Kostya (Ken) Ostrikov
Plasma Nanoscience

Related Titles

E.L. Wolf

Nanophysics and Nanotechnology

An Introduction to Modern Concepts in Nanoscience

2006

ISBN: 978-3-527-40651-7

G. Wilkening, L. Koenders

Nanoscale Calibration Standards and Methods

Dimensional and Related Measurements in the Micro- and Nanometer Range

2005

ISBN: 978-3-527-40502-2

R. Kelsall, I.W. Hamley, M. Geoghegan (Eds.)

Nanoscale Science and Technology

2005

ISBN: 978-0-470-85086-2

R. d'Agostino, P. Favia, C. Oehr, M.R. Wertheimer (Eds.)

Plasma Processes and Polymers

16th International Symposium on Plasma Chemistry, Taormina/Italy June 22–27, 2003

2005

ISBN: 978-3-527-40487-2

S. Reich, C. Thomsen, J. Maultzsch

Carbon Nanotubes

Basic Concepts and Physical Properties

2004

ISBN: 978-3-527-40386-8

G. Schmid (Ed.)

Nanoparticles

From Theory to Application

2004

ISBN: 978-3-527-30507-0

Kostya (Ken) Ostrikov

Plasma Nanoscience

Basic Concepts and Applications
of Deterministic Nanofabrication

**WILEY-
VCH**

WILEY-VCH Verlag GmbH & Co. KGaA

The Author

Prof. Kostya (Ken) Ostrikov
The University of Sydney
School of Physics
Sydney, Australia
and
Plasma Nanoscience Centre Australia (PNCA)
CSIRO Materials Science and Engineering
Lindfield, Australia

Cover illustration
This figure summarizes the Plasma Nanoscience effort to understand and use plasma-related effects such as electric charges and fields for the creation of building blocks of the Universe, nano-technology and, possibly, life.

Library of Congress Card No.: applied for

British Library Cataloguing-in-Publication Data
A catalogue record for this book is available from the British Library.

Bibliographic information published by the Deutsche Nationalbibliothek
Die Deutsche Nationalbibliothek lists this publication in the Deutsche Nationalbibliografie; detailed bibliographic data are available in the Internet at http://dnb.d-nb.de

100 c0o1 265

© 2008 WILEY-VCH Verlag GmbH & Co. KGaA, Weinheim

Printed in the Federal Republic of Germany
Printed on acid-free paper

Composition   Da-TeX Gerd Blumenstein, Leipzig
Printing   Strauss GmbH, Mörlenbach
Bookbinding   Litges & Dopf Buchbinderei GmbH, Heppenheim

ISBN: 978-3-527-40740-8

To Tina, with love and appreciation

Contents

Plasma Nanoscience: Basic Concepts and Applications of Deterministic Nanofabrication
Kostya (Ken) Ostrikov
Copyright © 2008 WILEY-VCH Verlag GmbH & Co. KGaA, Weinheim
ISBN: 978-3-527-40740-8

Preface

Applications of low-temperature plasmas for nanofabrication is a very
new and quickly emerging area at the frontier of physics and chemistry
of plasmas and gas discharges, nanoscience and nanotechnology, solid-
state physics, and materials science. Such plasma systems contain a wide
range of neutral and charged, reactive and non-reactive species with the
chemical structure and other properties that make them indispensable
for nanoscale fabrication of exotic architectures of different dimensionality and functional thin films and places uniquely among other existing
nanofabrication tools. By nanoscales, it is typically implied that the spatial scales concerned are above 1 nm ($= 10^{-9}$m) and below few hundred
nm.

In the last decade, there has been a strong trend towards an increasing
use of various plasma-based tools for numerous processes at nanoscales,
including plasma-aided nanoassembly of individual nanostructures and
their intricate nanopatterns, deposition of nanostructured functional
materials (including biomaterials), nanopatterns and interlayers, synthesis of quantum confinement structures of different dimensionality
(e.g., zero-dimensional quantum dots, one-dimensional nanowires, two-
dimensional nanowalls and nanowells, and intricate three-dimensional
nanostructures), surface profiling and structuring with nanoscale features, functionalization of nanostructured surfaces and nanoarrays,
ultra-high precision plasma-assisted reactive chemical etching of sub-
100 nm-wide and high-aspect-ratio trenches and several others.

In many applications (such as in commonly used plasma-assisted reactive chemical etching of semiconductor wafers in microelectronics),
plasma-based nanotools have shown superior performance compared
to techniques primarily based on neutral gas chemistry such as chemical vapor deposition (CVD). However, compared to neutral gas routes,
in low-temperature plasmas there appears another level of complexity
related to the necessity of creating and sustaining a suitable degree of
ionization and a much larger number of species generated in the gas

Plasma Nanoscience: Basic Concepts and Applications of Deterministic Nanofabrication
Kostya (Ken) Ostrikov
Copyright © 2008 WILEY-VCH Verlag GmbH & Co. KGaA, Weinheim
ISBN: 978-3-527-40740-8

phase, which is no longer neutral. Furthermore, in many cases uncontrollable generation, delivery and deposition of a very large number of radical and ionic species, further complicated by intense physical (physisorption, sputtering, etc.) and chemical (chemisorption, bond passivation, reactive ion/radical etching) plasma-surface interactions substantially compromise the quality and yield of plasma-based processes.

This overwhelming complexity leads to a number of practical difficulties in operating and controlling plasma-based processes. In many cases, instead of nicely ordered arrays of nanoscale objects one obtains poor quality and very disordered films nowhere near having any nanoscale features. Moreover, improper use of plasmas may lead to severe and irreparable damage to nanoscale objects already synthesized. On the other hand, plasma-based processes can be used to create really beautiful nanostructures and nanofeatures such as single- and multiwalled carbon nanotubes and high-aspect ratio straight trenches in silicon wafers. These common facts give us a lead to think that certain knowledge and skills are required to operate and use plasma discharges to synthesize and process so delicate objects as nanoscale assemblies.

In our daily life we always use a broad range of appliances and tools. Some of them are so simple to operate so that no one even reads a user's guide. However, the more complex the tool or appliance becomes, the more options it offers, to everyone's benefit. On the other hand, as the complexity increases, it becomes increasingly difficult to operate them. Some of the new and uncommon features are very difficult to enable merely relying on the already existing knowledge and experience. It is of course possible to enable some of these features via trial and error but a chance of damaging the (presumably expensive!) tool or appliance becomes higher after each unsuccessful attempt. The more complex the object of our experimentation becomes the larger number of trials we need to undertake. Above a certain level of complexity, trial and error simply become futile and way too risky and the best way in this case would be simply to read the user's manual. Fortunately, it is a norm nowadays that manufacturers of household appliances and related tools and devices provide handy user's instructions and manuals.

The situation changes when one tries to experiment and create something uncommon and unusual, by suitably modifying the commonly used tools. This is a typical situation in nanotechnology, which aims to create exotic ultra-small objects with highly-unusual properties compared with their bulk material counterparts. Apparently, creation of such small objects would most likely require different tools, approaches and techniques. Since the nanoscale objects are usually more complex than their corresponding bulk materials, they also require more complex fabrication tools and processes. Moreover, the costs involved in nanoscale processing are usually substantially higher compared to treatment of similar bulk materials. For example, multi-step nanostructuring of silicon semiconductor wafers (which may involve pre-patterning, surface conditioning, etching, deposition, etc. stages) is far more expensive than its coating by a plain dielectric film. The complexity of processing and therefore, the associated cost continuously increase as the feature sizes become smaller and smaller. Taken that even a single faulty interconnect or a short-circuited gate of a field effect transistor (which is more and more difficult to fabricate as they reduce in size) may disable proper functioning of the whole microchip.

Hence, the price of even simple errors in nanoscale processing may be way too high to simply afford them! For example, a 45 nm-sized nanoparticle attached to the surface of a 5 μm-thick film will most likely make no difference in terms of the film properties and performance. However, the same particle can reconnect (and hence, short-circuit) the two gate electrodes of a field effect transistor (FET) fabricated using a 45-nm node technology. This particle can be mistakenly grown in the gate area (e.g., when a nucleus was formed in an uncontrollable fashion) or grown in the gas phase and then dropped onto the transistor's gate. In either cases the associated damage to the integrated circuit may become irrecoverable and the whole effort spent on fabricating a huge number of transistors, vias, interconnects, interlayer dielectrics, etc. may go to waste simply because of a single nanoparticle-damaged transistor!

Therefore, it becomes clear that as the complexity of nanoscale processing increases, the cost of a single error becomes higher and eventually any "trial and error" approach in adjusting the nanofabrication tool and/or process may become inappropriate. First of all, the more complex the tools and processes become, the more reliant the researchers, students, process engineers and technicians become on user's manuals and detailed process specifications. For precise materials synthesis and processinig these guides should be as precise as possible. But who is supposed to write these detailed instructions? Engineers should write such guides for technicians, researchers for engineers, but who is supposed to write these for researchers? In the sister monograph "Plasma-Aided Nanofabrication: From Plasma Sources to Nanoassembly" [1] published by Wiley-VCH in July 2007 we tried to give some most important practical advices to researchers how to develop plasma-based nanoassembly processes, select the right plasma type, design appropriate plasma tools and reactors, and provided specific process parameters that led us and our colleagues to the synthesis of a wide range of nanoscale objects. However, the number of recipes given in that book is limited to certain types of low-temperature plasmas and specific nanoscale objects. So, where to find advice what to do when, for example, a 45 nm–sized nanoparticle was found in the gate area of an FET?

A typical advocate of a "trial and error" approach would suggest to change some process parameters and see what happens. But what if this trial will not work or continue causing more problems? On the other hand, a typical advocate of strictly following the prescribed guidelines would suggest to check a troubleshooting guide. But what if there is nothing about which knob to turn to eliminate the above particles? Moreover, taken the huge number of nanoscale processes that involve higher-complexity environments such as low-temperature plasmas, how could one possibly develop suggestions to troubleshoot every possible problem? The more complex the system becomes, the more opportunities for better, faster, more precise synthesis and processing it offers; on the other hand, a chance that something will go wrong will increase substantially. No wonder, the system is *complex* and may cause even more complex problems!

There are no exhaustive recipes to eliminate and troubleshoot all possible problems in a myriad of plasma-based/assisted processes that either already exist or being developed. In fact, if the nanofabrication system is very complex, then it would be physically impossible to foresee everything that can go wrong... So, what to do in this case? There is only one clear advice in this regard: do research, find a cause of the problem and then eliminate it. Therefore, *the more complex systems we use* in nanofabrication (as well as in any other area of technology and everyday's life), *the more important is to understand how they work, how to make them operate smoothly and how to prevent and eliminate any potential problems at a minimum cost and effort.* The importance of this rather simple commonsense statement becomes crucial when dealing with nanoscale materials synthesis and processing and I hope that anyone involved in related research will agree with me without any major arguments.

We are almost near the point where it becomes very clear what is the main point of this book. It should already become perfectly clear that it is about plasma-based nanotechnology. This nanotechnology is based on low-temperature plasmas, which represent a significantly more complex nanofabrication environment as compared with neutral feed gases where such plasmas are generated. So, how to properly handle plasma-based nanoassembly, avoid costly errors and troubleshoot any potential problems? To do this, we have to understand which plasmas to use, which plasma reactors and processes to design, how exactly to operate the plasma and control the most important surface processes.

These are among the most important issues the Plasma Nanoscience deals with and this monograph primarily aims to introduce the main aims and approaches of the Plasma Nanoscience to a reasonably broad audience which includes not only experts in the areas of plasma processing, materials science, gas discharge physics, nanoscience and nanotechnology and other related areas but also other researchers, academics, engineers, technicians, school teachers, graduate, undergraduate and even high school students.

As we will see from this monograph, the "microscopic" key to overcome the above problems and ultimately improve the overall performance of plasma-aided nanofabrication tools is to control generation, delivery, deposition, and structural incorporation of the required building units (BUs) complemented by appropriately manipulating other functional species [hereinafter termed "working units" (WUs)] that are responsible, e.g., for preparing the surface for deposition of the BUs. This task is impossible without properly identifying the purpose of each species (that is, as a BU, WU, functionless, or even a deleterious species) and numerical modelling of number densities of such species in plasma nanofabrication facilities and their fluxes onto nanostructured solid surfaces being processed.

Thus, the fundamental key to the ability to properly operate and troubleshoot highly-complex plasma nanotools is in comprehensive understanding of underlying elementary physical and chemical processes both in the ionized gas phase and on the solid surfaces exposed to the plasma. This is one of the main objectives of this monograph.

In my decision to write this book I was motivated by the fact that even though basic properties and applications of low-temperature plasma systems and even a range of useful recipes how to use such plasmas have been widely discussed in the literature, there have been no attempt to systematically clarify and critically examine what actually makes low-temperature plasmas a versatile nanofabrication tool of the "nano-age". One of the aims of this work is to discuss, from different perspectives and viewpoints, from commonsense intuition to expert knowledge, numerous specific features of the plasma that make them particularly suitable for synthesizing a wide range of nanoscale assemblies, epitaxial films, functionalities and devices with nano-features.

Richard Feynman's visionary speech "There is plenty of room at the bottom" [2] and a recent rapid progress in nanotechnology gave me a source of additional inspiration and provoked a couple of simple questions:

- Is there a room, in the global nanoscience context, for atomic manipulation in the plasma?

- Since the plasma is an unique, the fourth (ionized) state of the matter associated in our minds to a collection of interacting charged particles, what is the difference between nanoscale objects assembled in ionized and non-ionized gas environments?

Moreover, as we will stress in Chapter 1 of this monograph, since more than 99% of the visible matter in the Universe finds itself in an ionized (plasma) state (and contains charged atoms and electrons), the formation of the remaining $\sim 1\%$ of the matter should have inevitably passed through the nano-scale synthesis process (termed nanoassembly hereinafter) step. The nanoassembly is basically a rearrangement of gas-phase borne subnanometer-sized atomic building units into more ordered macroscopic liquid- and solid-like structures. Thus, one can intuitively suspect (even without any specialist knowledge apart from the sizes of the atoms and macroscopic ordered structures) that the process of formation of solid matter in the Universe did include the nano-assembly step in the plasma environment. Meanwhile, our commonsense tells us that the Nature always chooses the best option for arranging the things! So, could the plasma environment was chosen by the Nature for a specific purpose? As we will see from this monograph, the plasma environment could serve as an accelerator of nanoparticle creation in stellar outflows. Amazingly, without the plasma, there might have been insufficient dust particles, which are essential to maintain chemical balance in the Universe!

Another interesting area where in-depth investigation of the elementary plasma-based processes may shed some light on many existing mysteries is possible creation of building blocks of life such as DNA, RNA, proteins and living cells. There is a number of theories suggesting that these building blocks might have formed from simple organic molecules through a chain of elementary chemical reactions in methane, hydrogen, and water vapor-rich atmosphere of primordial Earth. The most amazing related fact is that at that time electrical discharges in the Earth's atmosphere (e.g., lightnings, coronas and sparks) were so frequent so that they may have played a significant role in chemical synthesis of macromolecules that eventually led to the formation of DNA, RNA and more complex building blocks of life. Despite more than 50 years of intense research and related debates about the creation and the origins of life which involve an extremely broad audience, this issue is far from being complete. On a positive note, reactive plasmas have been used to synthesize, in laboratory conditions close to those in primordial Earth, many complex organic macromolecules whithout which the existence of more complex building blocks of life would be impossible.

Even though this particular issue is only briefly mentioned in this monograph, here we stress that creation of building blocks of life is as important for the Plasma Nanoscience as the plasma-assisted synthesis of cosmic dust (building blocks of the Universe) and various building blocks (nanostructures, nanoarrays, etc.) of modern nanotechnology. These seemingly very different and unrelated issues have one most important thing in common: plasma environment which is used for deterministic creation of the above building blocks.

Since the Nature's nanofab uses the plasma in the Universe and quite possibly used quite similar ideas to synthesize building blocks of life in the atmosphere of primordial Earth, it sounds quite logical that so many companies and research institutions presently use cleanroom and laboratory plasma environments to synthesize a variety of nano-sized objects and nanodevices. Indeed, if a so reputed authority as Nature uses low-temperature plasmas to create many useful nanoscale things, then why should not one use that in terrestrial labs and commercial fabs? However, human mind always aims to create something that the Nature either cannot create or creates way too slow and inefficiently.

It is remarkable that the number of nanofilms, nano-sized structures, architectures, assemblies, and micro-/nanodevices fabricated by using low-temperature plasmas, has been enormous in the last ten years. Amazingly, using catalyzed plasma-assisted growth, it is possible to synthesize carbon nanotubes which are not among the common products of the Nature's astrophysical nanofab, and moreover, at rates which are orders and orders of magnitude higher.

Interestingly, the competition for priority synthesis and improved performance of nano-objects has been very tough in the last decade and gave rise to currently prevailing "trial and error" (followed by a rapid dissemination of the results) practice in the nanofabrication area. Furthermore, there is presently a wide gap between the practical performance of numerous plasma-based nanofabrication facilities and in-depth understanding of fundamental properties and operation principles of such devices and tools and elementary processes involved at every nanofabrication step. Indeed, if a particular plasma tool works well and allows one to fabricate nanostructured wafers and integrated circuits with a huge number of nano-sized transistors and synthesize a myriad of different nanostructures and materials, what is the point to research why it does so? Does one really need to?

Yes, one has to do that, and for a number of reasons. The most important reason for in-depth study of elementary physical and chemical processes involved is the need to keep the pace with miniaturization and ever-increasing demands for better quality nanomaterials and high-performance functionalities and nanodevices. At some stage the existing pool of tools will fail to meet the requirements, and what shall one do next? Do the trial and error as we discussed earlier?

After several years of active and productive research in the area, my colleagues and myself realized that the capabilities of "trial and error" approaches will soon be exhausted and deterministic "cause and effect" approaches to nanofabrication will need to be widely used to achieve any significant improvement in the properties and performance of the targeted nano-assemblies, nanomaterials, and nanodevices, which was quite easy to achieve several years ago by "trial and error". Indeed, in early and mid-90s, after a pioneering discovery of carbon nanotubes by Iijima [3], almost every carbon nanostructure synthesized under different process conditions, might have had quite different properties. But it is very difficult to impress anyone by synthesizing a carbon nanotube in 2008, when such a work has become a routine exercise in the third year chemistry or nanotechnology undergraduate programs.

Therefore, there is a vital demand for the development and wider practical use of sophisticated, and yet simple, *deterministic* "cause and effect" approaches. It is important to mention that such approaches would be impossible without a comprehensive understanding and generic recipes on the appropriate use and control, at the microscopic level (and more importantly, both in the ionized gas phase and on the solid surfaces), of the "causes" to achieve the envisaged and pre-determined goals ("effects").

Evidently, in the nanofabrication context, one can use the building blocks (e.g., specific atoms or radicals) of the nanoassemblies as the "cause" and the nanoassemblies themselves as the "effect". Indeed, the "building block" has been among the most commonly used and popular terms of the nanoscience and nanotechnology in the last decade. This term usually encompasses both elementary building units of atomic and molecular assemblies and some nanostructures and nanoparticles that are in turn used to build more complex nanoscale functionalities and nanodevices. However, merely praising the building units of the plasma-aided nanoassembly would be unfair, since many other particles also serve for other, merely than as building material, purposes.

For this reason, in this monograph we introduced the expanded notion of "working units" that encompasses all the relevant plasma species that contribute to any particular nanofabrication step. For instance, without appropriate surface preparation by suitable plasma species, the deposition and stacking of the building units into a nanostructure would be impossible. Thus, one should be fair in acknowledging contributions from all working units and realize that every one of them has to do their specific job properly to achieve the overall success. This is how the "nano-team effort" work!

It should also be emphasized that despite an enormous number of research monographs and textbooks related to nano-science and nanotechnology, only a few of them report on and analyze superior performance of plasma enhanced chemical vapor deposition (PECVD) and other plasma-based systems in nanofabrication of a wide variety of common nanostructures, such as carbon nanotubes, quantum dots, nanowalls, nanowires, etc. Therefore, there is a significant gap between the knowledge and information related to basic properties and applications of low-temperature plasmas and numerous nanoassembly processes that merely use such plasmas as a tool. Thus, the question about the actual role of the plasma in a large number of relevant processes remains essentially open. This book is intended to fill this obvious gap in the literature.

This monograph introduces the Plasma Nanoscience as a distinct research area and shows the way from Nature's mastery in assembling nano-sized dust grains in the Universe to deterministic plasma-aided nanoassembly of a variety of nanoscale structures and their arrays, a base of the future nanomanufacturing industry. We also introduce a concept of deterministic nanoassembly together with a multidisciplinary approach to bridge the spatial gap of nine orders of magnitude between the sizes of plasma reactors and atomic building units that self-assemble, in a controlled fashion, on plasma-exposed surfaces. By discussing the results of ongoing numerical simulation and experimental efforts on highly-controlled synthesis of various nanostructures and nanoarrays we show potential benefits of using ionized gas environments in nanofabrication.

In this monograph, we systematically discuss numerous advantages of using low-temperature plasmas to synthesize various nano-scale objects, and also introduce basic concepts of Plasma Nanoscience as a distinctive research area. For consistency of illustrating the benefits of using the advocated "cause and effect" approach, the majority of the examples come from own research experience of the author and his colleagues. Nevertheless, we will also attempt to provide a reasonable coverage of relevant ongoing reserach efforts that ultimately aim at achieving the goal of plasma-based deterministic synthesis of various nanostructures and elements of nanodevices.

In a systematic and easy-to-follow way, this monograph highlights the fundamental physics and relevant nanoscale applications of low-temperature plasmas and attempts to give detailed comments on what exactly makes the plasma a versatile nanofabrication tool of the "nano-age". An initial attempt to answer this very intriguing and timely puzzle of modern interdisciplinary science was made in a Colloquium article of Reviews of Modern Physics published in 2005 [4]. This original effort was further supported by a Special Cluster Issue of the Journal of Physics D on plasma-aided fabrication of nanostructures and nanoassemblies. For more details about this Special Issue please refer to the editorial review [5] and a cluster of 19 articles in the same issue. This monograph continues this series of efforts and aims to consolidate, in a single publication, some of the most important bits of knowledge about the unique properties and outstanding performance of the plasma-based systems in nanofabrication, as well as about possible ways of controlling the plasma-based nanoassembly.

Main attention is paid to the conditions relevant to the laboratory gas discharges and industrial plasma reactors. A specialized and comprehensive description of the most recent experimental, theoretical and computational efforts to understand unique properties of low-temperature plasma-aided nanofabrication systems involving a large number of associated phenomena is provided. Special emphasis is made on fundamental physics behind the most recent developments in major applications of relevant plasma systems in nanoscale materials synthesis and processing.

This monograph covers a specific area of the cutting-edge interdisciplinary research at the cross-roads where the physics and chemistry of plasmas and gas discharges meet nanoscience and materials physics and engineering. It certainly does not aim at the entire coverage of the existing reports on the variety of nanostructures, nanomaterials, and nanodevices on one hand and on the plasma tools and techniques for materials synthesis and processing at nanoscales and plasma-aided nanofabrication on the other one. Neither does it aim to introduce the physics of low-temperature plasmas for materials processing. We refer the interested reader to some of the many existing books that cover some of the relevant areas of knowledge [6–20]. From the perspective of fundamental studies, one of the purposes of this book is to pose a number of open questions to foresee the future development of this research area and also urge the researchers to look into fundamental, elementary bits (and not merely limited to the building units!) that make their nano-tools work.

The author extends his very special thanks to S. Xu, his principal collaborator in the last 8 years and a co-author of the sister monograh [1] and I. Levchenko, a co-author of Chapter 6, who also made substantial original contributions to a large number of original publications used in this monograph and created many exciting visualizations of original computational results and excellent illustrations for this book.

I am particularly grateful to my co-authors (alphabetic order) Q. J. Cheng, U. Cvelbar, I. Denysenko, J. C. Ho, S. Y. Huang, M. Keidar, J. D. Long, A. B. Murphy, A. E. Rider, P. P. Rutkevych, E. Tam, Z. L. Tsakadze, H.-J. Yoon, L. Yuan, X. X. Zhong, and W. Zhou, who made major contributions to the original publications used in this monograph.

I greatly acknowledge contributions and collaborations of other present and past members and associates of the Plasma Nanoscience (The University of Sydney, Australia) and Plasma Sources and Applications Center (NTU, Singapore) teams Y. Akimov, K. Chan, J. W. Chai, M. Chan, H. L. Chua, Y. C. Ee, S. Fisenko, N. Jiang, Y. A. Li, V. Ligatchev, W. Luo, E. L. Tsakadze, C. Mirpuri, V. Ng, L. Sim, Y. P. Ren, M. Xu, and all other co-authors of my research papers and conference presentations.

I also greatly appreciate all participants of the international research network and Plasma Nanoscience enthusiasts around the globe, as well as fruitful collaborations, mind-puzzling discussions, and critical comments of (alphabetic order) A. Anders, M. Bilek, I. H. Cairns, L. Chan, P. K. Chu, K. De Bleecker, C. Drummond, C. H. Diong, T. Desai, C. Foley, F. J. Gordillo-Vazquez, M. Hori, N. M. Hwang, A. Green, B. James, H. Kersten, S. Komatsu, U. Kortshagen, S. Kumar, O. Louchev, X. P. Lu, D. Mariotti, D. R. McKenzie, M. Mozetic, A. Okita, X. Q. Pan, F. Rosei, P. A. Robinson, P. Roca i Cabarrocas, F. Rossi, Y. Setsuhara, M. Shiratani, M. P. Srivastava, L. Stenflo, R. G. Storer, H. Sugai, H. Toyoda, S. V. Vladimirov, M. Y. Yu, and many other colleagues, collaborators and industry partners. I also thank all the authors of original figures for their kind permission to reproduce them. I sincerely appreciate the interest of a large number of undergraduate and postgraduate students at the University of Sydney in our special and summer vacation projects.

Last but not the least, I thank my family for their support and encouragement and extend very special thanks to my beloved wife Tina for her love, inspiration, motivation, patience, emotional support, and sacrifice of family time over weekends, evenings and public holidays that enabled me to work on this book and a large number of associated original publications, review papers, and project applications. My special thanks to my beloved pet Grace The Golden Retriever, who inspired me on a number of occasions during long evening walks around the suburb where we live.

This work was partially supported by the Australian Research Council, the University of Sydney, CSIRO, Institute of Advanced Studies (NTU, Singapore), and the International Reserach Network for Deterministic Plasma-Aided Nanofabrication.

Sydney, June 2008   *Kostya (Ken) Ostrikov*

Acronyms

0D	zero-dimensional
1D	one-dimensional
2D	two-dimensional
3D	three-dimensional
ADI	alternative direction implicit
AFM	atomic force microscopy
ALD	atomic layer deposition
BN	boron nitride
BNSLs	binary nanoparticle superlattices
BUs	building units
CBD	cluster beam deposition
CCT	charged cluster theory
CdS	cadmium sulfide
CNFs	carbon nanofibers
CNS	carbon nanostructure
CNSs	carbon nanostructures
CNTs	carbon nanotubes
CPU	central processing unit
DFT	density functional theory
DLC	diamond-like carbon
EEDF	electron energy distribution function
FED	field-emission display
FESEM	field emission scanning electron microscopy
FM	Frank–van der Merwe (growth mode)
FTG	floating temperature growth
H	high-density inductive discharge mode
HA	hydroxyapatite
HOMO	highest occupied molecular orbital
HRTEM	high-resolution transmission electron microscopy
ICP	inductively coupled plasma
ICPs	inductively coupled plasmas

Plasma Nanoscience: Basic Concepts and Applications of Deterministic Nanofabrication
Kostya (Ken) Ostrikov
Copyright © 2008 WILEY-VCH Verlag GmbH & Co. KGaA, Weinheim
ISBN: 978-3-527-40740-8

ICT	information and communications technology
ILDs	interlevel dielectrics
IPANF	integrated plasma-aided nanofabrication facility
IR	infrared
ISNs	initial seed nuclei
KMC	Kinetic Monte Carlo
LBL	layer-by-layer
LEED	low-energy electron diffraction
LEEM	low-energy electron microscopy
LP	langmuir probe
LUMO	lowest unocuppied molecular orbital
LVCS	laser vaporization cluster source
MBE	molecular beam epitaxy
MC	Monte Carlo
MD	molecular dynamics
ML/s	monolayer/s
MOSFET	metal-on-semiconductor field effect transistor
MOVPE	metal-organic vapor phase epitaxy
MSS	mask-substrate system
MWCNTs	multiwalled carbon nanotubes
NAs	nanoassemblies
NEMS	nanoelectromechanical systems
NGRs	neutral gas routes
NP	nanoparticle
NPRDF	nanoparticle radius distribution function
NPs	nanoparticles
NRDF	nanotip radii distribution function
NSs	nanostructures
OEI	optical emission intensity
OES	optical emission spectroscopy
PACIS	pulsed arc cluster ion source
PALCVD	plasma-assisted laser chemical vapor deposition
PAPLD	plasma-assisted pulsed laser deposition
PEALD	plasma enhanced ALD
PECVD	plasma-enhanced chemical vapor deposition
PEM	proton exchange membrane
PET	polyethylene terephtalate
PL	photoluminescence
PLD	pulsed laser deposition
PMCS	pulsed microplasma cluster source
PT	porous template
QD	quantum dot

QDN	quantum dot nuclei
QDs	quantum dots
QMS	quadrupole mass spectrometry
RAM	random access memory
RBW	red brick wall
SBF	simulated body fluid
SED	surface-conduction electron-emitter display
SEM	scanning electron microscopy
SK	Stranski–Krastanov (growth mode)
SMS	surface microstructure
SMSs	surface micro-structures
SNMS	secondary neutral mass spectrometry
SOR	successive-over-relaxation
STM	scanning tunnelling microscopy
SWCNTs	single-walled carbon nanotubes
TAM	total annual market
TE	thermal evaporation
TEM	transmission electron microscopy
TG	temperature gradient
VACNs	vertically aligned carbon nanostuctures
VANs	vertically aligned nanostructures
VLS	vapor-liquid-solid
VW	Volmer–Weber (growth mode)
WUs	working units
XPS	X-ray photoelectron spectroscopy
XRD	X-ray diffractometry

1
Introduction

This monograph aims to introduce the basic concepts and applications of plasma nanoscience, a rapidly emerging multidisciplinary research area at the forefront of, the physics of plasmas and gas discharges; nanoscience and nanotechnology; astrophysics; materials science and engineering; surface science and structural chemistry [4], and to show the importance of plasma environments in nanoscale processes spanning from astrophysics to plasma-aided nanofabrication in the laboratory.

Plasma nanoscience is a multidisciplinary topic which involves knowledge, methods and approaches from a broad range of disciplines, ranging from stellar astrophysics and astro-nucleosynthesis through "traditional" nanoscience and nanotechnology, materials science, the physics and chemistry of plasmas and gas discharges, to various engineering, health-related and socio-economic and business subjects. At one extreme, a variety of nanoscale solid objects are produced in the plasmas of stellar environments, while at the other, plasma nanofabrication has had a marked impact on capital investment, economy, trade and other aspects of our lives [5]. As a consequence, one can find reports on plasma applications in nanoscience and nanotechnology in a wide range of publications; from electronic archives to Science, Nature, not to mention numerous monographs and edited books (see, e.g., References [1, 4, 6–8, 21, 22] and references therein).

We will begin this chapter by introducing the main concepts and issues of plasma nanoscience in Section 1.1, followed by a discussion of various reasons why a self-organized nanoworld should be created in a low-temperature plasma environment (Section 1.2). Section 1.3 explains how nature's nanofab works in generating cosmic dust and discusses the issues related to nanotechnology research directions. In Section 1.4 we introduce the concept of deterministic nanofabrication and briefly discuss some of the most important aims and approaches of plasma nanoscience. Section 1.5 explains the structure of the monograph and gives advice to the reader.

Plasma Nanoscience: Basic Concepts and Applications of Deterministic Nanofabrication
Kostya (Ken) Ostrikov
Copyright © 2008 WILEY-VCH Verlag GmbH & Co. KGaA, Weinheim
ISBN: 978-3-527-40740-8

1.1
Main Concepts and Issues

By "a plasma" one usually implies a fully- or partially-ionized gas with many unique properties attributable to long-range electromagnetic interactions between charged particles, interactions which do not occur in neutral gases. The plasma is usually composed of electrons and two other categories of species, termed "ions" and "neutrals" depending on their charging state. The intrinsic property of the plasma is to preserve its overall charge neutrality, that is, that the combined number of all negatively-charged species is equal to that of all positively-charged species. Species that belong to the "ion" and "neutral" categories are identical except for the presence of positive or negative charges in the case of "ions". Relevant species can range from individual atoms, molecules, monomers and radicals to chain and aromatic polymers and macromolecules, atomic and molecular clusters, small grains and nanocrystallites and even particle agglomerates and mesoparticles. Amazingly, all these objects can be charge neutral or otherwise charged positively or negatively. The electric charge of such particles varies from a single electron charge for most positive and negative ions to hundreds and even thousands of electron charges for solid nanosized clusters and micron-sized grains.

It is common knowledge that more than 99% of the visible matter in the universe finds itself in the plasma state. Therefore, plasma plays a prominent role in a variety of processes that take place over spatial scales as large as galaxy-scale turbulence, which can be of the order of tens of light years, and as small as atomic collisions and interactions, the latter occurring at distances comparable to the sizes of individual atoms (ca. 0.1 nm). Here we focus on the relatively narrow spatial range, namely ca. 10^{-10}–10^{-5} m, which covers atomic processes and most of the existing microscopic structures. The main attention here will be the assembly of nanoscale objects from sub-nanometer-sized atomic (and also other) building units (BUs) in plasma environments and the discussion of the role of the plasma environment in such processes [23].

The concept of building units is central to plasma nanoscience and is used throughout this monograph to denote all microscopic matter that can be gainfully used to create nanoscale objects. Depending on the specific situation BUs can vary from the most fundamental atoms to macromolecules, nanoclusters, nanoparticles, nanocrystallites and even nanoparticle aggregates [4]. There are numerous examples of plasma-grown nanoscale objects, for example, ultra-small solid dust particles in stellar environments, interstellar gas, cometary tails, the upper layers of the earth's atmosphere, industrial materials processing reactors, electrostatic precipitators and laboratory plasma devices [24–28]. Additionally there are a number of higher-complexity nanoassemblies of different dimensionality, such as quantum dots (0D), nanotubes, nanoneedles, nanorods, nanowires (1D), nanowalls, nanowells, nanoribbons (2D), bulk nanocrystals, nanocones, nanopyramids, nanoparticles and other nanostructures of complex shapes (3D) synthesized by using laboratory plasma-aided nanofabrication [4,21,22,29–31].

It is noteworthy that in the existing literature most of the above mentioned nanoscaled objects are often termed "nanoparticles" . In turn, the "nanoparticles" are also commonly, and arguably well-justifiably, referred to as the building blocks of nanotechnology. To avoid confusion and emphasize that the nanoassemblies are also built using the smallest bits of matter we use the notion of building units rather than building blocks. And since one of the main aims of this work is to advocate the deterministic approach for plasma-based nanofabrication, we try wherever possible to be more specific when referring to individual nanoassemblies. Nonetheless, in cases where the shape and internal structure are not important we also use the term "nanoparticle". Wherever unconventional terminology is used it is explained at the beginning of the relevant section.

It is interesting to note that carbon nanotubes, arguably the cutting edge research topic at the moment (at least judging by recent citation reports), were first synthesized using arc discharge plasmas [3]. However, the existing approaches for fabrication of exotic nanostructures and functional nanofilms in plasmas still remain process-specific and suffer from cost-inefficient "trial and error" practices. This is mostly due to the fact that the ability to control – in the plasma – the generation, transport, deposition and structural incorporation of the BUs of such films and structures still remains elusive. On the other hand, the idea of deterministic plasma-based nanofabrication is treated with a bit of a caution due to the fact that plasma is inherently unstable and is thus quite difficult to control as controlling tools may introduce fresh instabilities. Recently, advanced non-linear dynamic techniques suited for instability control in low-pressure cold plasmas through chaos control mechanisms have been developed [32]; however, most of the existing plasma nanotools still have relatively weak control capacities at the microscopic level. To this end our basic understanding of intimately interlinked elementary processes in the ionized gas phase and on solid surfaces during the plasma-based nanoassembly needs to be substantially improved [23].

This is one of the main issues plasma nanoscience deals with. In this monograph we discuss a broad range of problems related to the assembly of nanoscaled objects in various plasma environments ranging from stellar envelopes in astrophysics to nanofabrication facilities in research laboratories and commercial nanofabs of the near future. Further, we will elucidate the naturally occuring self-assembly of nanometre-sized particles in the universe and how to approach the problem of deterministic synthesis of exotic nanoassemblies in laboratory plasmas.

We will also address the important issue of how to challenge one of the previously intractable problems of *deterministic* plasma-based nanofabrication, namely the ability to create nanosized objects with the required composition, structure and properties for their envisaged applications. This level of determinism is based on the relation between the macroscopic process parameters and the eventual function and performance of the nano-object in question and can be termed macroscopic determinism.

On the other hand, from the viewpoint of fundamental science, the required level of determinism can be achieved by properly creating, manipulating and arranging elementary building units into nanoscale assemblies in a way that will eventually determine the highly unusual properties of such nano-objects. This is in essence the method for creating exotic, unusual forms of matter by arranging "the atoms one by one the way we want them" envisioned by R. Feynman in his speech "There is plenty of room at the bottom" at the Annual Meeting of the Americal Physical Society on 29 December 1959 [2]. This is exactly what we are aiming to discuss in this book, with the specific focus on the arrangement of atomic building units in various ionized gas environments of plasma discharges.

As will be seen from the following discussion, microscopic determinism can be achieved via bridging macroscopic and microscopic processes that are characterized by spatial scales that differ by nine orders of magnitude! Indeed, typical dimensions of plasma nanofabrication facilities (ca. 0.5 m) are more than a billion times larger than the sizes (ca. 0.1 nm) of adsorbed atoms (adatoms) that self-assemble into intricate nanoassemblies and nanopatterns on solid surfaces.

One possibility [4] is to manipulate the plasma-generated species in the plasma sheath that separates the plasma and solid surfaces and to control self-organization of nanostructure building units on plasma-exposed surfaces and their insertion into nanoassemblies (NAs). By nanoassemblies, we will hereinafter refer to any solid object with at least one dimension larger than approximately 1 nm. Nanoassembly can also mean the process of arrangement of subnanometer-sized building units into structures with at least one dimension exceeding approximately 1 nm. This concept involves appropriate preparation of building units and the actual synthesis of the NA and is illustrated in Figure 1.1. If

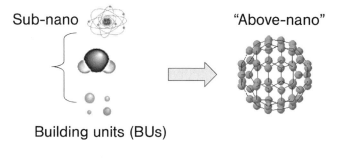

Sub-nano   "Above-nano"

Building units (BUs)

Nanoassembly (process) =
BU preparation + synthesis

Figure 1.1 Basic concept of nanoassembly.

appropriate, the process of nanoassembly can also involve removal or exchange of bits of matter.

The word "approximately" was added deliberately to this definition even though our commonsense suggests 1 nm as the most obvious lower size limit of nanoassemblies. However if we are dealing with a nanocluster of 0.5 nm diameter, it would be more accurate to consider it as a "sub-nanoassembly" (since it is constructed from more elementary building units) or as a building unit of larger nanostructures and nanoassemblies. Additionally, the diameters of surface-bound single-walled carbon nanotubes (SWCNTs), the most common nanostructures, which were always considered to exceed 1 nm, have in the last few years shrunk to approximately 0.6–0.7 nm. Does this mean that such ultra-thin SWCNTs with lengths well in the micrometer range should be excluded from the list of common nanoassemblies? Of course not! Instead, the lower limit for at least one size of nanoassemblies should be flexible and not necessarily be a fixed value of 1 nm. For example, to include micron-long and 0.7 nm thin SWCNTs in the list of nanoassemblies this lower limit should be reduced to below 0.7 nm. This might spark a discussion on the smallest diameter a single-walled nanotube can have yet having a length of excess of 1 μm. This is one of the as yet open questions in nanoscience; it will be addressed in the carbon nanotube-related section of this monograph.

By "nanofabrication" [5] one usually means the combination of a nanoassembly process and a suitable process environment; for example, synthesis of 1.5 nm-sized SiC quantum dots on a silicon surface in a thermally non-equilibrium low-temperature plasma of a $SiH_4 + CH_4$ gas mixture. However, common usage suggests that fabrication ultimately means producing some commercially marketable goods (otherwise this might be just a sophisticated academic exercise to satisfy scientific curiosity!). Therefore, at the very least, the above combination *nanoassembly + process environment* has to be complemented by one more component: *function* (ultimately related to the envisaged applications) to warrant serious consideration as *nanofabrication*. In simple terms, nanofabrication implies production of functionalities, elements, materials, and ultimately, coatings and devices (using just these examples for simplicity) that contain nanoscale features (e.g., size, nanostructure, nanopores) or have been made by using nanostructures or nanoassemblies as building blocks. Thus, synthesis of a carbon nanoneedle-like (at least potentially operational) microemitter mounted in a nanosized electron emitter cell or ordered arrays of luminescent quantum dots on stepped terraces on Si(111) surfaces are viable examples of nanofabrication.

Therefore, the ability to optimize the process environment and parameters to produce (at least potentially) the required function(s) of the nano-objects and show unique and unusual (intrinsic to the nanoscale only) properties is what differentiates between a simple process of nanoassembly (which often proceeds via self-assembly) and nanofabrication.

Plasma nanoscience is often understood as a bridge between plasma physics and surface science. Currently, there are enormous problems with the compatibility of *in situ* plasma diagnostics and surface science characterization techniques. Thererefore, researchers have to rely on quite separate experimental studies of the plasma processes and (in most cases *ex situ*) nanostructure characterization. On the other hand, there is a vital demand for reliable physical models and numerical simulations that could bridge the "unbridgeable" gap between gas-phases and surface processes separated in space by nine orders of magnitude.

In the following, we will discuss some advantages of using plasmas to generate, process and transport a variety of building units and then using them to synthesize nano-scale objects and, moreover, control "uncontrollable" atomic-level self-organization processes on plasma-exposed solid surfaces. We will also introduce basic concepts of plasma nanoscience and overview the ongoing reserach efforts aimed at achieving the ultimate goal of plasma-based deterministic synthesis of various nanostructures and elements of nanodevices. Finally, we will show that plasma nanoscience is a broad multidimensional notion that covers all situations in the universe and terrestrial laboratories wherein the nanoassembly process sketched in Figure 1.1 occurs in an ionized gas environment rather than merely the surface science of plasma-exposed surfaces.

1.2
Self-Organized Nanoworld, Commonsense Science of the Small and Socio-Economic Push

In the previous section we have mentioned self-organization and self-assembly as very useful and effective tools for nanoassembly. Both terms are crucial for nanoscience and nanotechnology and there exist plenty of definitions (see, e.g., Introduction to Nanotechnology [6]). However, such definitions generally do not reflect the overwhelming variety of different situations where self-organization processes play a role. Here we will only give working definitions to both of the terms; these definitions, although accurate in general, will mostly be related to those nanoassembly processes in ionized gas environments of our interest here.

Before giving the definitions we need to introduce the appropriate environment where self-organization and self-assembly take place. In this regard it will be prudent to introduce a broad term, "nanoworld", which will be used to denote various ensembles of nanoassemblies, with patterns or ordered arrays of individual nanostructures on solid surfaces as a typical example. This nanoworld is exposed to the plasma as shown in Figure 1.2. It is important to note that the nanoworld can have dimensions much larger than the sizes of individual nanoassemblies that compose it. In the example shown in Figure 1.2, the nanoworld on a plasma-exposed solid surface is made of small (1–20 nm in size) nanoislands, which can occupy large surface areas comparable to those of silicon wafers presently used by microelectronic industry.

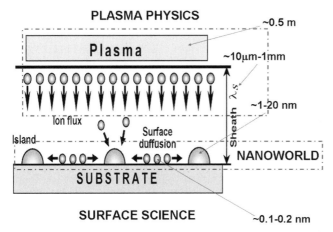

Figure 1.2 Nanoworld exposed to a plasma. Typical sizes of the plasma sources, transition layer (sheath) between the plasma and solid surface are shown.

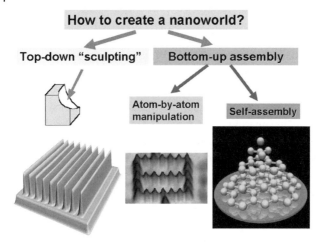

Figure 1.3 Two fundamental approaches of modern nanotechnology. Bottom-up approach has two basic possibilities: either atom-by-atom manipulation (nanomanipulation) or self-assembly.

In some cases the nanoworld can be limited to a single nanoassembly, this is the case for a single nanocluster levitated in a gas. It is also possible that the nanoworld can have macroscopic dimensions in all three dimensions yet having nanoscaled features. Dense arrays of micrometer-long single-walled carbon nanotubes with an average thickness of approximately 1 nm and bulk films with nanocrystalline or nanoporous features are especially good examples.

Some readers might find the introduction of this new term a bit artificial. The main reason we have introduced the nanoworld as a special term is the need to have the most generic notion that would be appropriate for virtually all objects that have any feature with at least one size ranging from sub-nanometers to the upper limit of approximately a few hundred nanometers. This generalization allows us to treat surface-bound dense SWCNT forests, ordered arrays of quantum dots, nanolayers and heterostructures, nanoporous and nanocrystalline films, films with nanoscale inclusions (e.g., nanocrystalline or simply cluster-sized defects), nanometer-sized trenches, vias and interconnects in nanoelectronic circuitry, complex assemblies such as nanoelectromechanical systems (NEMS), nanophotonic functionalities, as well as freestanding (e.g., gas borne) nanoassemblies from the same principles.

Once we have reached a convention on what the nanoworld is, the most obvious next step would be to identify plausible ways to create it. The two basic approaches of nanoscience are sketched in Figure 1.3. In the first, the "top-down" approach, smaller objects are carved from larger ones as sketched on the left side in this figure. For example, one can use energetic ion beams or reactive radicals to reduce the size of an initially micrometer-sized crystal to the nanometer range using the effects of physical sputtering and reactive chemical etching, respectively. The "top-down" approach based on masks, pattern transfer and reactive chemical etching is widely used in microelectronic manufacturing to fabricate patterns of two-dimensional trenches in silicon wafers or ordered arrays of high-aspect-ratio cylinders for two-dimensional photonic crystals. In this case reactive species etch holes through a mask placed on top of a bulk substrate; nanostructures are formed after a sufficient amount of matter has been removed from the bulk material.

It is worth noting that this technique requires pattern transfer and delineation, which is commonly achieved using microlithography approaches, which are based on micropattern transfer through natural templates or artificially created masks. Porous alumina with hexagonal nanopore arrays is perhaps the best example of natural templates used for creating ordered arrays of metal (e.g., nickel-based) catalyst nanoparticles required for carbon nanotube synthesis. This is also an example of a templated top-down nanofabrication approach, even though bits of matter are added to the substrate through the mask rather than removed. Artificial masks can be prepared, for example, by steering focused ion or laser beams about a solid surface; these beams can be used to drill small holes in thin and soft materials.

From the above arguments it becomes clear that "top-down" nanofabrication approaches critically depend on the ability to remove or add bits of matter along pre-delineated patterns. In simple terms, the resolution of this process strongly depends on the characteristics (hole patterns and sizes) of the masks involved in nanofabrication. Therefore, the smaller the nanostructures which are targeted, the smaller should be the mask holes. For example, using porous anodized alumina one can produce masks with tuneable pores of diameter ca. 10–500 nm, heights up to 6 μm, and nanopore densities of up to 10^{11} cm^{-2} (minimum spacing between the pore centres of ca. 30 nm), arranged in fine hexagonal arrays [33,34]. These holes can be used to fabricate, for example via a hot-filament evaporation process, hexagonal arrays of metal catalyst islands of sizes about the same as the sizes of the template nanopores. These catalyst islands can in turn be used to synthesize carbon nanotubes and related structures with diameters almost the same as the nanopores, which is 10–500 nm as mentioned above. Unfortunately, the sizes of nanopores in such templates are usually very non-uniform with the size dispersion reaching 100% and even more! This means that the carbon nanotubes will also be very non-uniform in size, and moreover, must be separated by at least 30 nm, the minimum inter-nanopore spacing.

However, what can be done as regards ultra-thin single-walled carbon nanotubes which require metal catalyst nanoislands as small as 0.6–0.7 nm in diameter? Moreover, how does one position such nanoislands very close to each other (inter-island spacing $\sim$ island diameter)? How should one design and create such a mask with holes so small and dense that they would be suitable for condensation of metal atoms? This size range is apparently far too small for the "top-down" nanofabrication despite very impressive recent advances in nanolithography [35] and more sophisticated nanopattern transfer techniques such as nanopanthography [36]. Generally speaking, "top-down" nanofabrication already experiences substantial problems in the sub-100 nm range [5]. Therefore, the global economic and technological demand for continued reduction in feature sizes in microelectronic devices (which as of mid-2007 are approximately in the 60–70 nm range in width and as thin as a few atomic layers) will inevitably move the top-down approach to the sidelines of industrial nanofabrication.

So, is there any other approach that can outperform and potentially replace the commonly used top-down nanofabrication techniques? If we consider the ultra-small metal (e.g., nickel) nanoislands required for the synthesis of single-walled carbon nanotubes, how many atoms do they contain? Such semi-spherical islands are generally constructed from approximately 15-25 atoms.

In such cases involving small number of atoms, would it not be wise to consider manipulating and stacking them one by one, the way Feynman suggested in his visionary speech? Yes, indeed for such a small number of atoms one could use another procedure, the "bottom-up" nanomanipulation approach, sketched in the middle of Figure 1.3. There are numerous reports on using nanomanipulators to displace and then reposition individual atoms into atomic chains or structures similar to the commonly known "atomic coral" [6]. At present, suitably adjusted scanning tunnelling and atomic force microscopes (STM and AFM, respectively) are extensively used for this purpose. By varying the amplitude, duration and sequence of voltage pulses applied between the tip of the microscope and the sample surface, one can induce electric charge on, or polarize otherwise charge-neutral atoms. In this way one can lift, move, replace, or otherwise manipulate individual atoms. Interestingly, this process involves ionization – the most important physical process that leads to the creation of a plasma! However, in nanomanipulation one ionizes only a very small number of atoms, which cannot qualify as a

plasma. Nevertheless, it is worth noting that ionization is used not only
in plasma-based nanotools!

We should also stress that the nanomanipulation technique is exactly
what Richard Feynman meant by stating "atom by atom, the way we
want them". In other words, to create a 50-atom nickel cluster on a silicon surface (or, for example, a Ge/Si quantum dot of a similar size), a
nanomanipulator device (e.g., STM) should repeat the

$$ionize/polarize \rightarrow lift/remove \rightarrow move \rightarrow stack$$

sequence at least 50 times (if everything works well) for each island.
Aiming to achieve any reasonable Si surface coverage by SWCNTs, one
would be looking at creating something in the order of ca. 10^{12} (or even
more!) nanoislands per square centimeter. This enormous number of
nickel clusters would thus require approximately 5×10^{13} atoms to be
ionized/polarized, lifted, moved, and then stacked *individually*! If every
move takes only 1 s, then the whole process of synthesizing the required
array of nickel nanoislands would take approximately 10 million years!

But what if the atoms do not want to be stacked where they are moved
by the nanomanipulator arm? What if the position they are put into is
not suitable or is unstable? Will the atoms remain firmly stacked in this
place or would they prefer to move further? These are just a few questions that need to be considered before committing time, resources and
effort to this arguably very precise and sophisticated technique, which is
commonly accepted as the best nanotool to manipulate very small numbers of individual atoms.

The most obvious and nature-inspired answer is just to do nothing and
let the atoms do what they want, in other words, self-assemble into nano-
objects of nature's choice. One of the most powerful of nature's tools in
this regard is the fundamental energy minimization principle

$$\mathcal{E}_{NA} = \mathcal{E}_{NA}^{min} < \Sigma\mathcal{E}_a$$

which states that the ensemble of atomic building units should self-
assemble to ensure that the resulting nanoassembly will have a total en-
ergy $\mathcal{E}_{NA}$ less than the combined energy of individual building units $\Sigma\mathcal{E}_a$.
Moreover, the assembly process will proceed along the minimum-total-
energy pathway, which means that $\mathcal{E}_{NA}$ will have the minimum possible
value $\mathcal{E}_{NA}^{min}$ under equilibrium conditions.

From this point of view, the ultimate crux of nanoscience is to create
unusual arrangements of atoms by whatever means, be it "top-down"
nanofabrication, nanomanipulation, or self-assembly. To illustrate this
concept, let us assume that there is some structure with a "regular" (reference) atomic structure and we want to create a similar structure but

with another arrangement of atoms, using one of the basic approaches of the nanoscience. It is noteworthy that "regular", nature-inspired structures are the simplest, the most stable and satisfy the minimum energy principle under equilibrium conditions.

Therefore, our commonsense would suggest that nature's approach is actually nothing else but the line of the least resistance. Indeed, the nature-preferred equilibrium conditions are normal for every particular environment; such conditions include room temperatures ($T = 20\,°C$) and gas pressure of 1 atm ($= 760\,mm$ mercury). Quite similar normal conditions exist elsewhere, outside the earth; moreover, such conditions are the most appropriate for the normal (line of the least resistance!) course of events and are chosen by the "lazy" yet "astute" nature.

For example, under normal terrestrial conditions, graphite is the most abundant and stable form (allotrope) of carbon. Carbon atoms are arranged in flat graphene sheets with a periodic hexagonal atomic network. Bulk graphite is made of parallel stacks of graphene sheets separated by a small interlayer spacing. Interestingly, the strength of atomic bonds between different graphene sheets appears lower compared with the inter-atomic bonds within each two-dimensional sheet. This is the reason why it is so easy to remove these sheets one by one, which is the way conventional pencils work! We can consider this atomic arrangement as a regular reference structure.

It is worth recalling at this juncture that creating exotic nanoassemblies implies applying some additional effort to create and use unusual, non-equilibrium conditions to rearrange the atoms in a different way than in the reference structure. Let us consider what that means in the context of carbon nanomaterials. If high pressures are applied and some other conditions are met, by using exactly the same carbon atoms one can synthesize diamond, a very different carbon material. This new material has a quite different crystalline lattice made of pyramid-like unit shells. These shells are interlinked three-dimensionally; this is why it is no longer possible to scrape off atomic carbon layers one by one as was possible in the case of graphite. It goes without saying that pure diamond and a range of diamond-like carbon (DLC) materials exhibit very different physical and chemical properties compared to graphite. We reiterate that diamond is usually synthesized under non-equilibrium conditions, such as high pressures, and once synthesized, remains stable at normal conditions. Even more non-equilibrium conditions are used to synthesize a very special diamond-like material – nanocrystalline diamond. More importantly, these non-equilibrium conditions are found in thermally non-equilibrium low-temperature plasma, a common environment for the synthesis of ultrananocrystalline diamond – a nanoworld made of ultra-small (ca. 1–3 nm in size) sp^3-hybridized (diamond-like) carbon [37,38].

Under different non-equilibrium synthesis conditions, which include relatively high temperatures and extrusion of carbon atoms through metal (e.g., nickel) catalyst nanoparticles, one can assemble carbon nanotubes. The same carbon atoms are now arranged in a similar graphene sheet but rolled into a graphitic tubule. Carbon nanotubes are also stable and also meet the energy minimum principle but under *modified* process conditions. Furthermore, their properties appear to be very different from the "regular" graphitic structure. Carbon nanotubes synthesized under non-equilibrium conditions (such as arc discharge plasmas in Iijima's pioneering experiments [3]) also remain stable under normal conditions and can be used for a variety of purposes including hydrogen storage, reinforced ceramic and polymer composites, electron field emission and wire-like interconnects in nanodevices to mention a just few.

Generalizing the above examples, we can state that nanoscience and nanotechnology aim at *using specific, non-equilibrium process conditions to create unusual and otherwise non-existing ultra-small nano-objects*! An important point to keep in mind is that these nano-objects must remain stable once returned to normal conditions.

Let us now return to the discussion of the possibilities offered by self-assembly and try to relate that to non-equilibrium process conditions. To begin with, let us pose a simple question: from the self-assembly perspective, what should one expect from a randomly chosen ensemble of atomic building units? Using the arguments we have already developed, it becomes clear that if the BUs are left without any external action and under equilibrium conditions, the BUs will simply self-assemble into the froms nature and the energy minimum principles prescribes under the given (in this case normal) conditions! Therefore, if one wants to create an exotic yet stable nano-object via self-assembly, suitable non-equilibrium conditions are required. In this case one can reasonably expect that self-assembly will proceed quite differently and will result in an exotic arrangement of atoms, otherwise non-existent under the equilibrium conditions. It is very important to stress that altering the process environment is perhaps the only way to control self-assembly, since the BUs are left without any external action and are not manipulated externally by any nanomanipulator arm!

We hope that the reader has become convinced that self-assembly can be effectively controlled by the nanofabrication environment. And with that we have just inadvertently revealed the fundamental concept of *guided self-assembly*, which is central to the entire nanoscience!

In this regard, it would be instructive to note that the ionized gas (plasma) environments of our interest here in most cases offer strongly non-equilibrium conditions. Capitalizing only on this point, we are now in a very good position to state a priori that low-temperature plasmas are excellent process environments for nanofabrication. For more background on why such plasmas can be regarded as a versatile nanofabrication tool please see [4]. These reasons will be discussed throughout this monograph and supported by relevant experimental and theoretical/computational results. More importantly, from the following consideration it will become clear that low-temperature plasmas have a number of effective (electric charge and field-related) tuning knobs to guide this self-assembly.

Moreover, as will be clarified in Section 1.3, nature creates non-equilibrium conditions (simply by adding a weakly ionized gas component) deliberately (or in other words, deterministically) to create a sufficient amount of solid dust particles in stellar environments. The ionization degree, one of the most important parameters of weakly ionized plasmas, turns out to be a very effective control of the formation of self-assembled nanoparticles.

Let us now complete the introduction of the main concepts used in this monograph and more specifically, in the context of plasma nanoscience. In this context, by self-assembly, we will imply a "bottom-up" process of arrangement of building units into subnanometer and nanometer-sized objects without any external action. In a sense, the nanoassemblies build themselves on plasma-exposed surfaces.

It is noteworthy that the terms self-assembly and self-organization are often used interchangeably in the literature; moreover, both terms are also frequently related to the formation of structured patterns such as quantum dot arrays. In this monograph we will try to avoid this ambiguity by using self-organization as a more global and generic term related to the nanoworld rather than an individual nanostructure. More specifically, self-organization phenomena considered in this monograph will also include the evolution of structural, size and positional order in nanoassembly patterns on solid surfaces from essentially non-uniform patterns, which cannot be merely attributed to self-assembly of individual nanostructures. In the following we will use the term "self-organized nanoworld", which encompasses any nanoscale objects that are formed exclusively via self-assembly and self-organization processes.

We hope that we have made our terminology and contextual issues more transparent to the reader. It should be emphasized, however, that many of the terms, although used commonly, do not have conventional definitions and their meaning may vary from one context to another. For example, many literature sources separate nanofabrication and self-assembly and attribute top-down and nanomanipulation processes to the nanofabrication. However, we believe that there is no good reason whatsoever why nanofabrication should not include a self-assembly step. Moreover, it is extremely important to note that to include self-assembly as a commercially viable nanofabrication approach, one should learn how to control it and so create exactly what is required. Thus, we have just arrived at the new, important notion of controlled (ultimately deterministic) self-assembly (and more globally, self-organization) – something which is still remains elusive despite the enormous efforts of a large number of universities, research and development institutions and industrial laboratories worldwide! However, if the level of understanding of how self-assembly works is poor (as it presently is!), nanofabrication and self-assembly are indeed quite separate issues and this is reflected in the existing terminology.

After this seemingly long discussion of basic terms and relevant issues, we will now try to answer one of the central questions of plasma nanoscience:

why should the nanoworld be self-organized and created in a plasma?

To answer the first part of this question let us consider a strong socio-economic push for miniaturization and nanotechnology. In the mid-1960s the introduction of computer and IT technologies transformed virtually every sector (manufacturing industry, transport, agriculture, finance, trades, government, defence, etc.) of our society and revolutionized the way we live. Many economists refer to this as the computer and communications revolution of the mid-1960s. Computer-based technologies received a rapid boost, which after a certain period of time slowed down and reached saturation in the mid-1990s. This behavior is commonly referred to as the "S-curve of technology".

We are currently entering the Information Age, when everyone (including developing countries) will (hopefully!) have a wireless broadband access to global information networks, and all information can be retrieved and processed almost instantly using palmtop computers with the capabilities of powerful present-day workstations. This is just one impression of what the Information Age can bring to society and how it can dramatically change our lives. In fact, the actual possibilities of what new technologies will be able to do (e.g., store and process) with enormous amounts of information go beyond our imagination.

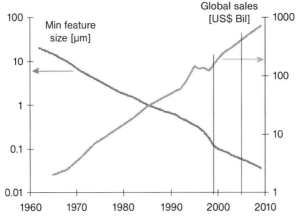

Figure 1.4 Minimum feature sizes of integrated circuitry and global sales of semiconductor microelectronic products (after [39]).

These new Information Age technologies require more powerful and faster computers in ever greater numbers. This results in a rapidly growing demand for better, faster computers and eventually in an exponential increase of the global market for computer-related products.

At this stage it would be instructive to recall that the main component of any computer is its motherboard which includes a central processing unit (CPU) and random access memory (RAM) on a semiconductor chip platform. Thus, to satisfy the demands of the Information Age, computer CPUs should work much faster and RAMs store a lot more information without any substantial increase in the microchip surface area. This in turn has led to an *exponential* increase of the total annual market (TAM) of semiconductor-based microelectronic products from only a couple of billion $US in the mid-1960 to more than $US 500 Billion in 2005 [39]. As can be seen from Figure 1.4, this amount is set to rise further to $US 1 trillion in 2010 and has excellent prospects of reaching $US 1.5 trillion in the foreseeable future.

However, careful size-cost-function-demand calculations show this can only become possible if the cost per electronic function falls at a rate of at least 18% per year to drop below 1 microcent per transistor within the next few years. To sustain this significant cost reduction while maintaining the cost of one square centimeter of a silicon wafer in the few $US range, the feature sizes (which determine the number of field effect transistors (FETs) and ultimately the number of logic operations a computer can perform per second) should reduce in size as shown in Figure 1.4. As can be seen from Figure 1.4, to reach $US 1 trillion in sales (thus, satisfy

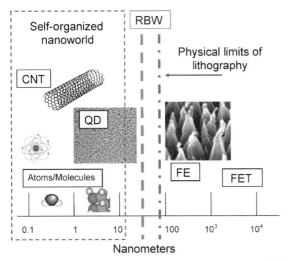

Figure 1.5 Self-organized nanoworld, physical limits of lithography, and red brick wall (RBW) of semiconductor technology. CNT = carbon nanotube; QD = quantum dot; FE = field emitter; FET = field effect transistor.

the demand of the Information Age), the feature sizes must be reduced to at least 20–30 nm by 2010–2012.

Therefore, there is a very strong socio-economic push to develop semiconductor features with sizes in the 20–30 nm range and below. If we have a look at the typical sizes of most common building units and nanoassemblies depicted in Figure 1.5 we will immediately notice that these sizes are comparable with those of carbon nanotubes and semiconductor quantum dots. Moreover, upon reduction of the nanostructure sizes to approximately 10 nm (which is of the order of the exciton's Bohr radius for some common semiconductors) quantum confinement effects begin to manifest. These effects lead to unique electronic properties of low-dimensional nanostructures not available in bulk materials. Therefore, the "true" nanoworld begins at spatial scales of approximately 10 nm as shown in Figure 1.5.

Another feature of Figure 1.5 is that it shows the present-day limits of lithographic tools (dash-dotted line) and also the ultimate physical limits the top-down nanofabrication approaches can achieve in the foreseeable future. A set of these ultimate physical limits is often referred to as the red brick wall (RBW) of semiconductor technology. Even the 20–30 nm features the semiconductor market demands to be achieved by the year 2010–2012 find themselves on the other, "nanoworld's", side of the RBW. Therefore, the strong socio-economic push forces one to develop means to fabricate 20–30 nm and smaller features and eventually "barge" into the "true" nanoworld with outstanding electron confinement capabilities.

It needs to be added at this stage that there is another group of breakthrough technologies with a rapidly expanding multi-billon dollar market that also demand substantially reduced sizes of nanoscale objects. As the reader may have guessed we are talking about biotechnology, which is becoming increasingly reliant on sophisticated nanotechnology products. The examples are numerous, for instance, quantum dot-based luminescent biomarkers, nanoparticle-based drug delivery systems, biocompatible and bioactive nanofilms and various biosensors on nanostructured film platforms. Synergy of nanostructured materials and biology gave rise to a rapidly emerging field of bionanotechnology. It is commonly accepted nowadays that the transition to the new industrial age will be marked by a synergetic triangle formed by information and communications technology (ICT), biotechnology and nanotechnology [39].

To be a bit more specific, let us concentrate on the link between the ICT and nanotechnology and consider how to challenge the problem of ultra-small size range of the "true" nanoworld. Some may say that this is easier said than done. Indeed, by which means are we supposed to achieve this? Since the required size range is on the other side of the RBW, the top-down nanofabrication approaches may not be applicable anymore. The other remaining choices are thus nanomanipulation and self-assembly (Figure 1.3). However, due to the extremely large number of atoms that make even a tiny interlayer in a single metal-on-semiconductor field effect transistor (MOSFET), the nanomanipulation approach should be immediately taken off the list.

Therefore, we are left with the only one option:

> *"True" nanoworld is self-organized and we must "barge" into it to satisfy the socio-economic push for better, faster, cheaper computers!*

Now the question is, what does this have to do with plasma and plasma nanoscience in particular? To answer this seemingly non-trivial question one should note that computer microchips are commonly produced in semiconductor fabs equipped with sophisticated plasma microfabrication facilities. In fact, the semiconductor industry widely uses inductively coupled RF plasma devices as sources of low-temperature thermally non-equilibrium plasmas. A representative plasma source of this type is shown in Figure 1.6.

The examples of plasma-based processes used in semiconductor microfabrication are numerous: reactive highly-anisotropic and highly-

Figure 1.6 A source of high-density, highly-uniform inductively coupled RF (~460 kHz) plasmas. Plasma sources of this type are widely adopted in semiconductor industry as a benchmark plasma reactor. Almost 50% of process steps in USLI micromanufacturing use low-temperature, thermally non-equilibrium plasmas. The insets show plasma glows around the magnetron sputtering targets which serve as sources of solid precursors. Photo courtesy of the Plasma Sources and Applications Center, NTU, Singapore.

selective chemical etching is used to fabricate deep high-aspect-ratio trenches in semiconductor wafers; plasma enhanced chemical vapor deposition is used to deposit ultra-thin (with the thickness approaching a few atomic layers) interconnect and copper diffusion barrier layers as well as surface activation and passivation; electric-field guided ion fluxes in the plasma-assisted physical vapor deposition (commonly known as i-PVD) are used for metallization of deep semiconductor features where neutral species cannot penetrate, just to mention a few practical applications. For a detailed coverage of the most important aspects of applications of plasma-based processes in microelectronics the reader should refer elsewhere [40]. Here we should stress that the total cost of plasma facilities used by the semiconductor industry worldwide is enormous and is clearly in the multi-billion range.

It is now a good time to move to the next step and pose another important question: is it possible to create a self-organized nanoworld made of nanoassemblies smaller than 10 nm in a typical plasma environment as currently used in semiconductor microfabrication? So far plasma-based nanotools, although extremely successful in the syntheis of carbon nanotubes and related structures, have not shown particularly impressive results in nanoassembly of low-dimensional semiconductor structures, which are of utmost importance for the creation of nanodevices based on quantum confinement effects. The reasons for this will be analyzed elsewhere in this monograph by using the arguments of balancing the demand and supply of plasma-generated building units. This is perhaps the most likely reason why sophisticated and extermely expensive nanotools such as molecular beam epitaxy (MBE), where one can precisely control incoming fluxes of neutral species, have shown a clearly better performance compared with the plasma nanotools.

Therefore, one might be tempted to start replacing plasma microfabrication facilities with non-plasma-based nanofabrication tools to enable production of "self-organized" computer microchips as soon as possible and no later than in 2012. However, the cost of such a replacement of entire microchip production lines may be completely unsustainable, taken that the yearly demand will be well above $US 1 trillion at that time. And one should also not forget about possible disruptions of computer production cycles, which may cost many billions of dollars.

But why is this radical change needed? To allow the use of tools which can create those self-organized nanoworlds so badly needed to satisfy the demand of the Information Age for better, faster, smarter and cheaper computers! And these tools need to replace the existing multibillion dollar pool of plasma facilities currently used by the semiconductor industry worldwide.

However, before committing such enormous resources and efforts one should make absolutely sure that it is not possible to create the self-organized nanoworld with the existing production lines, which, as we emphasize, are at present largely plasma-based. Indeed, why replace the existing production lines without first trying to create self-organized nanoworlds in the existing plasma-based microfabrication facilities!

Therefore, we have arrived, again inadvertently, at the conclusion that if we want to avoid huge losses because of major disruptions in the microchip (actually, nanochip!) production we need to learn

how to create a self-organized nanoworld in a plasma

and, moreover, in a *deterministic fashion*. Amazingly, this is what plasma nanoscience is all about!

From the above it becomes perfectly clear that there is a very strong socio-economic push to further develop plasma-based nanofabrication approaches and techniques and make them versatile nanotools of the new industrial age dominated by a synergy of information and computer technology, biotechnology and nanotechnology. Without a successful synergy of the three breakthrough technologies the S-curve of technology may not rise quickly enough and the age of transitions may stretch to quite a number of years, thus significantly delaying the much expected new industrial revolution.

What is even more amazing is that nature also encourages one to more widely use plasma-based environments for deterministic nanoassembly. In the following section we will discuss how nature's mastery works in the self-assembly of nanometre-sized particles in the universe. We will also comment on the nanotechnology research directions of the U.S. Nanotechnology Initiative.

1.3
Nature's Plasma Nanofab and Nanotechnology Research Directions

Let us now discuss how exactly nature uses plasma environments to create nano-sized objects. As we have already mentioned above, plasmas constitute more than 99% of the visible matter in the universe. The most striking example of how nature creates solid nanoscale objects from atomic building units is condensation and nucleation of cosmic dust in stellar environments. This process involves structural transformation from atomic (less than 1 nm in size) to nanocluster/nucleate stage (exceeding 1 nm) and, according to the convention we introduced in the previous section, qualifies as a nanoassembly process. In the following we will briefly discuss how the plasma nanofab works in the universe-based nanoassembly of dust grains and comment on the unique and specific roles of the plasma environment [23]. An astrophysical setting where cosmic dust nucleation takes place is sketched in Figure 1.7.

It is a common knowledge that in the universe most of the visible matter exists as a fully- or partially-ionized gas composed of subnanometre-sized particles such as atoms, molecules, radicals and ions. Therefore, synthesis of any bits of matter with sizes exceeding 1 nm (such as interstellar solid dust and other particles of increased complexity) necessarily involves the nanoassembly stage!

First of all, we note that this "above-nanometer" matter is solid and therefore cannot be made of hydrogen or helium atoms. Therefore, something should be done to create atoms of stable solid elements, which will be suitable for nanoassembly purposes. What is remarkable is that hydrogen is far more abundant in the universe than any heavier element such as helium, lithium, beryllium or carbon. Therefore, any heavier elements should be created by nuclear fusion of hydrogen nuclei. Nuclear fusion requires extra-high temperatures of the order of tens of million (or even more, depending on the required energy release) degrees to get the fusing atoms close enough for nuclear forces to come into play. Such high temperatures also serve the purpose of stripping interacting atoms of their electrons to eliminate atomic repulsion at distances com-

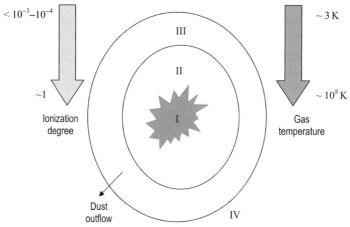

$< 10^{-3}-10^{-4}$

III

$\sim 3\,K$

II

~ 1

I

$\sim 10^8\,K$

Ionization
degree

Gas
temperature

Dust
outflow

IV

Figure 1.7 Schematic of nature's nanofab [23].
I – star; II – area of nanoparticle nucleation; III – area of further
dust growth and expulsion; IV – interstellar space.

parable with the radii of electron orbits (or more precisely, with the sizes
of electron clouds as accepted in quantum mechanics). In other words,
the gaseous environment should be hot enough to find itself in a hot and
fully ionized (plasma) state. Such conditions are met in the interiors of
stars. For example, temperatures inside the sun can be as high as 15 million degrees. Nuclear fusion reactions in hot and fully ionized plasmas
result in release of enormous amounts of energy and as such sustain the
entire existence of the stars and very possible, the universe.

However, here we are not interested in the energy generation processes
and refer the interested reader to relevant astrophysical and nuclear fusion literature. What we emphasize is that stellar nucleosynthesis proceeds via chains of nuclear fusion reactions which start from the most
elementary fusion of two protons into deuterium (heavy hydrogen) and
eventually result in the formation of a large number of elements that are
present in stellar environments.

Having a quick look at the Periodic Table of Elements, one would immediately work out that carbon would probably be the best candidate to
serve as a building unit of solid nanoscale matter. What is the shortest
and most effective way to create carbon using nuclear fusion and starting
from protons and neutrons, the elementary building blocks of subatomic
matter? One of the possibilities is to combine protons and neutrons into
an alpha particle ^4_2He, and then fuse three such particles to form a carbon atom $^{12}_6\text{C}$. These reactions can be accompanied by the creation of
other particles such as neutrons, positrons, and neutrinos and also by
the release of substantial amounts of energy. The actual reaction chains,

particles and amounts of energy involved depend on the mass of the star. For example, for stars comparable with the sun the nucleosynthesis processes proceed along a quite different scenario than in heavy (e.g., red giant) stars.

Therefore, the first building units (carbon atoms) suitable for the nanoassembly of solid matter can be generated in fully ionized plasmas of hot star interiors (denoted I in Figure 1.7) as a result of nuclear fusion reactions involving three alpha particles, such as the following "triple-alpha" reaction

$$^4\text{He} +^4\text{He} \rightarrow^8 \text{Be};$$

$$^8\text{Be} +^4 \text{He} \rightarrow^{12} \text{C} + 7.367\,\text{MeV}$$

which results in the synthesis of atomic carbon ^{12}C. Interestingly, this element is more effectively synthesized in the interiors of large stars evolved to the red giant and later stages. In fact, red giant stars are commonly recognized as primary sites of carbon synthesis in the universe [41].

New elements created as a result of nucleosynthesis are then carried away in stellar outflows to the star envelopes and then to the interstellar space. Therefore, carbon (and also other species synthesized in hot star interiors via nuclear fusion) condense and nucleate in the relatively cold and partially ionized plasmas of star environments. This happens in the area of primary nucleation denoted II in Figure 1.7. As a result, basic nanoassemblies such as critical clusters are formed [24]. A quite similar process is also possible in planetary atmospheres.

Thus, nature's "nano-mastery" proceeds in three stages. First, building units are generated via nucleosynthesis in hot fusion plasmas of star interiors. This is followed by expulsion of the as-created atomic building units in stellar outflows (in other words, they are delivered where they are required for nanoassembly). Finally, during the third stage the BUs condense into larger assemblies, this time in a much cooler and less dense low-temperature plasma environment. Thus, the whole process can be split into creation, delivery and assembly stages. Let us bookmark this point and compare it later with the building unit-based "cause and effect" nanofabrication approach, which we will introduce in Chapter 2 and commend throughout this monograph.

More observant readers should have already noticed that for some reason nature has chosen cold and weakly ionized plasma as the most suitable environment for the most elementary nanoassembly! Exploring alternative possibilities for nanoassembly in astrophysical environments, we should immediately exclude atom-by-atom nanomanipulation. Indeed, where are nanomanipulator arms (e.g., STM tip) in the universe that could ionize/polarize and move atoms from one place to another? Top-down "sculpting" is, in principle, possible; for example, non-thermal sputtering of amorphous carbon grains can reshape them. However, such processes have very low reaction rates under typical conditions of stellar environments. Therefore, self-assembly remains the only realistic choice.

One can thus write the basic formula that governs the operation of nature's nanofab:

Nature's mastery = self-assembly of building units + plasma.

As was mentioned in the original article [23], this statement might spark philosophic arguments on the appropriateness of the terminology used. For instance, there are numerous examples of natural self-assembly of biological objects on earth, which do not require plasma as a nanoassembly environment. Thus, one possible alternative term for the phenomenon depicted in Figure 1.7 would be the "universe's nanofab", with a multidimensional notion of the "nature's nanofab" attributed to all natural nanoassembly processes in the universe, in space and on earth. However, some other philosophers would offer counter-arguments based on the fact that earth itself takes its origin from cosmic dust created earlier in the "universe's" nanofab! Moreover, as we have mentioned in the preface, there is a possibility that the most basic building blocks of life were created in atmospheric gas discharges under primordial earth conditions. These and all other philosophical issues of the relevance of the nanoscale processes in the universe and on the earth to the awe of creation are outside the scope of this monograph.

The basic solid nanoassemblies mentioned above may grow further through other mechanisms, such as collection of atoms/ions from the adjacent plasma, and eventually form dust matter [28]. The area where the dust formation process proceeds by this method, is denoted III in Figure 1.7. The dust matter may be expelled into interstellar space, denoted IV in Figure 1.7. More importantly, the dust expulsion may serve a specific purpose, such as synthesis (via reactions on solid surfaces of dust grains) of molecular hydrogen H_2, much needed to maintain a proper chemical balance of the universe [25]. This and similar mechanisms lead to the appearance of various (mostly in simple nanoparticle forms) nanoassemblies in low-density partially ionized interstellar and interplanetary plasmas. To transport such nanoassemblies where they are actually needed (for example, to deliver atomic and molecular carbon for the synthesis of the solar system [42]), nature's nano-factory uses various "conveyor belts" such as high-velocity dust streams or comets [23,42,43]. We will return to the issues related to cosmic dust creation and its role in the universe later in this book.

A straightforward conclusion from the above arguments is that if the laboratory-based nanotechnology aims to be truly nature-inspired, it should ideally be plasma-based. Moreover, as we have already mentioned, the laboratory-based plasma-aided nanoassembly follows the same sequence of steps as the universe-based model, namely, generation of building units in the plasma environment and their transport and assembly into nano-objects [4]. Unfortunately, despite remarkable progress in the plasma-assisted synthesis of nanomaterials and functional nanostructures, the current use of the plasma-based techniques in nanotechnology is still quite limited, and mostly used for the synthesis of relatively simple nanoparticles, nanometer-thick functional coatings, nanocrystalline films and post-processing of nanostructures. Even though each of these are in most cases state-of-the-art on their own, none of them really deal with plasma-controlled self-assembly, the most effective driving force of self-organized nanoworld discussed in the previous section.

At this stage the reader might ask about the actual role of the plasma in nucleation of nanometer-sized nanoclusters and dust particles in the astrophysical situation depicted in Figure 1.7. In other words, why does the nucleation not happen either inside the stars (Area I) or in the interstellar space (Area IV)? The answer to the first part of the question is obvious: star interiors are suitable for generating the first solid atoms as a result of nuclear fusion reactions, but are way too hot for their nucleation. On the other hand, at the periphery of stellar gas envelopes and in the interstellar space (zones III and IV in Figure 1.7) the atom density is too low for the efficient nucleation. It is remarkable that dust nucleation actually takes place in the areas where the gas density is still reasonably high, the temperature is low and a weakly-ionized plasma is present (Area II in Figure 1.7).

It has been suggested that the nano-sized protoparticles appear as a result of ion-induced nucleation, which significantly increases the rates of generation of new solid grains [24]. Amazingly, a very similar conclusion was also made for laser ablation plasmas with the parameters different by many orders of magnitude [44]. Therefore, nature's nanofab actually uses plasma to increase the efficiency of the dust growth process and make it faster. And this sparks some extra optimism to pursue the plasma nanoscience research even further!

Unfortunately, despite all the apparent advantages and existing experience of the nature's nanofab, plasma-based nanoassembly routes have not been highlighted in the Nanotechnology Research Directions of the US Nanotechnology Initiative [45]. This provokes a reasonable question: since our major nanoscience and nanotechnology programs did not adequately follow nature's plasma nanofab mastery (which, as we have seen from the above discussion, explicitly prescribes one to use cold and weakly-ionized plasmas in nanoassembly processes), how did it affect the overall progress in the "nano-area"? Some might even consider this as one of the major reasons for the significant delays of the much expected industrial revolution that would lead us to the "IC-Nano-Bio"-Age.

However, extra care should be taken when assessing what plasma nanotools can and what they cannot do [23]. Let us recall one of the earlier remarks that the basic nature's nanofab formula usually leads to relatively simple, mostly nanoparticle-like nanoassemblies. Such nanoscale objects feature the minimum possible energy, are most stable, and hence, are the easiest to synthesize under equilibrium conditions. If a more complex nanostructure is targeted, some additional effort is required, such as using masks, catalysts, delineated patterns, and so on, which are not readily available in nature's nanofab. Thus, the basic formula should be complemented by the "minimum effort" principle. In this particular "ugly" nanoparticle-making process nature is indeed quite "lazy"! However, nature did spend some effort to ionize the background gas, which effectively leads to higher rates of "ugly" nanoparticle production. Here we stress that if the synthesis of more "beautiful" nanostructures is a goal, then more complex, non-equilibrium processes should be used.

Let us now recall that the main aim of modern nanotechnology is to create *complex* and unusual nano-objects such as quantum dots or nanowires and arrange them into intricate arrays and/or integrating into nanodevices. The relative simplicity of the nanostructures fabricated in nature's plasma nanofab is a possible reason for the common belief that other (e.g., chemical, lithographic, template-directed assembly, etc.) ways to create complex nanostructures and their nanopatterns had a better appeal for their inclusion and better highlighting in the Nanotechnology Research Directions [23,45].

This leads to another couple of concerns. The first and the most obvious is the level of competitive advantage of plasma-based nanoscale processes, techniques and facilities over the most commonly used non-plasma-based ones? Indeed, can plasma nanotools and processes compete with leading atomic-precision techniques and ensure a better quality of the resulting nanoassemblies and a higher process efficiency? Such high-precision, non-plasma-based routes include, but are not limited to, atom-by-atom nanomanipulation (e.g., by using the tip of a scanning tunneling microscope, STM), metal-organic vapor phase epitaxy (MOVPE), atomic layer deposition (ALD) and various modifications of molecular beam epitaxy (MBE) [23].

The future of plasma-based nanotools will critically depend on how realistic the prospects are of them winning this competition. If such prospects are not so optimistic, then it is very likely that plasma-based nanotools and processes will remain in the sidelines of modern nanoscience and nanotechnology and perform only certain steps (even though the number of such steps can be quite large) in nanomanufacturing, in a similar way to how plasma is currently used in microelectronics.

Shall we settle for this, or is there a better, more prominent role for plasma nanotools and approaches? Resolution (and, hopefully, positive) of this vital dilemma is one of the main aims of plasma nanoscience. Research endeavors in this area focus, in particular, on competitive advantages and disadvantages of using plasma-based tools and processes as compared with the leading and most established nanofabrication techniques [23].

At the momment it looks like the only way to resolve the "plasma-or-no-plasma" dilemma is to carry out a detailed investigation into how exactly the majority of nanoscale synthesis processes work. As has already been highlighted above, such processes rely in most cases on guided or controlled self-organization (building units into nanostructures, nanostructures into ordered patterns, etc.) in a specific nanofabrication environment.

Because of the extreme importance of the issue, let us briefly summarize what we have already learned form the previous sections. First, the role of self-assembly processes becomes even more prominent as the sizes of nanoassemblies shrink. Indeed, when the sizes of typical nanostructures become smaller than the presently achievable feature sizes of lithorgaphic patterns and nano-templates, self-assembly becomes the only possible way to control the formation of nanostructures and their self-organization into ordered patterns, the fundamental processes that lead to the formation of the self-organized nanoworld. This is particularly important for nanofabrication of ordered arrays of tiny ($< \sim 10\,\text{nm}$) quantum dots (QDs) or ultra-thin and high-aspect-ratio single-walled carbon nanotubes, which have been successfully fabricated by non-plasma synthesis techniques, such as MBE, MOVPE or CVD.

Therefore, the discussion about the suitability of plasma nanotools for the next-generation of nanofabrication is at the level of their ability to guide self-assembly of building units on solid surfaces, and eventually to create a self-organized nanoworld of "beautiful" (and ultimately properly functioning and useful in applications) nano-assemblies. If the plasma-based methods of controlling self-assembly of building units and nanostructure growth turn out to be competitive in terms of quality, cost efficiency, economic viability and investment risk assessment, plasma nanotools will have a bright future. The anticipated rapid expansion of the nanotooling market, which is expected to exceed $US 1.2 billion in 2008 [46] and is set to rise even further beyond that, makes us quite optimistic in this regard.

Having said that, one should use plasma-based tools and approaches to create a self-organized nanoworld, we should now try to specify how exactly to approach that in a highly-controlled, ultimately deterministic, fashion. Another important aspect is to properly identify the research area of plasma nanoscience and the main issues it deals with. These issues are clarified in the following section.

1.4
Deterministic Nanofabrication and Plasma Nanoscience

Previously, it has been stressed that the uniqueness of any plasma-based nanofabrication environment is the presence of a highly unusual layer of uncompensated space charge that separates the charge-neutral plasma bulk and a nanostructured solid surface. Referring to Figure 1.2, one sees that the typical dimensions of plasma nanofabrication facilities (ca. 0.5 m) differ by at least nine orders of magnitude from the sizes of the atomic building units (ca. 0.1 nm). In this section we will discuss how to challenge one of the previously intractable problems of bridging this nine order of magnitude spatial gap and systematically approach the problem of deterministic plasma-aided nanofabrication. As was suggested earlier [4], one of the possibilities is to manipulate the plasma-generated species in the plasma sheath that separates the plasma and solid surfaces and to control self-assembly of building units into nanostructures on plasma-exposed surfaces or their direct incorporation into growing nanoassemblies.

Owing to enormous problems with the compatibility of *in situ* plasma diagnostics and surface science characterization techniques, researchers have to rely on quite separate experimental studies of the plasma processes and (in most cases *ex situ*) nanostructure characterization. However, there is a vital demand for reliable physical models and numerical simulations that could bridge the "unbridgeable" gap between gas-phases and surface processes separated in space by nine orders of magnitude and generate recipes that can be used in nanofabrication process development.

From the previous section, it becomes clear that plasma-based environments are beneficial for creating solid particles. If such particles can have nanometer dimensions, they can be termed nanoparticles and as such become suitable as building blocks of nanotechnology. Here, by nanoparticles we mean solid grains in a purely crystalline or amorphous phase or a mixture thereof. This is why there is such a large, and continuously increasing, number of reports on nanoparticle synthesis in various plasmas, ranging from low-pressure glow discharges to atmospheric-pressure arc discharges and "submerged" discharges in water. Considering the importance of nanoparticles, which feature large surface-to-volume ratios (which in turn increases their surface reactivity and makes them particularly attractive for applications in chemical catalysis), relevant processes are already state-of-the-art on their own. For a comprehensive review of nanoparticle synthesis in thermally non-equilibrium and thermal low-temperature plasmas the reader can be referred elsewhere [5]. By using plasmas it becomes possible to significantly increase the concentrations and reactivity of assembling species, which eventually gives rise to very high nanoparticle production rates. Moreover, high gas temperatures in thermal plasma discharges are very favorable for the effective and rapid crystallization of solid particles in the ionized gas phase. Highly-crystalline and perfectly shaped nanoparticles can also be synthesized in thermally non-equilibrium plasmas, see for example [47,48].

At this point it would be worthwhile to shed a reasonable doubt on the applicablity of low-temperature plasmas for the fabrication of more delicate individual nanoassemblies of higher complexity and differing dimensionality. A few examples of such more complex nanoscale objects are zero-dimensional (0D) quantum dots (QDs) and tiny nanopores, one-dimensional (1D) nanorods, nanowires, nanohelixes, nanosprings, nanoneedles, two-dimensional (2D) nanowells and nanowall-like structures, periodic heterostructures and superlattices as well as three-dimensional (3D) nanostructures of complex shapes (e.g., pyramids, cones, multifaceted crystals, etc.). Moreover, are low-temperature plasmas appropriate for the fabrication of more complex assemblies of individual nanostructures, such as spatially-ordered patterns and arrays, mixed-dimensionality assemblies (a multilayered 2D heterostructure with zero-dimensional nanodot inclusions is a good example of such an assembly), interlinked networks of nanostructures arranged in ordered nanoarrays and eventually integrated nanodevices? What is even more important, all these nanoscale objects ranging from individual nanostructures to nanodevices should be fabricated at the minimum cost and maximum efficiency, which necessarily demands a substantially reduced number of experimental trials. We also recall that the way of creating such objects should ideally be through controlled self-organization on plasma-exposed surfaces.

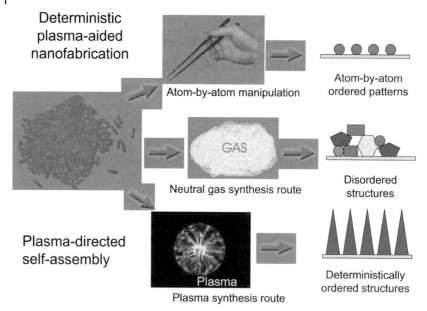

Deterministic plasma-aided nanofabrication

Atom-by-atom manipulation

Atom-by-atom ordered patterns

GAS

Neutral gas synthesis route

Disordered structures

Plasma-directed self-assembly

Plasma

Plasma synthesis route

Deterministically ordered structures

Figure 1.8 Deterministic concept of plasma nanoscience [23].

This leads to the need of pursuing a more efficient, deterministic approach, which is central to plasma nanoscience and is sketched in Figure 1.8. Generally speaking, full determinism means the ability to reach the targets (e.g., the shape, size, ordering and other parameters of the nanoassemblies concerned) using the absolute minumum number of experimental trials. In fact, Figure 1.8 illustrates one of the major aims of plasma nanoscience. It shows a process wherein self-organization of a "handful" of building units (for simplicity visualized here as rice grains) in a plasma environment results in a much more efficient (compared to the extremely time-consuming atom-by-atom nanomanipulation) and better-quality (compared to a neutral gas route) nano-sized product [23]. Identifying such processes and elaborating specific conditions when such a clear advantage of using plasma-based tools, approaches and techniques can be achieved is one of the major thrusts of plasma nanoscience.

Figure 1.9 summarizes the main aim of plasma nanoscience, which in other words is to generate suitable species and in some way control their self-organization in a suitable ionized gas environment. This needs to be done in a highly-controlled, ultimately deterministic fashion. To do this, nature's recipes (e.g., how to create solid nanoparticles in weakly ionized plasmas) should be rigorously followed, modified and optimized to achieve the required determinism not only in nanoscience research

Figure 1.9 Main aim of plasma nanoscience.

but also eventually in industrial nanomanufacturing, which as we have discussed in the previous section, will be based on the plasma-created self-organized nanoworld.

Therefore, the

> *Plasma Nanoscience is a multidisciplinary research area which aims at elucidating specific roles and purposes of the plasma environment in assembling nano-objects in natural, laboratory and technological situations and find ways to bring this plasma-based assembly to the deterministic level in nanofabrication.*

Some of the most commonly asked questions in this research area are: "should plasma be used?", "if so, why?" "what sort of plasmas to use?", "how exactly to use it?", and "what competitive advantages over non-plasma-based routes can one gain?". These scientific enquiries are expected to be directly related to a specific pre-determined goal, such as a nanoassembly with the desired characteristics.

Plasma nanoscience is intimately linked to the physics of plasmas and gas discharges, interdisciplinary nanoscience, surface science, astrophysics, solid state physics, materials science and engineering, structural chemistry, microelectronic engineering and photonics and some other areas [23]. These links naturally come about owing to the intrinsic ability of low-temperature plasmas to generate all sorts of building units ranging from atoms and ions to nanoclusters and nanocrystallites as depicted in Figure 1.10. The processes of building unit generation, transport and self-assembly or incorporation into growing nanoassemblies, accompanied by a suitable surface preparation (Figure 1.10) by other plasma-generated

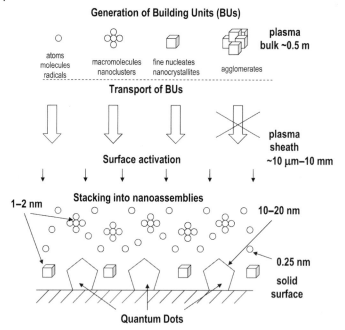

Figure 1.10 Schematics of a typical plasma-aided nanofabrication environment [23].

species (termed "working units" in this monograph) is what necessarily requires input from the above mentioned research areas.

As we have already mentioned above, in a typical plasma environment for nanoscale synthesis and processing (shown in Figure 1.10) the plasma bulk and the nanoworld are separated by the plasma sheath. A typical size for the plasma bulk area can be taken approximately as the typical dimensions of conventional plasma reactors, for example ca. 0.5 m for the integrated plasma-aided nanofabrication facility (IPANF) used in nanofabrication experiments within our research network [49,50]. On the other hand, the width λ_s of the charge non-neutral area (plasma sheath) in the vicinity of the surface critically depends on the plasma (e.g., electron density and temperature) and process (e.g., DC substrate bias) parameters. Nonetheless, λ_s typically ranges from ca. 10 μm to 10 mm as can be seen in Figure 1.10.

It is instructive to note that the ionized gas within the plasma sheath contains uncompensated positive charge. In this case the quasi-neutrality condition ($n_e = n_i$, where n_e and n_i are the number densities of electrons and ions, respectively), the most essential requirement for a plasma to exist, does not hold. Hence, the plasma sheath is no longer charge neutral and cannot be termed plasma in its usual sense.

We can now make a very important conclusion:

Plasma is separated from surface nanoassemblies by several orders
of magnitude.

In other words, the plasma bulk ends several orders of magnitude before
one can even "touch" nanoscale objects on a solid surface! In the example
depicted in Figure 1.10, a representative size of quantum dots is in the
range from a few to a couple of tens of nanometers. Here we recall that
the nanodots most desired for the self-organized nanoworld should have
the size less than 10 nm. Comparing this size with the typical dimensions
of plasma sheaths (ca. 10 μm to 10 mm), one can immediately notice a
huge difference of 3 to 6 orders of magnitude.

This interesting finding immediately prompts the curious reader to ask
a very reasonable question: *Since the plasma does not directly "touch" the*
nanostructures like a common neutral gas, then what is its actual role in the
nanoassembly process?

Let us briefly clarify this issue. If a plasma is partially ionized, it
contains two components, namely the ionized (electrons and ions) and
neutral (all neutrals) gas components. The above conclusion about the
multi-order of magnitude separation applies to the ionized component
only. Therefore, the actual contact of the environment with the surface
nanoassemblies in fact critically depends on the ionization degree of the
plasma

$$\xi_i = \Sigma n^+ / \Sigma n_{tot}, \tag{1.1}$$

where n^+ is the combined number density of all positively charged
species ($n^+ = n^-$ in quasineutral plasmas) and $n_{tot} = n^+ + n_n$ is
the total (combined) number density of neutral and positively charged
species. If $n^+ < n_n$, we have partially ionized plasmas, most common to
laboratory- and universe-based synthesis of nanoscale objects. In cases
where $n^+ \ll n_n$, the plasma is commonly termed weakly ionized. Equation (1.1) is also valid in case of fully ionized plasmas ($n_n = 0$) and yields
$\xi_i = 1$.

Let us now consider the issue of the contact between a plasma-based
environment and a solid surface in more detail. The two components
of the plasma contact the surface in their own ways. The neutrals are
not affected by electric fields in the plasma sheath area and deposit on
the surface via random thermal motion. In a two-dimensional geometry sketched in Figure 1.2, the flux of neutral species impinging on the
surface can be written as

$$j_n = \frac{1}{4} n_n V_{Tn},$$

where V_{Tn} is the thermal speed of neutrals. On the other hand, positively charged ions are accelerated by the electric field and, if ion-ion and ion-neutral collisions are not very frequent, are driven towards the surface. Even though the concentration of ions is usually less than that of the neutrals, the ion flux

$$j_i = n_i V_i$$

can in many cases exceed the neutral flux j_n. This can happen because ions can be easily accelerated to velocities V_i, which can be much larger than V_{Tn}. Therefore, even though the ions may be much less abundant than the neutrals, they can still arrive on the surface in larger amounts. Moreover, the energies of the ions impinging on the surface are usually very different from neutrals; this can cause very different effects discussed elsewhere in this monograph.

Therefore, if the plasma is weakly ionized ($\xi_i \ll 1$, which is the case in many low-temperature gas discharges), the neutral component of the plasma "touches" the surface nanostructures in almost the same way as in thermal, non-plasma-based chemical vapor deposition. However, the ions and the surface charges interfere with this process and eventually make the nanoassembly process quite different. It is amazing that the ionized gas species which in many cases constitute an overwhelming minority, can make a dramatic difference at virtually every growth step of nanostructures. In a sense, this effect is "remote"; intuitively, it is mostly related to electric fields, electric charges, and ionized atoms/radicals not otherwise available in neutral gas environments.

If the ionization degree is of the order of unity (which can be the case in various i-PVD schemes), then the surface is mostly exposed to intense ion fluxes rather than the neutral fluxes. In this case the ion- and charge-related effects lead to the creation of a very unique nanofabrication environment, impossible in any neutral gas-based routes.

As we can see from these very basic arguments, the presence of ionized species, uncompensated space charge, charges on solid surfaces and nanostructures and electric fields can make a dramatic difference in a very large number of processes that involve nanoassembly synthesis and processing. Most importantly, this difference can be quite substantial even if the fraction of ions among all atomic/radical species in the gas is very small!

Plasma nanoscience aims to shed some light on this issue and quantify the related effects. As was proposed earlier [4], this important issue can be approached systematically by following the sequence of events that occur when plasma-generated species cross the near-surface sheath area and self-assemble on (charged) plasma-exposed solid surfaces or incorporate into already existing nanoassemblies. Let us now turn our attention to the "plasma-building unit" nanofabrication approach that involves a range of specific working units as a primary cause of the growth of nanoassemblies ("effect") [4] and describes the above mentioned events. This approach will be considered in detail in Chapter 2 of this monograph. We emphasize that fully deterministic synthesis can only be achieved by following the cause and effect sequences involved in any particular nanoassembly process. In addition to the already introduced concept of building units, the "cause and effect" approach uses another notion of working units (WUs) to reflect the fact that some of the plasma-generated species (BUs) work as a primary building material for the nanoassemblies whereas others serve different purposes such as surface activation or passivation, reactive chemical etching, physical sputtering, and so on and for simplicity are referred to as WUs.

It is noteworthy that while building units are being transported to the surface nanoassemblies from the plasma bulk through the plasma sheath, the solid surface is being suitably prepared (by specific working units, e.g., argon ions or reactive radicals) to accommodate the deposited building units. Depending on the specific requirements, these working units can activate or passivate surface dangling bonds, alter the surface temperature, modify surface morphology via chemical etching or physical sputtering processes and perform some other functions. The last step discussed in the original publication [4] was to appropriately control the fluxes and energy of building units that tend to stack into nano-patterns being assembled.

However, as it turns out, in many cases this is just the beginning of the surface stage of the story. Indeed, it is extremely important where and how exactly the building units land onto the nanostructured surface. Depending on prevailing surface morphology and temperature, as well as the energy and incidence angle of impinging species, there can be an overwhelming variety of different possibilities.

Let us briefly consider some of these possibilities and begin with low-energy species that land on surface areas unoccupied by nanoassemblies. Such species are usually adsorbed at the surface and show the ability to migrate from one site to another. The notion used for such species is formed by adding the prefix "ad-" to their names. For example, the term adatom means an adsorbed atom, adradical denotes an adsorbed radical, and so on. In most cases it is implicitly assumed that any electric charge the species may have had in the gas phase is completely dissipated/neutralized upon adsoption on the surface. In neutral gas-based chemical vapor deposition and related processes the surfaces are charge neutral. However, this is not so obvious for plasma-based routes. Indeed, surface charges and/or currents make interaction of the incoming species with the substrate quite different. In some cases plasma-generated ions can retain their electric charge (or at least remain charged) upon deposition and therefore become adions. Another possibility is that plasma-generated neutral species become polarized either upon a close approach or chemisorption to the (possibly charged) nanostructured surface. In this case, microscopic electric fields in the vicinity of the substrate can substantially redistribute the polarized species about the surface as compared to the purely thermal chemical deposition case.

For simplicity, we will deal mostly with low-energy neutral adatoms. Migration of such adatoms is primarily controlled by the substrate temperature and material, morphology, chemical structure and other properties of the solid surface. Phenomenologically, these properties are reflected by the diffusion activation energy, which is usually calculated using atomistic simulation approaches, such as the density functional theory.

Upon migration from one surface site to another, adatoms can collide with other adatoms or surface features (such as defects, dislocations, bunched terraces, etc.) and form small clusters which in turn can serve as seed nuclei for nanoassemblies being created. It is imperative to note that adatoms migrate to where it is most energetically favorable for them to move. For example, consider a crystal with a few different facets; all these facets (numbered using superscript i) have quite different diffusion activation energies ε_{da}^i. Therefore, adatoms will prefer to move towards a facet with the lowest ε_{da}^i. Indeed, it saves a great deal energy to hop to a site where less energy is required to enable surface diffusion.

Adatoms can also leave the surface by desorbing and/or evaporating back to the bulk of the gas phase or join the two-dimensional vapor which remains in the immediate vicinity of the surface. Eventually, adsorbed species can find a suitable surface site to form chemical bonds with the surface atoms; this process excludes them from any further migration about the surface. Interactions between the plasma-generated species and the surface also include chemical etching which happens, for example, when highly reactive working units extract volatile species from the solid surface. We emphasize that the exact scenario ultimately depends on the relative chemical reactivity and affinity (which reflects elemental compatibility from the energetic point of view) of the building units and the host surface.

Using plasmas for nanofabrication has another indisputably attractive feature; this feature is related to the possibility of ion acceleration to relatively high energies and using highly energetic ions in surface modification and processing. If the ion energy upon landing is high, some sort of surface damage is inevitable. Physical sputtering is perhaps the simplest effect caused by energetic species; more complex phenomena may include ion subplantation as well as more substantial structural transformations in the material exposed to such fluxes. It is amazing that high-energy species can also do useful things when they crash onto the surface. Substantially improved crystallinity and structural transformations of amorphous materials to the crystalline state under the action of reasonably enertgetic ions is one salient example of this effect. We reiterate that this option is unique to ionized gas environments and in most cases involves appropriate substrate biasing.

From the above simple considerations, it becomes clear that using plasma environments in nanofabrication or surface processing does make a substantial difference as compared to the neutral gas processes. On the other hand, it is commonly understood that the plasma is a more complex environment than an equivalent neutral gas. The main evidence of a

higher complexity of the plasma environment is in the presence
of the ionized component

otherwise non-existent in charge neutral gases. Moreover, the presence of even a small fraction of ionized component and associated electric fields dramatically improves the plasmas ability to generate the entire range of building and working units in atomic, molecular, cluster and other forms [4].

Therefore, for the purpose of deterministic nanoassembly it is crucial to selectively generate and manipulate the required BUs and WUs. It goes without saying that different nanoassemblies and nanofabrication processes require very specific control strategies; some of them may be appropriate for one sort of plasma-generated species and completely ineffective for another. Therefore, which recipes should one use to fabricate the desired nanoassembly in a plasma?

Apparently this question has no general answer, with the number of possible solutions exceeding the number of presently known nanoassemblies. It was therefore proposed that the problem of choice of the appropriate building units (to be generated in the plasma) can be based on the "cause and effect" logic sequence:

precursor → building unit(s) → nanoassembly,

which also requires a feedback/optimization procedure, which will be discussed in detail in Chapter 2. The above choice should be supported by the existing knowledge from other areas. For example, sophisticated surface science experiments or atomistic simulations can shed some light on what species are most suitable for each particular purpose.

Otherwise, an infinite sequence of trials (these trials can for example be aimed at generating larger densities of specific precursors) can go to nowhere due to a huge number of abundant species (including highly-reactive ones) and polymerization, clustering, and nucleation scenarios in a plasma. Therefore, research efforts in the plasma nanoscience area are usually based on assumptions on the specific building units that are needed for the desired plasma-synthesized nanostructured materials.

Without trying to provide exhaustive recipes for the appropriate choice of building and associated working units (this in fact deserves to appear in the near future as the Encyclopedia of nano-assemblies and their building units), one can state that the relevant choice should be motivated by the structural considerations of the nanoassemblies being created. At this stage, it would be reasonable to appeal to the established theories of growth kinetics of specific nano-sized objects.

For example, in the assembly of open-ended carbon nanotubes (with a simple chiral structure) or ultrananocrystalline diamond, one can use re-active dimers C_2, which can appropriately insert into carbon atomic networks on reconstructed surfaces. Other details of the building unit-based "cause and effect" approach of the plasma nanoscience will be discussed in Chapter 2 (see also the original article [4]).

As was mentioned above, stacking or incorporation of plasma-generated building units into a developing nanoassembly can proceed via two major routes. The first route involves landing of the plasma-generated building units onto open surface areas followed by their surface migration from the deposition site to the nanoassembly site. The other pathway for the BUs to stack into the nanopattern being synthesized is via their direct incorporation from the low-temperature plasma [51, 52]. From Figure 1.10 one can clearly notice a huge (up to nine orders of magnitude!) difference between the spatial scales of the area where the building units are generated (ca. 0.5 m, which is a typical dimension of plasma reactors), the nanoassembly sites (ca. 5–20 nm, which is a typical size of quantum dots), and atomic/ionic/radical building units themselves (ca. 0.1–0.25 nm).

Therefore, in an attempt to achieve a fully deterministic plasma-based synthesis of surface-bound nanoassemblies, one needs to "bridge" the spatial gap of nine orders of magnitude to be able to generate, manipulate, and insert the building units into the nanoassemblies being grown. And all this needs to be done in a highly-controlled (ultimately deterministic) fashion enabling one to reduce the number of experimental trials and errors to the absolute minimum.

Figure 1.11 shows the sequence of events involved in the process of bridging the processes occuring in the plasma bulk and on solid surfaces;

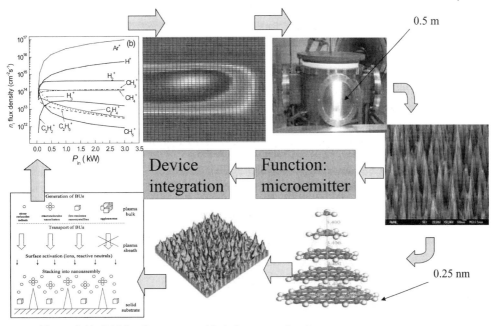

Figure 1.11 Bridging the macroworld of plasma reactors to atomic arrangement in nanoassemblies [23]. An example of carbon nanotip microemitter structures is shown.

some of these processes are characterized by spatial scales that differ by up to nine orders of magnitude and even more [23]. Let us describe the sequence of research steps involved in such a synthesis. First and foremost, one should be very clear on exactly what nanostructure is required, what are its sizes, shape, structural properties, and so on. In the example considered, high-aspect-ratio (sharp) conical nanotip-like nanocrystalline structures are of interest.

The first step in this direction is to figure out possible chemically stable atomic structures with the required shape and aspect ratio [23]. This can be achieved by using the *ab initio* atomistic density functional theory (DFT) simulation of downscaled (to within the acceptable number of atoms the most advanced computations can handle; at present this is a couple of hundred atoms) carbon nanotips [50]. One such atomistic structure of a downscaled carbon nanotip is shown in the bottom right side of Figure 1.11. Using energy minimization principles, one can work out stable configurations of the nanoassemblies concerned. In particular, it turns out that single-crystalline carbon nanotips are most stable if their lateral surfaces are terminated by hydrogen atoms as can be seen in Figure 1.11. Using these results, it is possible to work out specific aspect (height to radius) ratios, which the nanostructures may have.

We stress that such atomistic simulations do not describe the process of synthesis of the nanoscale objects concerned. However, they can specify the numbers of atoms located in the bulk or on the surface of the nanoassemblies. These data can then be used in the modeling of controlled and site-specific delivery of building units. Once this is done, it is possible to proceed with the modeling of the actual growth process which involves the two main routes of building unit incorporation into the nanotips being grown. As has already been mentioned above, this can be done via BU diffusion about the surface or their direct insertion upon deposition onto nanostructure surfaces. The next thumbnail figure in Figure 1.11 shows a three-dimensional microscopic topology of ion fluxes distributed about an ordered two-dimensional pattern of carbon nanotips [51].

Having estimated the rates of arrival of different species to specific nanoassembly sites on nanostructured surfaces, it is possible to formulate the process conditions for the optimized delivery of the plasma-generated building units to where they are actually needed. The fluxes of ionic species are most effectively controlled by the parameters of the plasma sheath, such as the potential drop across it. Moving backwards from the desired characteristics of nanostructures, one can elaborate the parameters of the plasma sheath (shown in the next thumbnail figure in the bottom left corner of Figure 1.11), such as the value of the DC substrate bias. It is prudent to mention here that the electric field magnitude, sheath width and the energy of the plasma ions significantly affect the surface temperature, which in turn dramatically influences the nanostructure growth. For example, additional heating and activation of the surface of nickel-catalyzed silicon substrates by intense ion fluxes turns out to be a decisive factor in low-temperature synthesis of carbon nanotubes and related structures.

Meanwhile, the fluxes of the building units are intimately linked to the plasma parameters, such as the electron temperature, number density of electrons/ions, neutral gas temperature, species composition and some others. This logic link is reflected by the next thumbnail figure in Figure 1.11 which shows a representative composition of thermally non-equilibrium plasmas sustained in a mixture of argon, hydrogen and methane gases. In the same figure, the dependence of the surface flux of cationic species on the input power applied to sustain the discharge is also shown [53].

The next logical step in this direction is to use the information on the composition, number densities, energies and fluxes of the required building units as input conditions in two-dimensional fluid modeling of the species and energy balance in the plasma discharge. Such modeling can generate detailed spatial maps of the densities/temperatures/energies of the most important charged and neutral species inside the plasma reactor. A typical two-dimensional distribution of neutral radical species in the integrated plasma aided nanofabrication facility (IPANF) [49] is shown in the next thumbnail figure in the middle of the top row in Figure 1.11 [54].

In the above, we have mapped the way from the atomistic carbon nanotip structure on the bottom right in Figure 1.11 (which, in fact, is much smaller than the actual carbon nanotip microemitter structure) to the spatial profiles of the main plasma-generated species in a macroscopic (with ca. 0.5 m dimensions) plasma reactor (the third figure from the left in the top row in Figure 1.11) used in nanofabrication of the carbon nanotip microemitter structures in question [49, 50, 55]. The numerical results mentioned so far can be used to optimize the parameters of trial laboratory experiments and eventually commercial nanofabrication processes [23].

This parameter optimization can be implemented through experimental verification of numerical results on spatial distributions of neutral and ionized atomic and radical plasma species in the plasma reactor concerned. Relevant experimental approaches can include Langmuir probe (LP), optical emission spectroscopy (OES) and a range of mass spectrometry diagnostic techniques. Application of these plasma diagnostic techniques to monitoring various plasma-based nanoassembly processes is discussed in detail in a recent monograph [1]. For example, one can match the experimentally measured and computed values of the electron number density by placing the probe tip at various spatial points and adjusting the gas pressure, gas flow rates and the RF input power. If the densities of negatively charged species other than electrons (e.g., anions or dust grains) are low, the electron and ion number densities will be approximately the same. The value of the substrate temperatures can be estimated by considering several factors that include external heating sources, heat conduction of the substrate material and the gas ambient, radiative losses, as well as the intensities of the plasma ion fluxes onto the surface. The calculated/measured changes of the surface temperature due to the ion bombardment may allow one to quantify the effect of the plasma environment on the deterministic nanoassembly process being developed. In the case of the ordered patterns of vertically aligned carbon nanotips shown in the thumbnail scanning electron micrograph on the far right in Figure 1.11, the relations between the computed and experimental values of the ion/radical densities and fluxes have been used to substantially reduce the number of experimental trials [49, 50, 55].

The outcomes of nanostructure synthesis are commonly investigated by using a range of analytical tools of materials science and surface

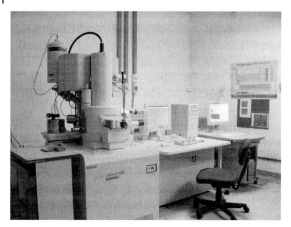

Figure 1.12 A Field Emission Scanning Electron Microscope,
a common high-resolution tool for surface analysis.
Photo courtesy of the Plasma Sources and Applications
Center of Nanyang Technological University, Singapore.

science such as scanning electron microscopy (SEM) (a field-emission
scanning electron microscope is shown in Figure 1.12), atomic force microscopy (AFM), scanning tunnelling microscopy (STM), high-resolution
transmission electron microscopy (HRTEM), Raman spectroscopy, low-
energy electron diffraction (LEED), low-energy electron microscopy
(LEEM), X-ray photoelectron spectroscopy (XPS), X-ray diffractometry
(XRD), secondary ion mass spectrometry (SIMS), secondary neutral mass
spectrometry (SNMS) as well as several other techniques.

At the end of the relatively long chain of processes (with every one of
them occurring at quite different temporal and spatial scales) depicted in
Figure 1.11 one should expect an array of shaped and structured carbon
nanotip microemitters arranged in ordered spatial arrays. If the quality of the final product meets the expectations, it is possible to proceed
with testing the nanoassemblies for their performance in microemitter
devices. After such tests are completed and if the results are encouraging, one can move on to the final process step, namely, nanodevice integration. Thus, the arrays of crystalline nanotips need to be properly integrated into microemitter devices. From the practical perspective this step
can be implemented by the growth of the carbon nanotip arrays directly
in the specified device locations. In this case, low-gas-temperature conditions of gas discharges are extremely useful for direct *in situ* processing
of nanoelectronic features with interlayers and interconnects, which can
melt very easily because of their ultra-small thickness (down to ca. 1 nm
and even thinner).

However, as was mentioned in the original article [23], if the quality of the nanostructures does not meet the required standards, the entire process or some of its cycles need to be repeated and new, even better optimized, process parameters should be used. In this way the process stages and parameters can be optimized within any segment of the nine-order of magnitude "bridge" in Figure 1.11 until the desired outcome is achieved [56]. Some would argue that this is often easier said than done in practical applications because of the huge number of processes in the ionized gas phase and on the surface. However, such practical difficulties can be overcome by studying a lumped effect of any change in a single process parameter, one-by-one, in order of their decreasing importance (e.g., the surface temperature first, then the substrate bias, the working gas composition, the input power, and so on) [23].

To conclude this section, we emphasize that the process milestones of Figure 1.11 quite accurately reflect the main essence of research endeavors in the plasma nanoscience area. Finally, practical implementation of this approach (which is applicable to virtually any nanoassembly and can be used to deterministically create the nanoworld we want) requires well-coordinated and concerted experimental, theoretical and numerical simulation efforts; each of these efforts can focus on processes that occur at specific spatial scales.

1.5
Structure of the Monograph and Advice to the Reader

Let us now make it more clear which specific material one should expect in this monograph. Structurally, the monograph consists of this introductory chapter (Chapter 1), 7 main chapters, a concluding Chapter 9, and two Appendices A and B.

In the introductory Chapter 1, we have already introduced the main aims, notions and concepts of plasma nanoscience. It has also been stressed that the most viable direction for the future development of nanoelectronics, as well as the nanoscale materials synthesis and processing should be based on plasma-guided self-organization of ultra-small nanoassemblies, creation of their ordered and interlinked networks and eventually fully functional nanodevices. The plasma-based approach to creating nanoparticle matter is common in nature's nanofab, which relies heavily on the plasma environment to substantially increase the nucleation rates of cosmic dust which is in turn essential for maintaining the chemical balance in the universe. We have also introduced the determinism, one of the most important concepts of the nanoscience and have discussed a conceptual pathway to achieve it through bridging, via a chain of various plasma-assisted processes, a spatial gap of up to nine orders of magnitude between the sizes of the plasma reactors and atomic building units.

In Chapter 2, we will introduce the concept of plasma-based nanoscale assembly based on the "plasma-building unit" approach. This approach considers plasma-generated building units as a "cause" and the desired nanoassemblies (or a specific nanoworld made of such NAs). A range of important issues related to salient features of low-temperature plasma environments (such as the plasma sheath) is also discussed. We also specify how the plasma environment can affect some elementary processes on solid surfaces. In Chapter 2, the reader can find an answer to what exactly makes low-temperature plasmas a versatile nanofabrication tool of the nano-age.

Chapter 3 explains how the "plasma-building unit" approach introduced in Chapter 2 may be used in the nanofabrication of a range of nanofilms and low-dimensional nanostructures made of most common semiconducting and carbon-based materials. The details of the sophisticated numerical and computational approaches that can be used to bridge the spatial gap of up to nine orders of magnitude (see Figure 1.11) are introduced. This practical computational framework is used in different sections of this monograph.

In Chapters 4 and 5 we continue the study of specific plasma-based effects within the "plasma-building unit" approach. The main focus of this approach is to generate appropriate building and working units. For this reason the whole Chapter 4 is devoted to the consideration of different possibilities for generating the required species in low-temperature plasmas. The plasmas considered also include reactive plasmas, where it is possible to create a broad range of reactive radical species and also nanoclusters and nanoparticles. The main focus of Chapter 4 is on thermally non-equilibrium low-temperature plasmas of silane- and hydrocarbon-based gas mixtures. However, examples of plasma-assisted nanoparticle nucleation and growth in different environments (such as in very low-density stellar ouflows and very high-density pulsed laser ablation of solid targets) are also introduced and discussed.

Chapter 5 focuses on various aspects of the delivery of a range of plasma-generated building units to the nanoassembly sites on solid surfaces. In particular, it describes a way to control microscopic ion fluxes with subnanometer precision and deposit ions onto specified areas on nanostructured surfaces. Electric fields sustained in the plasma sheath and created by surface nanostructures turn out to be powerful control tools. By properly using a combination of forces representative of plasma environments, one can effectively manipulate the plasma-grown nanoparticles in the plasma sheath area and in this way control their deposition onto selected areas on nanostructured surfaces.

In Chapter 6, written jointly with I. Levchenko, we discuss the basic ideas and approaches of surface science of plasma-exposed surfaces. The main accent here is placed on the explanation of highly unusual features the elementary processes acquire owing to the presence of electric charges, ion fluxes and other plasma attributes. The material in this chapter is centered around the demonstration of the possibility of formation of self-organized arrays of size- and position-uniform arrays of quantum dots on solid surfaces facing the plasma environment. Practical ways to implement a range of other requirements that are essential for the eventual applications of such nanodot arrays in nanodevices are also discussed. In this chapter, the reader will also find a range of plasma-specific effects on nanoscale self-organization on solid surfaces.

Chapter 7 is devoted to a specific class of nanoscale objects that show a strong ability to focus ion fluxes. The examples of such objects include single- and multiwalled nanotubes, nanotips, nanoneedles, nanocones, nanorods and some other one-dimensional nanostructures. The plasma and, in particular, ion fluxes exert a significant effect on the growth of such nanostructures and make it very different compared to similar neutral gas-based processes. The examples of advantages offered by the plasma-based fabrication routes considered in Chapter 7 are higher growth rates, better size and positional uniformity of nanostructure arrays, vertical alignment, controlled reshaping and several others. One of the most exciting examples introduced in Chapter 7 deals with the unique possibility of using plasma-controlled self-organization to synthesize uniform arrays of carbon nanocones from essentially non-uniform nickel catalyst nanoislanded films.

Examples of using various plasma-generated building and working units in nanoscale applications are shown in Chapter 8. In the first example, it is shown that a suitable variation of the plasma and sheath parameters can enable electric field-related control of ion-assisted post-processing of arrays of nanotubes and nanorods with different densities. It is demonstrated that ion- and plasma-assisted processes offer a great deal of advantages (compared to the neutral gas routes) in terms of charged species penetration into the areas inaccessible by the neutral species. In another example, we demonstrate the possibility of synthesizing ordered arrays of gold nanodots using nanoporous template-assisted ionized physical vapor deposition (i-PVD). Reactive plasmas can also be used to generate building units on solid surfaces as is the case in the synthesis of metal oxide nanostructures such as nanopyramids and nanowires. Nanocluster building units can be successfully used to synthesize nanostructured titanium dioxide films with excellent biocompatible properties. In particular, by capitalizing on size-dependent properties of such nanoclusters one can control the relative presence of rutile and anatase phases in the film.

The monograph concludes in Chapter 9 with a brief summary of current issues of the plasma-aided nanofabrication and an outlook for future directions in this exciting research area. In particular, Chapter 9 further elaborates on the issues of determinism and complexity, summarizes some of the most salient benefits and advantages in nanoscale assembly offered by the plasma-based processes and approaches, as well as providing a concise outlook for the future developments in the area.

Despite a very large number of relevant works cited in this monograph we did not aim to provide an exhaustive coverage of the current status of the major research efforts in the area of plasma-based nanoscience and nanotechnology. Such was clearly impossible to implement given the limited size of this work, and even more importantly, because of extremely limited time budget of the author. Nonetheless, Chapter 9 contains a link to Appendix B, which briefly outlines a large number of other reasons why nanoscale synthesis and processing should be ultimately plasma-based.

As we have stressed in the preface, this monograph is primarily based on personal research experience of the author and refelects his personal views on a range of relevant issues. Most of the results discussed in this work have been published in high-impact international research journals. These results have been put in the context of the "plasma-building unit" approach advocated by the author in his earlier publication [4]. This makes this work a little specialized and more suitable for researchers, academics, engineers and postgraduate students. However, tertiary college and school teachers and undergraduate students may also be interested to understand how the advocated generic nanofabrication approach works in a large number of applications. Moreover, anyone interested in general science is encouraged to browse this work to see how intricate phenomena in very complex systems can be eventually explained using commonsense approaches supplemented by solid scientific findings. This is why the level of presentation varies from a very simple, commonsense-based to highly-technical with multiple formulas and graphs. Moreover, a large number of visualizations and illustrations should make the basic concepts and ideas of this monograph easily understood by a broad audience with a very limited specialist knowledge. Above all, this monograph can also serve as a textbook or a reference manual for third-year undergraduate, Honours and postgraduate courses.

Readers are also highly recommended to familiarize themselves with the contents of the sister monograph "Plasma-aided nanofabrication: from plasma sources to nanoassembly" [1] which gives a number of essential practical hints on how to appropriately choose the plasma and develop processes and facilities suitable for the envisaged nanoscale applications. Of particular importance, especially for broad audience, is the introductory section [1] which explains what is a plasma, what are the most important, from the nanofabrication perspective, issues in nanoscience and nanotechnology and how to choose the right plasma type with certain features for the envisaged nanoscale applications. The lists of references in these two monographs are complementary and should be appended to each other. However, even this will not cover the whole range, and exponentially increasing, number of publications related to applications of low-temperature plasmas at nanoscales. To convince yourself and to observe what is happening in the area, the reader is strongly encouraged to do the subject search "nano and plasma" using any major research database such as the ISI Web of Science or Scopus.

Finally, all the best with the reading (which is expected to be enjoyable and relatively easy yet not effortless) and feel free to ask any questions!

2
What Makes Low-Temperature Plasmas a Versatile Nanotool?

This chapter introduces a generic "cause and effect" nanoassembly approach, which is centered around plasma-generated and processed precursor species (building units) and their interaction with nanostructured surfaces ("nanoworld"). The main concepts of this approach have already been reported in the original work [4]. In this chapter the main points of this approach are summarized and clarified. Recent advances in our basic understanding of how this approach works in various nanoassembly processes are also highlighted alongside the most recent developments in this very active and highly topical research field.

This chapter introduces several unique features of low-temperature plasmas that make them an indispensable nanofabrication tool in a number of common nanoassembly processes. The main focus here is in the building units (BUs) and their deposition onto the surface and stacking/self-assembly into the nano-scaled object being targeted. In a sense, the plasma-generated building units can be interpreted as the most fundamental building blocks of nanotechnology. Associated working units (WUs), not highlighted in the original publication [4], and their interactions with ultra-small features, such as micro-/nanopores, defects, dislocations and submicron-sized trenches on nanostructured surfaces are also considered. From this point of view, working units can be considered as the elementary bits of matter responsible for processing of matter at nanoscales. These species have a specific function, yet they do not act as primary building material of nanoassemblies.

In Section 2.1, the most basic ideas and major issues are highlighted. The next section (Section 2.2) introduces general considerations and explains the main points of the plasma-generated building unit-based assembly of various nanoscale objects. In Section 2.3 we consider some of the features of low-temperature plasmas most useful for nano-scale fabrication and processing. In the next section (Section 2.4) the best way to choose and generate appropriate building and working units is discussed. Section 2.5 reveals the main effects of the near-surface plasma sheath on the nanoassembly processes of our interest. This chapter con-

Plasma Nanoscience: Basic Concepts and Applications of Deterministic Nanofabrication
Kostya (Ken) Ostrikov
Copyright © 2008 WILEY-VCH Verlag GmbH & Co. KGaA, Weinheim
ISBN: 978-3-527-40740-8

cludes with Section 2.6 where the basics of the influence of the plasma process environment on surface processes are introduced.

2.1
Basic Ideas and Major Issues

Management and control of the assembly of various forms of nanostructured matter – involving numerous global pathways, specific scenarios and building units – is the ultimate crux of modern nanoscience. Various liquid, gasous, solid, colloidal systems (and their combinations) and fabrication methods have to date been successfully used for the fabrication of a myriad of nanoscale assemblies. The choice of any particular precursor, synthesis method and process environment ultimately depends on the desired nano-scaled object or assembly.

The first step in this direction is to generate the required building units with the desired size, shape, morphology and in the correct energetic, chemical and bonding states. It is also important to work out which specific working units are needed, for example, to prepare the surface to accommodate the BUs when they arrive. Depending on the actual process, either BUs or WUs may not even be necessary. For example, top-down nanofabrication based on sculpting of an already small object does not rely on any building units and only requires appropriate working units. Suitable working units can be in the form of plasma-generated reactive and/or energetic species which can effectively interact with the surface.

Some specific working units may be needed when BUs are not readily accepted by a host surface, which usually happens because of insufficient reactivity or very short lifetime to form chemical bonds with the atoms of the surface material. In some cases, working units are needed to suitably heat the surface to prepare it for the landing of building units. This structuring is required, for example, for the synthesis of carbon nanotubes on nickel-catalyzed silicon surfaces. In this case it is commonly accepted that the size of nickel islands determines the nanotube diameter. Also, these nickel nanoparticles need to be hot enough to support the extrusion of carbon material needed to build the nanotube walls. The required level of heating can be achieved using bombardment by the plasma ions, and in some cases even without external substrate heating.

On the other hand, if a surface can accommodate building units without any specific pre-treatment or, alternatively, when freestanding (non-surface-bound) nanoassemblies are created, then additional working units may not be required. In some situations, building units are reactive enough to interact with the surface and to create to dangling bonds with which to attach. Moreover, if the energy of the BUs is high enough for their subplantation into the bulk of the surface material, then additional working units are not needed.

The next step is to choose the most appropriate process parameters for the gas-phase generation of the desired building units and their transport to the deposition surface. Subsequently, special attention should be paid to finding the optimal process conditions for stacking the BUs into the assembly (which in general can be airborne, floating, freestanding, or solid substrate-bound) via direct or WU-assisted incorporation into the nanoassembly or via nucleation or condensation into a new nanoscale object. Whenever a nano-feature on a solid surface needs to be processed (e.g., etched, activated/passivated or sputtered), the optimal conditions for the creation and manipulation of suitable working units also need to be calculated.

Here we stress that the control of the generation and assembly of the building units into the nanofilms and nanostructures, alongside the generation of the desired reactive species (WUs) that can process nano-features with ultra-high precision, remains a vital condition for the development of advanced nanofabrication processes. However, the question of how exactly one chooses the most appropriate plasma species remains essentially open. The answer to this question is essentially process-, assembly- and feature-specific: different nanoassemblies may require quite different building units even if they are made of the same material. Moreover, exactly the same nanoassembly may be built using quite different combinations of building and working units in different nanofabrication processes. This is why worldwide a large number of international teams are conducting research into finding the most suitable and effective precursors in various nanofabrication processes. Using the terminology of this chapter, these efforts are centered around unwinding a very complex chain of interlinked processes involved in the BU creation → transport → deposition → nanoassembly sequence.

The examples considered in this monograph highlight the unique ability of low-temperature plasmas to generate the desired building and working units (encompassing the entire range of species from atoms and molecules to nano-sized clusters, particulates and agglomerates) through numerous reactions in the gas phase which include, but are not limited to, ionization/dissociation and polymerization processes. These processes commonly occur during plasma enhanced chemical vapor deposition (PECVD), which is perhaps one of the most widespread processes in plasma processing and synthesis of solid materials. In specific case studies we consider basic physical phenomena in plasma-based nanofabrication facilities, and try to elucidate the most important competitive advantages and benefits of using the plasma-assisted processes as compared to most commonly used neutral gas routes.

One such case study is related to the PECVD of carbon nanotubes (CNTs) on Ni-catalyzed Si surfaces, arguably the most commonly known nanostructures. Apparently, CNTs require carbon-bearing building units, which can be carbon atoms, hydrocarbon radicals, small carbon clusters or graphitic nanofragments. Most of the required species can be generated by either thermal or plasma-assisted dissociation of hydrocarbon precursor gases such as methane, acetylene, and so on mixed with other inert or reactive gases.

In thermal chemical vapor deposition, hydrocarbon precursors can only dissociate upon landing on sufficiently hot solid surfaces. Substrate temperatures required for thermal dissociation of precursor species are usually very high (typically ranging form ca. 800–1200 °C and even higher), which is far above the present-day demands of semiconductor micromanufacturing. Indeed, metal interconnect layers in the existing 65 nm ULSI technology node already reach 1 nm in thickness. Such ultra-thin layers can easily melt if heated to 500–600 °C. Therefore, the integration of carbon nanotube-based elements into ultra-large scale integrated circuits becomes a major issue of thermal compatibility of CNT synthesis processes with the established silicon-based microelectronic technology.

On the other hand, in plasma-based processes, substantial dissociation of hydrocarbon precursors into ionic and radical species can be achieved in the ionized gas phase. Moreover, ion bombardment can significantly reduce thresholds for a number of surface processes, which may, for example, enhance surface mobility of adsorbed species. As a result, process temperatures (of both working gases and substrate surfaces) can be significantly reduced, whereas deposition rates can become much higher.

However, one should be extremely cautious not to overdo species generation. Indeed, the number of generated precursor species can be very high and reach several hundred in reactive plasmas. Moreover, concentrations of dissociated species, and hence, their deposition rates can be undesirably high.

In most applications high rates of film deposition is a definite advantage. However, in nanoassembly of delicate nanoscale objects such as single-walled carbon nanotubes, arrays of quantum dots or ultra-thin quantum wires only a limited amount of building units are actually required. Uncontrolled overproduction of BUs can lead to large accumulations of unconsumed species, or, alternatively, growth of other structures or features on the surface. It goes without saying that large amounts of other species that do not participate in the synthesis process can easily compromise the whole effort. Because of the high complexity and very large number of elementary processes in the gas phase, it is generally believed that plasmas are a lot more difficult to control than neutral gases.

In part due to the substantial lack of appropriate control tools, approaches and techniques, a lot of time and effort was required before it was possible to synthesize surface-bound single-walled carbon nanotubes in low-temperature plasmas; SWCNTs had earlier been assembled using thermal CVD methods. Until the early 2000s, there existed a common disbelief that plasmas could even be used to fabricate single-walled nanotubes. However, with substantial advances in the control capabilities of plasma nanotools and techniques it is nowadays very common to perform nanofabrication processes ranging from "traditional" high-precision etching of sub-100 nm features in silicon and deposition of ultra-thin and highly-conformal nanofilms to nanoassembly of ordered arrays of tiny quantum dots, single-walled carbon nanotubes and even subplantation of extra-small amounts of phosphorus ions for quantum computing applications.

Therefore, it looks like there are almost no nanoscale objects for which plasmas are not suitable. Nonetheless, most of the achivments still rely heavily on "trial and error" and certainly, good luck. This is why the quest for deterministic nanofabrication should be continued towards complete elimination of trial and error practices and introducing the appropriate level of certainty in the plasma-based processes.

As mentioned above, the choice of a particular precursor medium and assembly method is dependent on the desired nanoscaled object or assembly. Nonetheless, according to the commonsense "bricklayer's approach", most of the building processes (including nanoassembly) proceed in the following sequence [4]:

1. choice and preparation of appropriate building units;

2. preparation of the surface where the BUs will be deposited;

3. transport of the building units to the assembly;

4. appropriate stacking of the BUs into the assembly.

Philosophically, no matter what the assembly is (a conventional brick wall or an exotic nanostructure), it commonly requires a sequence of appropriate manipulations of the building units, which appear to be the most essential parts of the assembly.

Following the arguments of the original article [4], let us consider what this means in the context of the plasma-assisted nanofabrication. Thus, our focus here is on the plasma-assisted generation of the building units and their deposition onto the surface and subsequent stacking into the desired nanoassemblies on solid surfaces. However, this approach is also applicable for freestanding nano-scale objects that do not require any substrate support.

In this chapter, by following the above approach for deterministic nanoassembly, we will focus on some of the most typical plasma-based systems and discuss the unique features of such systems that make them particularly attractive for advanced nano-scale applications. We will also attempt to systematically address, among others, the following issues:

- When exactly should plasma-based processes be used and when has PECVD a competitive advantage over thermal CVD?

- How can we develop effective nanofabrication processes to synthesize the desired nano-films, nanostructures, nanoarrays and elements of nanodevices?

- In which situations should one avoid using plasma-based processes?

- What is the actual role of the ionized gas environment in plasma-assisted assembly of nanofilms and nanostructures and ultrahigh-precision reactive anisotropic etching of nano-features on nanostructured substrates?

- What is the specific (and, hopefully, positive) role of the plasma sheaths, precursor dissociation and plasma polymerization in the nanoassembly processes being considered?

- How can we predict which mechanisms for creating building units (e.g., in the gas phase or on the surface) prevail?

- How can we transport the plasma-grown BUs to the solid surface and stack into the required nanoassembly?

- How can we selectively manipulate the rates of building unit delivery via two different channels: directly from the ionized gas phase and via migration about nanostructured surfaces?

- How can we control the fluxes and energy of reactive species in a broad range of plasma-aided processes, including nanoparticle synthesis, nanopattern and nanofeature development, and so on?

- How can we develop suitable plasma nanofabrication facilities?

- How can we optimize the plasma parameters in view of the targeted process outcomes?

- How can we tailor the plasma composition and other parameters to control nucleation and growth of nanoassemblies with the required characteristics?

Referring to any particular examples of plasma applications for nanomaterials synthesis or processing, we will try to compare the basic physical processes involved and compare the nanotool performance with other common approaches to nanoscale processing, such as thermal CVD. The applicability of the "cause and effect" approach will also be illustrated using some examples of the most common nanofilms and nanostructures. The ultimate aim of this discussion is to define the most striking features, benefits and challenges of using the plasma-based systems in several typical examples of nanofabrication.

The cutting-edge nanoscale applications we use as examples include, but are not limited to, ordered patterns of single- and multiwalled carbon nanotubes and related structures such as carbon nanorods and nanotips; nanostructured Si and SiC-based films and heterostructures; low-dimensional semiconductor quantum confinement structures, such as self-assembled quantum dot arrays; and ultra-high-aspect-ratio quantum wires. Furthermore, we also consider nanostructured bioceramic films such as hydroxyapatite- and titania-based biocompatible films, ultra-fine highly-anisotropic reactive ion etching of submicrometer features on semiconductor wafers, synthesis and processing of micro- and nanoporous materials, as well as new and high-precision means of nanopattern transfer to solid surfaces. Special attention will be paid to the identification and control strategies of the main building units both in the ionized gas phase and on plasma-exposed solid surfaces. Possible roles of related working units will also be identified.

2.2
Plasma Nanofabrication Concept

Following the original publication [4], we will now discuss the building unit-based "cause and effect" approach in more detail. First and foremost, one should create (in the plasma environment) the required building units with appropriate attributes. These attributes include, but are not limited to, chemical structure/organization, elemental composition, size, shape, surface morphology, electric charge, energetic (e.g., excited atomic/molecular states) state and availability of dangling bonds. It is amazing that low-temperature plasmas are uniquely able to generate the entire range of building units (Figure 2.1), from atoms, molecules, ions and radicals through macromolecules, nanosized clusters, nucleates, par-

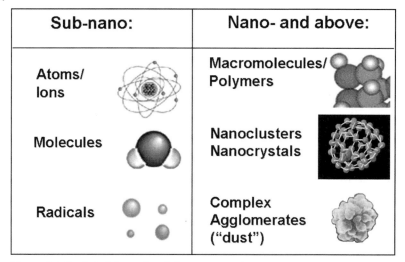

Sub-nano:	Nano- and above:
Atoms/ Ions	Macromolecules/ Polymers
Molecules	Nanoclusters Nanocrystals
Radicals	Complex Agglomerates ("dust")

Figure 2.1 Basic building units of plasma nanoscience.

ticulates to complex aggregates, agglomerates and even basic nanostructures.

These BUs can range in size from fractions of a nanometer to a few microns and are usually created through numerous gas-phase ionization, dissociation, polymerization, clustering, nucleation and other processes that occur in the ionized gas phase. It should be emphasized that following the basic definitions in Chapter 1, we will be primarily interested in subnanometer-sized building units which can participate in the most elementary stage of nanoassembly processes. Nonetheless, in some cases we will also consider uses of plasma-grown nanoclusters and nanocrystallites, for example to create specific nanostructured films such as amorphous films with nanocrystalline inclusions.

Taken that larger building units are made of smaller and more elemetary species, one would be tempted to introduce two separate notions of primary and secondary BUs, primary for those that can be used in the most elementary nanoassembly (when an object with the size exceeding 1 nm is created) and secondary for those which cannot. On the other hand, no matter how big any particular BUs is, it is always made of atoms. Therefore, it would be to consider atoms as primary building units and all larger atomic assemblies as secondary BUs. However, this emphasis is trivial since even common bricks are made of atoms at the most elementary structural level! Interestingly, no one has ever raised the question what the main building units of a common wall of a building are: a clay brick or a silicon atom? In this case, the answer would be a brick rather than an atom since building a common macroscopic house atom by atom would be extremely inefficient. This is why we do not separate the primary and secondary BUs and instead use BU for any reasonably small microscopic object created in a plasma that can be readily, most effectively and most efficiently used in a nanoassembly process.

For example, in the sythesis of a carbon cluster with a small number of atoms (e.g., C_{15}), the most likely building units are simple carbon atoms. Nucleation of 15 carbon atoms into this small cluster is perhaps the most energetically favorable process. Therefore, in this process one should use carbon atoms as building units. However, depending on the specific process conditions, one could also use carbon dimers C_2 or trimers C_3 or fuse two smaller clusters C_7 and C_8 to create C_{15}. Hydrocarbons C_xH_y, especially those with a large number of hydrogen atoms y are much less likely to be effective BUs for the desired carbon nanocluster. And certainly, no one would use a micron-sized diamond crystal to create such a small nanocluster! There is a theoretical possibility to use a top-down approach (e.g., sputtering) to carve the micron-sized crystal down to a 15-atom cluster, but such an exercise would just be a waste of time, effort and material!

Let us consider another example, the growth of amorphous silicon films with nanocrystalline inclusions in reactive (also commonly referred to as chemically active) plasmas. Such films feature small silicon nanocrystals embedded in an amorphous silicon matrix. One of the most effective approaches would be to use ultra-small nanocrystals, which can be readily prepared (e.g., via plasma polymerization and nucleation) in the gas phase, and transport them to the surface where these BUs will be covered with amorphous silicon.

It is commonly known that the most effective growth conditions for hydrogenated amorphous silicon ($a-$Si:H) requires hydrosilicon radicals Si_xH_y with a single silicon atom and a small number (typically 1–3) of hydrogen atoms. In this case using pre-prepared and readily available nanocrystals as building units is worthwhile. Indeed, one can cover such nanocrystals with a layer of amorphous material as they land on the growth surface. The grain inclusions are then already in the crystalline form as they are deposited – thus making it possible to dramatically increase the film growth rates.

Alternatively, one could avoid using plasma-grown Si nanocrystallites and try to deposit such a material on a layer-by-layer basis: first deposit radical/atomic building units on the surface, then wait until they self-assemble into Si nanocrystals, cover the layer of Si nanocrystals with amorphous silicon, and finally repeat the whole cycle as required depending on the required film thickness.

It turns out that the first approach is more efficient and that using plasma-grown silicon nanocrystals (as well as higher hydrosilanes with a larger number of silicon and hydrogen atoms) enables much higher deposition rates than achievable otherwise. Therefore, in this particular process we can use silicon nanocrystalline "bricks" as effective building units of amorphous silicon films with nanocrystalline inclusions alongside hydrosilicon radicals SiH_y with $y = 1 - 3$. We recall that in the first process, the nanoassembly of C_{15} clusters, the best choice of suitable building units would be simple carbon atoms.

It is now clear how to approach the problem of the choice of the most appropriate building units. However, similar to many processes in our everyday lives, the choice depends on the availability – and the availability depends on the "manufacturer", which in our case is the sort of a plasma one uses. Depending on the sort of plasmas, there could be different possibilities to create the same building unit. For example, metal atoms can be obtained through dissociation of metalorganic gaseous precursors in reactive (chemically active) plasmas, from metal plasmas of cathodic vacuum arcs, using plasmas of laser plumes in laser ablation, or simply sputter a high-purity metal target in a discharge in an inert gas such as argon.

Reactive (chemically active) plasmas are probably most effective in terms of producing large varieties of building units. However, such plasmas also bring more complexity into the process. To better understand the issues involved it would be very reasonable to pose the question: How does one differentiate between the chemically active and "usual" plasmas? Without trying to offer a comprehensive definition, it is usually quite accurate to state that reactive (chemically active) plasmas are usually composed of multiple reactive species that continuously transform into each other, and also generate new species as a result of numerous chemical reactions in the ionized gas phase [4, 57]. To this end, mutual transformations and chemical reactivity of the species is what makes such plasmas very different from conventional multi-component plasmas. We recall that chemical reactivity of the species is often associated with dangling (activated, or available for bonding) chemical bonds. Thus, one can conclude that such reactive plasmas can be useful for various assemblies (including nanostructures) because of their intrinsic ability to generate species with sufficient reactivity and ability to stack into the assembly being created.

However, the species should be reactive at the very moment they approach and stack into the nanoscale object being grown. If an individual building unit was created with a dangling bond in the plasma bulk, it does not necessarily mean it will reach the assembly site in the same reactive state. Indeed, the only dangling bond of a CH_3 radical can easily be terminated by a hydrogen atom while the radical moves and experiences numerous collisions with other plasma species. The resulting CH_4 molecule is not readily available for bonding. However, it can loose one hydrogen atom as a result of interaction with the surface or other particles (such as the plasma ions) and then stack using the just-formed dangling bond.

Therefore, creation of building units in the ionized gas phase and on surfaces is an extremely complex process. In fact, it goes through a very large number of elementary reactions both in the plasma and solid surfaces. For example, lower atomic mass units (e.g., atoms, molecules, radicals and ions) are usually generated through gas-phase electron-impact or heavy particle collisional ionization or dissociation of feedstock gases. Alternatively, they can be released from solid surfaces exposed to plasmas (or intense laser or particle beams) as a result of physical sputtering, chemical etching, ablation and evaporation processes. Another possibility is that building units can be generated on the deposition surface as a result of the breakup of larger objects into nanofragments, clusters, smaller grains and/or atoms/radicals. Furthermore, the already deposited building material can be re-released back into the plasma bulk as a result of desorption or re-evaporation.

Larger building units are usually created via quite different mechanisms, such as assembly of smaller BUs. We recall that if the size of the larger BUs is above 1 nm and they are usually created via assembly of atomic/radical units, this process is nothing else but a nanoassembly as discussed in Chapter 1. More specifically, such building units can be synthesized as a result of complex polymerization or clustering processes in the gas phase, or, alternatively, the release of various nanofragmets from the solid surfaces [4]. Polymerization processes in a plasma are usually triggered by certain reactive precursor species (e.g., anion SiH_3^- in silane-based plasmas) and proceeds, through a chain of polymerization reactions, to macromolecules (which can be either neutral or charged) and critical clusters, large enough to trigger the nucleation process [58,59].

It is important to note that critical clusters (also frequently termed seed nuclei in this monograph) can also be formed in the gas phase from species released from the surface or directly on the deposition surface. The latter is the case of quantum dot formation on plasma-exposed surfaces considered in Chapter 6. When the number density of such clusters reaches a critical threshold, nucleation becomes more pronounced. In most cases the sizes of critical clusters depend on specific plasma parameters and surface conditions (if applicable) and are typically less than a few nanometers.

As suggested in the original publication [4], generation of gas-phase-borne building units can be managed by adjusting the rates of major elementary reactions in the plasma, which in turn can be achieved by simply altering the discharge control parameters. The simplest way is to adopt a "trial and error" approach and achieve the required composition and other properties of these larger building units through an endless number of trial experiments. On the other hand, a fully self-consistent and systematic approach to this problem would require a comprehensive knowledge of all reactive species and chemical reactions (very often exceeding several hundred and even more) involved in the generation of the building units. Alternatively, one can use a somewhat simpler, "cause and effect" approach, which requires adequate knowledge of the main precursors and other conditions appropriate for creation of the desired building units. With both these methods, one can optimize the process parameters to obtain the required number of suitable BUs.

Once the building units have been created, they need to be transported to where they will either self-assemble or stack into a nanoassembly being created. The surface of a substrate or the nanoassembly in question needs to be suitably prepared to accommodate the incoming building units. Here we recall that the BUs can arrive at the nanoassembly site either directly from the plasma or via the surface as sketched in Figure 1.2. Regardless the arrival route, the deposition surface must be suitably prepared before the building units can land on it. Again, the requirements for surface preparation are process-specific [4]. For example, if a nanoassembly requires certain reactive radicals with a single dangling bond (such as SiH_3), the adequate number of compatible dangling bonds should be available at the required surface sites (e.g., where the nanoscale object is being assembled). On the other hand, in the growth of epitaxial quantum dots or other low-dimensional semiconductor nanostructures, surface temperatures should be high enough to enable efficient migration of adatoms about the surface and their stacking into the nanoassemblies being grown. In this case surface preparation can also include adjusting surface conditioning by using, for example, ions of an inert gas such as argon.

As another example, deposition of gas-phase nucleated nanocrystals or nanoparticles (NPs) would require a suitable (e.g., amorphous) matrix, or otherwise prepared surface with the required adhesive properties to secure embedding into or attachment of these building units to the surface. Some nanoassemblies often require specific (e.g., thermal) activation of catalyst layers. For example, growth of commonly known carbon nanotubes (CNTs) is usually preceded by thermal fragmentation of thin (a few to a few tens of nanometers) Ni/Fe/Co catalyst layers into nanosized particles forming a wetting contact with the substrate (e.g., silicon or glass) surface [3, 60]. This fragmentation can be achieved via external heating of the substrate or using plasma-related effects such as ion bombardment or reactive chemical etching.

Low-temperature plasmas offer a great range of options for the required surface preparation. The possibilities include for example, surface heating, activation and physical sputtering by intense ion fluxes accelerated in the plasma sheath to plasma-enhanced chemical vapor deposition (PECVD) of passivating, adhesive, conformal, functional and so on nanolayers, and microstructuring of growth surfaces via selective and highly anisotropic reactive chemical etching. It is relevant to mention that some of the above processes affect the nanostructures being grown. For example, reactive chemical etching plays a major role in reshaping (e.g., sharpening) of various carbon nanostructures grown in a plasma. More specific details will be considered in the following sections of this monograph.

As mentioned above, plasma-grown building units need to be appropriately transported to the nanoassembly site. In this regard, one has to keep in mind several basic possibilities and challenges. First and foremost, one can transport ionic, atomic and radical species onto the substrate surface and then rely on their surface migration from the deposition point to the nanoassembly site. Another basic possibility is to transport the building units directly to the nanoassembly being grown. In this case one should appropriately stack them directly from the gas phase, without the need of intermediate landing on open surface areas followed by migration. As will be discussed below, the importance of the second channel of building unit transport is greater in low-temperature plasmas compared with non-plasma-based routes. One of the reasons is that intense fluxes of positive ions are accelerated within the plasma sheath and converge towards sharper tips of some high-aspect-ratio nanostructures, such as nanoneedles, nanotips, nanotubes, nanorods and nanowires.

In the above, we have mentioned that the building units should participate in the nanoassembly process in the most suitable structural and energetic state. Put in simple words, if a BU is created in a plasma bulk exactly in the same form as required for its insertion/self-assembly on the surface, this form should not be in any way altered while the building unit is being transported towards the nanoassembly site. Indeed, a radical should remain the same radical and keep the same number of dangling bonds; a nanocluster should retain the same number of the same atoms; a small nanocrystal should not break into smaller nano- and subnanofragments. This issue is not only about preserving structural integrity but also about avoiding any unwanted alteration of the structure.

For example, if one wants to deposit a pure silicon nanocrystal grown in a plasma, intense incorporation of hydrogen while the crystal moves towards the surface will lead to a hydrogenated nanocrystal instead of a pure one and so should be avoided at all costs.

Let us briefly consider how to transport relatively large (e.g., crystalline grains) building units to suitably prepared surfaces without them breaking apart into smaller fragments. In low-temperature plasmas this process can be controlled by manipulation of the fluxes of the plasma species impinging onto them as well as the driving forces (e.g., electrostatic and ion/neutral drag forces) in the near-substrate areas. Appropriate manipulation of these forces makes it possible to deposit nanoparticles on pre-selected areas of microstructured surfaces, and, more importantly, in a highly-controlled fashion.

In some situations it is beneficial to conduct deposition of plasma-grown nanoparticles in afterglow, that is, swich a plasma discharge off and let the particles reach the substrate under plasma-off conditions. In this case one can separate the growth and transport stages and so avoid any electrostatic effects during the transport and deposition. Such effects include strong repulsion of negatively charged particles by the like charges on solid substrates as well as various effects associated with charge transfer upon close approach to and contact with the surface.

The last step of the building unit-based "cause and effect" approach is to ensure that the BUs are appropriately stacked or self-assembled into the required nanoassembly pattern [4]. As we have mentioned earlier, once the building units have reached the nanoassembly site, their further integration is controlled by the self-assembly (self-organization) processes at nanoscales [6,61].

It is a quite common opinion that at this stage the building units grown and transported in a plasma can no longer be controlled by the plasma conditions, and indeed, many surface conditions and features do come into play. These include surface morphology, activation energies of various processes such as diffusion, evaporation, bond formation and so on, defects, dislocations, surface instabilities, deformations, surface temperature, phase conditions of the surface material and several others. Generally speaking, these features and conditions always exist in any nanofabrication environment such as a neutral gas in thermal CVD.

However, a plasma can significantly alter almost any surface feature and condition. The ability of the plasma to control these comes about through surface charges and currents that originate in the plasma and may flow through the substrate. A cautious "may" was added in the above sentence deliberately to reflect the two basic possibilities depending on substrate conductivity. If the substrate is conducting, the current can flow through it and cause its Ohmic heating as is common to conventional conductors. Alternatively, dielectric surfaces can stop the ion current, in this case positive charge can build up on such surfaces. There is another fundamental possibility where the substrate is electrically floating, that is, disconnected from an external circuit. In both cases, intense ion fluxes carry significant energy which is converted into heat and eventually results in higher surface temperatures.

Therefore, it becomes obvious that low-temperature plasma environments can provide important background conditions for nanoscale assembly [4]. For example, if the discharge is run continuously, intense fluxes of ions and neutrals dynamically maintain the equilibrium substrate temperature, which is the key control factor at this stage. It is important to mention that substrate heating due to ion bombardment is often sufficient to grow a number of nanostructures even without any additional surface heating. Moreover, if the plasma discharge is stable (which unfortunately is not always the case), the surface temperature quickly reaches its equilibrium value and remains unchanged during the nanoassembly process. Therefore, ion bombardment is often what makes external substrate temperature control superfluous [1].

As we have mentioned above, the plasma environment can significantly affect the features and state of the solid surface. In fact, depending on the surface activation performed using specific working units, the development of nanostructured films or nanoassemblies can proceed either through the nano-island (Volmer–Weber) or layer-by-layer (Frank–Van-der-Merwe) growth scenarios [62]. The latter mechanism usually requires a greater precision in homogeneous activation of surface dangling bonds over large areas. Under certain conditions a mixed growth mode (Stranski–Krastanov) is also possible. These conditions will be discussed in more detail elsewhere in this monograph. Here it is sufficient to mention that plasma-related process conditions can significantly affect the surface conditions and as a result, the actual nanoassembly growth mode.

In the following sections of this monograph we will consider how to implement the aforementioned nanofabrication process by capitalizing on unique properties of low-temperature non-equilibrium plasmas. We will highlight the unique ability of such plasmas to generate the desired building units (encompassing the entire range of the species from atoms and molecules to nanosized clusters, small grains and agglomerates) through numerous gas-phase ionization/dissociation and polymerization processes in the plasma-assisted deposition of selected most common nanoassemblies. We will also discuss the most important physical phenomena involved in plasma-based processes. Furthermore, by comparing common plasma- and neutral gas-based nanofabrication routes we will be in a very good position to comment on competitive advantages and benefits of using plasma-assisted techniques.

For example, carbon nanotubes are commonly synthesized by either thermal or plasma-assisted chemical vapor deposition (CVD) using carbon-bearing precursors such as methane, acetylene and so on, mixed with other inert or reactive gases [3,63]. Thermal CVD usually requires very high gas temperatures (and also quite often high pressures) to decompose the feedstock gas, which adversely affects the suitability of this process for temperature-sensitive technologies such as micromanufacturing of metal interconnects and vias in semiconductor integrated circuits. It is already commonly accepted [4] that by using low-temperature non-equilibrium plasmas, one can achieve substantial precursor dissociation into a large number of ionic and radical species. In this case, one can noticeably lower the process temperatures and also achieve much higher deposition rates. However, this is not straightforward and substantial care should be taken to achieve this. This is particularly the case in the nanoassembly of delicate nanostructures such as ultra-small quantum dots and ultra-thin single-walled carbon nanotubes. It is indeed interesting that single-walled carbon nanotubes easily assembled by thermal CVD methods had not been synthesized via plasma routes until 2003–2004 [21].

Throughout the monograph, we focus on typical low-temperature plasma-based systems and figure out the unique features of such systems that make them particularly attractive for numerous nanoscale applications. In particular, we endeavor to clarify some of the following issues [5]:

- the salient features of plasma-based environments that make them versatile nanofabrication tools;

- the competitive advantages of the plasma route(s), with examples of superior performance of plasma-based processes compared to neutral gas-based (e.g., chemical vapour deposition (CVD), molecular beam epitaxy (MBE), cluster beam deposition, etc.) and other (e.g., wet chemical, electron beam, microlithography, flame) fabrication routes; and

- the optimum parameters for, and the physical mechanisms governing, deterministic (highly controlled and predictable) plasma-aided nanoassembly.

Furthermore, we also attempt to shed light on the following problems, more specifically related to ionized gas-assisted processes and approaches [5]:

- the importance of thermal equilibrium (e.g., thermal versus non-
  equilibrium plasmas);

- the effect of the degree of ionization and pressure range of operation;

- the role of the specific precursor (the building units of nanostructures) and other functional (e.g., surface preparation) species;

- the interaction of ions and reactive species with nanostructured
  surfaces and nanoscale features (e.g., pores and trenches);

- the various effects of low-temperature plasma environments on nu-
  cleation/clustering in the ionized gas phase and on solid surfaces;

- the control of surface temperature by intense ion fluxes; and

- the use of plasma environments to control the self-assembly of
  nanopatterns on solid surfaces (plasma-directed/guided self-assembly),

as well as other related problems [4], such as:

- when exactly should one use either thermal CVD or the PECVD?

- in which cases one should definitely use plasma-based tools and
  techniques and in which cases one should avoid using plasmas at
  all?

- how to develop process specifications to fabricate the desired
  nanofilms and nanoassemblies deterministically?

- what is the actual and specific role of the plasma in PECVD of
  nanofilms and nanostructures?

- what is the role of specific plasma features such as plasma sheaths,
  multiple species and plasma polymerization in the envisaged
  nanoassembly process?

- how to control creation, transport and structural incorporation of
  plasma-generated building units? In other words, how should one
  implement the BU-based "cause and effect" approach in practice?

Having addressed these issues, it will be possible to critically compare the underlying physics of most common thermal and plasma enhanced CVD systems. In doing so, we will attempt to highlight the most striking features, benefits and challenges of using low-temperature plasma-based systems in nanoassembly of ordered carbon nanotips and nanotubes, nanostructured silicon-based films for photovoltiac applications, low-dimensional quantum confinement structures such as quantum dots and nanowires, as well as a range of nanostructured films with biocompatible properties. Wherever possible, special attention will be paid to the identification and control strategies of the main plasma-generated precursor species-building units of the nanoassemblies being targeted.

2.3
Useful Plasma Features for Nanoscale Fabrication

In this section we will identify and discuss some of the most important features of low-temperature plasmas which can be gainfully used in the plasma-assisted assembly of a broad range of nanoscale objects. In doing so, we will follow the same sequence of nanofabrication steps and other important considerations outlined in the previous section, as well as attempt to compare the performance of the plasma- and neutral gas-based approaches.

As already mentioned above, nanoscale objects are usually deposited on a solid substrate as a result of the plasma-surface interactions in the gas-solid environment shown schematically in Figure 1.10. The entire near-surface area can be separated into three distinctive regions; namely, the plasma bulk, plasma sheath and solid substrate with the outer layer (growth surface) facing the plasma. As has already been discussed in Chapter 1, the sizes of each of these areas and the processes involved are very different. For example, spatial scales of adatom migration over facets of crystalline quantum dots and ambipolar diffusion in a plasma can differ by up to 9 orders of magnitude! More importantly, every area, be it a plasma bulk with typical sizes of ca. 50 cm, near-surface plasma sheath with the width λ_s ca. 10 μm–10 mm, or the surface itself with a typical root-mean-square roughness of ca. 3 nm, plays its unique role in nanofabrication processes of interest here.

As a point of reference, we recall that in the common thermal chemical vapor deposition process a hot reactive gas interacts with suitably pre-processed and additionally heated deposition substrates. In most cases, the temperatures in the gas phase and on the surface are quite different. Moreover, the commonly used temperatures of the neutral gas are substantially lower than the ionization threshold. Thus, gas feedstock usually remains neutral in the reactor chamber. For example, thermal dissociation of a $CH_4 \sim$ gas commonly used in the synthesis of various carbon-based nanostructures, is negligible even at gas temperatures of ca. 900 °C [64]. However, depostion substrates heated externally open an additional channel for the generation of ionized species; this happens through Saha–Langmuir ionization [57].

One of the main concerns in thermal chemical vapor deposition is that very often deposition rates are unacceptably low. This is usually attributed to the low number densities of suitable building units and their low reactivity and hence, inadequate interaction with the solid surface. Indeed, due to low rates of gas-phase precursor dissociation, generation of reactive radicals turns out to be rather inefficient. This can be improved in one of two ways: (i) either increased dissociation rates; or (ii) create other, larger, building units which are still suitable for the film growth. Nanoclusters are ideal candidates to fulfil this role. However, as was suggested by Hwang and colleagues [65], using electrically charged nanoclusters can bring in many exciting and effective opportunities for the formation of a variety of nanostructured films non-existing or ineffective otherwise.

More specifically, if a working pressure in a CVD reactor chamber exceeds a certain threshold, typically of the order of 13.33 Pa (100 mTorr), gas-phase clustering processes can lead to the generation of nanosized clusters [66]. These clusters most likely originate as a result of homogeneous condensation and nucleation of atomic species in the (essentially neutral) gas phase. These clusters are in most cases neutral; hovever, they can become ionized upon landing on a hot solid surface. Another more effective possibility for the generation of charged nanoclusters is through nucleation induced by positive ions that can be released from the surface. Interestingly, the size of charged carbon-based nanoclusters generated in a thermal CVD system can be effectively controlled by process parameters such as concentration of carbon-bearing gas feedstock. Figure 2.2 shows a representative size/mass distribution of carbon-based nanocluster building units in a hot filament-assisted chemical vapor deposition of diamondlike and other carbon films from a $CH_4 + H_2$ gas mixture [67]. The nanoclusters are negatively charged and their size and mass intrinsically depend on the concentration of carbon precursor gas in the gas mixture. When the CH_4 content is lower (ca. 1 %), the maximum of the size distribution is within the range between 200 and 300 atoms; when the concentration of CH_4 is increased to 5 %, significantly larger clusters of typically ca. 1000–1500 atoms dominate the mass spectrum as can be seen in Figure 2.2.

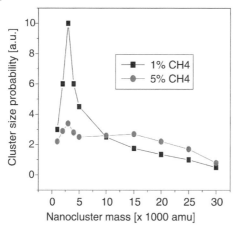

Figure 2.2 Mass distribution of negatively charged carbon clusters in hot filament CVD of diamond in $CH_4 + H_2$ gas mixtures. After Reference [67]

Thus, chemical vapor deposition systems usually feature a limited number of building units; in this case the most likely BUs are thermally activated atoms and molecules and, alternatively, neutral or charged nanoclusters. As suggested by the charged cluster theory [65], in many cases only nanoclusters stand a chance to be the main building units of various nanofilms and nanostructures, since (charge neutral) atomic and molecular species usually deposit with very low deposition rates. It is important to note that smaller and charged nanoclusters are more favorable for epitaxial recrystallization upon landing on the surface, whereas larger and neutral clusters usually coagulate to form cauliflower and porous skeletal structures [4,65].

On the other hand, in the PECVD, the gas phase is a two-component system and comprises the neutral and ionized gas (plasma) components as shown in Figure 2.3. In weakly-ionized plasmas the relative population of the ionized component $\Sigma_j n_i^j / \Sigma_k N_n^k$ is low, where n_i^j and N_n^k are the number densities of j-th ionic and k-th neutral species, respectively. Thus, the presence of the neutral component makes the PECVD quite similar to most of the commonly used thermal chemical vapor deposition systems. However, the presence of the additional ionized phase makes a remarkable difference at all four stages of the nanofabrication process. Let us recall that finding the difference the plasma can make is exactly what plasma nanoscience research is after.

The striking difference between the plasma and the neutral gas-based nanofabrication environment is in the presence of the plasma sheath, which is a non-neutral layer of space charge separating the plasma bulk

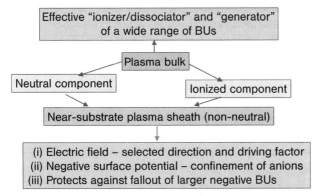

Figure 2.3 Components of a typical PECVD system and their roles. Because of significant precursor dissociation/ionization and nanocluster creation in the gas phase, surface ionization is less likely than in thermal CVD.

and the solid substrate. Because of the much higher mobility of electrons, the surface potential is negative with respect to the plasma bulk [68]. The resulting potential distribution sustains intense positive ion fluxes onto the surface and impedes the fluxes of negatively charged species such as anions, negatively charged nanoclusters and nanoparticles, and also leads to the commonly known Boltzmann's density distribution for electrons. In most situations, the plasma bulk acts as an efficient ionizer of neutral gas precursors, where a large spectrum of positively and negatively charged species is generated. As a result of numerous electron-impact and heavy particle collisions, neutral molecules are dissociated into reactive radical fragments.

It is also relevant to note that some of common reactive plasmas (e.g., $SiH_4, C_4F_8, SF_6, O_2, Cl_2$, and so on) feature quite large populations of anions (negative ions). The latter species can trigger polymerization and clustering processes, which eventually result in the creation of large nanoclusters and macromolecules [57–59]. It is important to note that the negative surface potential repels such particles from the surface; therefore, anions and negatively charged clusters, macromolecules and small nanoparticles can be held by the resulting potential and hence, reside somewhere in the plasma *pre-sheath* (a relatively large transition area between the plasma bulk and the near-substrate sheath [68]) longer than if they had no electric charge at all or were charged positively. The negative species confinement effect is one of the salient features of plasma-based nanoassembly environments which makes them very different from neutral gas-based thermal chemical vapor deposition. We emphasize that this happens even if there is no external bias applied to the solid surface.

As mentioned in the original publication [4], the plasma sheath protects the deposition surface against the uncontrollable fallout of negatively charged species such as anions, clusters, or nanoparticles [28]. It is amazing that this is the case even when a positive bias is applied to the substrate, as the potential distribution in the plasma is uniquely rearranged to maintain lower surface potentials with respect to the plasma bulk [68].

Reasonably long residence times of reactive plasma species favor assembly of primary macromolecules, nanoclusters and some other nanofragments (e.g., two-dimensional graphene sheets) into amorphous or (under certain conditions) crystalline nanoparticles. It needs to be stressed that this can happen even at very low pressures of gas feedstock when homogeneous nucleation, arguably one of the main mechanisms in thermal CVD systems, is inefficient.

This possibility serves as a further example that in low-temperature plasma-based processes, nanofilms can be synthesized using a larger variety of building units than in neutral gas-based approaches. Moreover, intense processes in the ionized gas phase significantly reduce the importance of thermal dissociation and ionization of precursor gases on deposition surface – a primary source of building units in thermal chemical vapor deposition. Even though charged nanoclusters can still be formed in a quite similar way as in thermal CVD, the relative importance of such a process is significantly reduced. As a rule of thumb, the more reactive the plasma is, the less important is the formation of larger building units on the surface. Strictly speaking, this conclusion is reasonable when the rates of clustering and polymerization in the plasma bulk are reasonably high and the effects of surface etching (which can also lead to the release of reactive radical precursors) are weak.

On the other hand, if the feedstock gas is non-reactive, then the importance of the release of building units from solid surfaces becomes very high. For example, in the case of a discharge operation in a low-reactive (e.g., inert) gas, building units need to be created via the plasma-surface interactions or injected otherwise. Ion sputtering of a solid target surface in DC/RF sputtering systems, evaporation of a wire material in hot filament systems, ablation of surface material in pulsed laser deposition systems, or erosion and evaporation of cathode material in vacuum arc plasmas are just a few posiblities to this effect.

Examples of external injection of building units into a plasma are also numerous. Plasma spray deposition is a good example when externally synthesized fine solid grains are dispersed downstream into an atmospheric-pressure plasma torch. Another example is related to plasma-assisted atomic or molecular beam epitaxy (ABE/MBE) systems,

Neutral gas

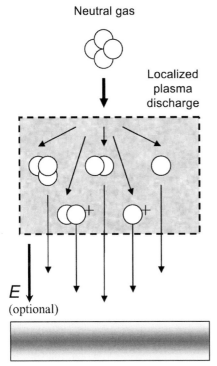

Localized
plasma
discharge

E
(optional)

Figure 2.4 Schematics of a typical remote plasma system
primarily aimed for effective and controlled dissociation of neutral
gas feedstock and avoiding ion bombardment and surface charge-
related effects unavoidable in the example shown in Figure 1.10.

where an externally formed atomic/molecular beam is subjected to a localized plasma discharge; the latter is usually sustained in an inert gas. In
many related cases, the plasma is solely used for the purpose of creation
of building units. Whenever it is required to generate reactive radicals
but direct effects of the plasma on the surface should be avoided (e.g., intense ion bombardment or surface charging are undesired), one can use
remote plasma systems, that is, move the plasma discharge away from
the surface as sketched in Figure 2.4.

Density of neutral and ionized components of the plasma is also an important factor that determines the relative rates of building unit creation
on the surface and within the plasma bulk. Indeed, the likelihood of collisions in denser plasmas is obviously higher; moreover, this usually results in higher nucleation and clustering rates. Higher process pressures
can lead to larger supersaturation of condensing gas. Moreover, above a
certain pressure, gas-phase nanoparticle growth can proceed largely via
the homogeneous nucleation channel, similar to neutral gas-based CVD.

In the following section we will consider the issues of appropriate choice and effective generation of suitable building and working units in more detail. This will significantly facilitate our consideration of specific features of plasma-aided nanoassembly due to the presence of the plasma sheath (Section 2.5).

2.4
Choice and Generation of Building and Working Units

The main concern of this section is to try to better understand how to choose and generate the most appropriate building and working units. As has become clear from the previous consideration, low-temperature plasmas can generate the entire range of species in atomic, molecular, cluster and other forms. To this end, it is imperative to be able to work out which specific units to generate and which specific control strategies, suitable for each specific sort of plasmas, are the best to use in synthesizing any desired nanoassembly. As was stressed in the original article [4], this question has no general answer, with the number of possible solutions exceeding the number of nanoassemblies ever fabricated by plasma-based methods. Therefore, the problem of choice of the appropriate building units (to be generated by the plasma in any particular experiment) can be based on the "cause and effect" logic sequence [4]: precursor → BU(s) → nanoassembly shown in Figure 2.5.

The chosen sequence should rely on the existing knowledge from other areas of nanoscience, structural chemistry, surface science, materials science, and so on. It also has feedback and optimization links, which enable one to adjust the choice of precursor gases, appropriate building units and methods of their manipulation and incorporation. This is particularly important for reactive plasmas where a very large number of abundant reactive species and polymerization/clustering scenarios can

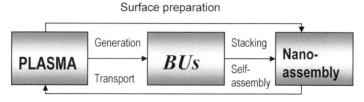

Surface preparation

PLASMA — Generation / Transport → *BUs* — Stacking / Self-assembly → Nano-assembly

Optimization of process parameters

Figure 2.5 A concept map of the low-temperature plasma-based "cause and effect" approach to deterministic nanoassembly (after [4]).

easily result in an inappropriately large and virtually infinite number of "trial and error" sequences.

This is the reason why in this chapter our considerations will mostly be based on specific assumptions concerning the building units needed for certain nanostructures or any other BU-assembled nanoassemblies such as nanostructured materials. Without trying to provide exhaustive recipes for the appropriate choice of building units, (which deserves to appear in the near future as the Encyclopedia of nano-assemblies and their building units), it is very clear that the relevant choice should be primarily based on structural considerations of the nanoassemblies being created. At this stage, using established theories of nanoscale synthesis and growth kinetics at nanoscales is very helpful [4,6,61,62].

We now discuss a few simple examples of when specific assumptions of some particular species as a building unit can be used to explain growth kinetics or some important features of common nanostructures. Considering the synthesis of carbon nanotubes, we note that it is widely accepted that carbon atoms are the simplest and the most effective building units. Indeed, any carbon nanotube is made of carbon atoms arranged in hexagonal lattices of wrapped graphene sheets! This is very obvious but this argument only confirms the final structural state of the nanoassembly of interest. Which building units were actually used to create this ordered stack of carbon atoms? This question, in fact, is very far from trivial.

Let us consider a carbon nanotube with a nickel catalyst particle either on the top or at the base. Such nanotubes are grown by the insertion of carbon atoms into graphene sheets. However, the carbon atoms can stack into the nanotube walls only through the bulk or the surface of the catalyst particle. This process is commonly known as carbon extrusion, wherein carbon atoms are pushed through a metal nanoparticle, which can be in a solid or a liquid state. Therefore, a building unit that lands on the surface of the catalyst nanoparticle (in the top growth scenario) or gets into it (base growth mode) may undergo significant changes on the surface or in the bulk of the catalyst particle. For example, hydrocarbon radical CH_2 or methyl cation CH_3^+ may be stripped of hydrogen atoms; a carbon trimer C_3 or a carbon cluster C_{15} may be broken into separate carbon atoms, and so on. Assuming that only carbon atoms can insert into the growing nanotube walls, which of the CH_2, CH_3^+, C_3, or C_{15} species would be the most effective building unit of the nanotube in question?

In fact, we have just arrived at one of the most difficult problems plasma nanoscience deals with! In a general form it can be formulated as *which specie(s) should one create in the plasma bulk so that they could appropriately insert into the desired nanoassembly, in the most suitable form, which may even be quite different from the original specie(s)?*

As we have mentioned earlier in this monograph, there is no general solution for this very complicated problem. The choice certainly depends on specific types and arrangements of the nanotubes as well as on the process details. However, within the framework of the "cause and effect" approach advocated here, we can reformulate the above question from the "effect" end and apply it to the nanotubes grown through the catalyst particle on top: *which species created in the plasma bulk is most efficient in terms of their arrival at the catalyst nanoparticle and most effective in terms of creating carbon atoms (on the surface or bulk of the catalyst) that can incorporate into the growing graphene sheets in the fastest and most effective way?* In simple terms, which species should we deliver so that after a chain of transformations on the surface or in the bulk of the catalyst nanoparticle, the resultant carbon atoms will stack into the growing CNT walls in the required order (e.g., number and density of walls for multi-walled carbon nanotubes and chirality for single-walled nanotubes) and ensure the highest possible growth rates of the nanostructures? It is prudent to note that this statement also contains some deterministic flavor, especially concerning the desired order and the maximizing of the growth rates.

In this example, if any of the above species can satisfy the above requirements better than others, then one can arguably call this particular species the best/most appropriate BU in the nanoassembly process concerned. If, for example, the effective rates of the process that include the delivery of carbon radicals CH_2, their conversion into carbon atoms, diffusion over the surface and/or through the bulk of the Ni catalyst nanoparticle turn out to be larger (even better if much larger!) than for other carbon-bearing species, then one should focus their attention on this specific radical species and try to create it in larger amounts and suppress generation of other species.

However, in several cases reactive dimers C_2 should be used for the assembly of carbon nanotubes [63]. We note that the chemical structure and reactivity (usually characterized by the availability of dangling bonds and energy barriers for specific interactions such as structural incorporation into a graphite sheet) of the C_2 molecule are also quite suitable for its insertion into graphite hexagonal lattice patterns which build the nanotube walls [62]. This building unit may be particularly important for the growth of open-ended (uncapped) nanotubes, which grow by attachment of incoming species to the open end of a nanotube where dangling bonds are available.

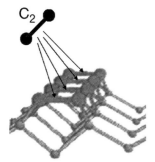

Figure 2.6 Schematics of insertion of carbon molecules C_2 into dimer rows on reconstructed surfaces of ultrananocrystalline diamond (after [37]).

It is also relevant to stress that the carbon dimer also plays a prominent role in the plasma-assisted synthesis of ultrananocrystalline diamond [37]. Physically, C_2 is an energetic and chemically active molecule which can insert directly into carbon-carbon and carbon-hydrogen bonds, often without any intermediaries, such as the reactive hydrogen atoms frequently used for activation of dangling bonds on hydrocarbon-based surfaces. For example, this mechanism is responsible for the self-assembly of dimer rows of the reconstructured surfaces of ultrananocrystalline diamond shown in Figure 2.6 [37]. Presumably, the rates of incorporation of carbon dimers appear to be higher than those of atomic carbon [37]. Therefore, one can conclude that in this particular example carbon dimer C_2 is the most effective building unit of ultrananocrystalline diamond. Apparently, C_2 represents a subclass of subnanometer-sized molecular building units.

We now turn our attention to the next group of plasma-generated building units (Figure 2.1); for convenience they will be termed "nanoclusters" below. This group includes nanosized clusters, macromolecules and nanofragments, which typically range in size from fractions of to a few nanometers. We note that building units of this category usually participate in nanoassembly processes concurrently with other species and often appear as a result of condensation and clustering of atomic/molecular species or, alternatively, complex chains of polymerization reactions in the ionized gas phase [58,69].

Interestingly, the size of nanocluster BUs strongly affects the structure and properties of nanocluster-assembled solid films and nanoscale objects. For example, if carbon clusters are relatively small (d ca. 2–3 nm) and contain only a few hundred C atoms, they appear to be ideal building units for epitaxial growth of crystalline carbon-based films. Moreover, if such clusters are negatively charged, one can synthesize dense and smooth single-crystalline diamond [70,71].

Electric charge on such nanoclusters enables them to recrystallize epitaxially on hot growth surfaces. Moreover, because of strong electrostatic repulsion, carbon nanoclusters do not agglomerate/coalesce while landing and experience contact epitaxy upon landing individually, without being greatly affected by other clusters. Another important reason for the successful transformation of such small carbon nanoclusters into nanocrystalline diamond is that small nano-sized objects melt at surface temperatures significantly lower than the melting points of the corresponding bulk materials. If the processes of full contact epitaxy upon landing and rearrangement into sp^3 networks could proceed instantaneously and with complete consumption of nanocluster BUs, one could potentially grow very large single-crystalline diamonds.

On the other hand, if one targets amorphous graphite-like carbon films with nanocluster/nanocrystalline inclusions, one should use relatively large (typically containing more than 1000 atoms and featuring sizes in the ca. 10 nm domain) carbon clusters [70,71]. Such clusters can be positively charged, which also makes it possible to avoid their agglomeration before reaching the surface or in the process of nanofilm growth. When larger nanoclusters are involved their contact epitaxy may be incomplete.

For example, if the surface temperature is not high enough to melt the landing cluster completely, atoms from molten outer shells of the cluster may epitaxially recrystallize on the growing surface, while the inner core will be left intact. In this way, a nanofragment inclusion can be formed within the growing matrix of the main film. Generally speaking, deposition of relatively large clusters may result in the development of the disordered structure in the films. In some cases, such disordered inclusions can actually be beneficial. For example, third-generation photovoltaic solar cells are based on amorphous silicon films with ultra-small nanocrystalline/nanocluster inclusions. Some of these conclusions have been confirmed by measurements of charged nanocluster mass and size distributions in hot filament CVD reactors and hydrocarbon-oxygen flames and computations within the framework of the charged cluster theory (CCT) [72,73].

One would possibly ask what could happen if nanocluster building units agglomerate before they land on the surface. This can happen, for example, when the clusters are not electrically charged and gas pressures are reasonably high and so allowing for strong collisions. If the surface temperature is very high, one could expect relatively fast agglomerate melting and recrystallization on the surface despite a large number of small nanoclusters that compose it. In the range of moderate temperatures, it is most likely that films with quite strong structural disorder will be formed from partially molten nanocluster agglomerates. Finally, at low temperatures the films will feature an even higher degree of disorder and also a porous structure with relatively large voids.

These conclusions have been made assuming that the energy of the nanoclusters upon landing is not sufficient for their structural disintegration into smaller fragments. This situation is quite common for neutral gas-based approaches and implicitly assumes soft nanocluster deposition on the surface. As we will see from the following discussion, the presence of electric charges on nanoclusters can make it easier to control their energy in a plasma. Even if nanoclusters are not electrically charged, their polarization ability (polarizability) can make them ideal building units that are capable of selectively attaching to elongated silicon nanostructures such as silicon nanowires, carbon nanoneedles and some others [49,50,74].

The arrangement of building units into nanoassemblies can proceed quite differently depending on the energetic preferences of the nanoassemblies being created. For example, the same carbon atoms can arrange themselves into a large number of structural networks attributed to different allotropes. In bulk graphite, the sheets of hexagonal patterns are flat, whereas in carbon nanotubes they are rolled, twisted or bent. This makes structural and other properties of carbon nanotubes very different from those of bulk graphite.

A striking example of how a specific arrangement of building units can result in an exotic nanostructured organization of the matter, is a recently discovered allotrope of carbon, the "carbon nanofoam" [75] with unique ferromagnetic properties uncommon to carbon in general. The carbon nanofoam, with its very unusual properties, is made of a web of randomly interconnected carbon-atom clusters, with an average diameter of 6–9 nanometers as shown in Figure 3 of Reference [4]. The new material also features pronounced semiconducting properties. The unexpected magnetic properties can be attributed to the unusual nanostructural organization of carbon atoms in the clusters. Indeed, carbon atoms in carbon nanofoam form heptagonal structures with an unpaired (and thus not involved in chemical bonding) electron. This unpaired electron has a magnetic moment which, in turn, may lead to the magnetism. It is remarkable that this form of nanostructured carbon can only be formed using carbon atoms as building units.

Let us now turn our attention to larger molecular building units which may contain quite a large number of atoms. Higher silanes (macromolecules) and hydrogenated silicon clusters also can create a certain degree of structural disorder in hydrogenated amorphous silicon films and as such play a pivotal role in the plasma-assisted deposition of nanostructured amorphous films for solar-cell applications [4]. Interestingly, concentration of higher silanes in silane-based discharges used for PECVD of *a*-Si films determines the film deposition rates and directly correlates with the microstructure and performance of the solar-cell elements [76]. Moreover, the amount of small (less than a few nm in size) hydrogenated silicon clusters $Si_x : H_y$ in silane-based plasmas is directly related to the microstructure parameter, which is one of the indicators of the device quality of hydrogenated amorphous silicon films [77].

We emphasize here that one of the major advantages of using larger than atomic/molecular building units is the possibility to significantly increase the film deposition rates. Indeed, such BUs carry a much larger amount of matter than atoms. This is why cluster beam deposition has recently been so popular in nanofabrication, materials synthesis and surface modification [78]. In particular, using supersonic cluster beam deposition [79], one can achieve quite high rates of deposition of nanostructured carbon films. However, one should always be careful not to exceed a certain threshold for the nanocluster flux. This threshold can be calculated using the recrystallization upon landing arguments we used above in the discussion of charged cluster deposition. If the incoming flux of nanoclusters is too high, it is very easy to compromise some of the film performance/structure specifications, critical in industrial applications.

As can be seen in Figure 2.1, larger clusters, nanocrystallites and complex aggregates (which for the sake of simplicity are collectively termed nanoparticles in this section) can be regarded as the third distinctive group of building units that can be generated in a plasma. In many applications, such nanoparticles can be intentionally grown in the ionized gas phase and then incorporated in the growing film. For example, silicon films grown with a significant contribution of nanoparticles coming from the plasma have been found to exhibit improved transport and stability properties, as well as a wider optical bandgap as compared to nanoparticle-free *a*-Si:H [4]. A larger bandgap primarily arises as a result of electron quantum confinemet effects in small nanocrystals (quantum dots) embedded in amorphous matrix of hydrogenated silicon or related materials such as SiC.

Structural incorporation of the plasma-grown nanoparticles (usually a few nm in size) makes it possible to synthesize polymorphous silicon (pm−Si:H) films, an unique form of nanostructured matter composed of a relatively uniform amorphous matrix with ultrasmall nanocrystalline inclusions [80]. It is noteworthy that recent results show that the performance of PIN (p-type semiconductor − insulator − n-type semiconductor) solar cells can be significantly improved through nanocrystalline (usually 1–2 nm) inclusions deposited with high rates using plasma enhanced chemical vapor deposition in silane-based reactive gas mixtures [81–83].

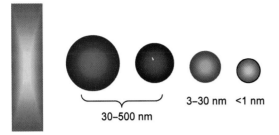

3–30 nm <1 nm

30–500 nm

Figure 2.7 Changes in color of gold nanoparticles when their size decreases (after [86]).

It is also interesting to note that charged clusters can be generated in the gas phase of non-equilibrium low-temperature plasmas at much lower pressures than are typically required in thermal non-plasma-based processes. For example, in the chemical vapor deposition of diamond and diamond-like films (DLC), gas-phase generation of charged clusters requires supersaturated reactive gas feedstock, and, hence, higher working pressures, typically exceeding ca. 13 Pa (100 mTorr) [66]. In comparison, numerous reports suggest that polymerization and clustering processes can be quite efficient in plasma discharges at working gas pressures as low as a few Pa [58,59].

Furthermore, single-crystalline nanoparticles in the gas phase of reactive plasmas have been favorably considered as effective building units in the fabrication of quantum information, molecular electronics, data storage and light emitting devices [49, 50, 84, 85]. Some elements of these devices, such as field emitting arrays of vertically aligned single-crystalline carbon nanotips/nanoneedles, ordered $Al_xIn_{1-x}N, Si_{1-x}C_xN$ quantum dot structures, silicon-based nanowires, as well as a number of other nanostructures can also be fabricated in such plasmas by using smaller atomic/radical and nanocluster units [49,50].

In recent years there appeared a new class of unique materials with very unusual properties, quite different from those of bulk materials made of the same chemical element(s) or compounds. These materials are built using nanoparticles and nanoclusters as primary building units and exhibit a prominent dependence of their electronic, thermal, mechanical, chemical and other properties on sizes and size distributions of the BUs. Indeed, it is commonly known that quantum size effects can strongly affect the arrangement of crystal lattices and electronic spectra of building units with sizes in the 1–10 nm range. One of the most remarkable manifestations of this effect is the changes in color gold nanoparticles experience when their size decreases. These changes are illustrated in Figure 2.7.

As can be seen in Figure 2.7, gold particles change their color from the yellowish color of conventional bulk gold to blue, purple and reddish blue when their size is reduced to 30–500 nm [86]. With a further decrease in size down to 3 nm, the reddish blue color of gold nanoparticles changes to orange and even red. Gold nanoparticles in this size range are metallic. However, in the sub-1 nm range, gold nanoclusters are non-metallic; their color is more likely to be light orange than dark orange or red. The color fades as the size is further reduced. Indeed, atoms are commonly treated as colorless.

It is relevant to mention here that small nanoparticles (e.g., nanoclusters or nanocrystallites) feature quite different (strained) chemical structures compared to bulk materials, which has numerous implications for nanocomposites and nanoelectronic systems assembled using such nanoparticles as building units [87].

To end this section, we note that complex aggregates (Figure 2.1), usually appear as a result of agglomeration of smaller building units and are very rarely used in plasma-aided nanofabrication. Moreover, for the past couple of decades they have been widely considered as unwelcome side-products. However, unique properties of low-temperature plasmas enable one to charge (usually negatively) gas phase-borne nanoparticles, confine them in electrostatic potential wells above the substrate and use this unique opportunity to deposit conformal coatings or functional monolayers on nanoparticle surfaces. Moreover, other means of post-processing, such as etching or plasma-mediated annealing also become very effective because of the longer residence times of the nanoparticles in the plasma. For example [4], decomposition of reactive precursor ATI $(Al(i - OC_3H_7)_3)$ in Ar-based RF plasmas has been used to deposit thin alumina films on fine particles of barium magnesium aluminate used as high-brightness phosphors in tri-color fluorescent lamps [88].

Having discussed the major basic possibilities in terms of possible building units that can be generated in the plasma, we now turn our attention to examining what effects one could expect from the plasma sheath – a unique near-surface area of uncompensated space charge otherwise non-existent in any neutral gas-based process. Moreover, we will try to elucidate how exactly the plasma sheath can affect transport of building units generated in the plasma bulk and their interaction with nanostructured surfaces.

2.5
Effect of the Plasma Sheath

As mentioned in the original publication [4], the plasma sheath plays an important role during the surface activation and building unit transport stages. Let us now clarify this statement and discuss possible effects of the plasma sheath on building units, their motion in the near-surface area and eventually deposition on the solid surface where nanostructures are assembled. The most basic effect comes about owing to the presence of uncompensated space charge (in the plasma sheath the number density of ions is higher than that of the electrons) and associated electric fields. The surface potential is lower than in the plasma bulk; the resulting electric field accelerates positively charged species (e.g., positive ions/cations) towards the solid surface and slows down the motion of negatively charged species (e.g., electrons and negative ions/anions). First of all, such a polarity of the near-substrate electric field strongly enhances surface fluxes of positively charged building units, otherwise quite weak in thermal CVD systems. Meanwhile, negatively charged species can be suspended by the sheath potentials and reside in the plasma discharge longer than neutral and positively charged species. This creates a unique opportunity for extended interaction with other plasma species and electromagnetic fields. For example, some of the long-term resident species can trigger and effectively mediate clustering, nucleation and polymerization processes in a plasma [57,69]. Under certain conditions this results in a pronounced generation of dust particulates.

It is very important to stress that in a plasma, almost any building or working units move quite differently from a neutral gas environment under similar conditions such as gas composition and working pressure. First of all, the presence of an ionized gas component brings about a very different collisional environment for all species, ranging from neutral atoms to relatively large nanocrystals. In the simplest example, let us consider a gas with only one neutral species. It is very common that under plasma conditions the rates of collisions of neutral atoms/molecules of such a gas with each other (the dominant collision process in a neutral gas) can be much less than the rates of electron-neutral and ion-neutral collisions.

Energetic electrons and ions can affect the momentum of neutral species. In this case even the interaction of the neutral species with the surface will be quite different. Electron-neutral and ion-neutral collisions also affect energetic states of the atoms/molecules and can lead to their ionization and/or dissociation. For example, if we need any particular radical (say, CH_3) to be transported to the deposition surface, in a plasma it will: (i) experience quite different collisions due to the presence of the ionized gas component; and (ii) may become ionized and, therefore, will be transported to the surface much faster because of the action of the electric force in the near-surface area.

In the simplest terms of newtonian mechanics, we can say that the forces acting on building units in a plasma are quite different from the neutral gas case. This difference depends on the nature, size and charge of the BU concerned and can be even more pronounced for larger (mesoscopic) species. For example, directed ion fluxes can drag nanoparticles towards the surface; this (frictional by virtue) force is very common in the physics of complex (dusty) plasma as an ion drag force.

Whichever forces come into play and dominate, depends on the size of building units and the prevailing experimental conditions. As we have discussed above, motion of BUs in the sub-nanometer range (atoms, molecules, radicals), can be substantially impeded as a result of collisions with other (usually mostly neutral in weakly ionized plasmas) species in the plasma sheath. The impulse associated with such collisions can divert the species from their straight paths towards the deposition surface leading to chaotic and oblique deposition, after a number of collisions. This effect is particularly important for collisional sheaths, when

$$\lambda_s > \lambda_{mfp},$$

where λ_s and λ_{mfp} are the sheath width and mean free path of the BUs concerned in collisions with other plasma species, respectively. In other words, the above condition means that once a particle enters the plasma sheath, it is most likely that it will experience a collision and its deposition on the surface will be strongly affected by the collision(s).

It is imperative to stress that for charged species the rates of collisions as they cross the sheath area can be quite different than for the neutral particles. For positive ions this may mean that because of their electric field-controlled directional velocity (which is in most cases larger than their thermal velocity) towards the substrate, the rates of their collisions with background neutral gas can be even lower. Therefore, collisions within the sheath can be very important for neutrals yet insignificant for ions. In this case we have $\lambda_s^n > \lambda_{mfp}^n$ for the neutrals and $\lambda_s^i < \lambda_{mfp}^i$ for the ions.

Therefore, in collisionless sheaths ($\lambda_s \ll \lambda_{mfp}$), the BUs are unaffected by collisions and move (more or less) in the same way as they entered the sheath. For example, if a neutral radical entered the sheath at an angle of 45°, it will continue its motion until it hits the surface with the same angle of incidence. On the other hand, a cation (positively charged BU) normally incident on the plasma sheath, will accelerate and move smoothly along the electric field lines directed normally to the surface. The latter case is most common in low-pressure PECVD reactors used for plasma-assisted nanofabrication.

This electric field-controlled motion of ions (certainly together with charge-related effects on the surface) appears to be a factor in the growth of nanotubes and related high-aspect-ratio nanostructures. Indeed, such nanostructures are aligned along the direction of the near-substrate electric field. Therefore, in plasma-assisted processes, the alignment is perfectly normal to the substrate [89, 90] since the electric field associated with the plasma sheath is usually perpendicular to the surface. We will discuss these issues in the section related to ion-focusing nanostructures in Chapter 7.

However, the effect of the degree of collisionality of the plasma sheath on the properties of the plasma-grown nanostructures and nanoassemblies still needs further study [4]. From practical considerations, collisionless sheaths remain most attractive from the point of view of transporting BUs from the plasma bulk to the surface. This is because neutral BUs (e.g., carbon dimer C_2), which are unaffected by the sheath electric field, can be transported through such a sheath without any significant change in their energetic state, the energetic state which was acquired in the plasma bulk through intense collisions with other neutral and ionized species. Indeed, the electron-depleted plasma sheaths effeively exclude electrons from electron-impact excitation and ionization of neutral building units. Furthermore, collisions which involve heavy particles (atoms, ions, molecules) are rather inefficient in collisionless sheaths when $\lambda_s \ll \lambda_{mfp}$.

Therefore, the energetic state, and hence, reactivity of neutrals can indeed be controlled by adjusting the rates of collisions in the plasma bulk; these are not expected to change within the collisionless plasma sheath. Here we reiterate that sufficient reactivity of the building units is essential for their insertion into nanoassemblies or reconstructed surfaces of bulk materials. From the practical perspective, the easiest way to control collision rates is to vary the pressure and composition of the working gas mixture.

In addition to the electric force, larger (>1 nm) building units are subject to some other forces, such as the ion and neutral drag forces, thermophoretic force and gravity [28]. Some of these forces (e.g., neutral drag and gravity) exist even without a plasma, while the ion drag force cannot exist without a plasma environment. The thermophoretic force is determined by temperature gradients, which appear to be quite different in a plasma because of intense ion bombardment of the surface.

What is even more important is that the above forces that act on meso-scopic BUs vary in scale according to the particle size. Physically, this creates a unique possibility of manipulating such building units by adjusting the force balance on them. This possibility will be considered in more detail in Chapter 5. It is noteworthy that the force of gravity is usually weak for those nanometer-sized particles that are suitable for meaningful nanofabrication purposes.

Electric charges on such particles open yet another possibility of manipulating them in the vicinity of deposition surfaces. On the other hand, this introduces another level of complexity associated with the charge and electrostatic potential of the surface itself. These features are primarily determined by several factors such as conductivity of the surface material, surface morphology, as well as whether the electric current flows through the surface or not. For example, if the surface material is conducting but the substrate is floating (disconnected from the external circuit), the charge on the surface will in most cases be negative. This is because of a much higher mobility of the plasma electrons which quickly deposit on the surface and set up a negative potential which in turn repels and slows down the electrons and attracts and accelerates positive ions so that the total current through the floating substrate is zero. Therefore, negatively charged building units (such as reasonably large nanoparticles in low-temperature plasma discharges) will be repelled by the negative substrate potentials. However, under certain conditions (when the electrostatic force cannot stop the particles with a large momentum) they can overcome repulsive potential barriers and still deposit on the solid surface.

After the nanoparticle BU lands on the surface then the whole range of other physical phenomena comes into play. There are many different possibilities of their interaction with the surface. For example, depending on the energy of the particles upon landing, they can either deposit without any significant structural transformation (soft landing with a very small or no shape change) or change their properties and structure after the interaction with the surface. In the extreme case of a nanoparticle deposition with high energy, a complete structural disintegration can take place. There is also a further plethora of different possibilities depending on types of interacting materials, their physical state, the presence of electric charges on the surfaces of the substrate and nanoparticles being deposited and several other important factors. These and several other possibilities are considered in more details in Sections 5.2 and 5.3 of this monograph.

Therefore, electric fields within the plasma sheath and especially in the vicinity of any microscopic textures on the deposition surface can be used to control the fluxes and energies of the charged BUs impinging on the surface. In some applications it is necessary to deliver large building units such as nanoclusters of a relatively large number of atoms or small nanocrystals intact, without any significant disintegration into smaller nano- and sub-nanofragments. This requirement can also be crucial in processes where some radicals or functional groups need to be attached, or, at least delivered, to the surface "as is". If the charge of such nanocluster/nanocrystal or radical/functional group BUs is different (e.g., positive) from the surface charge, the electric field can significantly enhance their fluxes onto the deposition surface and, therefore, increase the deposition rates.

However, the electric field should not be too strong so as to avoid undesired effects such as the above mentioned structural damage/disintegration of the BUs as well as any changes that might affect the ability of the surface to host the incoming building units. For example, surface sites suitably prepared to host the incoming particles should not be damaged. These factors include available dangling bonds, chemical potential of the surface, distribution of stress, strained sites, defects, dislocations and several others.

Let us now consider this issue in a bit more detail using an example of cluster beam deposition, a versatile tool for nanoscale science and technology, which is currently widely used for a variety of purposes [78,79]. These, and related deposition techniques are extremely sensitive to the landing energy and orientation of the clusters [65,91].

In the example considered by Frantz and Nordlund [91], if a cluster is small enough (e.g., has a size of a few nanometers), the substrate surface is hot enough (e.g., 600–1000 °C), lattice mismatch between the cluster and the substrate is small and the cluster is incident normally to the surface with a low energy (e.g., fractions of eV to a few eV), epitaxial recrystallization of the cluster atoms in the substrate lattice is likely, with the initial stage shown in Figure 2.8(a). In this case the lattice of the cluster can ideally fit into the substrate lattice and rearrange through vacancy migration along the cluster-substrate interface [91].

However, clusters with larger size, higher energy and oblique incidence, have less chance for epitaxial recrystallization and usually only partially burrow into the substrate. As a result, distinct grain boundaries within the substrate material and island-like structures on the surface are formed. This situation is sketched in Figure 2.8(b). In this case, the rearrangement proceeds through disordered motions of atoms along the cluster-substrate interface [91]. Furthermore, very small clusters (1–2 nm in size), with no fixed crystalline orientation and featuring lower melting points, compared to bulk materials, are able to adopt the crystal struc-

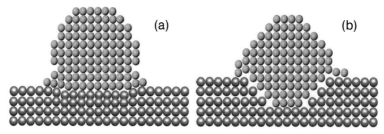

Figure 2.8 Schematics of initial stages of epitaxial recrystallization (a) and burrowing (b) of a nanocluster deposited on a solid
surface. The latter case is usually accompanied by formation
of well-resolved grain boundaries.

ture of the substrate or film already deposited. Hence, perfectly crystalline films can be grown by a relatively rapid epitaxial recrystallization of small nanoclusters [65]. On the other hand, larger clusters and
nanocrystallites tend to retain their ordered structure upon deposition;
this eventually leads to the formation of nanostructured films with well-
defined grain boundaries [4].

Therefore, the electric field in the plasma sheath turns out to be an additional powerful tool in helping control the energy and incidence angle
of various building and working units. These landing parameters are
extremely important in the process of interaction of a range of plasma-
generated BUs with arrays of various nanostructures considered elsewhere in this monograph. For example, normal incidence is certainly
the best option for plasma immersion treatment of nanostructured surfaces, for example, via ion implantation/subplantation. As we have seen
above, landing normally to the surface also improves the chance of direct contact epitaxy of plasma-generated nanoclusters. However, there
are situations when a glancing angle incidence is preferred. For example, if an as-deposited adatom needs to move quite a significant distance
along the surface (e.g., of a high aspect ratio nanotube) before incorporating into the nanostructure being assembled, an additional momentum in
the required direction of motion (e.g., towards the catalyst particle) can
be an advantage. A parallel component of the BU's velocity conserved in
the process of oblique incidence provides this momentum.

We should stress, however, that when selecting the process parameters
one should be very careful in choosing the appropriate potential drop
across the plasma sheath (this can be done by varying the bias voltage or
plasma parameters such as the plasma density or the electron temperature). Indeed, excessive voltage drops across the plasma sheath can be
detrimental to the integrity of charged nanocluster building units. For
example, nitrogen clusters break up upon landing on graphite surfaces

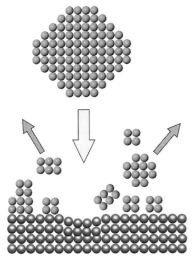

Figure 2.9 Structural disintegration of a nanocluster upon landing with an excessive energy. Newly formed nano- and sub-
nanometer fragments can be used as secondary building units
which under certain conditions re-deposit onto the surface and
form nanofilms or nanostructures.

if the potential difference between the surface and the plasma bulk is of
the order of 25 V, which is the estimated nanocluster cohesive energy
$\mathcal{E}_{nc}^{cohes}$ [92]. If the energy of the nanocluster $\mathcal{E}_{nc}$ upon landing is slightly
higher than $\mathcal{E}_{nc}^{cohes}$, smaller fragments released as a result of this collision
will spread about the surface and eventually form unwanted amorphous
deposits. On the other hand, if $\mathcal{E}_{nc}$ is well above the cluster breakdown
threshold, most of the deposition material carried by the clusters will go
back to the gas phase, with a small fraction embedded as defects in the
crystalline structure of the substrate. This situation is sketched in Figure 2.9.

Thus, in nanofabrication processes which rely heavily on charged nanoclusters, one should very carefully adjust the potential distribution
across the sheath (and hence, the sheath width) to avoid any undesired
nanocluster disintegration and surface damage. On the other hand, the
sheath width can be related to the specific parameters of the plasma and
the bias. For example, if $\tau_b \gg \tau_i$, where τ_b is the duration of the applied
bias pulse and τ_i is the ion motion timescale, the sheath width

$$\lambda_s = (\sqrt{2}/3)\lambda_D(2eV_0/T_e)^{3/4} \tag{2.1}$$

and potential profiles $\phi(x)$ near the substrate (located at $x = -\lambda_s$)

$$\phi(x) = -\left(\frac{3}{2}\right)^{4/3}\left(\frac{J_0}{\epsilon_0}\right)^{2/3}\left(\frac{2e}{m_i}\right)^{-1/3}x^{4/3} \tag{2.2}$$

can be adjusted to achieve the required energy and flux of the building units upon deposition on the surface [68]. Here

$$\lambda_D = \sqrt{\frac{\varepsilon_0 k_B T_e}{n_e e^2}}$$

is the Debye length, V_0 is the bias potential, T_e is the electron temperature, m_i is the ion mass, ε_0 is the permittivity of free space, k_B is Boltzmann's constant and

$$J_0 = e n_{iS} V_B$$

is the ion current entering the sheath, where n_{iS} and $V_B = (T_e/m_i)^{1/2}$ (Bohm velocity) are the ion number density and velocity at the sheath edge, respectively. Equations (2.1) and (2.2) are, strictly speaking, applicable to the case when the bias applied to the deposition surface is not very large. In other cases, specific expressions for the sheath width and potential distribution can be quite different. For a variety of options the reader should refer to the reference materials on principles of plasma discharges and materials processing [68] and plasma immersion ion implantation and deposition [12]. It is also important to emphasize that no matter what the specific case is, the electric potential of the plasma bulk is always higher than that of the plasma-exposed surface.

Evaluating possible effects of the plasma environment on nanoassembly processes, one should always keep in mind that (sometimes quite significant) potential differences across the plasma sheath lead to intense fluxes of energetic ions onto the substrates. Physically, the most important ion-related effects are the major contribution ion fluxes make to:

- deposition of the required material onto the surface;

- physical sputtering of the surface material via direct ion bombardment;

- ion-assisted reactive chemical etching of the surfaces being processed;

- activation and passivation of surface bonds;

- surface modification as a result of the penetration of energetic ions through several atomic layers of the surface material, which is commonly referred to as ion implantation;

- additional surface heating as a result of direct and localized transfer of ion energy upon impact;

- extra heating of the feedstock gas via intense ion-neutral collisions;

- creation of radical, molecular and atomic species directly on deposition surfaces;

- charge transfer to the surface and neutralization of any existing negative charge;

- various polarization-related effects.

In fact, these interactions can make a drastic difference in the nanoassembly process. For example, direct heating of deposition substrates by ion fluxes and hot gas in the reactor can be used to maintain the required surface temperature, which can even make external substrate heating unnecessary. In the case of PECVD of the nanocone-like carbon nanostructures, temperatures of nickel-catalyzed silicon surfaces strongly depend on the power that sustains the plasma discharge and under certain conditions can reach 300–350 °C and even higher [55, 93–95]. Under such conditions, a range of carbon-based nanostructures has been grown under no-external heating conditions.

Plasma-related temperature effects can be even more delicate and amazing. For example, low-temperature plasmas can be used to dissociate molecular oxygen into oxygen atoms, which in turn combine with metal atoms to form oxide pyramidal and nanowire-like structures [96]. However, the most amazing aspect in this process is that the plasma conditions can control the surface temperature so delicately to be able to catch the moment of localized melting of metal foils, when intricate nanostructures can develop via one of the variants of the vapor-liquid-solid growth scenario. More details will be discussed in Section 8.3 of this monograph.

Among the list of possible contributions of ion fluxes to the nanoassembly there is one which is seemingly not directly related to surface processes, namely the extra heating of feedstock gas. This was included intentionally, to spark the reader's curiosity and also illustrate the not-so-obvious link between the processes in the ionized gas phase and on solid surfaces. Let us now consider one of the processes which benefits from a reasonably hot gas environment of plasma-based processes.

Plasma-grown nanoparticles often require significant post-processing before they can be used as viable building blocks in nanoscale applications. A common experimental observation is that such nanoparticles are amorphous and do not posses regular shape. In this case, plasma-related effects can be used to simultaneously confine and anneal such particles. More specifically, negative electric charges which these particles usually acquire in low-temperature plasmas, prevent them from falling out onto the surface, which is also in many cases charged negatively. Once confined (levitated) in near-substrate areas of low-temperature plasmas, the nanoparticles can be subjected to *in situ* "annealing" by hot ambient gas and ion fluxes. If the energy transfer in the interactions of the nanoparticles with the neutral and ionic species is effective enough, nanoparticle sintering and eventual spheroidization can take place. This effect is most pronounced in thermal plasmas where gas feedstock temperatures can reach a few thousand degrees and higher [5].

The fundamental mechanism involved in this nanoparticle transformation may be related to the plasma-enhanced surface self-diffusion of atoms [58]. In this way, originally unsuitable for nanodevice integration or nanofilm incorporation, individual nanoparticles can be refined and otherwise post-processed via temperature-related effects, physical sputtering, chemical etching and so on, to become viable building blocks in targeted nanoscale applications [97].

Another interesting plasma-related effect is due to the presence of the electric field in the vicinity of nanostructured surfaces. This electric field can substantially change the growth of nanostructures made of polarizable materials. Thus, if the building units being deposited have a dipole moment, their behavior in a neutral and ionized gas can be completely different. To mentally visualize this difference, let us imagine dipole BUs as "dumbells" made of two oppositely charged atoms linked together; the direction of the dipole moment of this molecule is from the negative to the positive charge. If the deposition is carried out in a neutral gas, the orientation of the dipole moments of different dipoles is random; thus, the building units stack into the growing structure also quite randomly. On the other hand, the dipoles tend to turn to line themselves up along the direction of the electric field **E** of the plasma sheath, which is perpendicular to the solid surface. Therefore, dipole building units tend to align their dipole moments in a direction normal to the substrate. Furthermore, since the electric field in the plasma sheath always points to the substrate, the positively charged atom will face the surface and encounter the interaction faster than its negatively charged counterpart. This perfectly aligned insertion of dipole building units will eventually create a unique atomic architecture of the reconstructed surface exposed to a plasma. Thus, the insertion of radical and molecular building units with dipole moments into nanoassemblies can indeed be quite different in plasma-aided processes.

The next example is related to the role of the plasma in the preparation of solid surfaces for the deposition of building units with different polarizability and ability to interact with the surface. Using the example given in the original report [98] and referring to Figure 2.10, let us consider

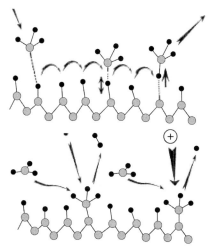

Figure 2.10 Reaction kinetics of SiH$_3$ (pyramidal structure, top panel) and CH$_3$ (planar structure, bottom panel) radicals with a-Si:H and a-C:H surfaces, respectively. SiH$_3$ is strongly physisorbed by the dipole interaction, whereas chemisorption of CH$_3$ is controlled by the creation of dangling bonds on the surface by either reactive hydrogen atoms or by ion bombardment [98].

the surface reaction kinetics of SiH$_3$ and CH$_3$ building units in the low-temperature plasma-assisted deposition of hydrogenated amorphous silicon (a-Si:H) and carbon (a-C:H) films. As was mentioned earlier [4], the difference in the specific atomic arrangement (which in turn causes quite different polarization response) turns out to be the main cause of remarkably different surface interactions of SiH$_3$ and CH$_3$ radicals with the growth surfaces. From Figure 2.10 one can see that a SiH$_3$ molecule has a pyramidal structure and a permanent dipole moment, whereas CH$_3$ is planar and has no dipole moment. In a sense, the dipole moment is a factor which substantially adds to the reactivity of SiH$_3$ radicals in their interaction with the amorphous silicon surface. In fact, the dipole interaction facilitates hydrogen abstraction from the growing a-Si:H surface. As a result, dangling bonds on the growth surface can be activated as shown in the upper panel in Figure 2.10. Therefore, the as-vacated dangling bond can be instantly occupied by another incoming SiH$_3$ radical. In this case, the film growth is supported by a vacancy migration along the plasma-silicon interface and is quite similar to the one involved in epitaxial recrystallization of small nanoclusters discussed above.

The mechanism of interaction of CH$_3$ radicals with the surface turns out to be quite different, as sketched in the bottom panel in Figure 2.10. To explain this very different radical-surface interaction, we note that methyl radicals exibit a planar structure (all four atoms lie in the same plane) and also do not have any dipole moment. Therefore, their inter-action with a-C:H surfaces cannot be effective if not facilitated by other species, which we term working units in this monograph. In this case the required dangling bonds can be created via direct ion bombardment or chemical interaction of reactive hydrogen atoms generated in the plasma bulk (Figure 2.10 (top panel)). To facilitate this process, a heavy dilution of hydrocarbon gas feedstock in argon and/or hydrogen is commonly used.

In this case, it is worth selecting the plasma discharge parameters when the efficiency of atomic hydrogen production and ion bombard-ment (e.g., by atoms of an inert gas) are at their highest. In this case the ion bombardment should be mild; the energy of impinging ions should exceed (but not to be much larger than) the hydrogen abstraction energy, which is approximately 0.34–0.4 eV for hydrogenated amorphous carbon and silicon materials [99]. Indeed, if the ion energy is too high, ion bom-bardment can cause significant damage to the growth surface, including its nanostructured textures. In this example, we can see that the ability to activate the deposition surface using a variety of working units (e.g., atomic hydrogen or ions of an inert gas) is one of the unique features of low-temperature plasmas and is not common to thermal chemical vapor deposition and other synthesis routes. Other examples of possible roles of the plasma environment and specific species (working units) in the preparation of solids for nanoscale synthesis are highlighted throughout this monograph.

As we have seen from the above discussion, low-temperature plasmas and, in particular, near-substrate sheath areas allow numerous unique and fascinating features in the fabrication of nanofilms and nanostruc-tures not available in neutral gas processes. We now discuss yet another interesting feature of low-temperature plasmas related to the possibil-ity of direct stacking of building units into nanoassemblies being grown without the requirement for intermediate deposition on the solid surface. Here it needs to be emphasized that most of basic concepts for fabricat-ing various nanostructures and thin films require the building units to stack into the film or the structure only at the required sites (e.g., attach to an activated dangling bond or adsorb to a specified facet of a crys-tal). Moreover, if the BUs attach inappropriately, they have to be able to migrate into place, and if possible, along the optimum path.

One immediately notices that the above two requirements pose strict constraints on the process, simultaneously demanding strong and weak binding between the precursor species and the growth surface. Amaz-ingly, by transporting the required building units through the plasma directly to the nanoassembly, one can substantially diminish the impor-

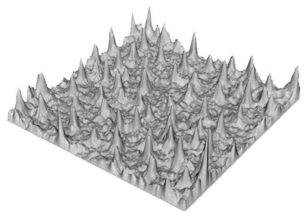

Figure 2.11 Representative distribution of ion current density over a nanostructured surface in the Ar + H$_2$ + CH$_4$ plasma-assisted nanofabrication of ordered carbon nanotip microemitter arrays [4, 51, 52, 100].

tance of the "move into place" (over the surface) factor. This is particularly important for charged nanocluster and radical building units which can be driven and focused towards the target sites by the sheath electric fields.

For example, in the deposition of high-aspect-ratio nano-objects such as nanoneedles, nanotips, nanopyramids, or vertically aligned carbon nanotubes, the electric field lines converge towards the sharp ends of these nanoassemblies. Moreover, microscopic ion fluxes focused by electric fields of sharp high-aspect-ratio nanostructures facilitate deposition and stacking of the building units in selected areas on their lateral surfaces. Moreover, by varying the plasma process conditions, one can effectively steer microscopic ion fluxes over the nanostructure surfaces and even concentrate them wherever more building units are needed, for example, near the nanostructure tips or bases. A variety of examples will be considered in detail in Chapters 5 and 7. A representative example of the distribution of the ion current density over the nanostructured surface in the PECVD of ordered arrays of carbon nanotip structures in Ar + H$_2$ + CH$_4$ plasmas, computed via a hybrid fluid/Monte Carlo simulation is shown in Figure 2.11, which is representative to numerical simulations by Levchenko et al. [4, 51, 52, 100].

Thus, in plasma-based systems one can effectively remove the requirement of weak binding to the substrate and focus on the enhancement of binding of the BUs to the nanoassembly upon landing [4]. One notices that various weakly adhesive buffer layers are frequently used in nanoscale synthesis processes to speed up the surface diffusion and therefore enhance coalescence of adatoms into three-dimensional nan-oclusters [6, 62]. As was mentioned elsewhere [4], the plasma also acts as a buffer layer, where nanocluster building units can coalesce, be electrostatically charged and, moreover, be transported directly to the nanoassembly site in a weakly collisional environment. In this re-gard, electrostatic charge prevents the plasma-grown nanocluster build-ing units from further coalescence in the gas phase or on the surface, which is ideal for fabrication of high-quality dense films and nanostruc-tures [65].

A further interesting point is that maintaining building units charged in nanocluster-based deposition processes usually prevents their unde-sired coagulation/agglomeration in the gas phase. This is particularly important for the synthesis of perfectly crystalline films by using nano-sized building units [4]. In this case, in addition to "intrinsic" charges owing to specific chemical and electronic structures of the nanoclus-ters/nanocrystals [101], continuous microscopic currents of the plasma electrons and ions (which are usually negligible in thermal chemical va-por deposition) also contribute to the charging process [28]. More specif-ically, surfaces of large enough nanoclusters/nanoparticles acquire the electric charge, which is just sufficient to balance microscopic electron and ion current so that the net current on such particles is zero. This negative charge slows down the electrons and accelerates the ions giving rise to the floating surface potential in a way similar to any mesoscopic bodies immersed in a plasma and not connected to any external circuit (e.g., not grounded).

When the clusters approach very close to the surface (distances of the order of a_c, where a_c is the nanocluster size), the further deposition pro-cess depends on the charging state of the surface and, moreover, relative conductivity of the gas phase and the surface [4]. Different (but certainly not all) possibilities of interaction of charged nanoclusters with solid sur-faces in neutral and ionized gas-based chemical vapor deposition are summarized in Figure 2.12.

As mentioned by Hwang et al. [65], in thermal CVD systems, charged nanoclusters deposit differently on conducting and insulating substrates. This often leads to selective deposition of the nanoclusters, especially when the gas phase is not conducting. This different behavior can be explained by quite different charge transfer rates (CTRs) of insulating and conducting surfaces. If a cluster approaches a conducting surface, the cluster's charge is easily dissipated during the deposition and the surface remains charge neutral. On the other hand, dielectric surfaces poorly dissipate the charge and remain charged. In this case, nanoclus-

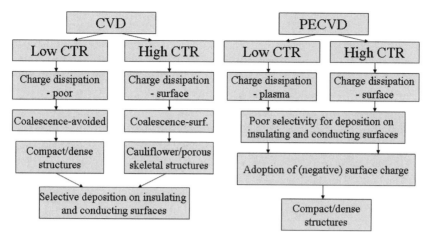

Figure 2.12 Various options to control interaction of charged nanoclusters with deposition surfaces in thermal and plasma-enhanced CVD.

ters charged similarly (same sign of the charge) to the solid surface may experience substantial difficulties during the deposition.

However, if the clusters are large enough, strong (short-range) polarization effects can diminish the (long-range) Coulomb repulsion and enable the deposition. Thus, smaller nanoclusters experience more difficulties in landing on insulating surfaces and are ideal to increase the selectivity of deposition. Interestingly, in neutral gas-based fabrication using substrates with high charge transfer rates, one can expect porous, skeletal, cauliflower and so on, structures. These voided structures usually develop as a result of coalescence of neutral clusters on the surface. On the other hand, in cases when the CTRs are low and small nanocluster BUs are used, perfect and highly compact three-dimensional assemblies, and eventually, dense nanofilms, can be synthesized.

On the other hand, in the plasma-enhanced CVD, the ionized gas phase is highly conductive and is favorable for efficient cluster charge dissipation upon contacting either insulating or conducting surfaces [4]. Moreover, as has been mentioned above, the substrates are usually charged negatively at the initial stage of deposition and remain negatively charged if the charge dissipation (e.g., in conducting substrates) is efficient. This favors the deposition of positively charged nanoclusters and small nanocrystallites. However, when an insulating material is deposited via PECVD, the charge is predominantly dissipated through the ionized gas phase. Therefore, low-temperature plasmas appear to be equally suitable for deposition on insulating and conducting substrates.

However, it is generally believed that is it quite challenging, if at all possible, to realize selective nanocluster deposition on insulating and conducting substrates in a plasma environment [102].

Nonetheless, as discussed above, plasma has additional effective controls that may enable selective deposition of various building units onto specific substrate areas. Indeed, surface areas with different surface morphology may feature quite different topology of microscopic electric fields [51, 52]. These electric fields are very non-uniform over spatial scales comparable with the sizes of individual nanostructures, which makes precise and selective deposition of a variety of building units possible. The non-uniformities exist even in the absence of any external substrate bias. Apparently, this electric field-related feature is not available in thermal CVD systems without any external substrate bias.

It is interesting to note that in some cases one can synthesize high-quality closely packed assemblies in the plasma, despite very high charge transfer rates in the ionized gas phase and without relying much on the effect of microscopic electric fields mentioned above. Let us consider two important factors that can shed some light on the issue [4]. First, the importance of reactive radical species in the plasma is much higher than in thermal CVD. Second, if the nanocluster deposition is the main process leading to nanoassembly, it is likely that while "burrowing" into the substrate material exposed to the plasma, the nanoluster BUs dissipate their original charge acquired in the gas phase and adopt the (usually negative) equilibrium charge of the local area at the substrate. Thus, such clusters remain charged when coming into contact with the substrate, which prevents them from coalescing on the surface, which would happen in thermal CVD systems which use substrates with high charge transfer rates.

Therefore, it may be quite advantageous to use PECVD to synthesize high-quality films on conducting substrates. Furthermore, it was stressed [4] that the benefits of selective deposition in thermal CVD systems are not so obvious because of larger (compared to plasma-based systems) numbers of neutral clusters which tend to agglomerate in the gas phase; the latter process eventually leads to various voided structures including porous skeletal or cauliflower morphologies.

It would be a remiss not to mention the highly unusual ability of low-temperature plasmas to support ordered assemblies of certain building units in the gas phase via long-range electrostatic interactions. For example, plasma-grown or externally dispensed negatively charged grains (typically in the sub-μm and μm size range) suspended in the near-sheath areas, frequently form ordered lattices commonly termed Coulomb (or dust) crystals [103, 104].

Such stable arrangements of the BUs can, in principle, be used for fabrication of ordered arrays of fine particles on the surface or epitaxial recrystallization on pre-patterned substrates. However, in the latter case the pattern size should match the lattice constant of the Coulomb crystal being deposited.

The idea of depositing, for example, a two-dimensional Coulomb crystal onto the surface and preserving the original gas-phase order sounds fascinating and can lead to this highly-unusual method of creating ordered arrays of nanoparticles on solid surfaces. However, it is extremely difficult to deposit such ordered structures during the discharge run because of strong electrostatic repulsion from the similarly charged surface. In some cases, individual nanoparticles can overcome this barrier and land on the surface. However, a number of conditions have to be met before such deposition can occur [105]. These conditions are extremely difficult to meet for all solid grains that make the Coulomb crystal.

An alternative, deposition of such ordered structures in a discharge afterglow, also turns out to be quite difficult. The reason is that the main causes of the existence of this ordered structure in the plasma are related to the electric fields, charges and ion fluxes, which all disappear quite rapidly when the plasma discharge is turned off. It was experimentally observed that ordered two-dimensional structures disperse quite rapidly in the discharge afterglow [106]. Therefore, these fascinating theoretical options still await their experimental realization.

To conclude this section, we reiterate that one can achieve reasonably high deposition rates (of the order of a few nm/s) in PECVD of various nanoassemblies. Moreover, the deposition rates can be further enhanced by controlling near-substrate electric fields and using positively charged radical, nanocluster, or small nanoparticle BUs. Interestingly, in typical neutral gas-based CVD systems reasonably high growth rates cannot usually be achieved using atomic/molecular species. The required film growth rates can nevertheless be achieved using charged nanoclusters [65].

2.6
How Plasmas Affect Elementary Surface Processes

As we have seen from the arguments laid out in the previous sections, a low-temperature plasma can indeed be considered as a very relevant and powerful tool for the synthesis and processing of various materials at nanoscales. Indeed, this nanofabrication environment features all the required attributes: suitable building units can be created, processed, transported and stacked into the nanostructures and nanofilms being created.

However, so far we have not attempted to address one of the major issues raised in the introductory section of this monograph, namely, *deterministic* synthesis of nanoassemblies with the required size, structure and other essential properties. Indeed, in the above, we have considered several critical issues related to creation, processing and delivery of building units to the solid surface where the remaining part of the nanoassembly process is expected to happen. We have also discussed different scenarios for the initial interactions of those building units depositing from the plasma. In the case of small (atomic, molecular and radical) BUs, these scenarios include stacking into appropriate surface sites by terminating available dangling bonds, adsorption and subplantation. Here we intentionally do not highlight other interactions (e.g., elastic scattering from the surface), which do not directly lead to the creation of new nanoscale assemblies or which assist but do not directly participate in the assembly processes (e.g., surface bombardment by ions of an inert gas). Thus, let us now concentrate on those interactions which lead to the successful arrival of suitable species at the nanoassembly sites – be it one of the surfaces (e.g., a facet) of the nanostructure being created or open surface areas between the developing nano-scale objects (see, e.g., a sketch in Figure 1.2). At this juncture it seems appropriate to pose the following questions:

1. What happens to those building units next?

2. How can we assemble them into the nanoassemblies we actually want?

3. What role does the plasma environment possibly play in this process?

Once again the number of possibilities in this regard is virtually infinite and depends on the specifics of nanoscale objects being created.

Therefore, we will need to move from the three-dimensional ionized gas environment and enter the totally different and unique two-dimensional world of *solid surfaces exposed to the plasma*. Generally speaking, these processes are studied in surface science and other relevant areas of materials science and chemical kinetics. We intentionally emphasized that the solid surfaces in which we are interested are unusual, in the sense that they are exposed to whatever the ionized gas environment brings forth, namely, electric fields, ionized species, surface charges, currents that flow on and through the surface, polarization and some others. Therefore, we must first attempt to develop the "surface science of plasma exposed surfaces", which is an extremely unexplored research area, and as such we most effectively start from scratch. Several attempts at developing suitable models for the most essential surface processes on plasma-facing surfaces will be presented in other chapters in this monograph. We emphasize that these attempts to bridge the gap between the plasma physics and surface science are still in their infancy, and in particular, in the understanding of basic physical and chemical processes in the process of nucleation and growth of the desired nanoassemblies under very unusual, non-equilibrium conditions imposed by the plasma environment.

Here we note that the foundations of surface science were laid decades ago and it is generally well understood what happens with certain species (e.g., carbon adatom) on some (e.g., Si(100)) solid surfaces. However, it should be stressed that the number of the "well understood" species and solid surfaces is quite limited even without immersion in the plasma environment!

One of the basic reasons for this effect is that the exact values of the energies and rate coefficients which are used to describe and quantify the overwhelming variety of all possible processes (just to mention surface diffusion, hopping from one facet to another, insertion into a specific atomic stack, evaporation from the surface, evaporation from nano-scale object, and so on) are not known and in most cases are calculated using sophisticated quantum mechanical tools such as *ab initio* density functional theory (DFT) [107–109] and associated packages of computational materials science such as DMol^3, SIESTA, and so on [110, 111]. These tools, while very accurate computationally, use a range of different assumptions which spark a general debate about possible optimum combinations of the physical models and computational tools, for example molecular dynamics (MD), DFT, Kinetic Monte Carlo (KMC), as well as various physical models based on specific nanostructure growth equations complemented by the species balance equations.

From the plasma nanoscience perspective, this already sophisticated toolbox needs to be redesigned and upgraded to accommodate at least the most likely effects of the plasma environment on the elementary processes on solid surfaces. This action will require more precise data on the characteristic energies which will account for the plasma-related effects. Thus, the quantum mechanics approaches used to calculate the specific energies of interactions of specific species with specific surfaces need to be modified to include at least some of the most essential ion- and plasma-related effects. No easy task, and one where no quick solution can currently be guaranteed.

The reader's disappointment will probably grow even further once they learn that experimental studies of such intricate surface phenomena in a plasma environment will prove to be an even greater challenge, even though quite similar phenomena on charge-free surfaces are routinely investigated using sophisticated surface science tools such as low energy electron diffraction (LEED), high-resolution transmission electron microscopy (TEM), or scanning tunnelling microscopy (STM). For example, high-resolution STM technique presently allows one to depict positions of individual atoms with temporal resolution of a few tens of snapshots per second. Using the frame-by-frame analysis of these snapshots, one can reconstruct the displacements of individual adatoms over selected surface areas. However, adapting this technique to harsh plasma-based environments is another challenge. One of the main reasons is that the pressure ranges suitable for the operation of plasma-based deposition tools and for the high-precision analytical tools of surface science differ by several orders of magnitude. Therefore, *in situ* STM monitoring of how adatoms move on plasma-exposed surfaces is presently out of the question.

On a more positive note, if plasma-based environments can indeed be used to create better-quality nanoassemblies then *ex situ* characterization using HRTEM, SEM, XRD, XPS, Raman spectroscopy and other tools is relatively straightforward. However, each of the techniques has its own set of technical recipes and "tricks" which are specific to different types of materials, nanostructures, and so on. For example, those who have worked with TEM know how time consuming it is just to prepare specimens for characterization. For more details on materials characterization tools, approaches and material-specific recipes and suggestions the reader is referred to suitable reference materials [112].

Therefore, we arrive at the amazing conclusion that plasma nanoscience demands us to "deterministically" create better (compared to neutral gas processes) nanoassemblies via proper understanding of how elementary atomic processes and assemblies work on plasma-exposed surfaces. However, as we have just inferred using quite simple arguments, these things presently cannot be either properly modelled or experimentally measured!

Therefore, development of adequate methods for theoretical modeling and experimental diagnostics/characterization of elementary processes on plasma-exposed surfaces is one of the major challenges for the coming years. Nevertheless, in this monograph we will present several advanced models developed by the Plasma Nanoscience team of the University of Sydney, Australia, complemented by the experimental results of our collaborators from the International Research Network for Deterministic Plasma-Aided Nanofabrication and other researchers.

However, some readers would argue that what we have discussed above is merely related to the major difficulties and challenges facing the surface science of plasma-exposed surfaces and that we have not even attempted to answer the question posed in this section's title, namely "How can plasmas affect elementary surface processes?" In other words, even if whatever is happening on such surfaces cannot be adequately described theoretically and characterized experimentally (at least at present) what can we actually expect? Some of the supporters of the idea of approaching everything deterministically (in fact, we have introduced this approach in Chapter 1) may even expect to be told upfront about any potential benefits of using plasma environments to control the "uncontrollable" elementary processes of self-organization on solid surfaces. This approach does make perfect sense since if there are no clearly foreseeable comparative advantages of using nanofabrication environments of a higher complexity such as a plasma, then one might not need to go any further beyond an exploratory stage with a relatively low consumption of time, efforts and resources.

Let us now use our commonsense and basic knowledge of physics and chemistry and try to elucidate how plasmas can affect adatom motion and self-organization into intricate nanoassemblies, one of the main goals of plasma nanoscience, which was highlighted in Chapter 1. Among the questions we need to ask are: how exactly shall we barge into the self-organized nanoworld on plasma exposed surfaces? How to use the plasma to help in achieving this as yet elusive goal?

Let us consider an example of quantum dots on a solid surface exposed to a plasma, sketched in Figure 2.13. To deternimistically create them and ensure that they indeed have a number of essential attributes (e.g., size, shape, crystalline structure, and so on) one should first of all deliver the appropriate amount of suitable building units. For example, our task is to create an array (ca. 10^{10} nanostructures per square centimeter) of silicon nanodots, each with 28 atoms (this would correspond to a size of approximately 1 cubic nanometer), within 1 second. First of all, we need to create and then deliver the appropriate amount of silicon atoms from the gas phase, in our case is 2.8×10^{11} atoms / $(cm^2 \cdot s)$. However, due to various losses of building units associated with detachment from the surface, inappropriate sticking/attachment to the surface, evaporation and collisional losses on the way to the substrate, the actual amount of silicon atoms needs to be larger, for example 3.5×10^{11} atoms / $(cm^2 \cdot s)$. Therefore, approximately 3.5×10^{11} silicon atoms should arrive at each square centimeter of the substrate within the one-second interval when

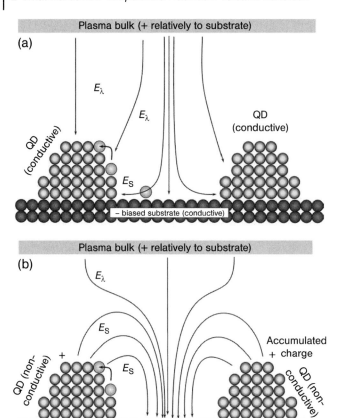

Figure 2.13 An illustration of a possible effect of a plasma environment on quantum dot (QD) self-assembly from adatoms. Panels (a) and (b) show the configurations of the electric field in the QD/surface system: (a) conductive substrate, conductive quantum dots, negative potential on the substrate; (b) conductive substrate, non-conductive quantum dots, negative potential on the substrate; positive charge on quantum dots.

the growth process is going to happen. Needless to say, before and after the growth interval no building units should arrive.

Now, it is very important to work out how to deliver this amount of atoms; simultaneously (e.g., before the growth process starts), in discrete batches or continuously during the growth process. It is noteworthy that this ability is extremely limited in thermal chemical vapor deposition, where the flux of neutral species (ca. $(1/4)V_{Tn}N_n$, where V_{Tn} and N_n are the thermal velocity and density of neutral species, respectively, and the coefficient $1/4$ appears due to four equally probable directions of motion) can only be controlled by the neutral gas temperature. In plasma, one can create silicon ions (e.g., using a hot filament along the way of the silicon atom beam; this beam can be created in any way, such as plasma-assisted magnetron sputtering and so on) and deliver them faster using the appropriate substrate bias. The associated potential difference can be tailored to draw the ions onto the growth surface in any desired way. For example, a pulsed bias can be used to deliver the whole amount of ions at the beginning of the growth process, split the ion flux into a few batches or gradually increase or decrease the ion influx depending on the specific requirements of the nanoassembly process.

Now, once the building units have been delivered to the clean deposition surface they need to move in place. In the case considered, we assume that silicon atoms adsorb to the surface at the point of landing and temporarily lose the ability to move until they are given some extra energy to overcome the potential barrier which holds them to the surface. In surface science, this energy is called the surface diffusion activation energy, ε_a. Its numerical value is mostly determined by the ability of the pair of materials to interact via adsorbtion of species and varies in a relatively broad range. For most materials considered in this monograph, $\varepsilon_a = 0.480 \times 10^{-19}$ J (0.3 eV)–1.440×10^{-19} J (0.9 eV).

To enable the adsorbed atoms (adatoms) to move, an additional energy supply to the growth surface is required. The simplest way to achieve this is to heat the substrate using an external heat source coupled with a temperature monitor and controller. Thus, the higher the surface temperature, the faster the adatoms will be able to reach the self-assembly sites (seed nuclei of quantum dots (QDs) in our case) and stack into place in a QD. However, to make this process effective, the surface temperatures should be increased quite substantially, for example to ca. 800–900 K, which is very undesirable for a broad range of applications which require substrates with much lower melting temperatures such as polymers for biodevices and ultra-thin (a few nm thick) interlayers in microelectronics.

Therefore, one should find methods to transfer the required energy locally within a very thin surface layer without heating the whole substrate. Interaction of the top surface with ionic and neutral species in a plasma can be used for this purpose. First of all, ions can transfer their energy and electric charge to the surface in the vicinity of the landing point, and yet still have sufficient energy to overcome the potential barrier (which can be higher for ions due to possible partial burrowing into the surface layer) and so start moving instantly on an already hotter surface! Thus, if adatoms arrive as ions, they may have a better chance of reaching the nanoassembly site (e.g., an appropriate facet) faster and, more importantly, of finding the most appropriate place in the growing crystalline lattice of the quantum dot.

Thus, the fact that adatoms can move and self-assemble faster on a thin surface layer locally heated by ion/neutral fluxes is a definite advantage of plasma nanotools in creating perfectly crystalline quantum dots. However, the adatoms can move faster not only because of the additional heating but also due to a reduced diffusion activation barrier. Thus, if a "normal" diffusion activation energy was 0.7 eV, then in a plasma it can become, for example 0.6 eV or even lower. Taken that the characteristic diffusion times

$$\tau_D = \tau_{D0} \exp\left(\frac{e\varepsilon_a}{k_B T_s}\right)$$

depend exponentially on ε_a, the actual diffusion rates can indeed be increased quite substantially. Here, τ_{D0} is the pre-exponential coefficient with the numerical value equal to the characteristic time scale of barrier-free ($\varepsilon_a = 0$) diffusion, k_B is Boltzmann's constant and T_s is the surface temperature.

The mechanism behind this reduction of the surface diffusion activation energy is quite complicated; in simple terms it is related to polarization effects in complex microscopic electric fields between the growing nanostructures sketched in Figure 2.13. In this figure, the microscopic electric field features a complex dipole topology which originates due to superposition of two electric fields, a "macroscopic" field of the plasma sheath $\mathbf{E}_\lambda$ and a truly microscopic field $\mathbf{E}_S$ due to nanoscale surface morphology. In the case depicted in Figure 2.13(b) the charge on the quantum dots is assumed positive. Generally speaking, the actual electric charge on a quantum dot depends on the rates of charge accummulation and removal/leakage. In the case sketched in Figure 2.13(b) it was assumed that the rates of positive charge leakage from QDs are small compared with the rates of ion deposition.

If adatoms/admolecules are polarizable, or, alternatively if not all of them lose their electric charge upon landing on the plasma-exposed surface, the above microscopic electric fields can facilitate the motion of the building units. In some cases, ions preserve their charge upon adsorption and can follow the electric field; in this case they are commonly referred to as adions. It is worth emphasizing that in the case depicted in Figure 2.13(a), there is a relatively small component of the field $\mathbf{E}_S$ which is parallel to the surface. This field faces the nanodot and can pull a (still positively charged) adion towards the growth site. On the other hand, a positive end of a dipole admolecule can also be driven towards the QD, apparently, at a rate higher than would otherwise be possible without the plasma exposure.

Furthermore, ions of an inert gas (e.g., argon) and hydrogen atoms (which appear as a result of dissociation of H_2 molecules in a plasma) can further improve the quantum dot crystallization process via a number of effects. These effects include, but are not limited to, localized energy transfer; compactification upon impact; appropriate balance of available dangling bonds via competition of bond activation by Ar^+ ions and passivation by hydrogen atoms; insertion of H atoms into strained Si–Si bonds followed by their structural relaxation.

Other plasma-related effects on the self-organized growth of ordered arrays of quantum dots include ion-assisted control of surface stress distribution, which eventually results in new preferential growth sites of quantum dots; delivery of appropriate amounts of plasma-grown seed nuclei to (deterministically!) obtain the required density of quantum dots on the surface; control of two-dimensional microscopic adatom fluxes on a "playground" surface; size uniformity; and on top of all that, even plasma-controlled self-ordering of QDs. Some of these examples will be considered in Chapter 6 of this monograph.

2.7
Concluding Remarks

To conclude this chapter, we stress that a low-temperature plasma can indeed be regarded as a versatile nanofabrication tool which can be used to:

- create appropriate amounts of suitable building units;

- deliver them to the deposition surface;

- suitably prepare this surface for BU landing (which is done by using appropriate working units);

- control surface processes including intricate self-assembly.

These processes lead eventually to the appropriate stacking of the BUs in the nanoassemblies being grown.

There are numerous challenges in this direction and a lot of combined and well-coordinated experimental and theoretical efforts are needed to fully realize the outstanding potential of plasma nanotools in nanofabrication. The ultimate aim of these research endeavours is deterministic control of self-assembly of ultra-small nanoscale and even sub-nanoscaled objects into functional arrays and eventually nanodevices. Even though the "cause and effect" approach based on self-assembly of plasma-generated building units is in its infacy, realization of its full potential will make it possible to eliminate the prevailing "trial and error" experimental practices, which will eventually lead to better, faster and cheaper ways of fabricating nanotechnology-enhanced products.

In this chapter we have discussed the basics of various physical processes in the plasma bulk and on nanostructured solid surfaces that need to be properly considered when applying the advocated "plasma-building unit" approach in practice. Some of these processes have neither a detailed and reliable theoretical description nor proper methods of *in situ* monitoring experimentally. This creates a range of exciting opportunities to expand these studies into the as yet unknown realm of plasma-exposed nanostructured solid surfaces. In the following chapters of this monograph we will continue this slow, step-by-step process of scientific enquiry and will try to reveal the optimum methods of their perception, quantification, and application.

In the following chapter, we will show some typical examples of the application of the generic plasma-based nanofabrication approach elaborated in this chapter. We will also introduce a theoretical approach and an associated computational framework of bridging the "unbridgeable" processes with characteristic spatial scales different by up to nine orders of magnitude.

3
Specific Examples and Practical Framework

We now turn our attention to a discussion of some of the most common nanoassemblies and nanofilms from the viewpoint of the building unit-based "cause and effect" approach introduced in the previous chapter. Wherever possible we will try to follow and comment on the most important issues related to the four main milestones of the above approach; namely, generation and transport of appropriate building units, suitable surface preparation, and incorporation into the nanofilms/nanostructures being grown.

In Section 3.1 several examples of using this approach in the synthesis of various (but primarily silicon-based) semiconducting nanofilms and nanostructures will be discussed. Section 3.2 focuses on the most common carbon-based nanofilms and nanostructures. Section 3.3 introduces a practical framework for using hybrid multi-scale numerical simulations to describe a range of elementary processes in the plasma bulk and on the solid surfaces, which are characterized by spatial scales that differ by up to nine orders of magnitude. Some of the most important points of this chapter are summarized and briefly discussed in the concluding Section 3.4.

3.1
Semiconducting Nanofilms and Nanostructures

In this section, following the pattern of the review article [4], we will concentrate on nanostructured silicon-based films which have an enormous potential for solar cell and several other applications and which are commonly fabricated using plasma discharges in reactive silane-based gas mixtures. The range of suitable materials for thin film-based solar cells is currently dominated by hydrogenated amorphous silicon a-Si:H, with a continuously increasing role of microcrystalline silicon. Unfortunately, even though silicon thin film technologies offer great potential for economically viable solutions in mass production, the share of thin films in the photovoltaic market still remains quite low. There are several

Plasma Nanoscience: Basic Concepts and Applications of Deterministic Nanofabrication
Kostya (Ken) Ostrikov
Copyright © 2008 WILEY-VCH Verlag GmbH & Co. KGaA, Weinheim
ISBN: 978-3-527-40740-8

reasons for this, for example, the quite low deposition rates, which results in prohibitively high production costs and inadequate photostability and power generation efficiency. Furthermore, the efficiency of solar energy absorption is substantially reduced with decreasing film thickness, hence, the relatively thick films that need to be used in solar cell devices.

Recently, it has been demonstrated that inclusion of nanometer-sized Si or hydrogenated silicon (Si:H) crystallites greatly improves performance of *a*-Si:H films in solar cell applications. Straightaway one can interpret that new building units, plasma-grown silicon nanocrystallites, can be considered as the cause of the observed improvement of the solar cell function ("effect"). Therefore, this is an excellent illustration of how the building unit-based "cause and effect" approach of the previous chapter works. The new sort of silicon-based films is thus a mixture of amorphous silicon and small Si crystallites and is termed polymorphous silicon. More precisely, since this material contains a significant fraction of hydrogen, it is called hydrogenated polymorphous silicon (*pm*-Si:H), see Figure 3.1. It is worth emphasizing that solar cell-grade *pm*-Si:H films feature substantially improved transport properties, reduced photo-induced degradation, smaller film thickness and can be deposited at very high growth rates (up to a few nanometers per second and even higher) [76, 80, 92]. Even more importantly from the nanoscience perspective, it appears that by varying the sizes of plasma-grown silicon nanocrystallites, one can effectively control the energy bandgap within the most important part of solar spectrum.

Numerous experiments suggest that plasma-nucleated nanocrystalline building units do indeed cause the remarkable improvement in solar-cell attributes and performance of *a*-Si:H films. To this end, it is imperative to figure out how to generate such BUs and then to incorporate them into the amorphous silicon matrix, following the four-milestone nanofabrication scenario introduced in Chapter 2.

It is remarkable that silane-based reactive plasmas have shown such a remarkable ability to generate a large number of reactive radicals and support gas-phase polymerization of macromolecules and generation of critical clusters [58, 59]. A cluster formation pathway is dominated by anion-neutral reactions, commonly referred to as the Winchester mechanism of ion-molecular cluster growth [57,113]. The essence of this mechanism is the thermodynamic advantage of the anion-induced clustering. A typical anion-supported clustering process

$$Si_nH_{2n+1}^- + SiH_4 \rightarrow Si_{n+1}H_{2n+3}^- + H_2$$

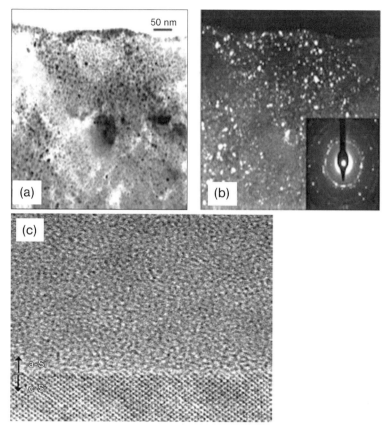

Figure 3.1 Polymorphous hydrogenated silicon films (pm-Si:H) [80], which contain a mixture of nanocrystalline and amorphous phases of silicon: (a) bright field TEM; (b) dark field TEM. HRTEM micrograph in panel (c) shows a clear, smooth, and defect-free interface between amorphous silicon and crystalline Si substrate.

involves the following sequence

$$\mathrm{SiH_3^-} \rightarrow \mathrm{Si_2H_5^-} \rightarrow \mathrm{Si_3H_7^-} \rightarrow \cdots \rightarrow \mathrm{Si_{n+1}H_{2n+3}^-}$$

of silyl anions, with electron affinities $\mathcal{E}_a^n$ increasing with the number of silicon atoms in the cluster and reaching the work function of bulk hydrogenated silicon [4]. As was mentioned by Fridman and Kennedy [57], each reaction step is exothermic. More importantly, exothermic anion-molecular reactions usually have a very low activation barrier and feature very high reaction rates. This fundamental conclusion explains the dominance of anion-induced clustering of hydrogenated silicon and has recently been confirmed by numerical simulations of particle generation mechanisms in silane-based discharges [114,115]. In fact, over 90 % of the critical cluster formation in silane plasmas is likely to proceed through the silyl anion pathway triggered by the SiH_3^- anion, whereas only about 10 % proceeds through the siluene anion ($Si_nH_{2n}^-$) pathway, initiated by SiH_2^- [115].

One can thus immediately conclude [4] that the best strategy for generating the required building units is to optimize the nanoparticle-loaded discharge [116, 117] operation and so enhance the gas-phase reactions leading to generation of SiH_3^- and SiH_2^- nanocluster precursors. One of the most effective ways to achieve this is to specifically tailor the electron energy distribution function (EEDF). In fact, the EEDF determines the rates of most of the electron-impact reactions that occur in the ionized gas phase; this significantly affects the balance, and hence, the abundance of reactive species in the discharge [118].

However, it appears quite challenging to ensure that critical clusters nucleate into the nanosized crystalline particles that are actually required for the PECVD of device-grade silicon films. A potential danger arises when the number densities of the primary nucleates exceed a certain threshold and trigger an (in most cases uncontrollable) agglomeration process. This condition is usually referred to as the nanopowder generation onset [119]. The powder particles usually have complex fractal, cauliflower, porous skeletal, as well as some other shapes and are usually larger than 40–50 nm [58,59]. These agglomerated particles, although viable for some other applications, should be avoided in this particular case of solar cell-grade polymorphous silicon films.

We have thus arrived at the important conclusion that since perfectly nanocrystalline building units are preferred for solar cell applications over the above mentioned nanoparticle agglomerates, then the plasma discharge should be operated away from nanopowder generation conditions. An example of such conditions is shown in Figure 3.2: RF plasmas of highly-diluted $SiH_4(2\%) + H_2(98\%)$, gas deposition temperatures of approximately 200 °C, relatively high deposition pressures (p_0 ca. 1.6 hPa (1.23 Torr)), and RF input powers (ca. 0.11 W/cm^3) [120]. Under such conditions, heavy dilution of silane in hydrogen is beneficial for the growth of the amorphous hydrogenated silicon matrix by SiH_3 radical building units via the hydrogen-mediated surface activation mechanism discussed in Section 2.5. We emphasize that the PECVD regime of interest here is based on selective deposition of the first population of small (1–2 nm) particles which appear in the ionized gas phase well before the agglomeration onset [80]. As has already been discussed, transport of such building units to the substrate critically depends on their

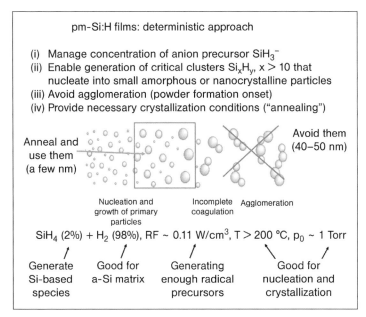

pm-Si:H films: deterministic approach

(i) Manage concentration of anion precursor SiH_3^-
(ii) Enable generation of critical clusters Si_xH_y, $x > 10$ that
     nucleate into small amorphous or nanocrystalline particles
(iii) Avoid agglomeration (powder formation onset)
(iv) Provide necessary crystallization conditions ("annealing")

Anneal and use them (a few nm)

Avoid them (40–50 nm)

Nucleation and growth of primary particles

Incomplete coagulation

Agglomeration

SiH_4 (2%) + H_2 (98%), RF ~ 0.11 W/cm^3, T > 200 °C, p_0 ~ 1 Torr

Generate Si-based species

Good for a-Si matrix

Generating enough radical precursors

Good for nucleation and crystallization

Figure 3.2 Summary of process conditions for deterministic plasma-assisted fabrication of polymorphous hydrogenated silicon films.

charge, on the distribution of gas temperature in the near-substrate areas and on other conditions.

Several experiments have been conducted to relate the nature and charge of specific building units to the surface roughness, microstructure and phase composition of the Si:H films, as well as to whether or not deposition of such particles causes any damage to the surface [4]. For example, surface roughness progressively increases with the size of the BUs, being ca. 2 nm when atomic/molecular units build microcrystalline silicon films, ca. 4–5 nm when the contribution of the plasma-grown nanocrystals is significant and ca. 10 nm under powder-generating conditions [92, 121, 122]. To study the transport to and the contribution of nanocrystalline building units to the properties of *pm*-Si:H, an independently biased "triode" mesh was placed in front of liquid nitrogen chilled (T_s ca. 80 K) substrates as shown schematically in Figure 3.3 [92, 122]. As was mentioned in the review paper [4], this arrangement substantially reduces the contributions from surface migration of atomic/molecular building units and allows one to control the impact energies of small (1–2 nm) positively charged, plasma-grown nanocrystals.

We emphasize that the charge on these ionized gas phase-borne nanocrystals is positive. In this series of experiments the potential in the

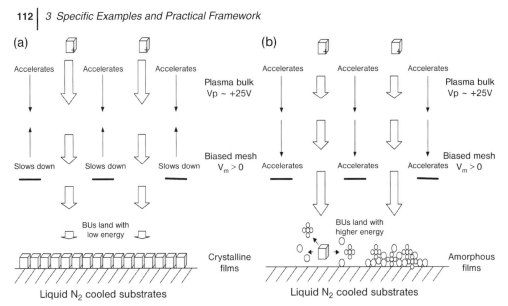

Figure 3.3 Schematics of the experiment on reactive plasma-based deposition of positively charged nanocrystals onto liquid nitrogen cooled substrates [4, 92, 122]. When the building units land with very low energy, there is no nanocluster disintegration and the films are predominantly crystalline (a). On the other hand, when the BUs are accelerated to higher energies, nanocrystals break up into atomic/radical fragments and the films are mostly amorphous (b).

plasma bulk was approximately 25 V higher than that of the substrate. Thus, applying a positive potential of +25 V, one can substantially reduce the energy of the building units upon landing to almost zero (Figure 3.3(a)). Under such conditions, Raman spectroscopy reveals that the films are purely crystalline, which implies that the nanocrystalline building units deposit in a perfectly non-destructive fashion.

However, by reducing the mesh potential down to zero or negative values, one effectively increases the impact energy, and hence, the disintegration probability of the BUs. In this case, the films feature a very high content of the amorphous phase. From Chapter 2, we recall that the amorphous phase predominantly grows from radical/molecular species. Moreover, when the mesh bias is reduced to −50 V, the amorphous phase content increases to almost 100 %. This is indicative of a complete breakdown of nanocrystalline particles into atomic/radical fragments as schematically shown in Figure 3.3(b) (see also Figure 2.9 where impact disintegration of a small nanoscluster is shown). This behavior has also been confirmed by the results of molecular dynamics simulations of the landing of small nanoparticles on various substrates under similar deposition conditions [92].

Another possibility for controlling the deposition dynamics of small nanoparticles is to use a gas temperature gradient-driven thermophoretic force [4, 123, 124]. Physically, when the substrates are externally heated from underneath, gas temperatures near the surface are usually higher than in the plasma bulk. In this case, the temperature gradient results in quite different rates of collisions between the gas species and the nanoparticles in the areas with different gas temperatures. This gives rise to a force directed opposite to the temperature gradient, which effectively pushes the nanoparticles away to the plasma bulk and eventually to the pump line. Thus, by using the thermophoretic manipulation of larger (> 1 nm) building units, one can enhance the probability of the film growth occurring predominantly by atomic/molecular units [4].

The next milestone of the plasma-aided nanofabrication approach (Chapter 2) is structural incorporation of building units into the nanostructured films. Under conditions of heavy dilution of silane in hydrogen, the amorphous silicon matrix grows at a rate comparable to the growth rates of the nanocrystalline BUs that embed into it. This has been confirmed by high-resolution transmission electron microscopy (Figure 3.1) [80], which suggests that while the nanocrystalline building units grow in the gas phase, the a-Si:H matrix forms the underlayer. Subsequent layers contain 1–2 nm nanocrystalline inclusions, as suggested by the presence of sharp and intense rings in Figure 3.1 superimposed on diffuse rings in selected areas of electron diffraction patterns, indicating the presence of ordered structures in an amorphous matrix.

As was noted by Ostrikov [4], there is a trade-off between the growth rates of the pm-Si:H films and the crystallinity of small silicon nanoparticles embedded in the amorphous matrix. Even though processes based on pure silane plasmas yield very high nanoparticle growth rates, which in some cases are up to ca. 100 times higher than in plasmas of heavily diluted SiH_4 gas mixtures, perfectly crystalline nanoparticles are very rarely observed in pure silane plasmas [125]. As a remedy, one can use an alternative approach based on initial growth of amorphous and irregular-shaped particles and their subsequent annealing in hot working gas mixtures of silane and an inert gas [84]. This approach has been successfully used by the University of Minnesota research group in the plasma-aided synthesis of silicon and germanium nanocrystals of various shapes, sizes and faceted structure [85, 126]. However, efficient thermal annealing of nanoparticles requires very high working gas temperatures, often exceeding 1000 °C. As was mentioned earlier in Chapter 2, intense ion bombardment and/or special crystallization agents (reactive atoms or radicals) can substantially help to achieve this goal. Another important issue is to minimize the size of those crystalline building units

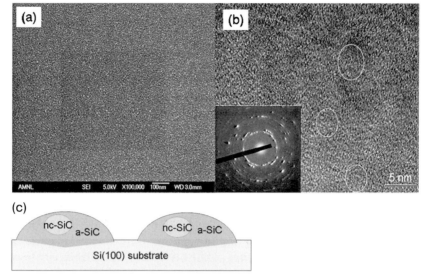

Figure 3.4 Typical SEM (a) plane-view HRTEM; and (b) EDP (inset) micrographs of nanoislanded nc-SiC films synthesized using high-density inductively coupled plasma-assisted RF magnetron sputtering of SiC with a surface temperature of $400\,°C$ and ICP power of 1200 W [129]. SiC nanocrystallites are encircled. Panel (c) shows a schematics of the film structure.

commonly generated by the ionized gas-phase growth technique (which are typically in the 20–80 nm range) to at least 10 nm, which would allow their applications in floating gate memory devices [4,127].

It is important to note that simultaneous integration of various building units in the same nanoassembly or nanostructured object can be very attractive from the applications point of view. For example, by using $SiH_4 + CH_4 + H_2$ gas mixtures, one can grow polymorphous hydrogenated silicon carbide $(pm - Si_{1-x}C_x : H)$, ideal for applications as a p-type nanolayer in PIN solar cells [92]. By varying the process parameters to adjust the relative production of silicon- and carbon-bearing species, one can control the value of x and also the crystallinity of the embedded nanoparticles. Under certain conditions (e.g., very small size, excellent crystallinity, and so on) these nanocrystals can act as buried quantum dots, which substantially alter tunnelling and photoabsorbing properties of functional layers in solar cells of the third generation [128].

An even more interesting sort of nanofilms reported recently is a nanoislanded SiC with embedded SiC nanocrystallites, as shown in Figure 3.4 [129]. In a sense, this is a more intricate, nanoislanded variant of polymorphous SiC films with the microstructure sketched in Figure 3.4(c). These nanostructured films were synthesized using radio-frequency magnetron sputtering of high-purity SiC targets in high-density argon plasmas created externally using inductively coupled plasmas sustained in the integrated plasma-aided nanofabrication facility [49]. By carefully manipulating the process parameters, it was possible to synthesize highly-stoichiometric (with the [Si]/[C] elemental ratio 0.96–0.98 close to a stoichiometric value of 1) and hydrogen-pure SiC films, which simultaneously feature a nanoislanded surface morphology and ultra-small (typical size of approximately 4–6 nm as shown in Figure 3.4(b)) nanocrystallites embedded in dome-shaped structures made of amorphous silicon carbide. We will revisit this issue later in this monograph and comment on the best strategies for deterministic plasma-aided synthesis of these and similar SiC nanoassemblies.

Sputtering of atomic and nanocluster building units from multiple targets in low-temperature non-equilibrium plasmas of reactive gas mixtures offers additional flexibility in controlling the growth of low-dimensional semiconductor nanostructures with the desired properties. In the example presented elsewhere [4], concurrent RF magnetron sputtering of Al and In targets in nitrogen plasma discharges at low pressures (in the 0.667 Pa (5 mTorr)–2.000 Pa (15 mTorr) range) can be used to synthesize $Al_xIn_{1-x}N$ quantum dot structures which uniformly cover large substrate areas (see Figure 3.5(a)) [49]. By varying the RF power supplied to Al and In sputtering targets, one can control the release of aluminum- and indium-containing building units into the ionized gas phase, and eventually, the relative elemental composition $x/(1-x)$ of Al and In in the films. In this way, one can control the size of the individual QDs and the energy bandgap of quantum dot structures in the range from ca. 5.680×10^{-19} J (3.55 eV), which is typical for InN (small x), to ca. 9.920×10^{-19} J (6.2 eV) (pure AlN, $x \approx 1$). This conclusion has been confirmed by the photoluminescence (PL) spectra of $Al_xIn_{1-x}N$ nanopatterns shown in Figure 3.5(b). Indeed, one immediately notices a marked blue shift of the PL maximum as the elemental presence of aluminium x and the quantum dot size decrease [49].

Figure 3.6(a) shows silicon nanowires synthesized in low-pressure (< 13.332 Pa (100 mTorr)) reactive plasmas of $SiH_4 + H_2$ gas mixtures in the presence of a Ni catalyst [4, 49, 130]. These nanoassemblies have an outstanding potential for the development of molecular nanoelectronic devices and have also been previously synthesized by charged nanocluster building units in thermal chemical vapor deposition systems [74]. In the example discussed by Hwang *et al.* [74], the assembly of one-dimensional (1D) silicon nanowires with a very high-aspect-ratio can be

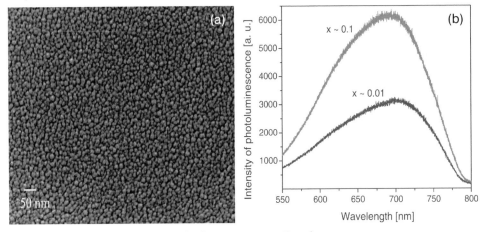

Figure 3.5 Synthesis and photoluminescence properties of ternary semiconductor quantum dots [4, 49]: (a) FESEM micrograph of $Al_xIn_{1-x}N$ quantum dot structures synthesized by the reactive plasma-assisted simultaneous RF sputtering of Al and In targets; and (b) the dependence of the photoluminescence intensity on QD size/composition. Quantum dot size decreases for smaller values of x.

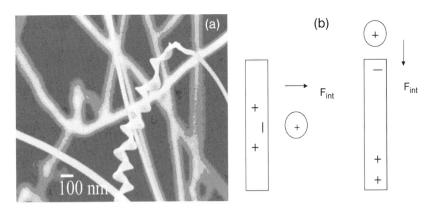

Figure 3.6 Silicon nanowire growth and interaction with charged nanoclusters [4]: (a) FESEM micrograph of silicon nanowires grown in low-pressure $SiH_4 + H_2$ reactive plasmas [49, 130], and (b) schematics of the charged nanocluster-nanowire interaction (after [65]). Here, F_{int} denotes the force of the electrostatic interaction between the incoming positively charged nanocluster and the nanowire.

explained by selective attraction of nanoclusters to the open end of 1D structures, which are schematically represented by rods in Figure 3.6(b). To be specific, it was assumed [74] that both the rod and the building units are positively charged. Here we note that in a similar situation involving low-temperature plasmas, the nanowires and nanoclusters will more likely be charged negatively.

Referring to Figure 3.6(b), if the nanocluster approaches the nanowire from the side, positive charges in the rod are repelled. However, since the diameter of silicon nanowires is very small (it does not exceed a few tens of nm in Figure 3.6(a)), the nanorod-cluster interaction still remains repulsive. However, when approaching the nanowire from its growth end (Figure 3.6(b)), positive charges are repelled to the opposite side of the rod, which results in an attractive electrostatic interaction between the cluster and the nanowire, and, hence, one-dimensional nanowire growth [65]. Interestingly, these arguments are also applicable to other charged species such as ions and also to building units which can be polarized in the electric field of the plasma sheath. However, one should be very cautious when using these arguments to explain ultra-high-rate catalyzed growth of semiconductor nanowires when a catalyst nanoparticle is anchored to the top of the one-dimensional structure. Thus, the low-temperature plasma-assisted growth of silicon nanowires has an outstanding potential to control the growth of high-aspect-ratio semiconductor nanostructures. However, as was mentioned in the review article [4], substantial efforts should be applied to enable deterministic control of crystallographic growth direction; this possibility has already been reported for the metal-organic CVD technique [131].

3.2
Carbon-Based Nanofilms and Nanostructures

In this section we focus our attention on some common carbon nanostructures which can be synthesized in low-temperature plasmas of mixtures of carbon-carrier gases (e.g., hydrocarbons C_xH_y, fluorocarbons C_xF_y, fullerenes, and so on) with other functional feedstock (H_2, NH_3, inert gases, and so on). We emphasize that in the "cause and effect" framework introduced in Chapter 2, the choice of working gases and process parameters should be driven by the required building/working units, surface preparation and specific building unit transport and stacking requirements [4].

Following the review article [4], let us now consider a few typical examples of plasma-synthesized carbon nanostructures (CNSs), such as carbon nanoparticles, nanotips, nanotubes, nanowalls and ultra-nanocrystalline diamond. With regard to these CNSs, the most commonly invoked building units are carbon dimer C_2 [37], graphitic nanofragments [132], charged nanoclusters [65, 70, 71] and carbon nanoparticles [133–135], in addition to carbon atoms already discussed in Chapter 2. Some other atomic and radical species have also been discussed extensively as mediators of clustering and film growth processes.

It is worth mentioning that in the nanofabrication of nanostructured films in silane-based plasmas of Section 3.1, nanoparticles larger than a few tens of a nm are usually only of limited interest for most common silicon-based nanoassemblies. However, some of common carbon nanoassemblies, such as carbon nanostructures, can self-assemble using large planar graphitic sheets which wrap themselves while in a gas phase. This is one particular example of carbon nanotube assembly from graphitic nanofragments [132]. In reactive plasma environments, graphitic nanofragments (which, generally speaking, are not merely limited to planar graphitic sheets) of various sizes can appear as a result of decomposition of larger carbon nanoparticles in a gas discharge or via sputtering of bulk graphite in a plasma. However, our existing knowledge on the origin and specific roles of graphitic nanofragments in low-temperature plasmas is very limited and needs further research efforts [4].

Regarding the carbon dimer and carbon nanocluster building units as well as other important species (e.g., growth precursors and mediators, some working units that condition or activate/passivate the growth surface), their abundance in various plasma systems can be predicted by numerical modeling of species production and ion-induced clustering in the ionized gas phase [53, 54, 136–138]. In particular, in RF plasmas of C_2H_2 highly diluted in argon, number densities of the carbon dimer in two excited states $X^1\Sigma_g^+$ and $a^3\Pi_u$ are higher at lower working gas pressures and argon concentration [137]. Therefore, optical emission spectroscopy (OES) appears to be a very powerful tool in monitoring the discharge species.

The strongest emission lines of carbon dimer C_2 generated in low-pressure (2.666 Pa (20 mTorr)–5.333 Pa (40 mTorr)) RF plasmas of Ar (75 %) + H_2(10 %) + C_2H_2(15 %) gas mixture are shown in Figure 3.7. These peaks can be used to trace the appearance of carbon dimers in various situations [4]. Amazingly, the strongest peak located at ca. 516.5 nm is characteristic of the Swan band visible absorption spectra of numerous protoplanetary nebulas, implying a possible important role of the carbon dimer in the evolution of red star-protoplanetary systems [139]. An interesting way to maximize the production of carbon dimer molecules is to dissociate purely carbon feedstock, such as gaseous C_{60}, a less stable (than diamond or graphite) carbon allotrope. In fact, by using plasma-assisted techniques one can achieve unusually high rates of conversion of the C_{60} feedstock into C_2 molecules [4,37].

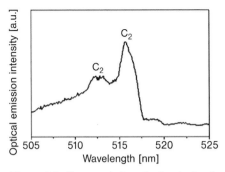

Figure 3.7 Characteristic optical emission lines of the carbon dimer in nanofabrication of carbon nanowall-like structures in $Ar + C_2H_2 + H_2$ RF plasmas [49].

Knowledge of the number densities of other (e.g., C_xH_y) neutral and charged species is crucial for the improvement of process control strategies. For example, as has already been discussed in Section 2.5, radical species CH_3 can be considered as a viable building unit in the growth of amorphous hydrogenated carbon (a-C:H) films [4,98]. On the other hand, atomic hydrogen serves to activate surface carbon bonds and simultaneously etch the amorphous carbon phase. Number densities of numerous charged and neutral species in the PECVD of carbon-based nanostructures in reactive gas mixtures of methane (CH_4) or acetylene (C_2H_2) with hydrogen and argon gases can be found elsewhere [53,54,138].

Similar to the Si:H clusters considered in Section 3.1, carbon clusters can also be formed in the gas phase, however, at significantly lower rates. The main reason for this is a relatively lower reactivity of C_xH_y radicals compared to Si_xH_y reactive species. Moreover, this process is usually believed to be faster in acetylene-based plasmas compared to gas discharges in methane-based mixtures. Interestingly, the most likely mechanism of carbon clustering in C_2H_2-based plasmas is quite similar to the Winchester mechanism of silicon hydride clustering and also involves the following chain

$$C_iH_j^- + C_mH_n \rightarrow C_{i+m}H_{j+n-1}^- + H$$

of anion-neutral clustering which proceeds via extraction of hydrogen and generation of higher anions [136]. Furthermore, carbon dimer C_2, featuring high electron affinity, can attach a plasma electron and so become a negatively charged C_2^- anion. This highly reactive radical has also been suggested as a possible trigger of anion-neutral clustering [65]. Relevant modeling results (that include charge neutralization, neutral clustering, diffusion loss of the plasma species and other effects) suggest

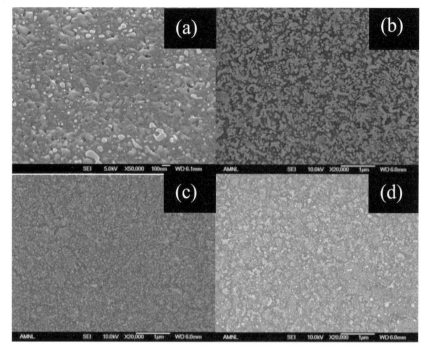

Figure 3.8 Surface morphology (top view) of (a) the catalyst layer after the etching stage; and (b–d) carbon nanostructure growth islands at T_s =500 ((a,b)), 350 (c), and $300\,^{\circ}\mathrm{C}$ (d), respectively [93].

that larger $C_m H_n$ clusters with $m > 10$ are negatively charged at higher gas temperatures and lower degrees of ionization and operating pressures [136]. Otherwise, one would expect a pronounced generation of neutral or positively charged nanoclusters. However, positively charged carbon-based clusters were not included in the clustering model [136], thus warranting their explicit consideration in the near future [4].

Having identified potential building units and important process mediators in hydrocarbon-based plasmas, we now discuss issues related to surface preparation and activation. Many nanofabrication processes, such as growth of carbon nanotubes and related nanostructures [21, 89, 140], require specifically activated thin catalyst (e.g., Ni, Fe, Co and their alloys) layers (with the thickness ranging from a few to a few tens of nm), which usually re-arrange into individual nanoparticles on the substrate surface, as can be seen in Figure 3.8. Figure 3.9 shows examples of most common vertically aligned carbon nanostructures, which rely on various metal catalysts in their growth. These examples include, but are not limited to, carbon nanotubes (Figure 3.9(a)) single-crystalline carbon

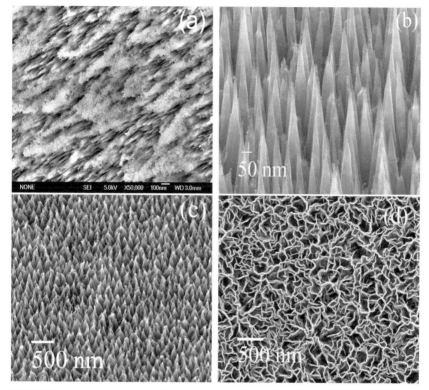

Figure 3.9 Examples of carbon-based nanostructures which are synthesized in hydrocarbon-based low-temperature plasmas on catalyzed silicon substrates [4, 49, 55, 93, 94, 140]: (a) vertically aligned carbon nanotubes; (b) carbon nanotips; (c) pyramid-like structures grown in $Ar + CH_4 + H_2$ plasmas without any external substrate heating; (d) carbon nanowall-like structures grown in $Ar + C_2H_2 + H_2$ RF plasmas.

nanotips (Figure 3.9(b)), nanopyramids (Figure 3.9(c)), and nanowall-like structures (Figure 3.9(d)).

Activation of the growth surface is usually achieved by externally heating the substrate to temperatures (typically 500–600 °C and higher) that exceed the melting point of the catalyst layer. It is notable that when the film thickness is in the nanometer range, the melting points are lower than those of the corresponding bulk materials [4]. To ensure efficient bonding of the nanostructure to the substrate, one should ensure adequate wetting of the substrate by the catalyst nanoparticles. In this case, one would expect the "base" growth scenario. However, sufficiently intense ion bombardment (controlled by the substrate bias) contributes to the loosening of the catalyst nanoparticles, thus leading to the "tip" growth pathway. In this case, plasma-assisted techniques often produce individual, free-standing, vertically aligned multiwalled carbon nanotubes (MWCNTs) [141,142]. This is one of the possible reasons why plasma-synthesized MWCNTs are more common than single-walled carbon nanotubes (SWCNTs) [21].

Here we emphasize that SWCNTs in most cases grow through the "base" growth mechanism. More importantly, the size of dome-shaped catalyst nanoparticles turns out to be a decisive factor which controls the nanotube's chirality, or in basic terms, the angle at which a planar graphene sheet is wrapped around its axis to form a nanotube [143–148]. Since typical diameters of single-walled carbon nanotubes can be as small as ca. 0.6–0.7 nm, the catalyst islands should be approximately of the same size. Therefore, if one aims at synthesizing a nanopattern with a large number of SWCNTs of the same chirality, then one has to create a somewhat larger number (this is required since some of them may turn catalytically inactive during the growth process) of metal catalyst particles and all of them should have nearly the same size! Even a minor difference in the catalyst island size can cause a significant variation of the resulting nanotube chiralities within the pattern. Moreover, the tube diameters, and also with high probability, their lengths, can also be affected by the catalyst nanoparticle size distribution. This eventually can result in electronic properties of the SWCNT patterns quite different from what was originally intended. This example is a clear illustration of the importance of proper catalyst choice and arrangement in the synthesis of delicate carbon-based nanoassemblies such as single-walled carbon nanotubes.

It is interesting that catalyst islands on solid surfaces can experience self-organized behavior and can restructure themselves over quite large deposition areas to improve spatial ordering of vertically-aligned carbon nanostructures such as nanotips and nanocones [93–95]. Indeed, the SEM analysis confirms that the sizes of the growth spots of the carbon nanotips correlate with the sizes of the reorganized (after reactive chemical etching using hydrogen) nickel-based catalyst islands (Figures 3.8(a) and (b)) and become smaller when the substrate temperature decreases to 300–350 °C (Figures 3.8(c) and (d)).

Amazingly, the actual growth process of carbon nanotip/nanocone-like structures can commence from quite non-uniform distribution of catalyst islands over the growth surface. Nevertheless, under plasma conditions carbon nanocones still manage to grow in size- and position-uniform nanoarrays [95]. This amazing phenomenon is attributed to the plasma-related affects on species behavior on the surfaces of the catalyst, silicon substrate and nanostructures themselves. Moreover, ion focusing

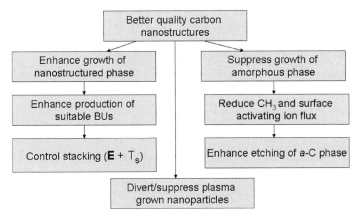

Figure 3.10 Practical approach to synthesize high-quality carbon nanostructures using low-temperature plasmas.

has a marked effect on the growth process [95]. This highly-unusual self-organized growth of uniform carbon nanocone arrays from very non-uniformly fragmented metal catalyst layers will be discussed in Chapter 7 of this monograph.

Plasma environments offer a range of effective options to activate and rearrange catalyst layers. For example, the Ni-based layer in Figure 3.8(a) was activated in low-pressure (ca. 5 Pa (40 mTorr)) RF plasmas of $Ar + H_2$ gas mixtures by using intense fluxes of argon ions, reactive chemical etching of metal surface by hydrogen atoms and heating by hot neutrals (with temperatures in the range 270–400 °C) [93, 94]. It is important to note that in some cases, such as plasma-enhanced chemical vapor deposition of nanopyramid-like structures in Figure 3.9(c), it was not necessary to additionally heat the substrates externally, a situation quite uncommon for thermal CVD systems [4].

While the catalyst nanoislands usually serve as the "base" for the growth of carbon nanostructures (CNSs), other areas, uncovered by the CNSs, are subject to deposition of amorphous carbon. Thus, the actual film growth process involves concurrent growth of two phases, namely, the nanostructured and the amorphous phases [4]. To synthesize better quality nanostructures (e.g., crystalline), one should thus inhibit the development of the amorphous phase and promote the growth of the nanoassemblies.

Figure 3.10 summarizes the requirements and some appropriate actions for better-quality carbon nanostructures. Again, by invoking the "cause and effect" approach of the previous chapter, we recall from Section 2.5 that efficient growth of amorphous carbon films requires plasma-mediated activation of dangling bonds, which is needed for the stacking of CH_3 radicals. On the other hand, preferential growth of the nanoassembly requires elevated abundance of carbon atoms, highly-reactive carbon dimers or suitable nanocluster BUs in the gas phase. To this end, it would be beneficial to maximize production of carbon atoms, C_2 and/or nanoclusters and minimize the presence of CH_3. Moreover, during the growth stage reasonably high densities of hydrogen atoms are desirable for preferential chemical etching of the amorphous phase. However, the hydrogen content cannot be made very high without detriment to the growth process. For example, large amounts of hydrogen can re-gasify embryonic nuclei during the synthesis of nanocrystalline diamond [37] and excessively activate carbon surface bonds, which are essential for the growth of *a*-C [4].

The effect of the ion bombardment on the plasma-based nanoassembly is one of the most controversial issues nowadays, as evidenced by strong debates in the literature, conferences, seminars, and so on. The main reason is that at present very little is understood about the actual effects of the plasma ions on *nanostructured* matter even though similar effects on bulk materials are much better understood. Some effects of the ion fluxes on the developing nanostructures have already been discussed in Chapter 2. As we emphasize in several places in this monograph, the answer depends on the specific ion species and nanostructure/nanopattern considered, structural/phase state of the nanoscale object, kinetic energy of impinging ions and some other factors. For example, is any particular ionic species a building unit needed to build the nanostructure, or a working unit used to facilitate the synthesis process?

Practical experience shows that the main difficulty in answering this question is to be specific! Indeed, it is extremely easy to ask the question whether ion bombardment causes severe damage to the nanoassemblies but it is not so easy to clearly state the input parameters to make the question more specific and direct! In some cases, it could be wise to ask a counter-question "why bombardment in the first place"? Well, in commonsense understanding it is sometimes similar to comparing the effects of a heavy rock and a feather thrown from the same height onto a range of different surfaces. Upon impact of a rock, some surfaces will be cracked instantly, whereas others will not even show a scratch - everything depends on the prevailing conditions. But who would seriously put together the words "feather" and "bombardment"?

To be specific, we will consider the effect of relatively heavy argon ions which definitely cannot be used as BUs. In this case we need to know exactly how structurally strong our films/structures are with respect to the impact of argon atoms. In most cases there will definitely be some sputtering of the building material off the surface. However, if the rates of deposition and structural incorporation of such materials are higher than the rates of physical sputtering caused by the ions, this effect is of a very little concern in practical experiments. A lot depends on the structural stability of the objects we actually deal with.

For instance, carbon nanotubes are usually very stable, both structurally and compositionally, against ion bombardment. Only high doses ca. 10^{13}– 10^{15} ions/cm^2 of high-energy ions (e.g., tens of kV) cause any noticeable structural modifications to multiwalled carbon nanotubes [149]. These harsh conditions can be used to selectively modify the structure of different layers throughout the MWCNTs; however, these conditions are not so common in experiments on nanostructure growth in low-temperature non-equilibrium plasmas. We should also stress that single-walled nanotubes are easier to damage by ion fluxes and extra care should be taken to avoid any undesirable effects.

Therefore, when using plasma-assisted nanofabrication methods one should always keep in mind that the ion dose and energy should be just sufficient for the desired effect, for example, surface activation or conditioning. There are reports in the literature that extra strong ion bombardment can cause irrecoverable defects to the nanostructures being grown and also destroy some nanocluster BUs in the gas phase. For example, if a nanostructure is made of amorphous material, then this effect is obvious. If the nanostructure is crystalline or a mixed-phase (e.g., polymorphous), substantially larger ion doses and energies are required to affect it. Furthermore, if the nanostructure is extra-stable, like the MWCNTs in the experiments discussed above [149], then there should be no appreciable damage under the normal conditions used for plasma-aided nanofabrication [5]. An example of how energetic ions impinging on the growth surface can compromise the integrity and ordering of vertically aligned nanostructures is presented by Hirata *et al.* [150], where additionally it is proposed to use strong magnetic fields (ca. 2T) directed parallel to the substrate to reduce this effect. Otherwise, one could reduce the potential difference between the growth surface and the plasma bulk and in this way control the ion density in the pre-sheath areas that separate the plasma bulk and the sheath.

Again, our commonsense logic tells us that if some ions are indeed undesired, then the best strategy would be not to create them in large amounts and not to accelerate them to high energies, rather than using complex means of confining them and preventing from landing on the substrate by using complex magnetic fields just like in nuclear fusion devices. Having said that, we nevertheless admit that magnetic enhancement of nanotube synthesis may bring about several important advantages. However, these advantages are not primarily related to the reduction of ion bombardment of nanotubes. Indeed, magnetic fields can be used for a variety of purposes ranging from the commonly known creation of higher-density plasmas (and therefore, building units!) to direct nanotube manipulation (via magnetic agitation and guiding ferromagnetic catalyst nanoparticles anchored to the tubes) in drug/gene delivery into mammalian cells and even impalefection (destroying by spearing) of cancerous cells [151].

It has been mentioned previously [4] that the transport of building units in hydrocarbon plasma environments has not yet attracted due attention when compared to similar research efforts on silane-based plasmas (see Section 3.1). However, the existing modeling results can be used to estimate the surface fluxes of numerous charged and neutral plasma species in the process of assembly of various carbon nanostructures [53,54,137,138,142]. Interestingly, larger building units (e.g., larger nanoclusters or nucleates exceeding 10 nm) can be effectively manipulated by using thermophoretic forces [105, 152, 153]. Indeed, when Ni-catalyzed silicon substrates are not externally heated, fallout of gas-phase nucleated carbon nanoparticles is frequently observed. However, by externally heating the substrates under the same operation conditions, one imposes an additional temperature gradient, and hence, the thermophoretic force that can completely remove carbon nanoparticles from the surface [4]. This approach makes it possible to implement the "divert plasma-grown nanoparticles" (from the nanostructures being synthesized) milestone shown in the practical approach diagram in Figure 3.10. The resulting nanoparticle-free assemblies resemble ordered carbon nanotip patterns in Figure 3.9(b).

We now consider the stacking of plasma-grown building units into carbon-based nanoassembly patterns [4]. The first example is the insertion of the carbon dimer C_2 into a dimer row of the reconstructed (100) surface of diamond, as sketched in Figure 2.6 [37]. Detailed density functional theory calculations show that the insertion of one of the carbon atoms of a gas-phase borne molecule C_2 into the dimer rows of the reconstructed (100) surface, leaves the other carbon atom free to react with incoming carbon dimers to form a new diamond crystallite, which grows larger and eventually forms a grain boundary [38]. Other examples of interactions of the carbon dimer with reconstructed (100) and (111) surfaces of diamond can be found elsewhere [154].

It is interesting that, depending on surface hydrogenation and other factors, the hybridization of carbon in the dimers on the reconstructed surface may not necessarily reproduce the sp^3 hybridization in the bulk. Thus, control of spontaneous reconstruction of carbon "dimer" surfaces is one of the current challenges of nanoscience. The specific

(a)
(b)

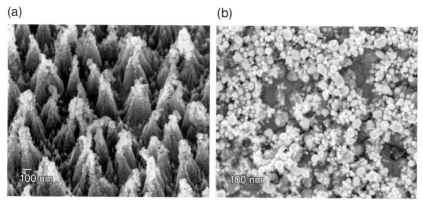

Figure 3.11 Nanoparticles (a) attached to the top of carbon nanotip structures; and (b) on a nickel-catalyzed silicon surface [4, 93, 94].

chemical structure and high reactivity of the carbon dimer have been argued as major factors in the reactive plasma-assisted fabrication of two-dimensional carbon nanostructures such as carbon nanowalls and nanowall-like structures (Figure 3.9(d)) [49, 50, 155, 156]. On the other hand, insertion of carbon into amorphous carbon films can be investigated by density-functional, tight-binding and empical simulation methods [157, 158].

Larger building units stack into carbon-based nanoassemblies quite differently [4]. For example, small nanoclusters can epitaxially recrystallize in a manner similar to that depicted in Figure 2.8(a) and discussed in Section 2.5. However, larger nanoclusters and nanoparticles can be driven by focused electric fields and eventually stick to the top ends of high-aspect-ratio carbon nanotip structures as can be seen in Figure 3.11(a). In some other cases (e.g., unbiased substrates) such building units can fall out onto the substrate, as can be seen in Figure 3.11(b). Interestingly, in some cases the top surface of the carbon nanocones (tapered or "truncated" carbon nanostructures) can be flat [93, 94], which can create ideal deposition spots for the nanoparticle building units.

Thus, adopting the "cause and effect" approach [4] introduced in Chapter 2, and following basic considerations during each of the nanofabrication steps, one can, in principle, grow and perfect the properties of various carbon nanostructures (such as single crystallinity of carbon nanotips in Figure 3.9(b)).

However, many important questions still remain open. For example, what is the cause of such pronounced vertical alignment of carbon nanotubes and carbon nanotips grown by the plasma-assisted techniques?

In fact, this also applies to a variety of other one-dimensional nanostructures such as nanorods, nanowires, nanocones and nanoneedles made of a variety of materials, including binary, ternary and even more complex compounds.

We recall that the electric field **E** is non-uniform in the plasma sheath, is focused on the growth spots from the initial stage of the growth island formation [159] and drives charged building units thus facilitating their stacking into the nanoassembly directly from the gas phase. Meanwhile, strong electric fields in the vicinity of nanotips can polarize neutral (e.g., nanoclusters or molecules/radicals of a polarizable material such as ZnO) building units and then align them to stack in the nanoscale object being created. This also increases the probability that the BUs will incorporate into the nanostructure while approaching from the top. Thus, one can conclude [4] that unidirectional precipitation of both charged and neutral building units favors the assembly of nano-sized structures aligned along the direction of **E**, on the unit-by-unit basis. However, stacking of neutral species with poor polarization response does not necessarily happen in the direction of **E**.

In particular, excellent alignment of carbon nanotubes and related nanostructures along the direction of the electric field is not fully understood. This is clearly the case in most plasma-based synthesis processes wherein the nanotubes align normally with respect to the deposition substrate (this can be seen, for example, in Figure 3.9(a)). This direction apparently concurs with the direction of the electric field in the plasma sheath. On the other hand, nanotubes and related structures in neutral gas-based processes form random networks somewhat similar to the network of interwoven nanowires shown in Figure 3.6. One would immediately speculate (in fact, this is one of the main privileges of physicists!) that this has something to do with the presence of ions and their focusing by the electric fields converging towards the sharp tips of the nanotubes. However, this needs to be complemented by the consideration of their structure and case-specific growth mechanism [160]. Interestingly, nanotubes also align along the electric field even without the presence of the plasma. For instance, the nanotubes can bend and form L-shaped structures if the direction of the electric field is changed during the growth process [161].

It was proposed that when nanotubes are subjected to an electric field, the electrostatic force creates stress non-uniformly distributed over the interface between the catalyst nanoparticle and the carbon nanotube structure grown through either the "tip" or "base" mechanisms. As a result, the carbon material precipitation rates become different in the areas with different stress, and vertical growth is dynamically maintained [160]. However, how exactly the stress non-uniformity over the catalyst nanoparticle-CNT interface translates into preferential stacking of the BUs still requires an adequate explanation [4].

Another interesting question is related to the thermokinetic choice of the growth mode, for example, whether to develop via the "tip" or the "base" growth mechanism, as a single-walled or multiwalled nanotube, or alternatively, as a nanofiber or a platelet-structured nanorod, and so on. From the plasma nanoscience perspective, it is particularly important to elucidate how exactly the plasma environment can affect this process. The bad news first, the latter issue is far from being resolved, even though there is plenty of experimental observations in specific cases which cover a very large number of possibilities; however, it is not possible to cover all the observations in a single monograph.

Let us instead approach this issue deterministically and discuss the structure and some properties of SWCNTs and MWCNTs in a bit more detail. First of all, from the point of view of applications in future nano-electronic devices, one should decide whether semiconducting or metallic nanotubes are required. Multiwalled nanotubes are conducting in the overwhelming majority of cases, except when they have a small number of walls. Thus, by controlling the number of walls, one can tune the electric conductivity properties of MWSNTs. On the other hand, SWCNTs can be either metallic (in approximately 1 out of 3 cases) or semiconducting (ca. 2/3 cases). This depends on the nanotube's chirality, which is determined by the chirality vector

$$C = n\mathbf{u}_1 + m\mathbf{u}_2, \tag{3.1}$$

where n and m are integers and $\mathbf{u}_1$ and $\mathbf{u}_2$ are the unit vectors of the planar graphene sheet [63]. The two numbers, n and m, therefore, determine the structure of the SWCNT. Moreover, if $n - m = 3k$, where k is also an integer, the nanotubes are metallic; otherwise, they exhibit semiconducting properties. If the SWCNTs are semiconducting, their electronic bandgap is [162]

$$\varepsilon_g = \gamma(2a/d), \tag{3.2}$$

where γ is the π matrix element between adjacent carbon atoms, a is the length of the carbon–carbon bond in the relevant nanostructure, and d is the diameter of the SWCNT concerned. In the case of carbon nanotubes made of graphene sheets, which contain sp^2-hybridized carbon, the estimated length of the C–C bond is approximately 0.135 nm. From Equation (3.2), one can see that the bandgap ε_g is inversely proportional to the nanotube's diameter d; thus, the bandgap increases as the nanotubes get thinner.

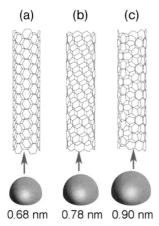

(a)   (b)   (c)

0.68 nm   0.78 nm   0.90 nm

Figure 3.12 Different chiral configurations of SWCNTs: (a) (5,5) "armchair" metallic CNT, with diameter of ca. 0.68 nm; (b) (10,0) "zig-zag" semiconducting SWCNT, with d of ca. 0.78 nm; (c) (9,4) semiconducting SWCNT, with diameter approx. 0.90 nm which may be formed from Ni catalyst nanoislands grown on a Si(100) substrate under the following conditions $T_s = 800\,°C$ and surface diffusion activation energy $\varepsilon_d = 0.47\,eV$ [163].

Therefore, the size of catalyst particles anchored to the SWCNT base appears to be of paramount importance and determines the structure, and hence, electronic properties of SWCNTs. Figure 3.12 shows three examples of numerical simulations of plasma-assisted growth of single-walled carbon nanotubes on semi-spherical nickel catalyst islands of different size [163]. One can see that a (5,5) "armchair" nanotube develops from a Ni catalyst island of diameter 0.68 nm. According to Definition (3.1), it is metallic as apparently $k = 0$ in this case. Slightly larger (with d = 0.78 nm) nickel islands give rise to the "zig-zag" nanotube depicted in Figure 3.12(b). In this case $n - m \neq 3k$; hence, this SWCNT is semi-conducting. The last nanotube of (9,4) chirality shown in Figure 3.12(c) is also semiconducting but features a slightly larger bandgap because of its larger diameter (d = 0.9 nm).

The primary role of the plasma environment in this case is to control the surface conditions (surface temperature and diffusion activation energy of nickel adatoms on Si(100) surface; the former was higher and the latter was lower than in neutral gas-based processes). As a result, larger densities or smaller and more size-uniform nickel islands can be grown. This will eventually lead to a denser and more uniform nanopattern of SWCNTs with a narrow distribution of chiralities and, hence, electronic properties.

Practical experience shows that plasma-assisted growth of single-walled carbon nanotubes is an extremely delicate process. One of the main reasons is that it requires precisely dosed amounts of carbon atoms to insert into the single wall of the nanotube through the catalyst particle at its base. Moreover, this process also routinely requires effective suppression of higher hydrocarbons [164], conditions that are quite difficult to meet in plasma-based processes [4].

Depending on the process conditions, multiwalled carbon nanotubes can grow either with the catalyst particle on the top or at the base. The most remarkable feature is that in plasma enhanced CVD the MWCNTs predominantly develop in the tip growth mode [21], with an additional possibility to evolve as carbon nanofiber structures [22]. In thermal CVD, both options are possible [162]. These amazing possibilities are not completely understood, especially in plasma-based processes.

Recent results of atomic-scale *in situ* TEM imaging of carbon nanotube and nanofiber growth shed some light on possible reasons for the formation of multiple walls and also demonstrate how exactly metal catalyst particles can detach from the growth surfaces and move on top of the nanostructures [144, 165]. In particular, it has been suggested that multiple graphene sheet-made nanofiber walls can be formed at mono-atomic step edges at the C–Ni interface [165]. The above steps continuously develop and disappear while an initially fairly spherical nickel catalyst particle periodically elongates, reshapes and eventually contracts to a spherical shape. The growth terminates when the graphene sheets encapsulate the Ni particle completely, indicating that metal-gas interaction is essential for the nanofiber growth.

If this process were conducted in a plasma environment, one should expect that electrostatic interactions between the similarly (usually negatively) charged catalyst nanoparticle and substrate surface can facilitate dynamic reshaping of the particle and its detachment from the surface. This effect has previously been used for electrostatic shedding of fine powder particles from solid surfaces in plasma discharges [166] and can be invoked for the explanation of predominant tip-growth of multi-walled carbon nanostructures in a plasma [4]. Atomic-scale, video-rate TEM [144] also convincingly confirmed that single-walled carbon nanotubes grow by lift-off of a carbon cap, which initially tightly covers the entire surface of the catalyst nanoislands. This mechanism had been proposed earlier by Reich *et al.* using DFT numerical simulations [143].

We will now follow the review article [4] and conclude that plasma-assisted techniques can be advantageous over thermal CVD when at least one of the following is required:

- control of densities and fluxes of the required building units in the gas phase, which is difficult (if possible at all) to do on the surface of catalyst nanoparticles (e.g., in the synthesis of carbon nanotubes);

- electric field-guided delivery of charged or polarizable building units straight to the nanoassembly directly from the ionized gas phase and without any intermediate landing on the substrate surface;

- specific substrate activation by ion/heat fluxes from the gas phase;

- preferential growth and alignment direction, such as the direction of the sheath electric field;

- nanostructures should be synthesized at low process temperatures to avoid any damage of easy-to-melt substrates.

This list is incomplete and a few more areas of competitive advantages of plasma-based approaches have been recently highlighted by Ostrikov and Murphy [5].

Nonetheless, the actual role of the plasma in the growth of carbon nanotubes and related structures is still a subject of intense discussion within the research community [4]. For example, it is commonly believed that since dissociation of precursor gas on the surface of catalyst particles is usually sufficient for carbon material precipitation, the ability of the plasma to dissociate the gas feedstock into reactive radicals should not be a factor in the growth of carbon nanotubes [21].

To this end, one should clearly understand the consequences of additional dissociation of the feedstock gas in the plasma. Extra radical building units produced in the gas phase reduce the need for their production on the catalyzed surface. Moreover, carbon-bearing species required for the nanotube/nanofiber growth can be produced by ion-assisted dissociation of hydrocarbon radicals directly on the surfaces of the nanostructures being grown [167]. Apparently, these processes are energetically favorable and could be one of the reasons for the lower substrate temperatures required to synthesize carbon nanostructures by plasma-assisted methods [4]. These processes will be considered in more detail in Chapter 7.

Another argument in favor of the importance of the gas-phase decomposition of working gas is the possibility of plasma-assisted growth of multiwalled carbon nanotubes without a catalyst [168, 169]. It is thus indeed likely that carbon-carrying species can precipitate into nanoassemblies directly from the gas phase, provided that the appropriate insertion conditions (e.g., dangling bond availability) are met.

Another interesting point is that the CVD of CNTs usually requires substrate temperatures of at least 550 °C, and the cold wafer scenario is quite unlikely [21]. By using thermally non-equilibrium low-temperature plasmas, growth temperatures of carbon nanotubes can be reduced to as low as 120 °C [170]. Furthermore, various carbon nanostructures can be grown in low-temperature plasmas without any external heating of the substrate [93,94]. In this case, the neutral and ionized components of the weakly ionized plasma environment can be responsible for the required heating of the catalyst layer.

To conclude this section, we note that more efficient species generation and plasma polymerization generally require higher chemical activity of the gas feedstock. Thus, using highly-reactive (also frequently termed chemically active) gases such as ethylene, acetylene, or propylene is considered beneficial to enhance the process yield. For example, nanoparticle generation and nanostructure growth is more efficient in acetylene-based than methane-based plasmas [4,49,133]. In the following section, we will approach the problem of tailoring the plasma environments for deterministic nanoscale assembly more closely and introduce the effective pratical theoretical/computational approach.

3.3
Practical Framework – Bridging Nine Orders of Magnitude

In Chapter 1, it was stressed that to achieve a fully deterministic synthesis of nanoscale objects, one needs to be able to understand a variety of processes that occur in the ionized gas phase and on the growth surfaces. These processes can be separated in space by up to nine orders of magnitude as sketched in Figures 1.2 and 1.10. A basic approach of how to bridge this spatial gap is sketched in Figure 1.11. In this section we will introduce the theoretical/computational approach, which is based on multiscale hybrid numerical simulations that bridge the nine-order-of-magnitude spatial gap between the macroscopic plasma nanotools and microscopic surface processes on nanostructured solids. In line with the original publication [171], we will also consider two specific examples of carbon nanotip-like and semicondictor quantum dot nanopatterns. We have chosen these two examples to align the consideration in this section to what has already been discussed in Sections 3.1 (semiconducting nanostructures) and 3.2 (carbon-based nanostructures). These simulations are instrumental in developing the physical principles of nano-scale assembly processes on solid surfaces exposed to low-temperature plasmas.

Table 3.1 Numerical models and spatial scales in the multi-scale hybrid simulations: M1 – generation of building units, M2 – building unit delivery, M3 – surface conditions, M4 – pattern self-organization and M5 – nanoassembly growth [171].

Module	Area	Scales	Comments
M1	Plasma bulk	$\lambda_s < z < L$	Uses multi-fluid global plasma models and provides densities and energies of plasma species as functions of operating parameters.
M2	Plasma sheath	$\lambda_s > z > h_{NA}$	Uses Monte Carlo simulation to trace ions in the sheath and their collisions with the surface.
M3	Solid surface	$x, y \sim l_p$	Models of plasma-surface interactions relate surface temperature, elastic stress and charge distributions to plasma process parameters.
M4	Solid surface	$l_{NA} < x, y \sim l_p$	Models of adatom migration, island nucleation, growth and coalescence on plasma-exposed surfaces relate nanopattern and plasma process parameters.
M5	Solid surface	$d_a < z \sim h_{NA}$	BU self-assembly and nanostructure growth models relate the parameters of nanoassemblies and the plasma process parameters.
		$d_a < x, y \sim l_{NA}$	

Let us again refer to Figure 1.2 which shows a typical plasma-aided nanofabrication environment, with the spatial scales and numerical models involved summarized in Table 3.1. From Figure 1.2 and Table 3.1 we can see that the modeling of building unit creation in the plasma spans the spatial scales of ca. 0.5 m (typical sizes of plasma reactors), whereas the scales involved in the modeling of surface self-organization processes are 7–9 orders of magnitude smaller. The BU delivery through the plasma sheath spans over ca. 10 μm–10 mm, which is a typical sheath width in low-temperature processing plasmas. This was already mentioned in Chapter 1. In Table 3.1, the following notations are used (same as in the original article [171]): L is a typical dimension of plasma reactors, λ_s is the plasma sheath width, l_p is the size of the simulation area on the surface, h_{NA} and l_{NA} are the NA sizes in vertical (in the z direction) and horizontal (in the x-y plane) directions.

To be more specific, here we focus on the description of various processes involved in the plasma-assisted synthesis of the two main groups of selected nanoassemblies (NAs) and their nanopatterns. For the con-

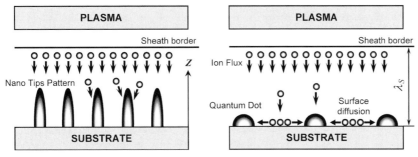

Figure 3.13 Two representative simulation geometries:
(a) plasma ions interact with an ion-focusing nanotip-like pattern;
(b) ions from the plasma are deposited onto the substrate and
quantum dot surfaces; surface diffusion enables adatoms to
contribute to the quantum dot growth. (x,y) plane (substrate) is
perpendicular to the z direction.

venience of the reader, we also summarize the links to the original papers where more details can be found. Group I includes (mostly carbon-
based) high-aspect-ratio, nanotip-like structures of different dimensionality, such as nanotubes, nanoneedles and nanowires (1D), nanowall-like
structures (2D), nanocones and nanopillars (3D) [34,51,52,100,172,173].
The relevant simulation geometry is shown in Figure 3.13(a). Group II
includes semiconducting (e.g., SiC on Si/AlN, Si on Ge, AlN/AlInN on
Si, and so on) quantum dot (QD) structures [174–176,178].

The plasma-sheath environmet used in simulations is sketched in Figure 3.13(b). In most cases a detailed comparison of the results obtained
by using plasma and charge-neutral fabrication routes (with the same
main process parameters) can be made. Nanotip-like structures of Group
I usually have h_{NA} ca. 100–900 nm and l_{NA} ca. 10–80 nm, whereas the
sizes of nanodots of Group II is smaller ($h_{NA} \approx l_{NA} \approx 5$–40 nm). The
choice of simulation geometries and other parameters is dictated by the
relevant experimental results [1,4,49,50,55,93–95,129,130,140].

Positioning of individual nanoassemblies in nanopatterns reflects the
most commonly used fabrication methods and patterning techniques.
For example, nanotip-like structures are usually grown on catalyzed substrates pre-patterned via pattern transfer by using ordered templates,
such as lithographic masks or porous materials. Hence, nanotip-like
structures should be seperated from each other by distances varying
from tens of nanometers to several micrometers. On the other hand,
tiny quantum dots are set in a pattern to reflect the commonly achievable surface coverage by the nanodots and some factors that can align or
order them. Other process conditions, such as the surface temperature,

DC substrate bias, working gas pressure, densities of plasma species, temperatures of electrons, ions and neutrals, as well as other parameters, are taken as typical values of recent experiments [1,4,49,50,55,93–95,129,130,140].

As stated in the original article [171], module M1 (building unit generation) uses multi-fluid global plasma models and provides densities and energies of plasma species as functions of operating parameters [53, 54, 116, 138]. The module is adjusted to the modelling of $Ar + H_2 + CH_4/C_2H_2$ plasmas, structures of Group I and $Ar + H_2$(optional) $+ N_2$ (optional) plasma-assisted sputtering of Al, In, SiC, and Ge targets for the nanostructures of Group II. This module generates fluxes of the plasma species at the sheath edge ($z = \lambda_s$) to be used in the next module, M2.

Module M2 (building unit delivery) uses Monte Carlo simulation to study the delivery of charged and neutral building units from the plasma to the surface. It enables one to (i) compute the distribution of microscopic ion fluxes in the vicinity of and on nanostructured surfaces, including open surface areas (areas not covered by the nanostructures) and lateral surfaces of individual nanoassemblies; (ii) relate the selective and targeted (e.g., to open surface areas or lateral surfaces) delivery of building units to the plasma parameters such as the densities of the plasma species, electron temperature, and so on; and (iii) obtain (e.g., argon) ion fluxes required in M3 for computation of temperature and stress distributions on the surface. In this model, the species are traced from the plasma sheath edge to the nanopattern surface [34,51,52,100,172,173].

Module M3 (surface conditions) [171] contains the models of plasma-surface interactions (heat, momentum and charge transfer) and specifies the effects of the plasma environment on the temperature, elastic stress and electric charge distribution on the growth surfaces. This module makes it possible to (i) obtain the dependence of the surface temperature as a function of the ion flux; (ii) compute surface charge and charges on individual nanoassemblies, and relate them to the plasma parameters. This module includes input from module M2, builds on relevant processes in the plasma sheath and on solid surfaces [34, 51–54,100,116,138,172,173] and provides the surface conditions required in M2, M4 and M5.

Module M4 (pattern self-organization) [171] describes the origin, development and self-organization of nanopatterns and contains two sub-modules [174–178]. The first sub-module M4-1 is used to model the nucleation of initial nuclei on surfaces subject to influx of species from the plasma and surface conditions (temperature, stress distribution and electric charge) imported from M3. The positions and surface density of the initial growth islands strongly depend on the rates of collisions between adsorbed species, which are controlled by the BU influx from the plasma, surface temperature and stress topography. The sub-module M4-1 enables one to (i) elucidate initial growth patterns; (ii) relate the size distribution and positioning of seed nuclei to the process parameters. It includes input from M2 and M3. Moreover, the module M4-1 generates initial nucleation sites required in M4-2, M2 and M5.

Sub-module M4-2 uses the advanced fluid-on-fluid technique [179], which is modified to describe migration of adsorbed adatoms and nanoisland growth and coalescence on plasma-exposed surfaces [171]. This model of the nanopattern development in the plasma takes into account (i) incoming building units from the plasma; (ii) surface diffusion of adatoms; (iii) evaporation from the substrate surface and island surfaces; (iv) adatom attachment to the island surfaces; (v) island growth, displacement and coalescence [174–178]. This module enables one to (i) describe density distributions of adsorbed species over nanostructured surfaces; (ii) consider the evolution of developing nanopatterns subject to intake of building units from over the surface and directly from the plasma; (iii) take into account surface conditions (surface temperature and surface charge) computed in M3; (iv) relate the characteristics of nanoassemblies and nanopatterns to the plasma parameters. This sub-module includes input from M2, M3 and M5 [171].

Module M5 (nanoassembly growth) [171] relates the growth kinetics of individual NAs from the initial nuclei to selective delivery of BUs (from over the surface and from the plasma) and surface conditions imported from M3 [34,51,52,100,172,173]. It enables one to follow the dependence of size and shape of individual nanostructures on the process conditions. This module is based on growth equations that relate the volume and shape of the structures to the incoming flux of building units; it includes input from M2, M3 and M4.

We will now discuss in more detail how the multiscale hybrid numerical simulation of this section works in modeling the growth of two selected nanostructures from Groups I and II. For more details of the relevant results please refer to Chapters 6 and 7 of this monograph and original publications. In the first example, let us consider carbon nanotip microemitter structures (Figure 3.9(b)), which should ideally feature high aspect ratios, vertical alignment and uniformity in sizes (both in the x-y plane and the z direction) across the entire nanopattern [171]. As the results of Levchenko *et al.* [51, 52, 173] suggest, deposition of ionic building units onto lateral surfaces of carbon nanotips can be effectively controlled by varying the plasma density, electron temperature and DC bias of the substrate. For example, in low-density plasmas, ionic building units deposit either onto open surface areas or on nanotip surfaces

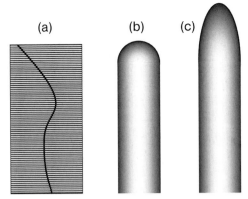

Figure 3.14 Example of a carbon nanotip reshaped by a microscopic ion flux [171]: a) ion flux distribution over a lateral surface; b) original nanotip; c) reshaped nanotip.

closer to their bases. On the other hand, when the plasma is dense, the ions tend to land on the upper sections of the nanotip structures [51]. Assuming that the ions can incorporate into the nanoassemblies being grown directly at the point of impact, the possibility of deterministic nanotip shape control (illustrated in Figure 3.14) has been demonstrated [100, 173]. In the example given in the original article [171], the plasma density is n_p ca. $5 \times 10^{11}\,\mathrm{cm}^{-3}$, electron temperature $T_e = 1.5\,\mathrm{eV}$, DC bias potential $V_b = -30\,\mathrm{V}$, initial nanotip height $h_{NA} = 300\,\mathrm{nm}$ and diameter $l_{NA} = 50\,\mathrm{nm}$.

We emphasize that the nanotip shape can also be controlled during the growth process. Moreover, by using a plasma-based approach, one can synthesize regular patterns of size-uniform high-aspect-ratio carbon nanotips, which turn out to be superior to similar nanopatterns synthesized by using a neutral gas with the same parameters. Specifically, the plasma-grown nanotips are taller and sharper, and more uniform within the entire simulation area, both in the surface plane and in vertical direction [52].

These results are also directly relevant for the post-processing of one-dimensional nanostructures, such as nanotubes, nanowires and nanoneedles. Indeed, relative positioning of such structures in a nanopattern strongly affects the quality of their coating/functionalization by an ion flux from the plasma. The best results are achieved when the ion flux is deposited uniformly over the lateral surfaces and is not lost to open surface areas. This can be done at intermediate plasma densities and bias voltages and when the nanostructures are sufficiently spaced (by Δx) away from each other ($\Delta x \sim h_{NA}$) [34]. For more details concerning

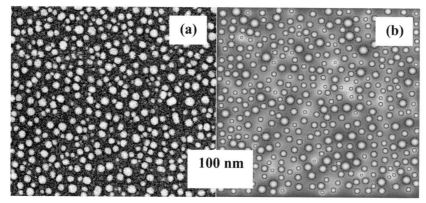

Figure 3.15 Experimentally synthesized (a) [49,50] and computed (b) [177,178] surface morphology of SiC/Si(100) quantum dots synthesized by plasma-assisted RF magnetron sputtering of SiC in Ar + H$_2$ plasmas [171].

the results of numerical simulations of carbon nanotip and other carbon nanostructures please refer to Chapter 7.

The second example is related to Group II nanostructures, more specifically, SiC/Si(100) quantum dots grown by plasma-assisted RF magnetron sputtering of SiC targets in Ar + H$_2$ inductively coupled plasmas, sustained in the integrated plasma-aided nanofabrication facility (IPANF) [49, 50]. Figure 3.15 shows the SEM micrograph of the experimentally synthesized (a) and numerically simulated quantum dot growth pattern (b). This simulation is mostly based on modules M2–M4 and accounts for neutral and singly ionized Si, C, SiC building units and involves two major stages [171]. During the first stage, by using module M4-1, the distribution of initial seed nuclei (ISNs, made of < 25 atoms) are computed depending on the incoming flux of building units from the plasma (taken from the experiments [49,50,129]) at different surface temperatures [174–176].

After optimization of size distributions, the initial seed nuclei are assigned random positions within the 1 μm×1 μm (or similar) pattern and their further growth is studied by using module M4-2 [177, 178]. The final simulation pattern with the surface coverage of ca. 0.4, which is within reasonable accuracy, recovers the experimentally achieved surface coverage [49,50,129], is shown in Figure 3.15(b). In numerical simulations, more complex plasma chemistries, such as reactive nanocluster-generating plasmas are also commonly used [53, 54, 116, 138]. The results of recent numerical simulation [176] of deposition of Ge/Si(100) quantum dot seed sub-monolayers from ionized and neutral germane

GeH_4 gas show the advantages of the plasma route in controlled (ca. 1–3×10^{-4} monolayers) growth of ultra-small (containing 3–25 Ge atoms) seed nuclei suitable for the growth of uniform quantum dot patterns with a high surface coverage. It is remarkable that the proposed technique challenges the conventional Stranski–Krastanov growth mode, is applicable for a broader range of epitaxial systems and is promising for deterministic synthesis of nanodevice-grade quantum dot arrays [171]. More details about plasma-assisted nanodot synthesis will be given in Chapter 6 of this monograph.

In this way, a multi-scale, hybrid "macroscopic" model, which is able to bridge the processes separated by spatial scales of up to nine orders of magnitude (Figure 1.11), can be developed. Here we note that these processes are normally considered by "three-dimensional" plasma physics and "two-dimensional" surface science, two seemingly unbridgeable areas of study. The model builds on the established, commonly used, well proven and justified principles and approaches of surface science to surface diffusion phenomena, island nucleation and growth [179]. These commonly used models have been advanced by individual treatment of each nanoassembly (up to ca. 10^3 in the pattern with each nanostructure containing up to ca. 1.5–2 million atoms or even more) arranged in ordered or randomly disordered arrays on plasma-exposed surfaces. Owing to limitations of the present-day atomic-level *ab initio* models (e.g., in the number of atoms they can handle), this "quasi-macroscopic" approach seems to be the most appropriate to describe nanopattern growth on plasma-exposed surfaces. Finally, the results generated by using this approach appear to be remarkably consistent with numerous experimental and computational results [4] and it has an outstanding potential to bridge the "unbridgeable" gap between plasma physics and surface science [171].

3.4
Concluding Remarks

In this chapter, we have focused on two main questions:

1. How can we use the building unit-based plasma-aided nanofabrication approach introduced in Chapter 2 in most common cases of semiconducting and carbon-based nanostructures and their patterns/arrays; and

2. How can we approach the problem of bridging the huge spatial gap between the scales of macroscopic processes in the plasma

bulk of plasma reactors and of elementary processes of atomic self-assembly on solid surfaces?

These issues are intimately related to the whole idea of the deterministic approach in plasma-aided nanofabrication.

We will now try to highlight some of the most important points discussed in this chapter:

- It is extremely important to be able to establish the logical link between the "cause" and the "effect". In the example considered in Section 3.1, ultra-small silicon nanocrystallites cause a remarkable improvement in the various functional properties of amorphous hydrogenated silicon films.

- This unique combination of plasma-grown silicon nanocrystallites and an amorphous matrix gives rise to a new class of materials commonly termed *polymorphous* materials.

- Figure 3.2 summarizes the main process requirements for the deterministic synthesis of polymorphous hydrogenated silicon films.

- The plasma-based approach is very effective in deterministic synthesis of nanoislanded nanocrystalline silicon carbide, a novel class of nanostructured materials in which nanocrystallites are embedded in dome-shaped nanoislands.

- The resulting nanostructure ultimately depends on the details of the interaction of plasma-generated building units and the surface. The examples shown in Figure 3.3 emphasize the importance of depositing nanocrystalline building units with appropriate energy to avoid their fragmentation upon impact on the substrate.

- Plasma-based environments can handle a variety of precursors in the gaseous, liquid and solid form. By using the appropriate combination of building units, one can synthesize a virtually unlimited number of binary, ternary, quarternary, and so on low-dimensional semiconductor nanostructures such as quantum dots and wires.

- In the synthesis of carbon-based nanostructures, the appropriate choice of suitable building units is one of the greatest and as yet unresolved challenges, mostly because of a very large number of possible options. Nevertheless, in most important cases (e.g., carbon nanotubes, nanocrystalline films, single-crystalline nanostructures), carbon atoms, carbon dimers and some other radicals can be effectively used to fabricate the desired carbon-based nanoassemblies deterministically.

- In a very large number of cases, carbon-based nanostructures can only grow if the process is catalyzed by metal nanoparticles such as Fe, Co, Ni, and various alloys.

- Size, shape and chemical activity of the catalyst are among the most essential factors that determine the parameters of the nanostructures and their arrays.

- Plasma plays a very important role at all stages of the synthesis of various carbon-based nanostructures, as discussed in Section 3.2.

- Plasma-based processes have been used to synthesize a large variety of carbon-based nanostructures, with some examples shown in Figure 3.9.

- By using relevant knowledge on precursor species of amorphous carbon films, one can tailor the nanofabrication process to enhance the development of the nanostructured phase and suppress the growth of the amorphous phase; this practical approach is summarized in Figure 3.10.

- Ion bombardment has always been, and still remains, one of the most controversial issues in plasma-assisted nanofabrication; depending on specific requirements and process conditions, it can be either a friend or a foe. A better understanding of the interaction of ion fluxes with various nanostructured films and nanoassemblies is therefore required.

- A range of high-aspect-ratio nanostructures grown in a plasma show pronounced alignment along the direction of the electric field within the plasma sheath. Even though the nature of this phenomenon is not fully understood, it is considered as one of the major competitive advantages of plasma-based processes over neutral gas-based routes.

- A properly balanced approach can enable the growth of very delicate nanostructures such as single-walled carbon nanotubes; one such example is shown in Figure 3.12.

- Typical situations when the growth of carbon-based (and in fact, many other) nanostructures can benefit from the involvement of low-temperature plasma environments are summarized at the end of Section 3.2.

- A practical approach, which bridges processes different by spatial scales by nine orders of magnitude, is based on multi-scale hybrid numerical simulations and is introduced in Section 3.3.

- This computational approach has been used to desribe plasma-based synthesis of a range of two major classes of nanoassemblies considered in Sections 3.1 and 3.2. The results are in most cases in remarkable agreement with the available experimental results.

We will now discuss the appropriateness of using the numerical approach of Section 3.3 to describe assembly of nanostructures, which takes place at ultra-small (atomic) scales over very short periods of time and therefore should involve quantum mechanical treatment heavily. Here we recall that the hybrid multiscale simulation that bridges the plasma and surface processes occurring at length scales different by several orders of magnitude involves a huge number of atoms and ions, which none of the presently available *ab initio* atomic-level techniques is able to simulate! In fact, the most advanced existing MD simulations (that require enormous computational resources) can typically handle just a few hundred to a few thousand atoms and, on the other hand, are not entirely reliable due to currently non-resolvable problems with choosing appropriate inter-atomic interaction potentials appropriate for chemical reactions and growth processes. On the other hand, relatively precise *ab initio* DFT models are also limited to a few hundred atoms but only are applicable to steady-state nanostructures and cannot be used to describe growth processes.

As we have mentioned in Section 3.2, and will be considered in more detail in Chapters 5–7, the microscopic topology of ion fluxes in the vicinity of selected functional nanostructures and the arrangement of adsorbed species into nanopatterns on solid surfaces are intimately related to the parameters of the plasma sheath separating the plasma bulk and the nanostructure growth substrate. More importantly, owing to selective delivery of ionic and neutral building blocks directly from the ionized gas phase and via surface migration, plasma environments can offer more options for deterministic synthesis of ordered nanoassemblies compared to neutral gas routes, such as thermal chemical vapor deposition (CVD), Molecular Beam Epitaxy, and so on. For example, the results of hybrid Monte Carlo (gas phase) and adatom self-organization (surface) simulation discussed in Chapter 7 of this monograph suggest that higher aspect ratios and better size and pattern uniformity of carbon nanotip microemitters can be achieved via the plasma route.

This is just a small piece of evidence that low-temperature plasmas have become a member of an elite club of versatile fabrication tools of the nano-age [4]. However, the level of understanding of how exactly the plasma nanotools work in various nanoscale applications, still remains far from perfect. One of the reasons is the enormous difference between the spatial and temporal scales involved in the main processes in the plasma and on solid surfaces. In fact, advanced multi-scale hybrid numerical simulations can be effectively used to bridge the gap between the gas-phase and surface phenomena, optimize nano-scale processes and eventually reach the so much needed deterministic level in nanofabrication, which will make the prevailing "trial and error" experimental practices extinct. In the following chapters we will consider more specific technical details of the main processes involved in the the plasma-assisted growth of various nanoassemblies.

4
Generation of Building and Working Units

As stressed in Chapter 2, the nanoassembly synthesis process starts with the creation of primary material, from which nanoscale objects can be built. This material comes in a variety of forms as summarized in Figure 2.1. In fact, from the numerous examples discussed in Chapters 2 and 3, it becomes clear that the choice of any particular building unit is dictated by the specific requirements of the "nano-building" process.

One might be tempted to suggest that since all nanoassemblies are ultimately made of atoms, one should create atomic species in the plasma bulk and then let them assemble into nanostructures where they are required. However, we have already learned that in many cases other atomic building units such as radicals, molecules, nanoclusters and nanocrystallites can be not only useful but also more effective and efficient in terms of better, faster synthesis of nanoscale objects. Furthermore a range of other species, which we term working units here, are required to suitably process the surfaces involved in the nanoscale synthesis.

Subsequently, as we have stressed several times above, the appropriate choice of suitable building units is one of the most difficult issues of plasma nanoscience. However, as our commonsense tells us, before choosing any particular species or a group of species, one needs to know, at least approximately, what the plasma environment can offer.

This particular question can be successfully solved using a range of plasma diagnostic tools such as optical emission spectroscopy, microwave interferometry, Langmuir probe diagnostics, quadrupole mass spectrometry and several others [180,181]. These diagnostic techniques should be complemented by the results of numerical modeling of plasma discharges, which can provide valuable information concerning the abundance and distribution of the plasma species in the active region of the plasma reactor. Several important issues and practical approaches to effective plasma diagnostics have been discussed in detail in a related monograph [1]. Here we will mostly concentrate on modelling and numerical aspects of this problem.

Plasma Nanoscience: Basic Concepts and Applications of Deterministic Nanofabrication
Kostya (Ken) Ostrikov
Copyright © 2008 WILEY-VCH Verlag GmbH & Co. KGaA, Weinheim
ISBN: 978-3-527-40740-8

In the following section we will show how discharge modeling complemented by experimental results can help in identifying the abundant plasma species and predicting possible effects of different sorts (e.g., neutral and charged) of species on the nanoassembly process in question. The discussion in Section 4.1 is based on the complementary modeling and diagnostics of radiofrequency plasma discharges in $Ar + H_2 + CH_4$ gas mixtures. The results presented in Section 4.1 have been obtained using a simplified, although very commonly used, spatially-averaged ("global") approach. This approach is particularly useful when it is necessary to estimate the concentrations of the plasma species averaged over the entire volume of the reactor chamber.

However, one should note that plasma is always non-uniform, with the species, densities higher in the plasma bulk and lower near the chamber walls or electrodes introduced into the chamber. If one wants to control the processing of a specimen immersed in the discharge glow, one needs to know the densities and energies of the plasma species exactly at the point where the specimen is located. And what if the surface area or the size of the specimen is large? In that case, different areas of the surfaces being processed may be subject to very different conditions, especially if the plasma is strongly non-uniform.

Therefore, a different approach which makes it possible to specify the two-dimensional topography of the densities and fluxes of the discharge species is required; such an approach based on the original work [54] is presented in Section 4.2. The third section of this chapter (Section 4.3) is devoted to an overview of the generation of larger (nanocluster, nanocrystalline etc.) building units in a plasma. As usual, this chapter concludes with concluding remarks (Section 4.4), which gives a brief overview of the main points discussed in Chapter 4.

4.1
Plasma Species in High-Density Inductively Coupled Plasmas for Low-Temperature Synthesis of Carbon Nanostructures

In this section, following the original report [53], we will discuss how numerical simulations complemented with the experimental results from optical emission spectroscopy (OES) and quadrupole mass spectrometry (QMS) can be used to calculate the abundance and surface fluxes of a variety of carbon-bearing species, which can potentially serve as building units in the plasma-asssited synthesis of vertically-aligned carbon-based nanostructures. The approach considered here is based on a spatially averaged (global) discharge model of inductively coupled $Ar/CH_4/H_2$

plasmas in the process of PECVD of vertically aligned carbon nanostructures (CNSs). This model makes it possible to estimate the densities and fluxes of the radical neutrals and charged species, the effective electron temperature and methane conversion factor under various growth conditions. The results of the numerical modeling show a remarkable agreement with the OES and QMS experimental results. The most interesting conclusion which follows from this study is that the incoming fluxes of cations (positively charged radicals) can exceed those of the radical neutrals, despite a low ionization degree of the plasma involved.

We now recall that plasma-enhanced chemical vapour deposition has recently become one of the most popular growth techniques of a variety of common carbon nanostructures, as can be seen from the large number of publications [1, 21, 90, 93, 94, 141, 182–184]. As has already been mentioned in previous sections, low-temperature PECVD offers a great deal of vertical alignment and positional ordering of carbon nanostructures, in particular, due to the presence of DC electric fields normal to the growth surface [90]. In many cases, this constitutes a significant advantage of PECVD processes compared with conventional thermal CVD techniques.

As we have learned from Chapter 2, the essential parameters for the PECVD growth of ordered carbon nanostructures are partial pressures of feed gases, input power, the nature and parameters of the catalyst used, the substrate temperature T_s and DC bias V_s. We also recall that because of the ever-decreasing surface temperatures acceptable in nanofabrication (film thicknesses shrink, which leads to substantially lower melting points compared to relevant bulk materials), many experiments have focused on the minimization of the synthesis temperatures of the carbon nanostructures [94, 185–187]. In recent years, a range of plasma sources have been used to synthesize carbon nanostructures at relatively low temperatures (ca. 300–500 °C). Relevant examples include capacitively coupled [188], microwave [186] and inductively coupled plasmas (ICPs) [21,93] of various hydrocarbon-based gas mixtures. However, capacitively coupled plasmas have not been widely used for this purpose due to a somewhat limited ability to control the substrate potential.

On the other hand, microwave and inductively coupled plasmas have recently been considered more attractive for this purpose. Some of the most important reasons are stable operation at low pressures (0.133 Pa (1 mTorr)–13.332 Pa (100 mTorr)), high electron and ion densities (10^{11}–10^{12} cm^{-3} and even higher) and controllable ion fluxes onto the surfaces [189].

As has also been already mentioned, ion bombardment significantly contributes to the fragmetation and activation of catalyst layers, both via direct physical impact and chemical reactions, as well as via additional substrate heating. Therefore, it is widely recognized that (properly controlled) ion fluxes play a key and a positive role in the PECVD growth of carbon nanostructures at substrate temparatures substantially lower than in thermal CVD processes with quite similar parameters [21, 187]. We now refer the reader to Figure 3.9 and the related monograph [1] and review articles [4,21,22] which provide plenty of evidence that numerous vertically aligned carbon nanostructures with high aspect ratios can be grown on large-area substrates on silicon substrates catalyzed by transition metals in the low-temperature range using radiofrequency plasmas of various mixtures that contain carbon-bearing precursor gases.

In order to control the deposition process and improve the emission and other properties of carbon nanostructures, as well as to explore the possibilities for the upscaling to larger growth areas, an insight into the gas-phase plasma processes is crucial [53]. Numerical modeling of the plasma discharges can reveal the underlying physics of the deposition process and contribute to the development of the future industrial process specifications.

In the last few decades, there have been a large number of reports on extensive and in-depth theoretical studies of plasmas in various hydrocarbon-based feedstocks used for plasma-enhanced CVD of diamond-like, graphite-like and amorphous carbon films [190–194]. Even though plasma-based techniques have been widely used to grow a large number of carbon-based nanostructures, the modeling of low-temperature plasmas in the carbon nanostructure growth has only recently attaracted the attention it merits [142, 184, 195, 196]. Some of the modeling works [142, 184, 195] are limited to high-pressure ($\geq$ 4.000 hPa (3 Torr)) regimes of carbon nanostructure growth in methane/acetylene carbon-bearing gases diluted with NH_3/H_2. The original work of Denysenko *et al.* [53] which we discuss in this section, refers to the properties of low-pressure ($<$ 13.332 Pa (100 mTorr)) plasmas in the PECVD of the ordered carbon nanostructures.

In this section we will follow the original article [53] and discuss the results of the diagnostics and numerical modeling of the low-frequency ICPs of $Ar/CH_4/H_2$ commonly used in the PECVD of ordered carbon nanotip structures. It is interesting that the results of the numerical study (using a spatially averaged (global) discharge model) of the effects of variation of the discharge control parameters on the methane conversion factor, electron energy distribution, densities of the neutral and charged species and their fluxes onto a nanostructured surface are in remarkable agreement with the results of the optical emission spectroscopy (OES) and quadrupole mass spectrometry (QMS) measurements.

4.1.1
Experimental Details

The experiments in the original work [53] were conducted using a source of low-frequency inductively coupled plasmas sketched in Figure 4.1. The results from the plasma and circuit diagnostics and the discharge operating parameters have been used for the discharge modeling by means of the spatially-averaged (global) model mentioned above. Below we give the most essential description of the setup and refer the reader to the recent monograph and other original experimental papers [1, 189, 197–202]. In brief, a cylindrical stainless steel reactor chamber of the plasma source has the inner diameter $2R = 32$ cm and length 23 cm. The chamber is cooled by a continuous water flow between the inner and outer walls of the chamber. The top plate of the chamber is a fused silica disk, with a diameter and thickness of 35 and 1.2 cm, respectively. The top surface of the stainless steel substrate holder, of diameter 17.5 cm, is located 11 cm below the bottom surface of the quartz window, as shown in Figure 4.1.

In the growth of a large number of carbon-based nanostructures, a DC voltage in the range of 0–300 V is typically applied to the substrate stage [93,94]. The plasma reactor sketched in Figure 4.1 is evacuated using a 450 L/s turbo-molecular pump backed by a two-stage rotary pump. The flow rates of the working gases (Ar, H_2, and CH_4, or acetylene in the growth of carbon nanowall-like structures) are regulated by MKS massflow controllers. In particular, the gas flow rates of argon and methane have been varied from 10 to 50 sccm and 3 to 7.5 sccm, respectively [53],

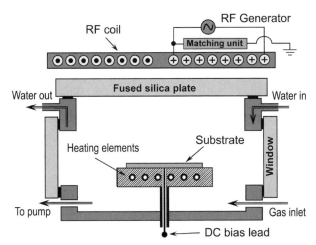

Figure 4.1 Schematics of the source of low-frequency inductively coupled plasmas used for plasma enhanced chemical vapor deposition of various carbon-based nanostructures [53, 95].

whereas the hydrogen flow rate is fixed at 12.4 sccm. As usual (see [1], Chapter 2), the total gas pressure in the discharge chamber is measured by a MKS Baratron capacitance manometer and is maintained in the range 2.666 Pa (20 mTorr)–9.332 Pa (70 mTorr). The plasma discharges were sustained by RF input power in the range from 1.8 to 3.0 kW. A typical scanning electron micrograph of the PECVD-grown ordered vertically aligned carbon nanotip structures is shown in Figures 3.9(a–c).

We note that argon, or another inert gas is commonly used to dilute mixtures of reactive gases. In the case considered, argon was used primarily for the purposes of surface conditioning and activation, as well as to maintain a stable discharge operation during the nanostructure growth at reasonably low input powers. The process begins with the initial pre-treatment of the Ni-based catalyst layer using argon ion bombardment, controlled by a DC substrate bias. As mentioned above, the dilution of the working gases with argon facilitates the discharge maintenance and operation due to the outstanding ionization/dissociation capacity of Ar in $CH_4/H_2/Ar$ gas mixtures [53,203,204]. Also, addition of an easily ionized inert gas can substantially improve the efficiency of the inductive coupling as compared with undiluted hydrocarbon/hydrogen mixtures.

For example, when the discharges are sustained in mixtures of hydrocarbons and hydrogen, the RF coupling can in some cases be predominantly capacitive [21]; in other words very high input powers may be required to run the discharges in the high-density inductive (H) mode. Moreover, the use of the pure hydrocarbon feedstock has previously been disfavored due to the enhanced deposition of amorphous carbon [53]. On the other hand, reactive hydrogen or ammonia are frequently used to enhance removal of the amorphous phase from nanostructured carbon-based films. The choice of a diluent gas also depends on the pressure range. For example, helium is more commonly used in the synthesis of carbon nanostructures in dielectric barrier discharges at atmospheric pressures [205].

The number densities of a range of neutral discharge species have been measured by a Microvision Plus LP101009 quadrupole mass spectrometer equipped with a Faraday cup detector. The operation and use of the QMS diagnostics to identify various plasma species are discussed in Chapter 2 of the related monograph [1]. A typical distribution [53] of the radical and non-radical neutral species in the PECVD of the self-assembled ordered carbon nanotip arrays is shown in Figure 4.2(a). The optical emission from the ICP discharge was collected in the radial direction using a collimated optical probe mounted 6 cm below the quartz window in the diagnostic side-port of the chamber [53]. The collected op-

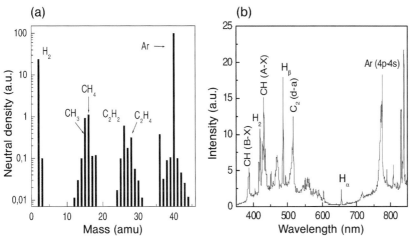

Figure 4.2 Mass spectrum of the neutrals measured (a) by QMS; and (b) OES of the excited species in the Ar/CH$_4$/H$_2$ plasma for the following experimental conditions: input power P_{in} = 2 kW, argon influx J_{Ar} = 35 sccm, hydrogen influx J_{H_2} = 12.4 sccm, methane influx J_{CH_4} = 7.5 sccm, DC bias V_s = −300 V, and substrate temperature T_s = 300 °C [53].

tical emission is transmitted via an optical fiber to a SpectroPro-750 spectrometer (Acton Research Corporation) with the spectral resolution of 0.023 nm. In this and similar experiments, the optical emission spectra of the excited neutral and ionized species are commonly studied in the wavelength range 350–850 nm. Further details of the optical emission intensity (OEI) measurements can be found elsewhere [1,189,197,198].

Figure 4.2(b) shows typical optical emission spectra in the process of PECVD of carbon nanostructures in methane-based gas mixtures. In the wavelength range 750–840 nm one can clearly observe argon lines originating due to $4p \rightarrow 4s$ transition. On the other hand, emission from the hydrogen Balmer line H$_\alpha$ is seen at 656.2 nm; the H$_\beta$ 486.1 nm emission common to gas discharges in hydrogen can also be identified. Furthermore, one observes the line at 420.5 nm attributed to optical emission form molecular hydrogen. From Figure 4.2(b), one can also observe intense emissions from the active hydrocarbon and carbon species, such as the molecular bands corresponding to the $B^2\Sigma \rightarrow X^2\Pi$ and $A^2\Delta \rightarrow X^2\Pi$ transitions (at 387.1 nm and 431.4 nm, respectively) [53]. The line belonging to the C$_2$ Swan band system, corresponding to the $d^3\Pi_g \rightarrow a^3\Pi_u$ transition with $\triangle v = 0$ is located at 516.5 nm. It is interesting to note that the role of carbon dimer C$_2$ as a building unit of nanocrystalline diamond and some other nanostructures and nanostructured films has already been mentioned in Chapter 2 of this monograph. Wherever possible, the experimental results showing how the external parameters (e.g., RF power and gas inlet) affect the internal discharge properties (densities of neutral species and optical emission intensities (OEIs optical emission intensities (OEIs) of the excited neutral species) will be presented below together with the computation results to enable a direct comparison.

4.1.2
Basic Assumptions of the Model

A spatially averaged (global) model has been developed to calculate the charged and neutral particle densities in the inductively coupled plasmas of $Ar/CH_4/H_2$ gas mixtures used in the PECVD of self-assembled ordered carbon nanostructures [53]. The two basic sets of equations of the model include the RF input power and species balance equations. We will now briefly go through the main assumptions, which successfully lead to reasonable results, in accordance with the relevant experimental results. The model of Denysenko *et al.* [53] includes the following species:

- Non-radical neutral species: Ar, H_2, CH_4, C_2H_2, C_2H_4, C_2H_6 and C_3H_8;

- Ions: Ar^+, H^+, H_2^+, H_3^+, CH_3^+, CH_4^+, CH_5^+, $C_2H_2^+$, $C_2H_4^+$ and $C_2H_5^+$;

- Radical neutral species: H, CH, CH_2, CH_3, and C_2H_5.

The anions (negative ions) are not accounted for here due to commonly recognized electropositive features of CH_4 and H_2 plasmas [206,207].

The results of modeling and diagnostics of plasma discharges strongly depend on the prevailing electron energy distribution function (EEDF) $f(\epsilon)$. In most cases, it is commonly assumed to be Maxwellian. However, in low-pressure plasma discharges the energy distribution function is often Druyvesteyn-like, as was the case in the original work of Denysenko *et al.* [53]. This assumption is supported by numerous experimental and theoretical studies of argon and CH_4-based RF plasmas [202,206].

Both Maxwellian and Druyvestein-like electron energy distribution functions can be explicitly presented in a similar form

$$f(\epsilon) = c_1 \epsilon^{1/2} exp(-c_2 \epsilon^x) \tag{4.1}$$

with different coefficients x [208, 209]. In this case $x = 1$ and $x = 2$ correspond to Maxwellian and Druyvesteyn EEDFs, respectively, ϵ is the electron energy, $c_1 = (x/\langle \epsilon \rangle^{3/2})[\Gamma(\xi_2)]^{3/2}/[\Gamma(\xi_1)]^{5/2}$, $c_2 = \langle \epsilon \rangle^{-x}[\Gamma(\xi_2)/\Gamma(\xi_1)]^x$, $\xi_1 = 3/2x$ and $\xi_2 = 5/2x$.

It is also assumed that the plasma is charge neutral $\sum_i n_i = n_e$, where n_e and n_i are the number densities of the electrons and cation (positive ion) species i, respectively. To be specific, the temperatures of the ions and neutrals have been fixed at $500 \, \mathrm{K}$, which is a typical gas temperature in argon-based inductively coupled plasmas [210, 211] under conditions similar to the PECVD growth of ordered carbon nanostructures [53,93,94].

4.1.3
Particle and Power Balance in Plasma Discharge

According to Denysenko *et al.* [53], the set of basic equations that describe the balance of radical and non-radical species in the discharge contains a set of rate equations for methane and hydrogen source gases as well as other radical and non-radical species

$$\frac{\partial n_{CH_4}}{\partial t} = I_{CH_4} - O_{CH_4} - \sum_i k_i n_e n_{CH_4} + \sum_{jkm} k_j n_k n_m - \sum_{ls} k_l n_s n_{CH_4} \quad (4.2)$$

$$\frac{\partial n_{H_2}}{\partial t} = I_{H_2} - O_{H_2} - \sum_i k_i n_e n_{H_2} + \sum_{jkm} k_j n_k n_m - \sum_{ls} k_l n_s n_{H_2}$$

$$+ \, 0.5 K_{wall}^H n_H \quad (4.3)$$

$$\frac{\partial n_r}{\partial t} = \sum_h k_h n_e n_h + \sum_{jkm} k_j n_k n_m - \sum_{ls} k_l n_s n_r - K_{wall}^r n_r - O_r \quad (4.4)$$

respectively, where $I_{C}H_4$ and I_{H2} are the incoming flow, and O_{CH4}, O_{H2} and O_r are the outflows of the CH_4, H_2 and other radical and non-radical neutral species r per unit time, respectively. A rate at which species $\alpha = \mathrm{Ar}$, CH_4, and H_2 enter the reactor I_α [cm^{-3}/s] $\approx 4.4 \times 10^{17} J_\alpha$[sccm]$/V$ is proportional to the gas inlet flow rate J_α, where V is the chamber volume in cm^3. The rate at which the molecules leave the discharge is $O_\alpha = v_{pump} n_\alpha / V$, where v_{pump} is the pumping rate in cm^3/s. The third terms in the right hand side of (4.2) and (4.3) account for the losses in the electron impact reactions, whereas the fourth and fifth terms account for the gains and losses from the neutral/ion-neutral reactions, respectively. The last term in Equation (4.3) reflects the fact that the atomic hydrogen is usually converted into the molecular state as a result of the surface reactions. The first term in the right hand side of Equation (4.4) describes the generation of species r as a result of the electron-impact reactions, while the second and third terms account for the gain and losses of the same species in the neutral/ion-neutral reactions, respectively. Here, $K_{wall}^r n_r$ is the number of radical neutral species r lost on the discharge walls per unit time per unit volume, where $K_{wall}^r = \gamma_r v_{thr} S_{surf} / 4V$ [53]. Furthermore, γ_r and v_{thr} are the model wall sticking coefficient and the average thermal velocity of radical species r, respectively, and S_{surf} is the chamber surface area. It can be further assumed that the sticking coefficients for CH, CH_2, ceCH3, C_2H_5 and H species are 0.025, 0.025 [212], 0.01 [213,214], 0.01 [215,216] and 0.001 [206], respectively. The sticking probabilities of all non-radical neutral species have been assumed to be zero [53]. Since argon is inert, its density n_{Ar} does not change during the discharge run. Moreover, in weakly ionized plasmas the density of neutral argon atoms is much higher than the combined ion density. In this case, it is reasonable to assume that $I_{Ar} = O_{Ar} = v_{pump} n_{Ar}/V$ [53].

The balance equation for the cation species i is [53,217]

$$V n_e v_{iz,i} = (h_L A_L + h_R A_R) n_i u_{B,i} + V \sum_{j=1}^{N_s} k_{cx,ij} n_i n_j, \tag{4.5}$$

where $k_{cx,ij}$ is the charge-exchange rate coefficient for asymmetric collisions between the ion species i and neutral species j, N_s is the number of neutrals that take part in the charge-exchange collisions with the ion species i and $v_{iz,i}$ is the ionization frequency for the production of the ion species i. Here, $A_L = 2\pi R^2$, $A_R = 2\pi RL$, h_L and h_R are the ratios of the densities of the cation species i on the outer surface of a cylindrical plasma column in the axial ($z = 0, L$) and radial ($r = R$) directions to the bulk averaged density n_i, respectively. In the low to intermediate pressure regime $[(R, L) \geq \lambda_i \geq (T_i/T_e)(R, L)]$ the above ratios are [218] $h_L = 0.86\{3 + L/2\lambda_i\}^{-1/2}$, and $h_R = 0.8\{4 + R/\lambda_i\}^{-1/2}$, where λ_i is the ion-neutral mean free path. For the EEDF (4.1), the velocity of ion species i at the plasma – sheath edge entering (4.5) is

$$u_{B,i} = \sqrt{2\langle\epsilon\rangle/m_i}\, \Gamma(\zeta_1)/\sqrt{\Gamma(\zeta_2)\Gamma(\zeta_3)},$$

where

$$\langle\epsilon\rangle = 1.5 T_{eff} = c_1 \int_0^\infty \epsilon^{3/2} exp(-c_2 e^x) d\epsilon,$$

$\zeta_3 = 1/2x$, and T_{eff} is the effective electron temperature [53].

The balance equations of the discharge species need to be complemented by the following power balance equation [53,217]

$$P_{in} = P_{ev} + P_w, \tag{4.6}$$

which completes the basic set of equations of this model. Here, P_{in} is the total RF power deposited to the plasma (discharge operating parameter). Furthermore, the energy lost in electron-neutral collisions is [53]

$$P_{ev} = e n_e V \sum_{i=1}^{q} v_{iz,i} \epsilon_{L,i},$$

where q is the number of cation species generated in the discharge, and

$$v_{iz,i}\epsilon_{L,i} = v_{iz,i}\epsilon_{iz,i} + \sum_{k=1}^{N_{exc}} v_{exc,k}\epsilon_{exc,k} + 3v_{elas}m_e T_{eff}/M_i.$$

Here, $v = \langle\sigma v\rangle n_n$ is the collision frequency, $\langle\sigma v\rangle$ is the rate coefficient obtained by the averaging of collision cross-section σ over the EEDF, n_n is the density of neutrals, N_{exc} is the number of excitation energy-loss channels, m_e and M_i are the electron mass and mass of i-th ion, respectively. Furthermore, $\epsilon_{iz,i}$ is the threshold ionization energy for the production of the cation species i, $v_{exc,k}$ and $\epsilon_{exc,k}$ are the excitation frequency and threshold energy for the k-th level of a neutral $\epsilon_{L,i}$ is the total collisional energy loss for the creation of the electron-ion (species i) pair. The sum over k includes all inelastic electron-neutral collisional processes that do not produce positive ions.

In addition, a number of rotational, vibrational/electronic excitation and dissociation processes have been taken into account here (see Appendix A and Table 10.1. The second term in (4.6)

$$P_w = \sum_{i=1}^{g} en_i v_{B,i}(h_L A_L + h_R A_R)(\epsilon_{ew} + \epsilon_{iw})$$

stands for the loss of kinetic energy of charged species to the discharge walls. The mean electron kinetic energy lost per electron lost to the walls is [53,209]

$$\epsilon_{ew} = \langle\epsilon\rangle\Gamma(\xi_1)\Gamma(\xi_5)/(\Gamma(\xi_2)\Gamma(\xi_4)),$$

where $\xi_4 = 2/x$ and $\xi_5 = 3/x$. The ion kinetic energy lost per ion lost to the wall ϵ_{iw} is the sum of the ion energy at the sheath edge and the energy gained by the ion as it traverses the sheath [53,209]

$$\epsilon_i = \langle\epsilon\rangle\Gamma(\xi_1)^2/(\Gamma(\xi_2)\Gamma(\xi_3)) + V_s,$$

where V_s is the sheath voltage drop. For argon ions and Maxwellian EEDF, $V_s \approx 4.68T_{eff}$; for Druyvesteyn distribution $V_s \approx 3.43T_{eff}$ [209]. Descriptions of the numerical method for the solution of the global model equations (4.2)–(4.6), the chemical reactions used in the computation and the rate coefficients for the reactions are presented in Tables 10.1–10.3 in Appendix A.

4.1.4
Densities of Neutral and Charged Species

The global model of the previous subsection is used here to show the effect of the discharge operating conditions on the main plasma parameters, including the number densities of the neutrals, electrons and ions, as well as the electron temperature originally computed by Denysenko *et al.* [53] for different powers P_{in} absorbed in the discharge, and the input flow rates of argon and methane. Wherever appropriate, the computation results are also compared with the QMS experimental data and are used to explain the variations of the OEI of selected CH (431.4 nm), C_2 (516.5 nm), hydrogen (656.2 nm) and argon (839.8 nm) emission lines.

4.1.4.1 Effect of RF Power

Let us now consider the effect of the input power on the densities of the major discharge species [53]. The methane, hydrogen and argon input flow rates are fixed at 7.5, 12.4 and 35 sccm, respectively, which corresponds to $n_{CH_4} = 1.37 \times 10^{14}$ cm^{-3}, $n_{H_2} = 2.27 \times 10^{14}$ cm^{-3}, and $n_{Ar} = 6.4 \times 10^{14}$ cm^{-3} under the "plasma-off" conditions, respectively. To study the effect of RF power on the plasma parameters, P_{in} is varied in the numerics in the range from 50 W to 3 kW, which is broader than was actually used in the PECVD of carbon nanostructures (1.8–3.0 kW) [93, 94].

Figure 4.3(a) shows the calculated densities of non-radical and radical neutrals in the Ar/CH$_4$/H$_2$ discharge of interest here. The variations of the electron and ion densities with P_{in} are shown in Figure 4.3(b). From Figure 4.3(a) one can see that both n_{CH_4} and n_{H_2} decrease dramatically with P_{in}. We note that in the power range suitable for the synthesis of carbon nanostructures, methane and hydrogen densities in the plasma are usually much smaller than in the absense of the discharge. Meanwhile, argon atoms appear to be the dominant neutral species in the discharge [53]. In fact, the density of hydrogen atoms is approximately 25 times smaller than the argon density and slightly decreases with power. As can be seen in Figure 4.3(a), the density of molecular hydrogen at low input powers is comparable to that of atomic hydrogen, and also decreases with P_{in}. The latter decrease can be attributed to the enhanced dissociation of hydrogen molecules at higher input powers accompanied by the rise in the electron number density (Figure 4.3(b)) [53]. Interestingly, under the process conditions suitable for carbon nanostructure growth, the hydrogen conversion factor [190] (degree of dissociation) can approach 99%. The enhanced dissociation of hydrogen molecules is naturally accompanied by the rise of the ratio of the densities of H and H$_2$ species, which can exceed 10 at elevated powers. Similar high n_H/n_{H_2} ratios have been reported by Bera *et al.* [194].

Meanwhile, the densities of methane and CH$_3$, C$_2$H$_2$, C$_2$H$_4$, C$_2$H$_5$, C$_2$H$_6$ and C$_3$H$_8$ species also decrease with RF power, which can be due to more intense electron-neutral collisions at higher electron densities [53]. On the other hand, CH and CH$_2$ densities increase with P_{in}. The latter

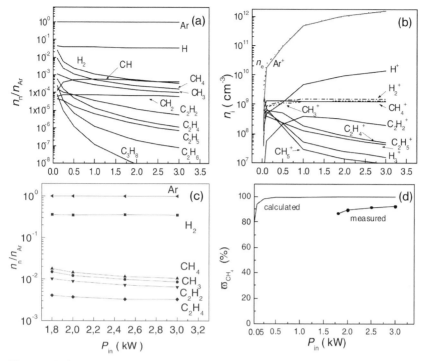

Figure 4.3 Computed (a) and measured by QMS (c) densities of neutrals (normalized on n_{Ar}), computed ion densities (b), and computed and measured by QMS CH_4 conversion factor $@_{CH_4}$ (d) as functions of the power input for the same gas feedstock as in Figure 4.2 [53].

can be attributed to the smaller collisional cross-sections of CH and CH_2 molecules compared to other hydrocarbon species. A similar tendency has also been reported elsewhere [191]. Therefore, it is probable that the above two molecules do not react with the plasma species as actively as CH_3, C_2H_2, C_2H_4, C_2H_5, C_2H_6 and C_3H_8 species do. Hence, higher hydrocarbons C_xH_y ($x \geq 2$) are more likely to break up into smaller radical fragments. This could be a reason why the densities of C_2H_6 and C_3H_8 in Figure 4.3 are quite low. However, the relative importance of the above two (as well as C_2H_5) species is higher at lower input powers, as can be seen in Figure 4.3(a).

The results of the quadrupole mass spectrometry in Figure 4.3(c) confirm that the densities of H_2, CH_4, CH_3, C_2H_2 and C_2H_4 species decrease with input power. Similarly, the CH_4 conversion factor $@_{CH_4} = 1 - n_{CH_4}/n^0_{CH_4}$, where $n^0_{CH_4}$ is the CH_4 density under the "plasma-off" conditions, appears to be very high at RF powers exceeding 1.8 kW [53].

A comparison between the experimental and calculated values of ω_{CH_4} at different RF powers in Figure 4.3(d) shows a consistent tendency of ω_{CH_4} to increase with the discharge power. In fact, there is a remarkable quantitative agreement between the computational and the experimental results. In the power range suitable for the growth of carbon nanostructures (1.8–3 kW), the conversion factor measured by the QMS varies from 86% to 92%, whereas the numerical value of ω_{CH_4} remains approximately 99% [53]. Similar high degrees of the methane dissociation (ca. 95%) in CH_4/H_2 inductively coupled plasmas have been reported by other authors [195]. High conversion rates of hydrogen-diluted reactive gas feedstocks is a common feature of many reactive plasmas. For example, ω_{SiH_4} can be as high as 99% in SiH_4/H_2 gas mixtures [219]. Apparently large conversion factors are due to the higher abundance of atomic hydrogen at elevated electron densities in hydrogen-containing plasmas, which in turn results in more intense chemical reactions between H and CH_4 (or SiH_4 in silane plasmas) [53].

The original global model of the plasma discharge also offers an explanation of the experimental tendencies of the optical emission intensities of neutral CH (431.4 nm), C_2 (516.5 nm), hydrogen (656.2 nm) and argon (839.8 nm) lines at different input RF power. The observation reported by Denysenko *et al.* [53] is a consistent increase of the OEIs of the above lines with RF power. This tendency can be explained by noting that the emission intensities of excited neutral species are proportional to the densities of the plasma electrons, n_e, and the corresponding neutral species, n_α, and also depend on the effective electron temperature [202, 220]. Since the electron temperature does not change (≈ 3.5 eV) within the power range of interest, the OEI is thus proportional to the product $n_e n_\alpha$. This product calculated at different input powers shows a remarkably similar tendency as the experimentally measured OEI [53].

4.1.4.2 Effect of Argon and Methane Dilution

The results of the original report [53] suggest that the densities of molecular and atomic hydrogen decrease with Ar inlet J_{Ar}, whereas the densities of hydrocarbon neutrals grow with J_{Ar} (RF input power is fixed). The majority of hydrocarbon species, however, show a clear tendency to increase with J_{Ar}. It was suggested that as the partial pressure of argon increases, the electron temperature decreases, which in turn results in a weaker methane conversion in collisional processes. As a result, the densities of CH_4 and other carbon-bearing (both neutral and ionic) species increase. This tendency is observed from both experimental (QMS and OES) and numerical results [53]. Thus, if a particular application requires

an increased concentration of carbon-containing species, one could consider an increase in the partial pressure of the argon feedstock gas.

The effect of methane inlet also appears to be quite straightforward
as the original report [53] suggests. A higher inflow of CH_4 is naturally
accompanied by a rise in the densities of methane and most of the hydrocarbon neutrals, which is confirmed by the QMS, OES measurements and
the results from the spatially-averaged model. An increase of the atomic
hydrogen density can be due to the enhancement of the electron-impact
dissociation of CH_4 summarized in Table 10.1 in Appendix A. On the
other hand, more molecular hydrogen is released as a result of the inten-
sified heavy particle collisions, see Table 10.1 in Appendix A. An interesting observation is that the electron density (and also that of argon ions)
slightly decreases. It was suggested that this can be attributed to higher
inelastic electron losses (e.g. for the vibrational, rotational, and electronic
excitations of hydrogen and hydrocarbon species) that inevitably accompany a higher inlet of CH_4 [53]. This decreased density of the plasma
electrons eventually leads to the lower intensity of optical emission from
most of the species involved.

4.1.5
Deposited Neutral and Ion Fluxes

In this subsection we consider the fluxes of the neutral and charged
species deposited onto the substrate under the growth conditions of carbon nanostructures [53]. To understand the role of each species in the
PECVD and to study the effect of the input plasma parameters on the
deposition process, one can compare the numbers of the radical neutrals
deposited per unit time per unit surface

$$\Psi_n^j = 0.25 n_j \gamma_j v_{thj}$$

with those of the plasma ions

$$\Psi_i = h_L n_i v_{Bi},$$

where the ion flux is calculated at the sheath edge rather than at the nanostructured surface where it can be substantially higher. Here, and below,
the indices j and i denote radical and ion species, respectively. Here we
emphasize that only H, CH, CH_2, CH_3 and C_2H_5 radical neutrals with
sufficient reactivity ("sticky") should be considered as being able to deposit on the film surface.

Figure 4.4(a) illustrates the effect of the input power on the deposited
fluxes of the radical neutrals. At low input powers ($P_{in} < 0.25\,kW$),
atomic hydrogen and methyl are the main radical neutrals deposited

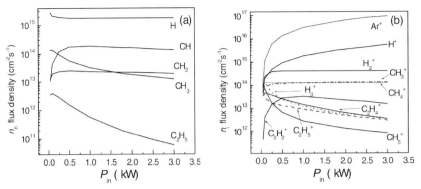

Figure 4.4 Deposited flux density of radical neutrals (a) and ions (b) as a function of P_{in} for the same conditions as in Figure 4.2 [53].

on the processing surface [53]. When the input power increases to P_{in} ca. 0.75 kW, the flux of CH increases. Furthermore, CH appears to be the major contributor to the carbon material deposition at relatively high input powers (≥ 0.5 kW). One of the most important observations from Figure 4.4 is that the flux of atomic hydrogen is about 10 times higher than those of the carbon-bearing neutrals. This finding supports the assertion of the crucial role of the etching of catalyzed surfaces by atomic hydrogen in the PECVD of carbon nanostructures [4,89].

The dependence of the deposited ion fluxes on P_{in} is shown in Figure 4.4(b). Under low input powers (ca. 0.05 kW), CH_5^+ flux is dominant. With an increase of the RF power the fluxes of Ar^+ and H^+ also increase due to the rise of the corresponding ion densities. Furthermore, at relatively high powers (≥ 0.4 kW) the fluxes of Ar^+ and H^+ cations become dominant. In this case, the H^+ flux is approximately 10 times smaller than that of Ar^+. Meanwhile, the fluxes of CH_4^+ and CH_3^+ are the strongest among the carbon-containing ions. However, the fluxes of CH_4^+ and CH_3^+ are in 2–3 orders of magnitude smaller than those of argon ions. Under the typical nanostructure growth conditions (P_{in} ca. 2 kW), the deposited flux of hydrogen atoms is comparable with that of H^+ ions [53].

The total flux density of hydrocarbon radical neutrals Ψ_n strongly depends on the input power. At $P_{in} = 0.05$ kW, Ψ_n ca. $1.63 \times 10^{14}\,\mathrm{s}^{-1}\mathrm{cm}^{-2}$ and further grows to $2.55 \times 10^{14}\,\mathrm{s}^{-1}\mathrm{cm}^{-2}$ at 0.5 kW. In the subsequent power range (0.5–3.0 kW) the flux density declines with power. For instance, $\Psi_n = 2.42, 1.98, 1.62\,\mathrm{s}^{-1}\mathrm{cm}^{-2}$ at $P_{in} = 1, 2,$ and 3 kW, respectively [53]. The total flux density of hydrocarbon ions onto the processing surface Ψ_i is slightly higher than Ψ_n and decreases

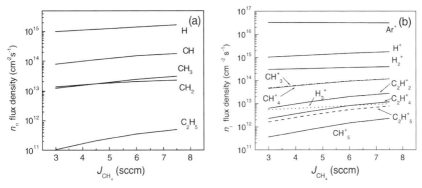

Figure 4.5 Same as in Figure 4.4 as a function of methane inflow rate for the same conditions as in Figure 4.2 [53].

with P_{in} in the entire power range of interest. Specifically, $\Psi_i =$ $3.66, 3.14, 3.02, 2.85, 2.75 \times 10^{14}\,\mathrm{s}^{-1}\mathrm{cm}^{-2}$ at $P_{in} = 0.05, 0.5, 1, 2$ and $3\,\mathrm{kW}$, respectively.

More importantly, under the prevailing carbon nanotip growth conditions ($P_{in} = 2$–$3\,\mathrm{kW}$) Ψ_i is ca. 1.5 times higher than Ψ_n, which is one of the most important conclusions of Denysenko *et al.* [53]. One can see that *despite the low degree of ionization, the rates of delivery of ionic species to the nanostructured surface can be higher than those of more abundant neutral radical species.* Moreover, as we have mentioned at the beginning of this subsection, the actual ion flux onto the surface can be much larger than the one calculated at the sheath boundary. Indeed, we recall that the ions enter the plasma sheath with the Bohm's velocity and are then accelerated by the negative substrate potential. At typical biases used in experiments on nanostructure synthesis, this difference can be a few tens and sometimes even hundreds of electron volts. This confirms the significance of nanostructure synthesis and processing using ion rather than neutral fluxes; therefore, ion-based processes are given a high priority in this monograph. In particular, as will be shown in the following chapters, ion fluxes play a pivotal role in the synthesis of a variety of nanostructures and nanostructured materials.

Methane gas inlet also appears as an important control factor of the densities of radical neutral/ion fluxes onto the surface. The dependences of the radical neutral and ion flux densities on J_{CH_4} are shown in Figures 4.5(a) and (b), respectively. One can conclude that the fluxes of all hydrocarbon species grow with the methane input flow rate. This can be attributed to the rise of the corresponding neutral and ion densities [53]. The total fluxes of ions and radical neutrals are almost the same at relatively low input flow rates of CH_4. Specifically, Ψ_i and Ψ_n are 1.56 and $1.51 \times 10^{14} \, s^{-1} cm^{-2}$ at $J_{CH_4} = 4.5$ sccm, respectively. With an increase of J_{CH_4} to 7.5 sccm, the total ion flux $\Psi_i = 2.85 \times 10^{14} \, s^{-1} cm^{-2}$ becomes approximately 20% higher than that of the radical neutrals $\Psi_n = 2.34 \times 10^{14} \, s^{-1} cm^{-2}$. This is yet another manifestation of the role of ion fluxes in the synthesis of the carbon nanostructures in which we are interested.

Variation of the argon input flow rate also affects the neutral and ion fluxes onto the processing surface. However, this dependence turns out to be quite similar to the variation of the number densities of the corresponding species with J_{Ar} and thus is not discussed in detail here. For further details the reader should refer to the original report [53]. Here we only note that the combined hydrocarbon ion flux on the processing surface diminishes with J_{Ar} as a result of the drop in the CH_4^+ and CH_3^+ ion fluxes. In contrast, the total hydrocarbon neutral flux is slightly smaller than Ψ_i and rises with J_{Ar}. Therefore, if the aim is to use ion fluxes for nanostructure/nanofilm synthesis or processing, argon inlet should be kept reasonably small. Indeed, as suggested by Denysenko *et al.* [53], at higher argon inlet flow rates the difference between the absolute values of the ion and neutral fluxes becomes very small.

4.1.6
Most Important Points and Summary

As we have seen from the previous subsections, the modeling and experimental results appear to be in good agreement, especially the variation trends of the number densities of the main neutral species H_2, CH_4, CH_3, C_2H_2, and C_2H_4 with the input power and methane/argon inlet flow rates. The global model also explains how the input parameters affect the OEIs of different emission lines in $Ar/CH_4/H_2$ discharges used for the PECVD of carbon nanostructures.

An interesting result is that methane and hydrogen conversion rates appear to be very high (ca. 99%); this conclusion follows from both experimental and numerical studies. This occurs because of the high electron number densities in low-frequency inductively coupled plasmas, which enhances the dissociation of molecular hydrogen and generation of large amounts of atomic hydrogen [53]. Hydrogen atoms, in turn, further accelerate the methane dissociation. More importantly, densities of neutral and radical species (with some of them being potential building units of targeted nanofilms and nanostructures) in the discharge can be effectively controlled by the input power as well as the flow rates of methane/argon gas feedstock.

Interestingly, the electron, Ar^+ and H^+ densities, as well as the methane conversion factor increase with power. On the other hand, the densities of most of hydrocarbon neutrals drop with P_{in} due to the enhancement of their collisions with the plasma electrons and ions. An increase of the input argon flow rate is accompanied by a rise of the densities of electrons, Ar^+ and hydrocarbon neutrals and cations. In contrast, the effective electron temperature and the densities of the atomic and molecular hydrogen neutrals and H^+ cations decline with J_{Ar}.

Higher inflow of the carbon source gas CH_4 naturally enhances the generation of C_xH_y cations and radical neutrals (here $x = 1-2$ and $y = 1-6$, and 8). Thus, various inelastic collisional processes intensify, which results in somewhat lower electron number densities. Furthermore, the electron-impact reactions involving a larger number of C_xH_y neutrals (see Table 10.1 in Appendix A) yield larger amounts of atomic hydrogen, thus contributing to higher densities of H_2 as a result of $C_xH_y + H$ reactions (see Table 10.2). In the model, some higher hydrocarbons are included despite their relatively low abundance in the discharge. The primary reason for this is that they play an important role in the gas-phase polymerization of carbon-based nanoparticles [134,136].

The computed T_{eff}, ion and radical neutral densities, can be used to analyze the fluxes of different species (building and working units) onto the catalyzed substrate in the PECVD of carbon nanostructures. The results of Denysenko et al. [53] reveal that total hydrocarbon neutral flux Ψ_n can be approximately 1.5 times smaller than the ion flux Ψ_i in a typical carbon nanotip growth process with $P_{in} = 1.8-3.0 \, kW$, $J_{CH_4} = 7.5 \, sccm$, $J_{H_2} = 12.4 \, sccm$, and $J_{Ar} = 35 \, sccm$. Therefore, by varying the inflow rates of argon and methane, one can control the ratio Ψ_i/Ψ_n, which characterizes the relative rates of arrival of positively charged and neutral building units to the surfaces of the nanostructures/nanofilms being grown. A reduced argon input flow rate can result in smaller Ψ_n and increased Ψ_i.

The total fluxes of the ion and neutral hydrocarbon species also increase with J_{CH_4}. At low methane inlets $\Psi_i \approx \Psi_n$, whereas at higher J_{CH_4} the deposited flux of hydrocarbon cations can be only ca. 20% higher than the neutral flux. Therefore, the ion fluxes onto the nanostructured surface in the carbon nanostructure growth process can exceed those of the neutral species and thus play a crucial role in the growth of nanonostructured carbon-based films. Their specific roles, in particular, in the growth and post-processing of high-aspect-ratio nanostructures, will be discussed in the following chapters of this monograph.

As was mentioned by Denysenko et al. [53], the effects of the plasma non-uniformity can play a significant role in the PECVD of carbon nanostructures and therefore a two-dimensional discharge model is required. Indeed, if one aims at synthesizing large-area nanoarrays, plasma non-uniformity can lead to very non-uniform growth of individual nanos-tructures in different areas within the pattern. In some cases these effects can be mediated and in some cases they significantly compromise the overall quality of the nanoarray. For a comprehensive discussion of the role of the plasma uniformity over large surface areas and processing volumes please refer to Chapters 2 and 3 of the recent monograph [1]. An example of a two-dimensional model of inductively coupled plasmas of acetylene-based gas mixtures is presented in the following section.

4.2
Two-Dimensional Distribution of Nanoassembly Precursor Species in Acetylene-Based Reactive Plasmas

In this section we follow the results of the original report [54] and discuss how two-dimensional fluid simulation of number densities and fluxes of the main building blocks and surface preparation species can help in improving control and predictability of nanoassembly of carbon-based nanopatterns in inductively coupled plamsas of $Ar + H_2 + C_2H_2$ reactive gas mixtures. These results suggest that the process parameters and non-uniformity of surface fluxes of each particular species may affect the quality of targeted nanoassemblies. In fact, the results of such predictive modeling can be used to control various plasma-aided nanofabrication processes and optimize the parameters of plasma nanotools.

It is important to note that despite a large number of experimental reports on successful applications of plasma-based tools for nanoscale applications and general-purpose plasma and discharge modeling results, the amount of currently available numerical results that specifically tackle the most important issues in applications of low-temperature plasma discharges for nanoscale processing and fabrication (such as two-dimensional fluxes of BUs and WUs on nanostructureed surfaces considered in this section) still remains limited. Moreover, existing relevant numerical efforts are mostly limited to the quite separate modeling of plasma discharges (similar to what was discussed in the previous section), thermal balance leading to additional substrate heating, or, on the other hand, *ab initio* simulations of atomistic configurations of relevant nanoassemblies [1,4,28,50,115,221–224].

More importantly, most existing plasma discharge modelling reports are based on "global" [53,68] (see also Section 4.1) or one-dimensional [117] models that provide only spatially averaged (or one-dimensional) densities and temperatures of the neutral and charged plasma species. Since laboratory and industrial low-temperature plasmas are intrinsically non-uniform, these models fall short of accurately describing fluxes of building and working units over large surface processing areas. In fact, this is a vital requirement of present-day micro- and nanoelectronic technologies. Another drawback of zero- and one-dimensional discharge models is their generic nature and, thus, lack of relevance to intrinsically process-specific nanoassembly processes. This is largely due to the overwhelming complexity of reactive plasma chemistry, complex geometries of plasma nanofabrication facilities [49] and a substantial lack of reliable knowledge on precursor species of most of the existing nanoassemblies [54].

In this section, we discuss the results of two-dimensional simulation of low-temperature inductively coupled plasmas (ICPs) of complex $Ar + H_2 + C_2H_2$ mixtures in a plasma-aided nanofabrication facility suitable for large-area processing of nanostructured surfaces [54]. In this simulation, the main focus is on number densities and surface fluxes of the most important species (building and working units, for definition please refer to Chapter 2) involved in the plasma-assisted synthesis of various carbon-based nanostructures. We will also relate the plasma non-uniformity to the distribution of fluxes of the BUs and WUs over a substrate with a large (ca. $300\,cm^2$) surface area. This will make it possible to elucidate the effects of the plasma non-uniformity on nanoscale processing and also of possible improvement pathways for the plasma-based nanotools and processes. The original report [54] is related to two-dimensional modelling of inductively coupled plamsas (ICPs) in complex $Ar + H_2 + C_2H_2$ mixtures in a realistic plasma reactor geometry, and in process parameter ranges relevant for plasma-aided nanofabrication of carbon nanotube/nanotip-based microemitter arrays.

4.2.1
Formulation of the Problem

The plasma reactor considered in this section is a metal cylinder of radius 15 cm and height 16 cm. A conducting, yet electrically insulated from the chamber walls, substrate stage of radius 10 cm is centered 8 cm below the chamber top. A 4-turn spiral antenna (inductive coil) is located at the top of the chamber and is connected to a 13.56 MHz generator. As is common to inductively coupled plasma (ICP) discharge configurations with a flat spiral coil, the RF power is coupled to the plasma in the reactor chamber through a dielectric window at the top. We recall that this RF coil configuration is typical for ICP plasma sources [1, 189] currently widely adopted by the microelectronic industry as benchmark plasma reactors.

The simulation area of the original report [54] is limited to the upper section ($z = 0$–8 cm) of the plasma reactor, where the plane of origin $z = 0$ is at the upper surface of the substrate stage.

The two-dimensional fluid model of interest to us here considers low-temperature plasmas of complex $Ar + H_2 + C_2H_2$ gas mixtures in high-pressure (1.333 hPa (1 Torr)–9.332 hPa (7 Torr)) and low-pressure (13.332 Pa (100 mTorr)) ranges. The above pressure ranges are taken from relevant experimental reports on successful synthesis of multiwalled carbon nanotubes, nanofibers, nanowalls, nanocones, nanotips/nanoneedles and pyramid-like structures [53,89,90,93,94,140,225–228] The most essential details of the 2D fluid plasma model can be found in Appendix A.

Radical and non-radical neutral and ionic (cationic) species include [54]

- 3 non-radical neutral species: Ar, H_2, C_2H_2;

- 7 radical neutral species: H, C_2H, CH, CH_2, CH_3, $C_2(X^1\Sigma_g^+)$, $C_2(a^3\Pi_u)$; and

- 11 ionic species: Ar^+, ArH^+, H^+, H_2^+, H_3^+, C_2H^+, $C_2H_2^+$, CH^+, CH_2^+, CH_3^+, C^+.

Details of chemical reactions in the gas phase are also provided in Appendix A. These species and chemical reactions have been chosen to include the most important and common building and working units and to incorporate the most essential channels of creation and destruction/sink of such species within a reasonable computational cost. Higher hydrocarbons, other complex radical species, as well as nanocluster species were not considered.

In the original work [54], the attempt was made to elucidate the possibility of controlling the number densities and fluxes of the most important building units CH, CH_3/CH_3^+, $C_2(X^1\Sigma_g^+)$ of carbon-based nanostructures and working units Ar^+ and H [4]. As mentioned above, the CH radical plays a major role in the synthesis of sp^3-hybridized carbon [229]; CH_3/CH_3^+ neutral/ionic radical species are ideal in creating hydrogen-terminated carbon nanostructures such as single-crystalline carbon nanotips, nanocones, or pyramid-like structures [4,230]; $C_2(X^1\Sigma_g^+)$ has been suggested as the most likely BU for open-ended carbon nanotubes [168] and ultrananocrystalline diamond [37]. We also recall that atomic hydrogen plays a vital role in termination of surface dangling bonds (surface passivation) and reactive chemical etching of amorphous carbon [4]. Likewise, initially hydrogen-terminated surface bonds can be activated as a result of the impact of energetic argon ions; this can enable attachment of radical species which would normally have difficulty to create these attachment sites for themselves. This ion-assisted process has already been discussed in the previous chapters of this book. Furthermore, impinging ions also transfer their kinetic energy to the processing surface, which can substantially elevate the surface temperature (under typical carbon nanostructure growth conditions this additional temperature increase can reach 100–150 degrees and even higher) [231]. This justifies the choice of CH, CH_3/CH_3^+ and $C_2(X^1\Sigma_g^+)$ building units and Ar^+ as well as atomic hydrogen working units as a primary focus in the simulations.

4.2.2
Number Densities of the Main Discharge Species

We will now focus on the main features of two-dimensional distributions of the electrostatic potential ϕ and electron temperature T_e in two different operating pressure ranges. Specifically, the high-pressure case ($p_0 = 1.333$ hPa (1 Torr)–9.332 hPa (7 Torr)) is relevant to the plasma-assisted synthesis of multiwalled carbon nanotubes [89, 228]. On the other hand, low-pressure ($p_0 = 13.332$ Pa (100 mTorr)) ICP discharges have been successfully used to synthesize carbon nanowall-like structures, pyramidal, conical and nanotip-like structures [4, 53, 93–95, 140, 258]

In the high-pressure case, the distribution of the electrostatic potential in the discharge bulk becomes more uniform when the pressure increases to 9.332 hPa (7 Torr); however, this change does not noticeably affect the potential profile in the vicinity of the (equipotential with a DC bias of -50 V) substrate stage. On the other hand, an increased gas pressure results in a lower electron temperature (from ca. 1.8–2.1 eV down to ca. 1.2–1.3 eV) above the substrate. We stress that this decrease of electron temperature results in notable changes in species production and destruction rates and the initial velocity with which the ionic species enter the plasma sheath (Bohm velocity) [54]. However, it does not significantly affect the deposition energy of charged species, which is mainly controlled by the negative DC bias applied to the substrate. In the low-pressure case, the electron temperature can be effectively controlled by varying the composition of the gas mixture. In fact, increasing the acetylene content from 5 to 60% results in a higher, and more uniform electron temperature, over the substrate surface. In this case the electron temperature can be maintained within the range 3–3.3 eV over the entire substrate stage.

The computed profiles of ϕ and T_e make it possible to map the number densities of the main BUs and WUs in the reactor chamber [54]. Changes

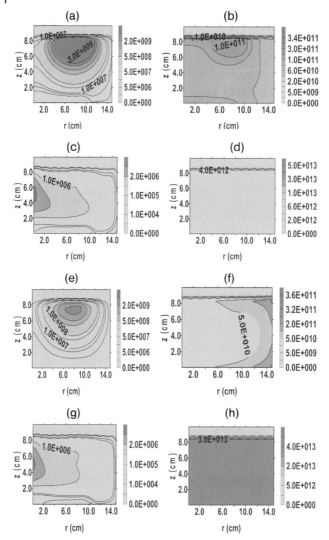

Figure 4.6 2D density in [cm^{-3}] profiles for: (a) Ar$^+$ at 1.333 hPa (1 Torr); (b) H at 1.333 hPa (1 Torr); (c) CH$_3^+$ at 1.333 hPa (1 Torr); (d) C$_2(X^1\Sigma_g^+)$ at 1.333 hPa (1 Torr); (e) Ar$^+$ at 9.332 hPa (7 Torr); (f) H at 9.332 hPa (7 Torr); (g) CH$_3^+$ at 9.332 hPa (7 Torr); (h) C$_2(X^1\Sigma_g^+)$ at 9.332 hPa (7 Torr). Other parameters: gas partial pressures 50% Ar, 10% H$_2$, and 40% C$_2$H$_2$; RF input power P_{in} =100 W; substrate bias V_S =-50 V; gas temperature T_g = 300 K [54].

in densities of charged and neutral species were computed when the gas feedstock pressure was increased from 1.333 hPa (1 Torr) to 9.332 hPa (7 Torr); these changes are visualized via two-dimensional density maps in Figure 4.6 and dependence of species densities on working pressure and relative densities of the species in 4.7.

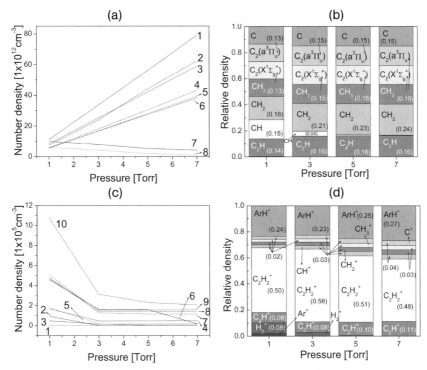

Figure 4.7 Densities (computed at $z = r = 0$) *versus* pressure: (a) Number densities of neutral species *versus* pressure: 1) 2 x [CH$_3$], 2) [CH$_2$], 3) 1.5 x [C$_2$(a$^3\Pi_u$)], 4) [C$_2$H], 5) [C], 6) [C$_2$($X^1\Sigma_g^+$)], 7) 10 x [H], 8) [CH]; (b) Relative densities of neutrals; (c) Number densities of charged species: 1) 10 x [H$^+$], 2) 2 x [Ar$^+$], 3) [CH$_2^+$], 4) [C$_2$H$^+$], 5) [H$_2^+$], 6) 10 x [C$^+$], 7) [ArH$^+$], 8) 10 x [CH$_3^+$], 9) 10 x [CH$^+$], 10) [C$_2$H$_2^+$]; (d) Relative densities of charged species. Other parameters are the same as in Figure 4.6 [54].

A general trend observed in Figure 4.7(a) was increased number densities of all neutral species except for CH and H. Interestingly, no corresponding trend for charged species can be observed. This suggests that when the pressure is increased, reactions involving the production of neutral species (such as abstraction and recombination) occur at higher rates than reactions (such as electron impact reactions) which favour the creation of charged species. It is the higher pressure (in the 1.333 hPa (1 Torr)–9.332 hPa (7 Torr) range) that makes neutral reactions particularly important. However, in low-pressure plasma processing such reactions are of less importance [54].

Looking specifically at Ar$^+$, CH$_3^+$, H and C$_2$($X^1\Sigma_g^+$) species, Figures 4.7(a) and 4.7(c) illustrate the increased C$_2$($X^1\Sigma_g^+$) and decreased Ar$^+$, CH$_3^+$ and H number density. One can also observe similar trends in the 2D density plots in Figure 4.6. Changes in relative densities are more subtle; the most striking feature, observed in Figure 4.7(b), is the decreasing relative density of CH with increasing pressure [54]. One sees that the CH density is the most sensitive to changing pressure. Since CH is one of the most important BUs of diamond-like carbon, which is commonly synthesised in a similar pressure range, a proper choice of the specific synthesis pressure is crucial for this process. In view of using CH_3^+ as a building unit, lower pressures would be favourable for synthesis techniques which require higher number densities of this BU. Similarly, if a carbon dimer $C_2(X^1\Sigma_g^+)$ is the intended building unit, an increased process pressure would be an advantage [54].

Indeed, if one requires a neutral BU species (other than CH and H) then higher pressures would be favoured [shown in Figure 4.7(a)] since the densities of all the neutrals increase with the working gas pressure. The number density of atomic hydrogen, in contrast to the other neutrals, decreases with increasing pressure. Thus, for an optimized plasma-based nanoassembly process a balance should be struck between the higher pressures required for carbon bearing neutrals and the lower pressure required for an adequate amount of hydrogen atoms [54]. We recall that atomic hydrogen is the main species responsible for surface termination and etching of amorphous carbon. Hence, one can expect that at lower pressures the surface termination should be better, in addition to more effective etching of amorphous carbon. Moreover, lower pressures also ensure a more efficient supply of carbon-bearing building units to the growth surface. On the other hand, higher pressures are beneficial for maintaining higher deposition rates.

A similar analysis of the densities of charged and neutral species was conducted in the lower-pressure regime when the percentage influx of C_2H_2 was varied between 0.1% and 60% [54]. This analysis generally shows quite similar trends as in the high-pressure case. In particular, the number densities of neutral species H and $C_2(X^1\Sigma_g^+)$ increase, and the densities of Ar^+ and CH_3^+ cations decrease when the acetylene inlet is increased. These results are consistent with the 2D density plots in the low-pressure case, which can be found in Figure 6 of the original report [54]. The most noteworthy feature is the significant decrease in the relative density of Ar^+.

It is possible that reduced amounts of Ar lead to reduced rates of the ionization or charge exchange reactions which destroy Ar in order to create Ar^+. In addition, an increased influx of C_2H_2 results in higher rates of the charge exchange reaction $Ar^+ + C_2H_2 \rightarrow C_2H_2^+ + Ar$, which destroys even more Ar^+, creating larger amounts of $C_2H_2^+$. Interestingly, a decrease in the relative density of Ar^+ is accompanied by a notable increase in the relative density of $C_2H_2^+$. This indicates that the aforementioned reaction is one of the most dominant mechanisms leading to a decreasing Ar^+ concentration [54]. Similarly, higher densities of atomic hydrogen may be attributed to the greater amount of acetylene which leads to higher rates of generation of H species as a result of heavy particle collisions.

Looking again at building and working units, both $C_2(X^1\Sigma_g^+)$ and H number densities increase with a larger percentage of acetylene precursor. Therefore, to maximize the amount of $C_2(X^1\Sigma_g^+)$ BUs and the amount of H available for surface passivation, the partial pressure of C_2H_2 should be increased. An increased C_2H_2 inlet leads to a decrease in the number density of CH_3^+. Therefore, higher densities of this important building unit can be achieved by reducing the pressure and partial pressure of acetylene feedstock. In terms of plasma-assisted nanoassembly, if CH_3^+ is to be used as a building unit and atomic hydrogen as a surface passivator a proper balance between the higher percentages of C_2H_2 (required for higher [H]) and the lower percentages (required for higher [CH_3^+]) should be established. Here, square brackets denote the species concentration. On the other hand, reduced densities of argon ions at higher partial pressures of C_2H_2 may lead to lower surface temperatures and also lower rates of activation of hydrogen-terminated surface dangling bonds by argon ions [54].

4.2.3
Fluxes of Building and Working Units

We now turn our attention to the dependence on process parameters and possible effects of *fluxes* of the major building and working units to the substrate [54]. For this analysis, only argon ion and atomic hydrogen working units (Figure 4.8) have been chosen. For comparison, the fluxes of carbon-bearing BUs CH, CH_3^+ and $C_2(X^1\Sigma_g^+)$ are shown in Figure 4.9. As can be seen from Figure 4.8, the fluxes of Ar^+ and H species behave quite similarly when the working pressure changes. Specifically, fluxes of both argon ion and atomic hydrogen species decrease when p_0 becomes higher. It is also important that the distribution of the fluxes of these working units over the substrate surface also appear to be similar. From Figures 4.8(a) and (b), one can observe that the fluxes of both Ar^+ and H are smaller in the central area of the substrate and increase towards the periphery. The density of Ar^+ in particular, is fairly uniform within the area $0 < r < 4$ cm, when $r > 4$ cm it begins to steadily increase. Atomic hydrogen flux, on the other hand, increases throughout the entire radial range. Under low-pressure conditions [Figure 4.8(c)

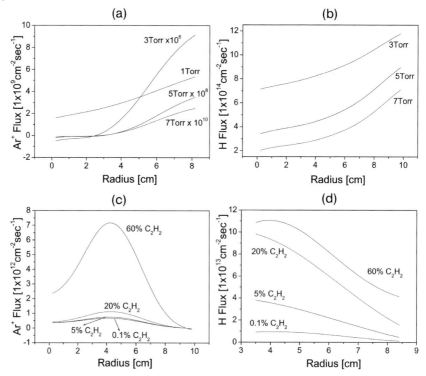

Figure 4.8 Fluxes of working units onto the substrate stage
($0 < r < 10$ cm): (a) Ar^+ varied pressure; (b) H varied pressure;
(c) Ar^+ varied C_2H_2 composition ($p_0 = 13.332$ Pa (100 mTorr));
(d) H varied C_2H_2 composition ($p_0 = 13.332$ Pa (100 mTorr)).
Parameters are the same as in Figure 4.6 [54].

and (d)], the fluxes of both species increase with the partial pressure of
the acetylene feedstock. Interestingly, the peak of the argon ion flux is
now located at the middle of the radial span of the substrate stage (at
$r \approx 4$–4.5 cm). The flux of atomic hydrogen is maximal in the central
area, in contrast to the high-pressure case [see Figure 4.8(b)] [54].

Let us consider what this means in terms of surface preparation in
plasma-assisted nanoassembly. The fluxes of both argon ions and atomic
hydrogen in the central area of the substrate stage are lower than they
are towards the edge, as Figures 4.8(a) and (b) suggest. Therefore, in
the outer reaches of a large-area substrate the rates of termination of
surface bonds by atomic hydrogen and the rates of activation of such
bonds by energetic argon ions will be higher than in the central substrate
stage [54]. However, here we stress that the actual availability of dangling bonds required for nanostructure growth, is controlled by the com-

petition between the fluxes of surface-activating and surface-passivating fluxes. Thus, the actual surface state will strongly depend on the *relative* changes of the fluxes of argon ions and atomic hydrogen along the radial direction.

Let us specify what this can mean in a real surface preparation process. If the rates of surface bond termination by hydrogen atoms are higher than the rates of their activation by argon ions, most of the surface bonds will remain terminated, thus unavailable for bonding. In the opposite case, surface bond activation by Ar^+ may outpace their termination by hydrogen atoms. In this case the availability of dangling surface bonds will certainly be much higher. As a result, one can expect significantly higher nanofilm growth rates [54].

In previous chapters of this monograph we have emphasized that the synthesis of ultra-fine nanostructures and their arrays is a very delicate process which usually requires quite low availability of surface dangling bonds. Exactly what is likely to grow in any selected substrate area, however, strongly depends on the area-specific delivery of the BUs and other surface conditions (surface temperature, defects, and so on). From Figures 4.8(a) and (b) one can deduce that the rates of surface bond activation (by argon ions) and passivation (by atomic hyrogen) change with approximately the same rates along the radial direction. Therefore, even though the fluxes of the main WUs are very non-uniform across the substrate, the actual surface bond availability may be quite uniform across the entire substrate area. With which we arrive at a counterintuitive conclusion: *a fairly uniform surface activation for the subsequent nanostructure growth can be achieved from essentially non-uniform plasmas!*

However, apart from the surface preparation, one still needs to make sure that the building units of the nanostructures are also uniformly delivered to the growth surface [54]. A further important requirement is that the surface temperature is maintained uniform throughout the substrate stage.

Figure 4.9 shows the details of spatial distributions of the main carbon-bearing BU species CH (panels (a) and (d)), $C_2(X^1\Sigma_g^+)$ (panels (b) and (e)), and CH_3 (panels (c) and (f)) [54]. In particular, from Figure 4.9 one can observe a decreased CH and CH_3^+ flux and an increased $C_2(X^1\Sigma_g^+)$ flux at higher pressures (in the pressure range 4.000 hPa (3 Torr)–9.332 hPa (7 Torr)). The dependence of the fluxes on the C_2H_2 partial pressure, however, appears to be more complicated. It is interesting that whilst one can observe a clear increase of the CH_3^+ flux with C_2H_2 inlet, the fluxes of CH and $C_2(X^1\Sigma_g^+)$ species behave quite differently. Indeed, these fluxes increase up to a certain point on the substrate, which is near to the substrate center, and is located at approximately

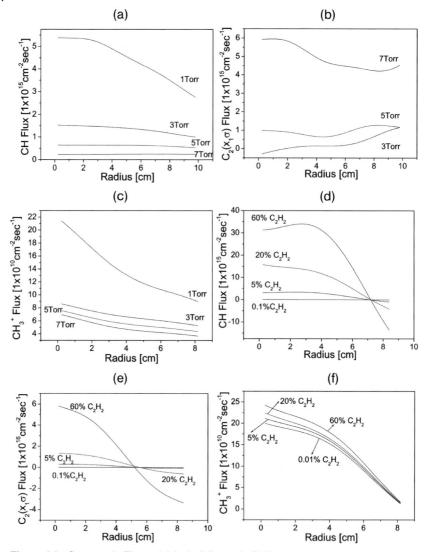

Figure 4.9 Same as in Figure 4.8 for building units [54]:
(a) CH varied pressure; (b) C_2 $(X^1\Sigma_g^+)$ varied pressure; (c) CH_3^+
varied pressure; (d) CH varied C_2H_2 composition ($p_0 = 13.332$ Pa
(100 mTorr)), (e) C_2 $(X^1\Sigma_g^+)$ varied C_2H_2 composition; (f) CH_3^+
varied C_2H_2 composition ($p_0 = 13.332$ Pa (100 mTorr)).

$r = 7$ cm for CH and $r = 5$ cm for C_2 species. At larger distances from
the substrate stage center these fluxes decrease, as can be clearly seen
from Figure 4.9. One of the important conclusions that follows from Figure 4.9 is that the rates of delivery of the CH and C_2 BUs towards the

central and outer areas of the substrate change differently with the gas feedstock composition.

From the point of view of applications, the areas around the center of the substrate stage are of particular importance. For this reason, let us now examine the behavior of fluxes of the most important carbon-bearing species in this particular area. From Figures 4.9(d–f), one can see that to increase the fluxes of CH, CH_3^+, and $C_2(X^1\Sigma_g^+)$ an increased partial pressure of C_2H_2 would be required. The optimized process pressure will then depend on the types of building units one wants to employ for the growth of specific nanoassemblies [54]. Indeed if CH or CH_3^+ is the desired building unit, then a decreased pressure directs more of those species towards the substrate where they can start building nanostructures. On the other hand, if $C_2(X^1\Sigma_g^+)$ is the building unit, then an increased pressure (at least in the range 4.000 hPa (3 Torr) to 9.332 hPa (7 Torr)) can provide the required greater fluxes [54].

The original report [54] also stresses that the fluxes of all carbon-bearing building units change significantly along the radius of the substrate stage. Apart from a few exemptions (such as fluxes of $C_2(X^1\Sigma_g^+)$ species in the 4.000 hPa (3 Torr)–6.666 hPa (5 Torr) range), a general trend is that the building unit fluxes are stronger in the areas near the substrate center. In this case, more energetic (e.g., CH_3^+ cations) or reactive (e.g., $C_2 (X^1\Sigma_g^+)$) species have a better chance to attach those surface dangling bonds available for bonding. Once all the available bonds are used, the excess species delivered from the ionized gas phase cannot form chemical bonds with the surface and, depending on the prevailing surface conditions, either physisorb and form irregular deposits on the surface – such deposits are usually amorphous – or migrate about the surface to occupy vacant dangling bonds available elsewhere. The first situation happens when the rates of supply of the building units exceed those for their surface diffusion. This situation can happen when the discharge is maintained with elevated input powers and surface temperatures are reasonably low.

Alternatively, when the incoming fluxes of the BUs are reduced and the surface temperature is increased, it is reasonable to anticipate that a large number of the BUs will diffuse radially and occupy the available dangling bonds. Therefore, under conditions when the dangling bond availability is fairly uniform across the surface, the nanostructures are expected to grow via the combination of the direct (from the plasma) and indirect (via surface migration) building unit delivery routes [54]. This may have important effects on the quality of the large-area nanopatterns being fabricated. As we have discussed above for the case of carbon nanocone structures, predominant delivery of CH_3/CH_3^+ building units to the nanoassembly site from the ionized gas phase results in the growth of high-aspect-ratio conical carbon nanostructures, which are ideal for electron microemitter applications [95]. However, increased rates of surface migration adversely affect the nanoassembly process and lead to shorter and wider nanotips [52].

Discharge modeling in quite similar gas mixtures has been reported by other authors. For example, Gordillo-Vazquez *et al.* [137] consider gas discharges in acetylene-based gas mixtures, albeit in the intermediate pressure range (0.133 hPa (0.1 Torr)–1.067 hPa (0.8 Torr)), in a simpler, spatially-averaged manner, and at much lower concentrations of C_2H_2 (ca. 1%) compared to the report of Ostrikov *et al.* [54]. It is interesting that these two papers report quite different trends for carbon dimers, C_2 ($X^1\Sigma_g^+$) and C_2 ($a^3\Pi_u$). In particular, the number densities of these species increase as the working pressure decreases [137]. Furthermore, the density of CH_3 remains fairly constant, whereas the densities of C_2H_2 decrease slightly [137]. This behavior, however, is not necessarily inconsistent with the results presented in this section, given the quite different pressure regimes, operation conditions and gas mixtures. Here we stress that according to Paschen's curves (dependence of densities of plasma species on working gas pressure), the density of plasma species shows quite a different response to pressure variations in low- and high-pressure ranges [68]. When the pressures are higher, electron impact reactions are weaker, with more heavy ion collisions occurring. At lower pressures, however, the effect of electron impact reactions are stronger, favouring the formation of lighter species such as carbon-bearing radicals [137].

One of the main points stressed in this section is that non-uniformity of fluxes of the main building and working units participating in the nanoassembly is a major issue in the application of plasma-based nanofabrication facilities for the synthesis of large-area nanopatterns. In the above, we have considered a conventional inductively coupled plasma source with an external flat spiral coil antenna with a small number of coil turns. Due to a specific spatial distribution of the induced azimuthal electric field (almost vanishing at the chamber axis), the maximum of the RF power deposition is shifted radially, typically by a few cms towards the walls. As a result, the rates of argon ionization (mostly by electron impact) are lower near the chamber axis [232]. This explains higher densities and fluxes of Ar^+ and atomic hydrogen to the outer areas of the large-area substrate. On the other hand, CH, $C_2(X^1\Sigma_g^+)$, and CH_3^+ species are predominantly generated via ion–neutral and neutral–neutral reactions. As a result, their fluxes are typically stronger near the chamber axis, where deposition substrates are usually located.

4.3
**Generation of Nanocluster and Nanoparticle Building Units
in the Ionized Gas Phase**

As we have seen from previous sections, building and working unit generation processes critically depend on plasma process conditions. For example, the range, densities and fluxes of major species produced in methane- (Section 4.1) and acetylene-based (Section 4.2) discharges are clearly different. In the above two cases our consideration was limited by atomic, molecular and radical species only.

In this section we will consider some typical situations when formation of larger species takes place. As was already mentioned in Chapter 2, examples of such species include polymeric chains and macromolecules, nanoclusters, primary nuclei of nanoparticles, developed nanoparticles in various (e.g., amorphous, crystalline or mixed) states, and nanoparticle agglomerates. Generally speaking, formation of such particles requires quite different conditions compared with creation of atomic, ionic, molecular, and radical species considered in previous sections.

Indeed, smaller species are most commonly created as a result of inelastic collisions between the species available in the gas phase. For example, various radicals are created through electron impact reactions such as ionization and dissociation. The probability of such reactions is strongly dependent on the plasma parameters such as electron density and temperature; specifically, these two parameters determine the ability of electrons to cause significant structural changes in the species they impact. As we have seen from the above discussion, plasma species can also recombine and form species with a larger number of atoms. It is worthwhile to stress that the main factor that determines the ability of particles to recombine are their densities in the ionized gas phase.

4.3.1
Nano-Sized Building Units from Reactive Plasmas

Let us consider an example of silicon nanoparticles generated in reactive silane-based plasmas and start from a deceptively trivial consideration. Suppose that densities of SiH_3 radicals and hydrogen atoms in such plasmas are large enough; accordingly, it is reasonable to assume that they may combine to form a silane molecule SiH_4. Our curious minds would immediately pose the question: What happens if one continues attaching larger numbers of hydrogen atoms and other available radicals to the SiH_3 radical and how fast can this process go? The answer is quite obvious: this process will eventually lead to the creation of heavier species Si_xH_y that contain larger (in some cases substantially larger!) numbers

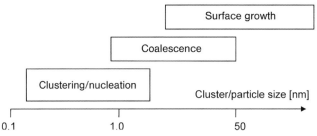

Figure 4.10 Parameter spaces for typical nanocluster/
nanoparticle size at three different stages of growth. All particles
require a nucleation/clustering stage while some particles may
continue their development via surface growth without
coalescence.

of silicon and hydrogen atoms. However, it is not so simple to work out
the products and rates of such reactions or to predict what happens next
with the Si_xH_y species.

Generally speaking, the rates of elementary reactions strongly depend
on the abundance and reactivity of the species involved. In particular,
this will determine the ability of specific species in a specific environment
to nucleate into larger species and choose the most appropriate pathway.
Let us consider nanoparticle nucleation in reactive silane-based plasmas
and compare this process with clustering of metal atoms in a plasma of
an inert gas such as argon or neon. The most striking difference between
the two cases is in reactivity of the species and the environment and also
in the actual triggers of the nucleation process. One can state that in the
first case the reactivity is very high and the nucleation process proceeds
via attachment of a larger and larger number of atoms that form chemical
bonds as they attach. This is a typical example of a chemical nucleation
pathway with different possibilities for further growth, as sketched in
Figure 4.10.

As we have mentioned above, the efficiency of nucleation in this case
strongly depends on the species that trigger the process. Sophisticated
numerical modeling and experiments [28, 233–235] suggest that silicon-
containing anions can be regarded as the most effective triggers of complex reaction chains that eventually lead to the formation of small silicon
hydride nuclei with a reasonably large number of Si and H atoms. In
addition to intrinsically high reactivity, SiH_x^- ($x \leq 3$) anions also show
remarkably longer residence times compared to similar neutral species.
This occurs because these negative ions can be effectively confined in
near-electrode electrostatic traps for reasonably long periods, until they
are pumped away. For comparison, positive ions stream towards negatively charged walls or electrodes and disappear from the active zone

of the discharge much faster than the anions. Thus, relatively long residence times and chemical reactivity make silicon-containing anions effective triggers of fast nanoparticle nucleation. This conclusion is supported by measurements of mass spectra which show a strong correlation between the abundance of SiH_x^- anions and particle nucleation rates in low-temperature silane-based plasma discharges [236].

Predictive numerical modeling of nucleation of hydrogenated silicon nanoparticles in silane-based plasmas [115, 224, 237] suggests that primary nanoparticles (nuclei) are primarily formed via a chain of reactions of SiH_x^- anions with silane molecules SiH_4. Such anions are usually created as a result of dissociative attachment of electrons

$$SiH_4 + e \rightarrow SiH_n^- + H_{4-n}, \tag{4.7}$$

where $n = 2$ or 3. Moreover, in 70% of cases, Reaction 4.7 leads to generation of SiH_3^- anions.

Interestingly, SiH^- anions, though reactive, are less abundant in mass spectra of nanoparticle-generating silane discharges and therefore do not lead to pronounced nanoparticle nucleation as compared to SiH_2^- and SiH_3^- cases. Therefore, there are two principal channels where the earliest stages of nanoparticle nucleation are triggered by SiH_3^- and SiH_2^- anions. The corresponding two distinctive pathways lead to the growth of larger silyl ($Si_jH_{2j+1}^-$) and siluene ($Si_jH_{2j}^-$) anions, respectively. The chains of polymerization reactions are

$$Si_jH_{2j+1}^- + SiH_4 \rightarrow Si_{j+1}H_{2j+3}^- + H_2 \tag{4.8}$$

for the SiH_3^--triggered nucleation pathway and

$$Si_jH_{2j}^- + SiH_4 \rightarrow Si_{j+1}H_{2j+2}^- + H_2 \tag{4.9}$$

for the SiH_2^--triggered nucleation pathway. The rate constants in these two cases are quite different but remain of the order of ca. $10^{-18}\,m^3/s$ [115, 237]. It is interesting that silane molecules vibrationally excited by electron impact can make a significant contribution to the nucleation process. Now the question is when does the polymerization process described by Reactions 4.8 and 4.9 saturate. In particular, until how many silicon atoms does the growth proceed in this way? At present, there is no clear answer to this question.

Elegant numerical simulations of De Bleecker *et al.* [115] only include anions containing up to 12 silicon atoms. It was assumed that immediately following $Si_{12}H_{25}^-$ formation, they transformed into spherical nanoparticles with a radius corresponding to a closely-packed system of 12 silicon and 25 hydrogen atoms. This rather simple approach makes

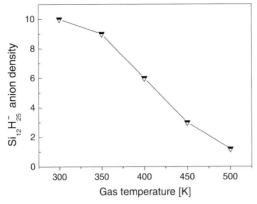

Figure 4.11 Density of $Si_{12}H_{25}^-$ anions *versus* background gas temperature, in units of $10^{15}\,m^{-3}$ [115].

it possible to move to the next step when nuclei (also frequently termed primary particles in the literature [58,59]) formed in such a way feature a well-shaped surface and grow further via attachment of plasma species to the surface, coalescence of primary nuclei into larger particles or a combination thereof.

At this stage, a very interesting phenomenon occurs. Indeed, before incorporating 12 silicon atoms, the protoparticles are simply anions and have electric charge -1. Note that in this case the charge on the particle does not depend on its size! However, when they transform into spherical nanoparticles, their charge is determined by the balance of microscopic electron and ion fluxes and strongly depends on the plasma parameters and the size of the nucleus. At present, one can only wonder at how exactly this amazing transformation takes place and how exactly the shape and size of $Si_xH_y^-$ anions change as the numbers x and y increase. This is just one of the examples of a challenge which is still awaiting its conclusive solution.

However, this approach makes it possible to estimate number densities of initial (ca. 1 nm-sized) nuclei from which larger Si nanoparticles are formed. These densities critically depend on the plasma environment and process parameters. For example, Figure 4.11 demonstrates the dependence of the concentration of anionic nuclei containing 12 silicon and 25 hydrogen atoms on the background gas temperature [115]. From this figure one can immediately conclude that variations in the gas temperature exert a profound effect on the nanoparticle growth from reactive silane-based plasmas. The trend shown in Figure 4.11 is that the density of primary nuclei decreases with gas temperature.

Therefore, lower densities of anions with the critical number of silicon atoms mean that the initial stage of nanoparticle growth proceeds slower when the temperatures of the silane-based working gas are lower. This conclusion is consistent with common experimental observations suggesting longer particulate ("dust") build-up times at higher process temperatures. For instance, at room temperatures (T_g ca. 300 K) small nanoparticles are generated within just a few milliseconds following discharge ignition. However, at $T_g \approx 500$ K, it typically takes 10–20 s to observe small nanopowders [238].

The number of primary particles (nuclei) is critical for the further development of the entire nanoparticle growth process. If the number densities of gas-borne solid nuclei are small, their further growth proceeds via relatively slow attachment (e.g. physi- or chemi-sorption) of reactive radicals to the surface of the growing particle. On the other hand, if the density of small nuclei is very high (which is often the case for silane-based plasmas), coalescence of primary nuclei into larger particles is commonly believed unavoidable. These possibilities are illustarted in Figure 4.10 which shows typical size ranges of nanoclusters and nanoparticles at nucleation, coalescence and surface growth stages. In both cases, reliable knowledge of reaction and growth rates is indispensable to achieve ultimate control of nanoparticle sizes.

In practice, it appears quite challenging to control the shape, sizes and structure of nanoparticles grown in the ionized gas phase. However, detection of the onset of rapid and uncontrollable coalescence of primary ca. 1 nm-sized nuclei into larger (typically 40–50 nm in size) particles gives a clear indication on the appropriate run of silane-based plasma discharges. In this case one can switch the discharge off when the primary particles have reached the desired size yet are not ready to coalesce in the plasma bulk. This is done by precisely following the discharge behavior and monitoring its parameters, which appear to be substantially different before and after the onset of particle coalescence. This transition is often accompanied by substantial changes of the electron temperature, plasma density, as well as the optical emission from the discharge and is commonly referred to as α–γ' transition [28].

The outcomes of such coalescence critically depend on how fast the particles can reach and stick to each other as well as on the gas temperature and the intensity of ion fluxes from the plasma on the particle's surface. In one extreme, if the rates of sticking are high, gas temperature is low and ion fluxes are weak, small nuclei stick to each other forming disordered (e.g., fractal-like) networks or cauliflower-shaped agglomerates. In most cases such particles feature a very high content of the amorphous phase. On the other hand, when the rates of sticking of individual particles to each other and ion fluxes are moderate, while the environment is very hot, one can expect formation of spherical nanoparticles; these particles quite often have a crystalline structure.

Under certain conditions, it appears possible to synthesize faceted (e.g., perfectly cubic) nanocrystals [239]. Interestingly, faceting arrangements are strongly dependent on the particle's size and surface conditions such as hydrogen passivation. All these interesting possibilities lead to the generation of a large variety of sub-nm and nm-sized particles ranging from relatively simple polyatomic molecules, complex polymers, clusters and nuclei to faceted nanocrystals, fractal chains of nanoclusters and nanoparticles and agglomerates with complex shapes and structures [4].

4.3.2
Nanoparticle Generation: Other Examples

Let us now consider some cases when the plasma is not as reactive as in the previous subsection. What one could expect is that the rates of nanoparticle production should be noticeably lower compared with "very reactive" plasmas. However, in several cases it is still possible to generate large amounts of nanoparticles at very high rates even from not-so-reactive plasmas like the SiH_4-based plasmas of the previous subsection. As we have already stressed above, a lot depends on the environment where nanoparticle nuclei are formed.

Another important factor which determines the rates of nanoparticle creation is the abundance of suitable precursor species. Indeed, under certain conditions a large amount of hydrosilicon anions can be created in silane-based plasmas; these anions induce polymerization and eventually initial nanoparticle nuclei (here we recall that they are also frequently termed primary particles in dusty plasma literature [28]) are formed. The reactivity of silane-based plasmas is high enough to create plentiful hydrosilicon anions via electron attachment to neutral radicals of the same composition or otherwise via complex heavy-particle reactions leading to charge exchange between the species. Under such conditions the chemical polymerization pathway dominates the nucleation of primary nanoparticles.

Another interesting example is using reactive titanium tetrachloride plasmas for production of Ti nanoparticles in arc discharges [240]. This process has long been used commercially. For example, the Tioxide company in the United Kingdom has used DC-arc gas heaters to produce TiO_2 pigment with nanoparticle sizes ranging from 200–400 nm [5]. We emphasize that reactive $TiCl_4$ gas is used in this process as a source of titanium atoms.

What if the plasma is not reactive and polymerization processes similar to those considered in the previous subsection are not possible or inefficient? Will the nanoparticle nucleation still be possible in this essentially non-reactive environment or an environment with a substantially weaker reactivity? The answer is yes, and of course everything is subject to prevailing process conditions. However, the strategy in such cases would be quite different. A viable option is to introduce a sufficient amount of the species of the desired solid material externally and create appropriate conditions for their condensation and nucleation. Such conditions should be chosen to ensure sufficient rates of collisions between the externally introduced species; as a result of such collisions nanoclusters of the solid material in question may be formed. In this regard, reasonably high densities of background (inert) gas is certainly an advantage. However, the precursor species of the solid nanoparticle should predominantly collide with each other and not with atoms and molecules of the background gas.

If we deal with plasmas of inert gases, there is no other way to create solid precursors other than to introduce them externally. For instance, atoms of some solid (metal, semiconductor and dielectric) materials can be evaporated from the solid state using intense electron beams (e-beam evaporation) or heating to very high temperatures (thermal evaporation). There are numerous other options for externally introducing nanoparticle precursors such as various modifications of sputtering (e.g., DC/RF/pulsed sputtering), current-assisted evaporation, ablation of solid targets as a result of action of intense beams of pulsed lasers, dispersion of ultra-small powders throughout the volume of the plasma reactor, injection of liquid droplets and several others. In most of these cases, nanoparticle precursors are created in atomic/molecular/radical forms with a reasonably small number of atoms. For example, titanium dioxide nanoparticles can also be produced by using liquid thermal precursors injected into thermal plasmas [5]. In all these situations, if the background conditions are appropriate, externally introduced precursors may condense and form a small nanocluster or a nucleus.

The above critical clusters and nuclei trigger spontaneous growth of the new solid phase from, for example, externally introduced gas-phase precursors. As we have already mentioned above, critical clusters/nuclei are formed through the process of nucleation, which can be described by the dynamic balance of clustering (condensation) of atoms or molecules from the gas phase and their re-evaporation back to the gas phase. Similar to chemical reactions, the difference between the free energies of the vapor and solid phases (the thermodynamic barrier) need to be overcome for the nucleation process to proceed. The nucleation rate ξ_n can be expressed as [241]

$$\xi_n = \xi_0 \exp(-\Delta\Phi^*/k_B T_g), \tag{4.10}$$

where k_B is Boltzmann's constant, T_g is the temperature of the gas-vapor mixture and ξ_0 is the so-called kinetic pre-factor. From Equation (4.10), it is seen that the nucleation rate critically depends on the free energy $\Delta\Phi^*$ required to overcome the above mentioned thermodynamic barrier. Apparently, the lower the barriers are, the easier it is to create the nuclei.

Such barriers can be lowered by the presence or active participation of external species or other objects such as surfaces. Indeed, it is common knowledge that it is often easier for small particles to nucleate on solid surfaces than in the gas phase. Some particles such as ions, species in excited states, clusters and small nanoparticles are known to be able to lower the thermodynamic barrier for nucleation and hence, increase the nuclei formation rates (4.10). Of particular relevance to plasma-based processes is to stress that ions reduce $\Delta\Phi^*$ and can thus substantially increase the nucleation rates. The reason behind this amazing phenomenon is that the ions, through their electric fields, can affect electric charge distribution within interacting atoms/molecules. Therefore, the presence of ions in a neutral gas can lead to induced strong dipole interactions between otherwise uncharged and non-polarized atoms/molecules.

This ion-related effect on nanoparticle nucleation can be quantified using the following expression for the free energy of formation of a cluster which has volume v and contains g molecules [241]

$$\Phi(g) = a_1 R^3 + a_2 R^2 + (\varepsilon_0/2) \left[\int_v EDdv - \int_v E^2 dv \right], \tag{4.11}$$

where R is the radius of the cluster, E is the magnitude of the electric field (e.g., in a plasma sheath), D is the induction field in the cluster, v is the cluster's volume, and ε_0 is the permittivity of free space. The first term in Equation (4.11) scales as $\sim R^3$ and quantifies the contribution of the bulk to the free energy. The coefficient a_1 is proportional to the vapor temperature and supersaturation. The latter is a measure of degree of non-equilibrium of the vapor-solid system and shows the partial pressure/concentration of the actual vapor relative to the equilibrium vapor pressure of the material in question under the given temperature. The second term in Equation (4.11) describes the surface contribution to the free energy; the coefficient $a_2 \approx 4\pi\sigma$, where σ is the surface tension. The last term in Equation (4.11) describes the effect of the electric field [241].

Using a simple Thompson's model which treats the ion as a uniformly charged sphere with charge q and located at the center of a cluster, Equation (4.11) can be simplified to

$$\Phi(g) = a_1 R^3 + a_2 R^2 + \frac{q^2}{8\pi\varepsilon_0}\left(1 - \frac{1}{\varepsilon}\right)\left[\frac{1}{R} - \frac{1}{r_i}\right], \qquad (4.12)$$

where ε is a dielectric constant of the condensing material, and r_i is the ion's radius. From Equation (4.12) one can immediately notice that since $\varepsilon > 1$ and $r_i < R$, the electric field-related addition to the free energy Φ is negative. Therefore, polarization-related effects lead to lower free energy of cluster formation. In simpler terms, the associated energy barriers may be substantially reduced and the nucleation rates significantly increased by the presence of ions.

Interestingly, the radius of a critical cluster R_{crit} nucleated in the presence of the plasma-related effects such as ions and/or electric field can be calculated as [241]

$$R_{crit} = R_0\left[1 - \left(\frac{R_1}{R_0}\right)^3\right], \qquad (4.13)$$

where R_0 is the radius of a critical cluster without the presence of the electric field and calculated using a classical nucleation theory and R_1 is the Raleigh radius which is determined by competition of polarization and surface tension effects. This radius scales as $\sim q/\sqrt{\sigma}$. Note that $R_1 < R_0$; therefore, the critical cluster size for ion-induced nucleation is always smaller than the critical cluster size for homogeneous nucleation at the same supersaturation [241]. In simple terms, in the presence of ions and/or electric fields, smaller clusters can induce pronounced nucleation in a vapor-gas mixture compared to the neutral gas case.

As stated in the original report [241], the differences between the ion-induced and homogeneous nucleation include the existence of a local minimum in the free energy of cluter formation, as well as the substantially increased collision cross-sections of the cluster ion caused by the polarizability and permanent dipole moment of the vapor-phase molecules. A very interesting conclusion that follows from this work is that the kinetics of vapor molecule condensation triggered by positively and negatively charged ions may be quite different, and is frequently referred to as the "ion-sign" effect. This effect plays a role when the electric field near a cluster ion is not spherically symmetric. In this case the dipole and higher-order moments of the cluster ion charge distribution provide an additional contribution to the electrostatic term of the free energy of cluster formation.

More importantly, the multipole expansion model [241] suggests a preference for nucleation on positive ions; this effect is expected to strengthen as the critical cluster size decreases. Therefore, *positive* ions have a higer chance of triggering nucleation in a vapor-gas mixture compared to anions. In fact, this result is quite different from the conclusions from the previous subsection, where negatively charged hydrosilicon radicals SiH_2^- and SiH_3^- triggered polymerization and nucleation of hydrogenated silicon nanoparticles in silane-based plasmas.

Another important point to mention is that the rates of ion-induced nucleation are strongly affected by the dipole moments of the molecules involved. In fact, dipole moments determine collision cross-sections of the ions with vapor molecules. Interestingly, this effect appears to be stronger at *lower temperatures* and may play an important role in stellar environments and interstellar medium.

Figure 4.12 shows a sequence of events involved in dust formation in the environments of relatively cool red giant stars. Some details of this process have already been discussed in Section 1.3 describing nature's nanofab. Slide 1 in Figure 4.12 shows that in the vicinity of the star the ionization degree is very high, approaching 100%, which is consistent with the schematic representation of nature's nanofab in Figure 1.7. In this area the temperatures are far too high for nucleation to occur. Hot gas of carbon atoms, created as a result of astronucleosynthesis processes discussed in detail in Chapter 1, expands into the stellar envelope areas, where the gas densities, ionization degrees, and background gas (hydrogen) temperatures are significantly lower.

In slides 2–4 the ionization degrees are 29, 2 and 0.001%, respectively. As the carbon gas cloud expands away from the stellar core, conditions for homogeneous nucleation of carbon nanoparticles become less favorable. Indeed, when the temperature of the medium and carbon atom densities are reasonably high, carbon nanoparticle nucleation may proceed through a homogeneous nucleation scenario. Under these conditions the contribution of the ion-induced nucleation channel can be smaller compared to the homogeneous nucleation. Indeed, the ratio of the rates of the ion-induced ς_n^{ion} and homogeneous ς_n^h nucleation is [241]

$$\frac{\varsigma_n^{ion}}{\varsigma_n^h} \approx \frac{n_i}{N_n} \exp\left(\frac{\Delta\Phi_h^* - \Delta\Phi^*}{k_B T_g}\right), \tag{4.14}$$

where n_i and N_n are the densities of ions and neutrals, respectively, and $\Delta\Phi_h^*$ is the free energy of the critical cluster for homogeneous nucleation. Usually, the number densities of ions are much lower than

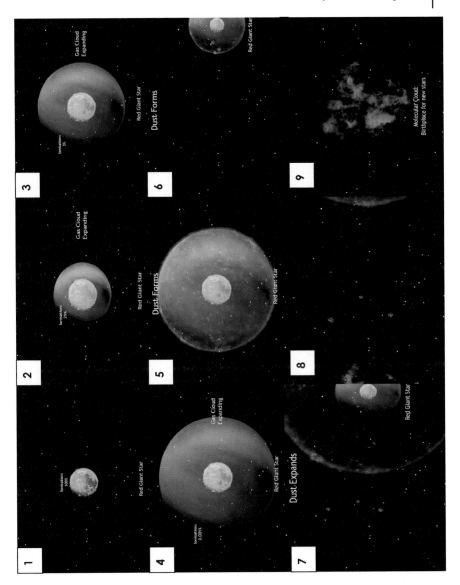

Figure 4.12 Different stages of dust formation in envelopes of relatively cool red giant stars.

those of neutrals, hence, the ratio $\zeta_n^{\text{ion}}/\zeta_n^{\text{h}} < 1$. However, as the temperatures decrease, the effect of ions on the nucleation process becomes much stronger. Thus, ion-induced nucleation effectively comes into play when the environment conditions become unfavorable for homogeneous nucleation.

Returning to Figure 4.12, we stress that dust nanoparticles nucleate where the environment (appropriate temperatures, ionization degree, level of exposure to UV radiation, and so on) is most suitable. This is sketched in slides 5 and 6 in Figure 4.12. The as-formed nanoparticle nuclei continue their growth in a free molecular regime by attachment of carbon atoms from the environment. This process is quite long and it may take up to 10^{19}–10^{21} s to reach micron sizes [242].

Thereafter, dust expands into the preriphery of stellar envelopes (slide 7 in Figure 4.12) and then into interstellar space (slide 8 in Figure 4.12). As mentioned in Chapter 1, dust particles provide small and reactive soild surfaces which are used to produce molecular hydrogen in interstellar medium. Eventually, this process contributes to the formation of molecular "dusty" clouds, from which new stars develop (slide 9 in Figure 4.12).

Thus, cosmic dust is used by nature's nanofab to create new stars far removed from where the dust itself is created. One can also speculate that the exact amount of dust created and expelled in stellar outflows into the interstellar medium is determined by the actual need to create new stars which are capable of creating more dust. In other words, nature's nanofabs reproduce themselves! As some of the nanofabs become inactive and no longer produce carbon atoms via nucleosynthesis, the other, more recently created, nanofabs take over the process and continue producing the required amounts of dust particles. Moreover, the amounts of such particles can be precisely controlled by creating (e.g., via UV irradiation) appropriate amounts of ions that significantly accelerate the nucleation process. This is a clear manifestation of nature's determinism (please refer to Chapter 1 for a detailed discussion of related issues) in reproducing its basic bits which can further reproduce themselves. And most amazingly, these basic bits are dust-generating stars – nature's nanofabs!

Interestingly, ion-induced nucleation plays an important role in the formation of small nanoclusters even in outer areas of stellar envelopes, where temperatures can be below 65 K. Simple estimates [243] suggest that under such temperatures and densities carbon atoms of approximately 1 atom per cubic meter (which is a typical value for interstellar medium), carbon supersaturation is huge and it is extremely unlikely that the dust nuclei formation can proceed via the conventional "macroscopic" condensation pathway. In this case, only small clusters can be formed as a result of collisions of carbon atoms in the presence of hydrogen. However, this process is extremely slow. Indeed, at $T_g = 10\,\text{K}$, $n_C = 1\,\text{atom}/\text{m}^3$ and $n_H = 100\,\text{atom}/\text{m}^3$, it may take up to 10^{31} s just to form one carbon dimer per cubic meter. A shorter but also very long period of time would be required to form a hydrocarbon radical CH. The latter species can be created as a result of collisions of carbon and hydrogen atoms in the presence of another hydrogen atom (three-body collisions). However, if a carbon atom is ionized, the rate of formation of such radicals can be increased by up to 15 orders of magnitude! Thus, ion–atom collisions significantly reduce the time necessary for the formation of carbon-containing species, which may further develop into larger molecules and eventually nucleate via a chemical pathway similar to what was discussed in the previous subsection for silicon nanoparticle nucleation in silane-based plasmas.

Nanoparticle nucleation can be strongly affected by external electromagnetic radiation such as light emissions from neighboring stars or laser radiation in laboratory experiments. In such cases, the effect of photo-induced nucleation can be quite significant. There are two main channels of photonucleation kinetics [244]: (i) direct heat impact of the incident radiation on the cluster of a condensed phase; (ii) resonant excitation (and possibly ionization/dissociation) of vapor molecules. It is interesting that the first channel usually leads to reduced rates of cluster nucleation, whereas the second channel results in enhanced rates of nanocluster formation.

It is interesting that quite similar conclusions regarding the effect of ions on nanoparticle nucleation also hold true in the case of pulsed laser ablation of solid targets. In this case the densities of the species involved are many orders of magnitude higher than in the astrophysical environment considered above. In fact, in laser ablation of Si targets in a low-pressure ambient using a pulsed 532 nm Nd:YAG laser radiation with a pulse duration of 8 ns, and intensities ranging from 0.05 to 5 GW/cm^2, typical densities of ablated silicon atoms and ions can reach ca. 3×10^{17} cm^{-3} and ca. 10^{20} cm^{-3}, respectively [44]. We recall that in the previous example in this subsection related to dust creation in the interstellar medium, the density of carbon atoms was of the order of just a few atoms per cubic meter! The plasma created in a laser ablation plume featured the electron temperature in the 0.5–1.0 eV range. Under such conditions, 5–50 nm-sized nanoclusters were synthesized [44].

Here we stress that despite a huge difference in process conditions, ion-induced nucleation also makes a major contribution even in the case of very dense plasmas of laser ablation plumes. Figure 4.13 shows the effect of ions on the Gibbs free energy for different numbers of atoms in nanoclusters. For illustration purposes, the ionization degree is assumed to be approximately 1%. From the upper curve in Figure 4.13, plotted without the presence of ions, one can see that the maximum of the free energy at ca. $24.5 k_B T_g$ corresponds to Si nanoclusters of approximately

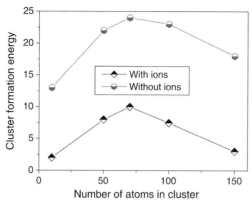

Figure 4.13 Free energy of silicon cluster formation (in units of $k_B T_g$) at gas temperature of 2000 K and $n_{Si} \sim 10^{20}$ cm^{-3} (upper curve) compared with the same conditions but with 1% ionization (lower curve) [44].

70 atoms. Introducing only one silicon ion among 100 atoms makes a dramatic difference: the peak of the Gibbs free energy reduces by 2.45 times and equals to only ca. $10k_B T_g$ for the same number of atoms! More importantly, the rates of nucleation can be significantly higher owing to the exponential dependence of the nucleation rates on the energy barrier for cluster formation in Figure 4.13.

The explanation given by Tillack *et al.* [44] is also quite similar to what we have used in the previous astrophysical example. Indeed, charged ions tend to be jacketed by surrounding neutral atoms/molecules. The electric field created by the ions causes polarization of the surrounding neutral atoms/molecules, which can now rearrange and attract to the ions. As a result, local minimum energy states, corresponding to these jacketed ions, are formed. Consequently, the critical radius, beyond which growth becomes energetically preferred, shifts to a lower value. Moreover, the nucleation energy barrier is lowered, which leads to significantly reduced rates of nucleation at the same process temperatures [44].

Another interesting consequence of the action of plasma-related effects is significantly improved size uniformity of silicon nanoclusters synthesized by pulsed laser ablation and deposited on single-crystalline Si wafers. Cluster size histograms for different intensities of laser radiation (under otherwise identical conditions) are shown in Figure 4.14 [44]. A striking observation from this figure is that an increase in the intensity of the laser beam led to a much narrower distribution in the sizes of silicon nanoclusters. In particular, as can be seen from Figures 4.14(b) and (c), nanoclusters with sizes exceeding 35 nm can be completely sup-

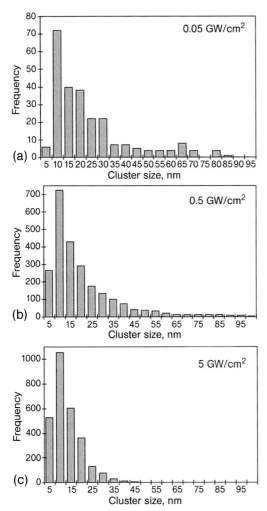

Figure 4.14 Size distributions of silicon nanoclusters
produced by pulsed laser ablation at different laser intensities:
(a) $0.05\,GW/cm^2$; (b) $0.5\,GW/cm^2$; and (c) $5\,GW/cm^2$ [44].

pressed by increasing the intensity of the laser from 0.5 to $5\,GW/cm^2$.
Higher laser intensity usually leads to higher plasma densities. Thus,
higher ion densities not only result in higher nucleation rates but also
in significantly more mono-disperse distributions of nanocluster sizes.
These conclusions are consistent with the modeling results of Tillack *et
al.* [44] that rigorously take into account the ion-enhanced nucleation and
with the experimental observations of other authors [245].

Therefore, plasma ions increase nucleation rates, which eventually results in a larger number of smaller nanoparticles in the final condensate

distribution. In fact, the ion presence causes a notable shift of the homogeneous nucleation pathway towards the heterogeneous mechanism wherein the ions act as condensation centers [44]. This interesting case shows that the presence of ions is central to nanocluster formation in laser plasmas and, more importantly, clearly shows advantages of using plasma environments in synthesizinig size-uniform ultra-small nanoclusters and nanoparticles.

Discussing the nucleation of carbonaceous dust, we have mentioned possible chemical pathways for further development of carbon nanoparticles. Below we will briefly discuss some of the main issues involved in the process of nucleation of such nanoparticles in low-temperature plasmas of acetylene discharges [224,246]. Sophisticated mass spectrometry experiments and detailed plasma modeling studies suggest that in this case carbon nanoparticcles may nucleate via quite different channels. Interestingly, both positive and negative ions may participate as precursors of the initial (polymerization) stage of particle growth in acetylene discharges. The positive ion pathway begins with $C_2H_2^+$ molecular ions and, after insertion of acetylene molecules and release of hydrogen atoms, leads to the formation of even numbered hydrocarbons $C_{2n}H_m$. The principal positive ion-triggered condensation pathway can be presented as follows

$$C_2H_2^+ \rightarrow C_4H_2^+ \rightarrow C_6H_4^+ \rightarrow C_8H_6^+ \rightarrow C_{10}H_6^+ \rightarrow \dots, \qquad (4.15)$$

where each step proceeds via addition of C_2H_2 and H_2 is released after the first and the fourth reaction steps.

On the other hand, the negative-ion supported reaction pathway is triggered by the primary C_2H^- anions that are formed through the electron impact dissociative attachment to acetylene molecules. These primary anions can trigger the following chain of polymerization reactions

$$C_2H^- + C_2H_2 \rightarrow C_4H^- + H_2$$

$$C_4H^- + C_2H_2 \rightarrow C_6H^- + H_2$$

$$\dots$$

$$C_{2n}H^- + C_2H_2 \rightarrow C_{2n+2}H^- + H_2$$

which is commonly referred to as the Winchester mechanism [4]. One can see that the repeated insertion of C_2H_2 molecules results in an anion sequence with nearly pure carbon anions that peaks at the $C_{2n}H^-$ species [224,246].

The results of comparison of the efficiency of both positive and negative ion pathways for nucleation of carbon nanoparticles in acetylene

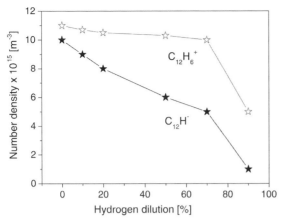

Figure 4.15 Number densities of highest polymerized positive and negative ions *versus* molecular hydrogen dilution [246]. Both positive and negative ion pathways give comparable densities of primary particles.

plasmas are shown in Figure 4.15 [246]. In this figure, the computed bulk densities of the highest polymerized ions of the above two pathways $C_{12}H_6^+$ and $C_{12}H^-$ are plotted as a function of hydrogen dilution. One can clearly see that for the same discharge conditions, a higher inflow of molecular hydrogen reduces the number densities of the primary nuclei. More importantly, both polymerization channels appear to be quite competitive and the concentrations of the $C_{12}H_6^+$ and $C_{12}H^-$ species always remain of the same order of magnitude. Thus, both positive and negative ions can be regarded as effective triggers of carbon nanoparticle nucleation in acetylene discharges. This appears to be quite different from the case of the silane-based plasmas of the previous subsection, where anionic pathways appear to be dominant.

Moreover, the clearly observed suppression of higher-mass hydrocarbons by increased hydrogen inflows can be used to control the rates of nanoparticle production in acetylene-based plasma discharges. This trend has also been observed experimentally under quite similar conditions [247]. Hydrogen can thus act as an inhibitor on the successive polymerization reactions and may thus be used as an effective suppression technique of undesirable dust contaminants in certain technological applications [248].

To conclude this section, we stress that we have discussed only a few examples of nanoparticle nucleation in various plasma environments. However, the chosen range of plasma environments encompasses dense and highly-reactive silane-based plasmas in one extreme and very rarefied plasmas in the interstellar medium in another one. We have also considered cases of very dense plasmas of laser ablated silicon material and also low-temperature RF plasmas of acetylene discharges. In every case considered, plasma-specific effects play a prominent role and in most cases make the nanoparticle production processes better and more efficient compared with neutral gas-based techniques with very similar operating conditions.

4.4
Concluding Remarks

In this chapter, we have considered an integrated approach for generation of specific species (e.g., building or working units) in various plasma environments ranging from extremely rarefied interstellar space to dense and very reactive nanoparticle-generating silane-based RF plasma discharges. In every particular situation, a target is set to optimize the generation of the desired species and suppress the growth of the unwanted ones. Basic requirements for the appropriate choice of the species (which, broadly defined, include atoms, molecules, radicals, macromolecules, clusters and nanoparticles) follow from the actual needs of the attempted nanoassembly processes. Some examples related to this choice have already been considered in Chapters 2 and 3. Once this choice is made, one has to decide exactly how to generate the required species in the ionized gas phase and also to ensure that their rates of delivery to the growth surface are appropriate. The key element in this regard is to properly choose the plasma environment and the way to generate such plasmas. Most of the discussion in this chapter has been centered around plasma-assisted synthesis of carbon-based nanostructures. Therefore, using methane- or acetylene-based plasma discharges would be the right choice in terms of providing carbon atoms – the most basic building units of carbon nanostructures. The first two sections were devoted to generation of atomic, molecular and radical species from methane- and acetylene-based plasmas, respectively. In Section 4.3, we considered a range of other examples of generation of nanoclusters and nanoparticles in diverse plasma environments and related basic processes such as ion-induced nucleation.

All the processes involved in the above examples occur in the ionized gas phase and lead to formation of the entire range of nanoassembly building and working units. Smaller species such as radicals are usually formed through different channels, with the most important of them being electron-impact dissociation and various heavy-particle collisions. These elementary processes may involve charge transfer; consequently, the radical species generated may be charge neutral of have either a positive or negative charge. On the other hand, clusters and nanoparticles are generated via association of atoms, molecules and radicals. This association can proceed via chemical pathways such as with the polymerization of silicon and carbon primary particles (initial nuclei) or through anionic and cationic pathways considered in Section 4.3. Smaller species can also condense to form larger nuclei; nucleation of metal clusters in plasmas of inert gases is a good example to this effect.

To understand the range of the processes involved and, hence, the outcomes in terms of generation of the desired species, numerical modeling and diagnostics are indispensable. The first approach to this problem can be undertaken using spatially-averaged (also commonly termed "global") models, which, despite their apparent simplicity, enable one to calculate the spatially-averaged number densities and fluxes of most important species in the plasma discharge. An example of using this approach to work out the abundance and possible roles of a range of cationic and neutral radical species in methane-based plasmas used in the assembly of arrays of sharp conical carbon nanostructures was presented in Section 4.1. The outcomes of numerical modeling were found to be in agreement with the results of low-temperature plasma diagnostics using optical emission spectroscopy (OES) and quadrupole mass spectrometry (QMS) of inductively coupled plasmas in $Ar + CH_4 + H_2$ gas mixtures. The global model of the inductively coupled plasmas made it possible to estimate the densities and fluxes of the radical neutrals and charged species, the effective electron temperature and the methane conversion factor under various growth conditions.

One of the most interesting conclusions following from Section 4.1 is that the rates of delivery of the nanostructure building units can be higher if they are deposited in their ionic, charged, form. Indeed, it was found that the fluxes of cationic radicals in the plasma enhanced chemical vapor deposition of carbon nanostructures often exceed those of the radical neutrals. This conclusion certainly favors extensive use of ion-based techniques in nanofabrication. More examples of successful applications of ion fluxes to fabricate a range of nanostructures is discussed in the following chapters of this monograph.

It would appear that the main limitation of spatially-averaged plasma models is in their inability to predict distributions of the species densities and fluxes in different areas of the discharge. Unfortunately, in most cases low-temperature discharge plasmas are very non-uniform throughout the volume they are sustained. This causes a problem when one aims to process or synthesize nanoscale-objects on large-area surfaces immersed in a plasma. Indeed, non-uniform deposition of specific building or working units over the area where the nanoassembly takes place may lead to location-dependent properties of individual nanostructures and their microscopic arrays. Therefore, discharge modeling and diagnostics should be space-resolved. Moreover, if one deals with non-stationary plasmas (e.g., pulsed discharges or afterglows), it is essential to know the temporal dependence of all the main characteristics involved.

Even though reactive plasmas can indeed be considered as a versatile nanofabrication tool [4], large-area synthesis of size- and property-uniform nanoassemblies is still quite challenging, especially if a plasma reactor and other process parameters are not properly chosen. If this is the case, the quality of the fabricated nanopatterns may be far below the quality expectations due to intrinsically non-uniform distribution of plasma densities and surface fluxes over large-area substrates. In the example of Section 4.2, spatially non-uniform inductively coupled plasmas were considered. This non-uniformity was created by an external inductive coil with a small number of turns. The uniformity of densities and surface fluxes of the main precursor species can be substantially improved by altering the reactor configuration or changing process conditions. One such effective way is to increase the number of coil turns to 17 and lower the operation frequency to ca. 460 kHz [189].

The "global" approach used in Section 4.1 has been extended to account for different conditions and rates of numerous elementary processes in different locations throughout the discharge chamber. More specifically, in Section 4.2 we have considered the problem of computing two-dimensional profiles of the major building and working units that may participate in the nanoassembly of carbon-based nanopatterns in reactive environments of $Ar + H_2 + C_2H_2$ gas mixtures. By using a two-dimensional fluid model, spatial distributions of the electron temperature, electrostatic potential and densities of the nanoassembly precursor species have been computed in a realistic plasma-aided nanofabrication facility with a large-area (with 10 cm radius) substrate stage [54]. In this facility, an RF discharge is sustained by means of inductive RF power coupling from a 4-turn flat spiral antenna at the top of the reactor chamber. In the chosen discharge configuration the distribution of the species densities and their incoming fluxes over the large substrate areas appears quite non-uniform.

Furthermore, the main discharge species have been separated by their functions: (i) Ar^+ and H as working units that serve for surface preparation and (ii) CH, C_2, and CH_3^+ carbon-bearing building units. In this way, it is possible to predict the effect of non-uniformities of fluxes of such species on nanofabrication of functional nanopatterns on large-area solid substrates and also elucidate how the variations of some of the process parameters (such as working pressure and composition of the reactive gas feedstock) can be used to mediate the effects of the plasma non-uniformities [54].

The results presented in Section 4.2 suggest that the "cause and effect" microscopic approach [4] (see also Chapter 2) and the detailed study of two-dimensional fluxes of the selected species towards specific areas on the substrate surface may lead to better control in the growth of nanostructures from reactive acetylene-based plasmas. One of the most amazing conclusions highlighted in Section 4.2 is that *quite similar surface preparation conditions can be achieved in different surface areas despite significant differencies in the working unit fluxes in these areas*. Thus, even though the BU/WU fluxes are indeed very non-uniform, in some cases their effect in different surface areas can be precisely engineered to prepare the substrate for uniform nanoassembly over its entire surface.

Examples of surface preparation are certainly not limited to those counteracting and concurrent dynamic processes of activation and passivation considered in Section 4.2 and may include etching, functionalization, doping, roughening, catalyst deposition, micropore formation, and so on in selected areas of large-area substrates. The original report [54] is just the beginning of a variety of possibilities to handle and gainfully use intrinsic plasma non-uniformities. The consideration in Section 4.2 leads us to the understanding that the key to achieving this goal is to properly identify, generate and manipulate the nanoassembly precursor and surface preparation species in every specific nanoassembly area. By doing so, one can dramatically improve the process outcomes.

Section 4.3 was devoted to various possibilities and issues related to nanocluster and nanoparticle synthesis in different environments. Without attempting to recap and summarize all the details given in Section 4.3, here we stress that plasma-based environments are truly unique and effective for nanocluster and nanoparticle synthesis. Some of the most interesting points to highlight in this regard are as follows:

- In a plasma, there are various pathways for nanocluster and nanoparticle nucleation due to an abundance of a large number of species in different energetic and charging states.

- There is a large number of effective options for physical and chemical synthesis of nanoclusters and nanoparticles which are unavailable in neutral gas environments. For instance, cationic and anionic routes to carbon and silicon nanoparticle formation are very inefficient in the bulk of the gas in thermal CVD systems where charged species can only be formed on hot surfaces.

- The presence of an additional ionized component introduces a range of unique ion-assisted processes such as ion-induced clustering/nucleation, ion-enhanced crystallization, and so on. These processes are involved in a broad range of phenomena spanning from ion-assisted nucleation of carbonaceous dust in stellar envelopes to nanoparticle production in ion fluxes generated in i-PVD, pulsed laser ablation and PECVD systems.

- Using ions as primary building units of nanoclusters/nanoparticles (which on their own can be building units of nanocluster/nanoparticle-made films) offers additional benefits in terms of effective control of nanoparticle shape, composition, crystallinity, structure, and so on These options will be considered in subsequent chapters of this monograph.

- Diverse gas temperature regimes in different plasmas (e.g., thermal or thermally non-equilibrium low-temperature plasmas) can be selectively used for *in situ* nanoparticle shaping, crystallization, and so on.

- Because of the broad range of available species and strong gas-phase dissociation processes, as well as several other factors, plasma-based environments offer very competitive rates of nanoparticle production.

To conclude this chapter, we stress that due to the overwhelming variety of abundant species and different growth pathways and options it is extremely important to choose the right plasma for any particular purpose. Generally speaking, if one desires very large quantities of spherical nanoparticles, relatively high-temperature environments of thermal plasmas of arc discharges would certainly be a viable option. On the other hand, thermally non-equilibrium low-temperature plasmas of radiofrequency discharges are more appropriate to synthesize surface-bound polymeric nanoparticles of complex shapes. For more details, the reader is referred to a recent review article [5]. We will continue discussing these issues throughout this monograph. Finally, low-temperature plasmas may and should be used to generate building and working units required for nanoscale processes!

5
Transport, Manipulation and Deposition of Building and Working Units

In the previous chapters we have considered the basic approaches of plasma nanoscience and several specific cases when these approaches can be successfully applied. The fundamental "cause and effect" approach discussed in detail in Chapter 2 highlights different stages involved in nanoscale assembly using low-temperature plasmas. One of the most important steps in this direction is to select suitable building units and then generate them in approriate amounts. Various possibilities, including creation of numerous atomic, radical, molecular, nanocluster and even nanoparticle building units, have been discussed in Chapter 4.

The next step is to transport the BUs from the plasma bulk and deposit them, ideally with atomic precision, onto specified surface areas on nanostructured surfaces. This chapter specifically deals with this problem. To illustrate various possibilities in this regard, we have selected a few representative examples. The first example discussed in Section 5.1 deals with the study of the precise deposition of positively charged ionic building units onto the lateral surfaces of conical carbon nanostructures intended for electron field microemitter applications. Knowledge of distributions of microscopic ion fluxes over the micon-sized patterns of the nanocones is indispensable in devising plasma-based nanofabrication processes ranging from high-precision nanoarray synthesis and post-processing.

The next three sections of this chapter are devoted to the investigation of various possibilities for controlling and manipulating small nanoparticles in hydrocarbon-based plasmas used for PECVD of various carbon nanostructures. In these particular examples, the main aim is to prevent the nanoparticles from contaminating the nanocone arrays, or to find other ways to deposit them into specified local areas on microstructured surfaces, in a way quite similar to what is done with ionic species in Section 5.1. In Section 5.2, we consider the possibility of manipulating carbon nanoparticles using thermophoretic forces and elucidate conditions when they can overcome replusive electrostatic barriers and deposit onto nanostructured surfaces. Section 5.3 introduces numerical simulations which enable one to predict the exact locations of nanoparticles landing onto a rough microstructured surface. In Section 5.4, the option of using electrostatic filtering to almost completely prevent undesirable fallout of the plasma-grown nanoparticles onto nanopatterns being synthesized is introduced. As usual, this chapter concludes with concluding remarks in Section 5.5.

5.1
Microscopic Ion Fluxes During Nanoassembly Processes

The results discussed in this section are relevant to the plasma en-hanced chemical vapor deposition of ordered large-area nanopatterns of vertically aligned carbon nanotips, nanofibers and nanopyramidal microemitter structures for flat panel display applications. Here we will follow the original report [51] and present the results of numeri-cal simulations of the three-dimensional topography of microscopic ion fluxes in the reactive hydrocarbon-based plasma-aided nanofabrication of ordered arrays of vertically aligned single crystalline carbon nanotip microemitter structures.

Let us begin the consideration by discussing the relevance of this prob-lem to the recent developments in the field of electron microemitter ar-rays. Surface-conduction electron-emitter display (SED) technology is one of the most recent advances in large flat-panel display manufactur-ing, a market area with currently high annual growth rates and a very positive long term prognosis [249].

Field-emission display (FED) technology, one of the derivatives of the SED, is based on ordered arrays of microemitters, each producing an in-dividual pixel. The combined multipixel emission from the microemitter array is expected to generate exotic color patterns with millions of in-dividual colors and superior color reproduction and darkness contrast. The FED technology is one of the most promising advances in the man-ufacturing of ultra-thin large-area flat-panel displays [51]. The emitter material usually contains ordered patterns of electron emitters on its sur-face.

Various quasi-one-dimensional carbon-based nanostructures, such as carbon nanofibers, nanopyramidal structures, multiwalled carbon nan-otubes and several others, are promising for microemitter applications owing to their outstanding size-dependent electronic properties, geo-metric field-enhancing factors such as high geometric aspect ratio and excellent conductivity [63, 250–254] and have recently been commonly recognized as an important alternative to conventional electronic materials [51].

Plasma-enhanced chemical vapor deposition (PECVD) is one of the most efficient and precise tools for the fabrication of the ordered arrays of carbon-based microemitters [21,89,90,170,225,227,228,255–257]. Recently, the possibility of plasma-aided nanofabrication of large-area ordered patterns of single-crystalline carbon nanotips [258], which are high-aspect-ratio nano-sized conical structures, has been demonstrated. These nanostructures have very good field emitting properties, which are characterized by high emission currents at low turn-on voltages [93].

However, relatively low emission currents, poor coherence in the emission pattern and difficulties in the integration into very-large-scale-integrated silicon-based microdevices, impede the overall progress in the carbon nanostructure (CNS)-based nanodevice fabrication and remain a major challenge [51]. Relevant problems can, for example, arise due to imperfections in the structure, sizes, orientations, alignment and ordering of the CNS patterns, and these become critical when upscaling the nanostructured patterns to larger surface areas. It is thus imperative to be able to control the plasma-aided nanofabrication process to ensure that the device quality of the individual nanostructures and their ordered arrays meets the continuously rising standards [4,5].

The main difficulty in this regard is to choose the ions that are best suited for the purpose. Fortunately, recent results of *ab initio* local density approximation-based density functional theory computations of the equilibrium chemical structure of the single-crystalline carbon nanotip microemitters revealed a crucial role of CH_3^+ cationic building units (BUs) which form an equilibrium carbon network peripherally terminated by hydrogen atoms [258,259].

Therefore, in nanoscale fabrication of such nanotip microemitters one must be able to control the distribution of ionic fluxes in the immediate vicinity (typically at distances $\leq 100\,nm$) of the nanotip surfaces required to maintain a high degree of three-dimensional positional uniformity of the growth pattern. Moreover, it is highly desirable to find ways to deposit ions with controlled energies and fluxes onto specified areas on lateral surfaces of individual nanocones.

In this section we use advanced numerical simulations to quantify the microscopic ion fluxes in the synthesis of carbon nanocones using low-temperature plasmas of $Ar + H_2 + CH_4$ or $Ar + H_2 + C_2H_2$ gas mixtures [1,93–95,258]. In the above experiments, the nanotips/nanocones were vertically aligned and typically had a base radius varying from 10–100 nm, and a heights from 100 nm–1 μm.

An interesting observation from the experiments is that the field emission characteristics strongly depend on the CNS geometrical parameters and their positioning in nanoarrays. Generally speaking, rarefied arrays of higher-aspect-ratio nanocones usually show stronger and more coherent field emission. Another important conclusion from the experiments [1, 53, 93–95, 258] is that under high-density conditions in Ar + H_2 + CH_4 plasmas, ion fluxes can even be stronger than the fluxes of neutral radical species. This conclusion has been highlighted in Section 4.1 and serves as an additional motivation to study ion-based nanoassembly processes in great detail.

In this section, we will follow the original report [51] and study the microscopic topology of the ion current distribution in the vicinity of ordered patterns of carbon nanotips/nanocones. In the case considered, the ion energy is controlled by the substrate bias, which is negative with respect to the plasma bulk and varies from a few volts for floating substrates to a few hundred volts for externally biased substrates. Using advanced Monte Carlo (MC) numerical simulations of the ion current distribution in the ordered carbon nanotip structures, we will show effective ways to control distributions of microscopic ion fluxes over the nanotip surfaces using the DC substrate bias, ion density and electron temperature.

5.1.1
Formulation and Model

In this section we consider an environment which consists of the plasma bulk and a biased substrate, on which carbon nanotip-like structures are grown (Figure 3.13, left panel). For simplicity, only a single cationic species CH_3^+ (assumed the main BU of the nanostructures in question, as discussed above) is considered in computations. A thin sheath separates the substrate and the plasma bulk. As the potential drop in the pre-sheath area is rather small (e.g., in the plasma facility [189] used for the PECVD of CNSs [93,94] it is ca. $T_e/2 \sim$1–2.5 eV, where T_e is the electron temperature), only the actual potential difference across the plasma sheath is taken into consideration [51]. The ions are assumed to enter the sheath at $z = \lambda_s$ with the Bohm velocity $V_B = (T_e/m_i)^{1/2}$, where m_i is the ion mass, and λ_s is the sheath thickness, and are accelerated in the sheath to an energy equal to the cross-sheath potential drop U_s.

When the cross-sheath potential drop is relatively high ($T_e \ll U_s$), one has an estimate for the sheath thickness [51]

$$\lambda_s = \frac{\sqrt{2}}{3}\lambda_D \left(\frac{2U_s}{T_e} \right)^{3/4}, \qquad (5.1)$$

where λ_D is the electron Debye length. On the other hand, when the external bias is low or the substrate is floating ($T_e \sim U_s$), it is quite accurate to assume that the sheath is of the order of a few Debye lengths

$$\lambda_s = \gamma_s \lambda_D = \gamma_s \left(\frac{\varepsilon_0 T_e}{e n_p} \right)^{1/2}, \tag{5.2}$$

where ε_0 is a dielectric constant, n_p is the plasma density (which in the single-ion-species approximation is either the electron n_e or ion n_i number density) and γ_s is a constant, typically in the range between 1 and 5 [68].

The microscopic topography of the ion current has been simulated for the nanotip pattern shown in Figure 2 of the original report [51]. The deposition surface is covered with an array of nanotips with the radius R varying from 20–70 nm (mean radius R_m is equal to 44 nm), height $h = 300$ nm and mean spacing (distance between adjacent nanostructures) $S_m = 100$ nm, representative to PECVD of carbon nanotip structures [49,93–95,258]. The distribution of the nanotip radii was Gaussian-inspired with the maximum at $R = R_m = 44$ nm. The entire simulated surface, including surfaces of the nanotips, is assumed conductive, which is appropriate for PECVD of carbon nanotips on nickel-catalyzed highly doped Si(100) substrates in the above mentioned experiments.

The electric field is strongly non-uniform in the vicinity of the nanostructures, and the ion motion is determined by the effective electric field $\mathbf{E}$ of the entire nanotip array. The ion motion in the field $\mathbf{E}$ is described by the following equations of motion

$$\mathbf{r}(t) = \mathbf{r}_0 + \int_{t_0}^{t} \left[\mathbf{v}_0 + \int_{r_0}^{0} \frac{\mathbf{E}(r,\tau)}{m_i} dr \right] d\tau, \tag{5.3}$$

where $\mathbf{r}(t)$ is the ion position vector, $\mathbf{r}_0$ is the initial position vector for an ion located at the sheath border $z = \lambda_s$, t_0 is the initial time moment of ion motion, and $\mathbf{v}_0 = V_B$ is the ion velocity at the sheath edge.

The surface charges on the nanotips have been calculated by approximating the nanotip surface as a hollow cylinder capped with a semi-sphere and using conventional electrodynamics for electrostatic potentials of the above structures. The effective electric field acting on an ion located at the space point $\mathbf{r}$ can be computed by integrating the surface charge of all nanotips in the pattern [51]

$$\mathbf{E}(\mathbf{r}) = \Sigma_{i=1}^{n} \int_{S_i} \frac{\sigma_i dS_i \mathbf{r}}{4\pi\varepsilon_0 r^3} + \mathbf{E}_S, \tag{5.4}$$

where σ_i is the surface density of electric charge on nanotip surfaces, S_i is the surface area of the i-th nanotip and $\mathbf{E}_S = -\nabla\phi$ is the electric field in the plasma sheath area.

The ion motion towards the nanopatterned substrate has been simulated in the simulation space bounded by the curved surface of the nanopattern and the plane $z = \lambda_s$. In the x and y directions (the substrate surface plane), the simulation space was bounded by the rectangular substrate area of $2.5 \times 2.5\,\mu m$ comprising 100 individual nanotips. The simulation set of the plasma and nanostructure parameters representative the low-frequency (ca. 460 kHz) inductively coupled plasma source [189] and the PECVD of CNSs [1,49,93–95,258] is summarized in Table 5.1. For further details of numerical method, boundary conditions and numerical methods the reader should refer to the original report [51].

Table 5.1 Parameters and representative values in computations [51].

Parameter	Notation	Value
Carbon nanotip radius	R	20–70 nm
Mean radius of nanotips	R_m	44 nm
Nanotip height	h	300 nm
Spacing between nanotips	S	50–150 nm
Mean spacing between nanotips	S_m	100 nm
Number of nanotips	N_{CNT}	100
Substrate dimensions	$l_s \times l_s$	$2.5 \times 2.5\,\mu m$
Cross-sheath potential drop	U_s	-20, −50 V
Mass of ionic building units	m_i	15 amu
Electron temperature	T_e	1.0–12.0 eV
Plasma density	n_p	10^{17}–$5 \times 10^{18}\,m^{-3}$
Number of ions in simulation	N_i	3×10^5
Relative enhancement of ion current density	ξ_i	1–5
Debye length	λ_D	8×10^{-4}–2×10^{-2} cm
Sheath constant	γ_s	1–5
Ion energy at the sheath edge	$\mathcal{E}_i$	1.0–12.0 eV
Distance between equidistant planes used for quantifying ion current distribution along lateral surfaces of nanotips	d_{lat}	6 nm
Magnitude of electric field	E	10^7 V/m
Ion velocity at the sheath edge	V_B	2.5–4×10^3 m/s

5.1.2
Numerical Results

Two distinctive cases of low and medium potential differences across the plasma sheath are considered. In these cases the potential drop is $U_s = 20$ and 50 V, respectively. Two typical ion current distributions onto the substrate surface are shown in Figure 5.1. From the figure on the left

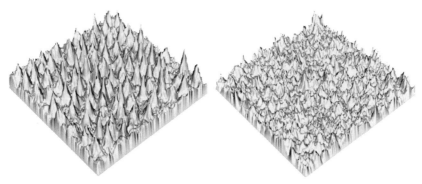

Figure 5.1 Representative three-dimensional topographies of ion
current distribution over the nanostructured surface for $U_s = 20$ V
(a) and $U_s = 50$ V (b). Other parameters are $n_p = 5 \times 10^{17}$ m^{-3}
and $T_e = 2.0$ eV [51].

in Figure 5.1, one sees that the ion current is strongly focused by the
nanotips (ion current spikes correspond to the nanotip positions).

Indeed, the electric field is non-uniform in the vicinity of sharp tips
and deflects the ions from straight trajectories causing their non-uniform
precipitation to the nanotip lateral surfaces in the areas closer to their
upper ends. We note that each ion from the ensemble is traced over an
individual trajectory with a randomly chosen starting point. This can be
regarded as one of the reasons for the stochastic-like pattern of the ion
current distribution. The figure on the left in Figure 5.1 shows a representative ion current distribution in the biased substrate case ($U_s = 50$ V).

A striking observation is that contrary to the narrow sheath case, the
ion current focusing is noticeably less effective. One can also see that
the ion current spikes at the nanotip positions become smaller and less
resolved compared to the results in the case shown in the figure on the
left. In the $U_s = 50$ V case the 3D topography of the ion current is also
quite stochastic and features numerous fluctuations in the inter-nanotip
gap areas [51].

The simulations make it possible to precisely record the z coordinates
of the points of the ion collisions with the nanotips. These data can
then be used to study the ion current density distribution along the nanotip lateral surfaces. To quantify this effect, each nanotip is split into 50
slices by equidistant parallel planes $z = kd_{lat}$, where $d_{lat} = 6$ nm and
$k = 1 \ldots 49$ is an integer. Accordingly, all the ions deposited on the nanotip lateral surfaces (excluding the ions precipitated to the substrate sur-

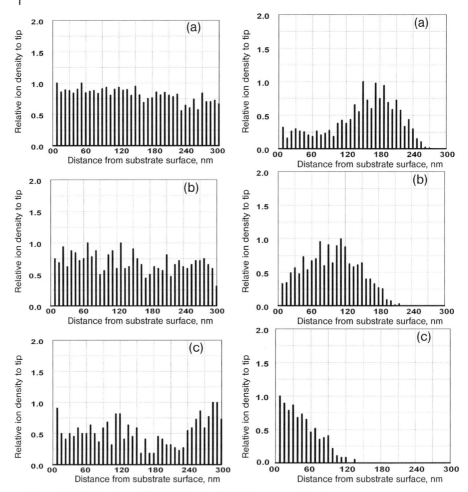

Figure 5.2 *Left column:* Distribution of relative ion current density along the nanotip lateral surface as a function of distance from the substrate surface z for $T_e = 2.0\,\text{eV}$ and $U_s = 20\,\text{V}$. Diagrams (a–c) correspond to the following values of the plasma density $1.7{\times}10^{18}$, $8.6{\times}10^{17}$, and $1.23 \times 10^{17}\,\text{m}^{-3}$, respectively. *Right column:* Same as above for $U_s = 50\,\text{V}$ and (a) $n_p = 2.4 \times 10^{18}\,\text{m}^{-3}$; (b) $1.0 \times 10^{18}\,\text{m}^{-3}$; and (c) $3.8 \times 10^{17}\,\text{m}^{-3}$ [51].

face free of nanotips) can also be divided into 50 groups according to their specific area of collision with the nanotip.

The histograms in Figure 5.2 suggest that the distribution of the ion current density over the nanotip lateral surfaces depends on the plasma density. The left column corresponds to the narrow sheath case ($U_s = 20\,\text{V}$) while the three histograms on the right correspond to the $U_s = $

50 V case. Let us now consider the results from the left column. From Figure 5.2(a), one observes that when the plasma density is higher, the ion current to the nanotip lateral surface is distributed fairly uniformly. In this case the ion flux is somewhat higher in the surface areas close to the nanotip base. In less dense plasmas, the ion flux distribution becomes less uniform as can be seen in Figure 5.2(b). Likewise, the position of the maximum of the ion flux still remains closer to the substrate surface. On the other hand, when n_p is reduced by one order of magnitude, two poorly resolved peaks located closer to the ends of the nanostructures can be observed (Figure 5.2(c)) [51].

Considering the histograms corresponding to the wider sheath case ($U_s = 50$ V, 3 panels on the right in Figure 5.2) we note that the distribution of the ion current density along the nanotip lateral surface strongly depends on the plasma density. It is seen that when the plasma density is higher (Figure 5.2(a)), the maximum of the microscopic ion current is located approximately 180 nm above the substrate surface, and so closer to the top of the nanostructure. However, the current density is much smaller in the upper part of the nanotip structures and diminishes near the tip. On the other hand, J_s is almost a quarter of the maximum value near the nanotip base. In lower-density plasmas, the maximum of the ion density shifts towards the substrate surface, while still remaining well-shaped. In this case the ion current density to the nanotip base increases as suggested by Figure 5.2(b). Indeed, if $n_p = 3.8 \times 10^{17}$ m^{-3}, the maximum of the ion current density is located in the immediate proximity to the substrate surface (Figure 5.2(c)). Furthermore, the ion flux to more distant from the substrate areas becomes very small [51].

Let us now consider the dependence of the mean relative current density to the nanotips on the plasma density and electron temperature, as is shown in Figure 5.3. As before, the left column is for the narrow sheath case, whereas the two graphs on the right show the results for the $U_s = 50$ V case. Let us examine the two graphs on the left first ($U_s = 20$ V). One can see that when the plasma density increases, the relative ion current density increases steadily and reaches 1.3 and 2.3 in the low- and high-density cases, respectively (Figure 5.3(a)). The curve plotted for lower T_e shows a higher ion current density. As can be seen from Figure 5.3(b), the relative current density decreases steeply in the range of 0.5–5 eV, and then decreases smoothly in the subsequent electron temperature range (5 eV $< T_e <$ 12 eV) levelling off at $T_e > 12$ eV. From Figure 5.3(b) in the left column, one can further observe that higher current densities can be obtained at elevated plasma densities [51].

In the wider sheath case ($U_s = 50$ V, two panels on the right in Figure 5.3), the dependence of J_s on the plasma density is quite similar to

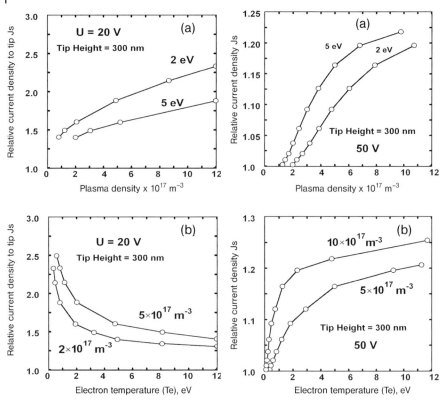

Figure 5.3 *Left column:* Relative ion current density to the nano-tips as a function of the plasma density (a) and electron temperature (b) for $U_s = 20\,\text{V}$. *Right column:* Same as in above for $U_s = 50\,\text{V}$ [51].

the $U_s = 20\,\text{V}$ case. In particular, the relative current density increases from 1 at lower plasma density to 1.2 for higher n_p (Figure 5.3(a)). The curve corresponding to the higher electron temperature shows a higher current density, which is opposite to the narrow sheath case. This is consistent with the results in Figure 5.3(b) which shows remarkable increase of J_s with increasing T_e, which is a different tendency compared with the narrow sheath case. The rates of increase of the ion current density appear to be different at lower and higher electron temperatures as can be clearly seen in Figure 5.3(b) in the right column. Meanwhile, the curve corresponding to the higher plasma density shows higher J_s [51].

5.1.3
Interpretation of Numerical Results

We will now interpret the results of the previous subsection and discuss their implications for the development of control strategies of the PECVD of ordered arrays of carbon nanotip microemitters. One of the key assumptions is that the nanotip pattern is conducting and there is no electric charge accumulated on the surface. This assumption is applicable to the case of conducting crystalline nanostructures synthesized through the "base growth" nanoassembly on metal-catalyzed highly-doped silicon substrates.

It is worth emphasizing that in other situations the presence of electric charges on the surface can significantly affect the microscopic ion flux distribution over the nanostructured surfaces. For example, dielectric islands on the surface can acquire a large positive charge, which can in turn cause quite irregular three-dimensional topographies of the ion flux onto the growth surface, as can be seen in the top panel in Figure 5.4 [260].

Graphs (a) and (b) in Figure 5.4 show the dependence of the ion trajectory displacement dX on the plasma density (with T_e and U_S as parameters) in the non-conducting nanopattern case. This dependence is quite similar to what is shown in the two panels on the right-hand-side of Figure 5.3. However, the absolute values of the ion deflection are larger for dielectric materials. This can be attributed to the dipole-like electric field present in the non-conductive pattern [260]. From Figure 5.4 one can see that the ion deflection by the nanotips is 35 nm under low-bias (20 eV) conditions and is only 15 nm when the bias is higher (50 eV). It is also notable that the displacement of ion trajectories increases steadily with the plasma density, as a result of the sheath shrinkage in denser plasmas [260].

Returning to the case of conducting carbon nanotips of the previous subsection, we note that the distribution of the ion current density onto carbon nanotip structures also shows quite significant irregularities. However, these irregularities are much less pronounced than in the process of growth of dielectric films. Physically, this is attributed to an excellent conductivity of the single crystalline nanotip structures and a substantial electric field enhancement in the proximity of the sharp ends of the carbon nanostructures, which strongly affect the ion motion and precipitation [51].

As we have seen from the previous subsection, the irregular electrical field created by the ordered nanotip pattern generally affects the distribution of the ion flux onto the substrate and nanotip surfaces. It is apparent that the presence of the nanotips on the substrate surface results in a remarkable redistribution of the ion current giving rise to non-

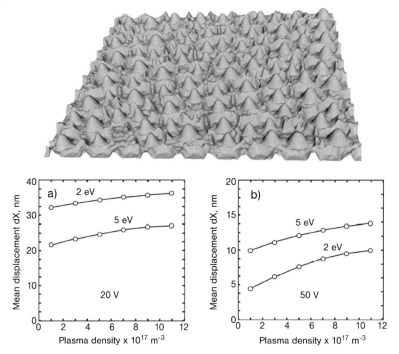

Figure 5.4 *Top row:* Distribution of microscopic ion current over a non-conductive nanotip pattern for $U_s = 50\,\text{V}$, $n_p = 5 \times 10^{17}\,\text{m}^{-3}$, and $T_e = 2\,\text{eV}$. The nanoislands do not focus the ion flux and are surrounded by the areas of strongly non-uniform ion density. *Bottom row:* Mean ion displacement in the vicinity of a nanostructured surface with a non-conductive nanotip pattern versus the plasma density with the electron temperature and bias voltage as parameters [260].

uniform and selective ion precipitation. The resulting three-dimensional non-uniform ion deposition patterns (representative examples shown in Figure 5.1) will inevitably affect any further nanostructure growth.

Even though both lower-U_s and higher-U_s conditions lead to quite substantial non-uniformity in the ion current profiles, there are remarkable differences between the two cases [51]. In particular, comparing the distributions of the relative ion current density along the nanotip lateral surfaces (Figure 5.2), one can clearly see that in the wide sheath case the distribution is strongly non-uniform, in remarkable contrast to the small sheath case featuring fairly uniform ion flux deposition. Physically, the nanotip growth critically depends on the location of the maximum of the ion current on the lateral nanotip surface. The displacement of the current maximum at higher plasma densities suggests that the actual shape and growth kinetics of the nanostructures are strongly affected by the value of n_p.

Let us now consider what that means for the growth conditions of carbon nanotips in a plasma. In dense plasmas, one could expect a preferential growth of the nanotip base and hence, the formation of wider and shorter nanoassemblies. On the other hand, the low-density plasma conditions favor the development of nanotips with a larger aspect ratio. It is notable that the narrow sheath case shows fairly uniform and regular distribution of the ion current over the entire simulation area. Therefore, the nanotips grown under such conditions should not change their shape significantly in the growth process. This conclusion is consistent with the experimental results on the CNS growth kinetics [95,258].

This clear difference in the ion current distribution can be explained as follows [51]. Under the wide sheath conditions, when the current is distributed non-uniformly along the lateral nanotip surface, the ion approaches the sharp end of the "host" nanotip (where the electric field is strongly non-uniform) with a relatively high velocity. Thus, J_s is less affected by the weak electric fields created by the neighboring nanotips, and the ion motion is mainly controlled by the electric field of the "host" nanotip. As a result, the ions precipitate locally on the surfaces of their "host" nanostructures, which results in somewhat diffuse but nevertheless pronounced maxima in the ion current distribution.

On the other hand, when the plasma sheath is narrow, the ions are slower when approaching the nanotips. In this case, weak electric fields created by the neighboring nanotips appreciably distort the trajectory, which diverts the ion flux and causes the fairly uniform ion current distributions observed in Figure 5.2. The dependence of relative ion current density on nanotip radius (Figures 6 and 10 of the original report [51]) suggest the possibility to control the distribution of nanotip radii in the process of nanofabrication of the ordered microemitter pattern. It appears that, preferential ion deposition on small nanotips can enhance their growth and, hence, contribute to the smoothening of the nanotip radii distribution function (NRDF). This effect is strongly pronounced for both low-U_s and high-U_s conditions [51].

Therefore, PECVD of CNSs in higher-density plasmas is advantageous for the assembly of microemitter arrays with highly uniform size distribution (e.g., NRDF) of individual nanostructures. Remarkably, in the experiments on the CNS synthesis in low-temperature RF plasmas of $Ar + H_2 + CH_4$ mixtures the combined ion number densities exceeded ca. $10^{18}\,m^{-3}$ [53] (see also Section 4.1). As has been discussed above, the most likely cause of the NRDF reshaping during the growth process is more effective focusing of the electric field and thus ion attraction by smaller nanotips. At higher plasma densities, this effect becomes stronger due to the smaller sheath thickness. One can thus expect a higher degree of the NRDF equalization under denser plasma conditions [51].

The dependence of the mean relative ion current density onto the nanotips on the plasma density (Figure 5.3) offers a possibility of controlling the nanotip density on the substrate. Indeed, the concentration of the adsorbed atoms on the substrate surface determines the rate of the new island nucleation [261]. Let us consider what that means in terms of CNS growth kinetics and assume that during the first growth stage a certain initial nanotip pattern (with the areas covered and free of nanotips) has been formed. When the ion current density onto the nanotips is high, one can expect that the nucleation rate of adatoms on the nanotip-free surface areas will be somewhat reduced; this may boost the growth of the already existing nanostructures. Furthermore, the surface areas between the nanotips should feature a smaller amount of irregular growth islands and thus a smoother surface morphology [51].

On the other hand, when the relative ion current density to the nanotips is close to J_s^m, one can expect a higher density of the adsorbed atoms in the nanotip-free areas, and hence, more efficient nucleation and growth island development. The resulting pattern will thus feature quite different surface morphology, with somewhat shorter nanotip structures and a number of unwelcome build-ups in the inter-nanotip areas. If the areas uncovered by the nanotips contain catalyst residues (this can happen when the Ni/Fe/Co catalyst layer fragmentation during the surface activation stage is incomplete), carbon nanostructures can also start growing in the inter-nanotip surface areas and can eventually interfere with the ordered nanotip pattern. Furthermore, this can result in an uncontrollable growth of the nanotips, with a large difference in height over the pattern [51].

Interestingly, the dependence of the mean relative ion current density onto the nanotips on the electron temperature (Figure 5.3) appears to be very different for low-U_s and high-U_s conditions. When $U_s = 20\,V$, the relative current density decreases with the electron temperature, whereas the inverse dependence is observed when the cross-sheath potential drop is $50\,V$. This phenomenon can be explained by noting that the dependence of the sheath thickness on T_e is quite different under the narrow- and wide-sheath conditions. Indeed, comparing Equations (5.1) and (5.2), one can notice that $\lambda_s \propto T_e^{-1/4}$ in the high-U_s case and $\lambda_s \propto T_e^{1/2}$ otherwise. Thus, different scaling of the sheath thickness with the varying electron temperature results in a quite different behavior of the relative ion current density onto the nanotips. The electron temperature can thus be regarded as another useful control knob of the growth kinetics of the ordered carbon nanotip structures.

To conclude this section, we stress that despite a close link to the specific plasma-aided nanofabrication process employing hydrocarbon-based reactive plasmas and specific building units (CH_3^+), the results presented are quite generic and the main conclusions can be applied to a broader range of ordered nanotip structures. Most importantly, this model can predict the exact locations where plasma-generated ionic building units can be deposited on nanostructured surfaces. The model can also be extended into different plasma chemistries, complex plasmas, deposition of multiple cationic species, denser ion fluxes and account for ion collisions within the plasma sheath, and broader ranges of substrate bias and other process control parameters to fit the demands of various reactive plasma-assisted nanofabrication processes [51]. The results of this section will be used in Chapters 7 and 8 in the studies of growth, reshaping, and post-processing of various one-dimensional nanostructures.

As we will see from the following sections, the dynamics of nanoparticles in the near-substrate areas can be substantially different from the ion case of this section. Again, as is often said in nanoscience, the size does matter!

5.2
Nanoparticle Manipulation in the Synthesis of Carbon Nanostructures

In this section, we will follow the original reports [105,152] and show the possibility of manipulation of the plasma-grown nanoparticle building units in the plasma-assisted deposition of various carbon-based nanostructures on Si substrates catalyzed using Ni-based alloys. The experimental results suggest that by varying the near-substrate temperature gradient (which can be achieved by applying variable external heating power to the substrate), one can selectively deposit or levitate the carbon-based nanoparticles grown in the low-temperature reactive plasmas of $Ar + H_2 + CH_4$ gas mixtures.

When the nanoparticles are levitated in the plasma presheath, the arrays of vertically aligned carbon nanotips (considered in Chapters 3 and 4 above) can be synthesized. On the other hand, when a large amount of plasma-grown nanoparticles precipitate from the gas phase, nanostructured mixed-phase (e.g., polymorphous) carbon films are fabricated. These experimental observations are supported by the basic one-dimensional model of the nanoparticle dynamics in the near-electrode area [105], which was later refined to account for a variety of important factors that may affect nanoparticle deposition during the discharge operation [152].

It turns out that the thermophoretic force can be crucial for manipulating the particle and, in particular, achieving their size-selective deposition onto plasma-exposed surfaces. In this section we also investigate specific conditions that enable plasma-borne negatively-charged nanoparticles to overcome the repulsive electrostatic barrier near the substrate and deposit onto the surface during the discharge run.

Let us now consider these interesting phenomena in more detail and as usual, let us briefly run through the significance and main aims pursued in the original reports. As we have already mentioned in Chapter 2, robust manipulation methods of the plasma-grown or externally introduced nano-sized clusters or particulates has been one of the hottest topics of research in recent years. These efforts essentially aim at achieving a reasonable level of creation, manipulation and stacking of a variety of building units during plasma-assisted synthesis of a range of nanoscale objects. It is particularly difficult to control the motion of such BUs in the immediate proximity of nanostructured surfaces.

We also recall that depending on the specific requirements of the nano-assemblies being targeted, different atomic, supramolecular, cluster, and mesoscopic species have been successfully used [70, 86, 132, 133, 152, 221, 262, 263]. For example, active (e.g., with activated dangling bonds, in excited or ionized states) radicals, macromolecules and small nano-clusters are ideal for the assembly of exotic nanostructures or single-crystalline advanced materials [21, 37, 74, 77, 152]. On the other hand, structural incorporation of larger units, such as larger nano-clusters, fine nano-crystals or aggregates, is often beneficial for the beam- or plasma-enhanced chemical vapor deposition (PECVD) of nano-structured (e.g., nano-crystalline) films [79, 80, 84, 114, 121, 152, 264–268].

Our focus here is on manipulation of the mesoscopic nano-sized building units in high-density, low-temperature plasmas of low-pressure glow discharges, by managing the dynamic balance of the plasma forces [105, 152]. The novelty of this research is that it goes beyond the widespread diagnostics and modeling of the nano-cluster and nanoparticle (NP) nucleation and growth, their applications for the PECVD of (mostly silicon-based) nanostructured films, and relevant discharge modeling (see, e.g., [28] and references therein). On the other hand, before the original report was published, the existing models of the solid particle dynamics in the near-electrode areas were mostly limited to micron- [269] and submicron-sized [270] particles and considered essentially particle motion-free cases. Indeed, it was a very common assumption in the studies of the force balance in the complex ("dusty") plasma discharges that a test grain moves from one point to another with a negligibly small directed velocity. At certain (often called "levitation point") positions in the vicinity of the deposition surface, the repulsive electrostatic and thermophoretic [124] forces balance the driving ion drag force and the particle is levitated (trapped) and cannot reach the substrate unless the discharge is turned off (deposition in the afterglow [80,262]). In the latter case, poly-disperse distributions of the particle size are frequently observed [152].

Here we note that the thermophoretic force is naturally present in the near-substrate areas of most low-temperature plasmas and is controlled by the difference between the gas temperatures in the plasma bulk T_g^b and near the substrate T_g^s. If $T_g^s > T_g^b$, F_{th} repels the particles from the substrate. On the other hand, if $T_g^s < T_g^b$, the thermophoretic force is attractive. If the substrate is additionally heated (which is a usual requirement in the PECVD of various functional thin films), the difference between T_g^b and T_g^s also changes [124,152,262].

However, substantial inertia of nanoparticles leads to significant variations of their velocities, which cannot be accurately accounted for by quasistatic models. This is why moving through the relatively long (compared to the plasma sheath) pre-sheath area, the (usually negatively charged) particle can acquire sufficient momentum to reach the surface even during the discharge operation. In this case, only particles of a certain size overcome the near-substrate potential barrier, as was frequently observed in the experiments on the plasma-aided nanofabrication of carbon-based electron field emitter arrays [50,53,93,94,105,152].

5.2.1
Nanoparticle Manipulation: Experimental Results

Recalling most important experimental tips on how to grow various carbon-based nanostructures [1] we stress that additional external (e.g., from underneath of the substrate stage) heating of metal-catalyzed substrates is one of the common requirements for the PECVD of a large number of carbon-based nanostructures (see also [21] and references therein). There are alternative schemes that merely rely on substrate heating by the hot gas in the chamber and the impinging ions and do not routinely require any external substrate heating [94]. The two regimes produce two remarkably different types of structures [105]. Specifically, in the external heating regime (also termed the fixed temperature growth regime [93]), arrays of vertically aligned carbon nanotip-like structures with no clear traces of particulate agglomerates on the surface have been synthesized.

Conversely, in the absence of any external substrate heating, numerous irregular-shaped nanoparticle agglomerates dispersed over large surface areas have been observed [94, 105]. Taken the small (in a few tens of nanometers range) size of the NPs and the fact that the only difference in the operating conditions of the experiments was in the temperature gradient, it was suggested that thermophoretic force [271] can be an important control factor in the deposition of the nanostructured carbon-based films in the low-pressure RF plasmas of $Ar + H_2 + CH_4$ gas mixtures [105].

The above hypothesis has been verified in a series of dedicated experiments [105, 152] wherein the heating power supplied to the catalyzed substrate was gradually changed from zero (no external heating, floating temperature regime [94]) to the maximum value used in the experiments with the fixed and externally controlled substrate temperatures [93]. In both series of experiments, the nanoparticles were either completely absent on the surface or exhibited a strong size selectivity of nanoparticle size distributions on nanostructured surfaces. In the earlier experiments [105], this observation was qualitatively related to external heating of the deposition substrate. This assumption was then quantified using direct *in situ* experimental measurements of the near-substrate temperature gradients using custom-designed temperature gradient probes [152].

Let us now briefly recap the most important experimental details. In the experiments, a high-density plasma of $Ar + H_2 + CH_4$ (typically with heavy dilution of carbon precursor gases in argon) gas mixtures was sustained, with RF powers of ca. 2 kW, in a low-frequency inductively coupled plasma (ICP) reactor pre-evacuated to base pressures ca. 10^{-5} hPa. The detailed description of the reactor, process control instrumentation, and plasma diagnostics can be found elsewhere [189]. Nanostructured carbon films (Figure 5.5) were deposited on Ni-coated highly-doped silicon substrates. The working pressures of the reactive gas mixture were maintained at ca. 5.333 Pa (40 mTorr)–6.666 Pa (50 mTorr). The deposition process was separated into three stages. First, high-purity argon was introduced for 30 min for the purpose of conditioning the deposition substrate and surfaces of the reactor chamber and substrate holder. At the end of the first stage, hydrogen was added to argon to perform the 20 min reactive chemical etching of the catalyst surface. Finally, methane was added to carry out the actual 40 min plasma-assisted deposition process in the $Ar + H_2 + CH_4$ RF glow discharge. An internal heating element, built into the substrate stage, was powered by a variable AC power supply. The above has been extensively used previously for the plasma-assisted nanofabrica-

(a)   (b)

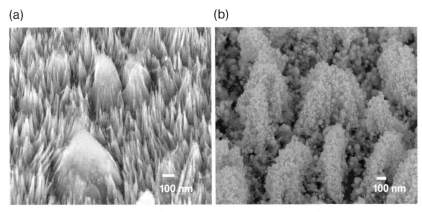

Figure 5.5 Scanning electron micrographs of the films grown
in different temperature regimes: (a) with the substrate heating (heating voltage and power are 30V and 17W, respectively);
(b) without the substrate heating [105].

tion of various carbon-based nanostructures [50, 53, 55, 93, 94] and later
optimized to obtain specific (with and without nanoparticles) surface
morphologies [152].

The surface morphology of the as-grown nanostructured carbon films
has been studied with a JEOL JSM-6700F field emission scanning electron
microscope (SEM) (Figure 1.12). When the external heating power was
applied to the substrate the deposition surfaces were free of nanoparticles as can be seen in Figure 5.5(a). However, when the deposition
substrates were not externally heated, fairly monodisperse nanoparticle distributions (Figure 5.5(b)) were observed (at exactly the same discharge parameters). These SEM results reveal that remarkable changes in
the surface morphology of the carbon nanofilms occur when the power
supplied to the substrate heater is varied. Specifically, at higher heater
voltages, exceeding a certain threshold value V_{thr}, the arrays of vertically aligned carbon nanotip structures were grown. A typical image of
the CNSs grown under heater voltages exceeding V_{thr}, is shown in Figure 5.5(a). For simplicity, this type of surface morphology was termed as
the "ordered nanostructured state" [105]. Here we note that the threshold heater voltage V_{thr} always remained within the range of 20–30 V,
showing an excellent reproducibility of the process.

When V_{heat} was decreased to V_{thr} and below, the resulting surface
morphology rearranged significantly resembling a (presumably amorphous [94]) matrix with the embedded nano-sized grains and particulate agglomerates. This state of the nanostructured carbon film was re-

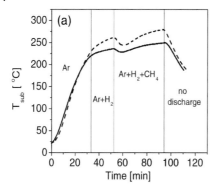
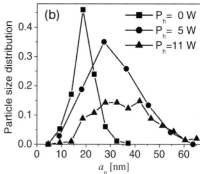

Figure 5.6 *Left panel:* Variation of the substrate temperature during different stages of the CNS deposition process for the following parameters of the experiment: heating voltage 20 V (dashed line) and no additional heating (solid line), total pressure 6.666 Pa (50 mTorr) –7.333 Pa (55 mTorr), RF input power ca. 2 kW, and 35, 35, and 100 sccm of Ar, H_2, and CH_4 inlet, respectively. The substrate bias was -4, -40, and -80 V during the conditioning, catalyst activation, and deposition stages, respectively [105]. *Right panel:* Effect of substrate heating power on nanoparticle radii distribution functions [152].

ferred to as the "disordered nanostructured state" [105]. In particular, Figure 5.5(b) shows that plasma-grown nanoparticles fall out onto the surface. In this case, no ordered nanostructure arrays have been observed. One of the most important conclusions of the original work [105] was that the *surface morphology transition from the ordered to the disordered nanostructured state occurs when the heater voltage reaches the threshold value* V_{thr}.

The variation of the substrate holder temperature during the three stages of the PECVD process is depicted in the left panel in Figure 5.6. From this figure one can see that in both growth regimes (with and without any additional substrate heating) following the rapid initial rise during the substrate conditioning stage and slower increase during the catalyst activation stage, the substrate temperature levels off after a few minutes into the actual film deposition process and remains stable, which indicates of the establishment of the quasi-stationary regime [105]. As can be seen from Figure 5.6 (left panel), the surface temperature does not change significantly under moderate values of the substrate heating power. For example, the difference between the substrate temperatures under conditions of the absence of any heating and when the heating voltage was 20 V (equivalent to ca. 11 W heating power), did not exceed 20 °C. This small difference ($< 7\%$) in the substrate temperatures is usually not sufficient to cause any significant changes in the shape and ordering of carbon-based nanostructures [105].

Furthermore, under the same conditions the plasma characteristics also did not change when the substrate heating was switched on and off, as suggested by highly-reproducible optical emission spectra of the plasma discharge (see Figure 4 in the original report [105]). The composition of the charged and neutral plasma species, reflected in the optical emission and mass spectra (the latter measured by the quadrupole mass spectrometry (QMS)) also remained unaffected by the changes in V_{heat}. Therefore, the powder generation process is also unaffected by the variations in the substrate heater voltage (and hence, the substrate temperature variations). In particular, in the case corresponding to the "ordered nanostructured state" of Figure 5.5(a), the nanoparticles homogeneously nucleated in the ionized gas phase presumably remain suspended in the plasma and are further transported to the pump line [105]. *Therefore, the temperature gradient changed, while other important parameters such as the plasma characteristics, gas pressure, species composition, substrate temperature and DC bias voltage did not.*

The effect of variation of the substrate heating power P_h on the nanoparticle size distribution over the selected areas of the deposition surface (e.g., shown in Figure 5.5) is quantified in Figure 5.6 (left panel). From this figure, one can see that the average size of the particles increases with the substrate heating power (even at moderate powers of several tens of Watts). Moreover, the nanoparticle radius distribution function (NPRDF) becomes more smooth and less monodisperse at higher P_h. This observation further supports the argument that by increasing external substrate heating power P_h, it is possible to vary the temperature gradient in the vicinity of the substrate, and hence, to modify the associated thermophoretic force acting on the nanoparticles.

Physically, an additional internal heating element produces a continuous additional upward heat flux emerging from the substrate stage. In the stationary stage, the above heat flux is compensated by an additional flux due to the thermoconductivity. Therefore, we arrive at the conclusion that the additional substrate heating changes the gradient ∇T_n. Therefore, in addition to any gas temperature gradient that exists in the absence of the external heating, one has the additional gradient ∇T_n^{add} controlled by the power applied to the internal heating element. Thus, if the additional temperature gradient is small enough, the gas-phase grown nanoparticles can reach the substrate surface and incorporate in the nanostructured film (Figure 5.5(b)). However, if ∇T_n^{add} is high enough, the thermophoretic force (originating due to the gas temperature gradient) may repel the nanoparticles to the plasma bulk. In this case, the resulting nanostructures resemble nano-needles ("nanotips") [50,105].

The additional gas temperature gradient can be estimated by assuming that the heat emerging from the substrate stage is transported through the gas homogeneously in all directions

$$\nabla T_n^{add} = \frac{P_{heat}}{\alpha \, S_h}. \tag{5.5}$$

Here S_h is the surface area of the substrate stage, P_{heat} is the power applied to the internal heater, α is the coefficient of the thermoconductivity of the neutral gas mixture [105].

This model was tested both experimentally and numerically [105,152]. One of the most important things in this regard is to experimentally establish the link between the temperature gradient in the near-substrate areas and the heating power, which was successfully done in Reference [152]. This made it possible to to link the nanoparticle size distributions (Figure 5.6) to the temperature gradients at fixed heating powers P_h. For further details of this procedure the reader should refer to the original report [152].

In the following, we will introduce the numerical model of the nanoparticle dynamics in the near-substrate area, which was successfully used to elucidate the dependence of the minimum nanoparticle size able to deposit on the surface during the discharge run on the value of the actual temperature gradient. Here we stress that a combination of the experimental and numerical results demonstrated the possibility of size-dependent deposition of the plasma-grown nanoparticles during the PECVD of carbon-based nanostructures in hydrocarbon-based plasmas.

5.2.2
Nanoparticle Manipulation: Numerical Model

In this subsection we will introduce the numerical model of the nanoparticle dynamics in the near-substrate area [105,153]. First of all, we note that in low-pressure laboratory plasma discharges fine particles are usually charged negatively due to a significant difference in electron and ion mobilities. For the same reason, solid surfaces are also charged negatively. Therefore, such nanoparticles have to overcome a repulsive electrostatic barrier before they can be deposited on the substrate (see e.g., [28] and references therein).

However, if the particle can reach the area of a significant depletion of the electron density (with respect to the density of the positive ions), the charge on it can change the polarity and the particle can be deposited onto the surface. This point is frequently referred to as the charge reversal point. Whether the NP is able to overcome the electrostatic barrier and reach the charge reversal point critically depends on the inertia the particle accumulated in the combined sheath-presheath area as a result of the action of the major plasma forces [105]. The nanoparticle is accelerated in the presheath area, and the final accummulated momentum depends on the presheath width. Generally speaking, the larger the presheath, the better the particle can be accelerated and, hence, the higher the chance it can overcome the electrostatic repulsion and deposit onto the surface.

To simulate the nanoparticle motion in the combined sheath-presheath area, a one-dimensional model was used [105,152]. The simulation area comprises the plasma bulk ($x < -l_{pr}$), presheath ($-l_{pr} < x < 0$), and sheath ($0 < x < l_s$) regions. In the first region, the plasma is assumed uniform with the zero electrostatic potential φ throughout. The width of and the potential distribution within the presheath were adjusted using relevant experimental results. The sheath width l_s was computed self-consistently using the basic set of equations and the appropriate boundary conditions [105,152].

The dynamics of a test nanoparticle (NP) in the near-electrode area has been studied for the following parameters: $n_0 = 2 \times 10^{11}$ cm^{-3}, $T_e = 2.0$ eV, and DC bias voltage of $V_s = -80$ V (with respect to the ground). The latter imposes the necessary boundary condition $\varphi(l_s) = V_s + V_{pl}$, where V_{pl} is the plasma bulk potential with respect to the grounded chamber walls. In high-density plasmas of inductively coupled plasmas $V_{pl} \sim 15$ V [189]. The pre-sheath width was varied from 0.5 to 3 cm. The equilibrium nanoparticle charge and the forces acting on it were computed using the orbit motion limited (OML) approximation [28]. It was assumed that the nanoparticle of a spherical shape takes its origin in the plasma bulk and participates in the film deposition process by moving through the sheath and presheath areas. Furthermore, the temperature distribution near the substrate has been modeled by a linear function of the coordinate x, which is justified in the case of a constant temperature conductivity of the operating gas in the near-substrate region [105].

The total force driving the test nanoparticle through the sheath can be expressed as

$$F_{tot} = F_{el} + F_{mg} + F_{ion}^{drag} + F_{neutr}^{drag} + F_{therm} \tag{5.6}$$

where the five terms in the right hand side represent the electrostatic, gravity, ion drag, neutral drag, and thermophoretic forces, respectively. The expressions for the above forces and the applicability limits are standard [28].

The motion of 10–50 nm sized nanoparticles in the sheath and presheath areas appears to be quite different. In the presheath, the ion drag and

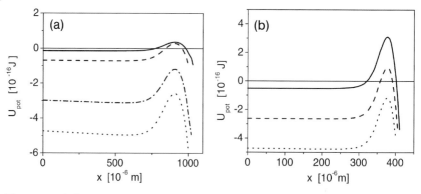

Figure 5.7 *Left panel:* The potential energy profiles for $T_e = 2.0\,\text{eV}$, $n_e = 10^{11}\,\text{cm}^{-3}$, $n_n = 10^{14}\,\text{cm}^{-3}$, $\nabla T = 0$, $m_i = 40\,\text{amu}$, and different values of particle radius r_p. Solid, dashed, dash-dotted and dotted lines correspond to $r_p = 10$, 20, 40, and 50 nm, respectively [105].

Right panel: Same as above for $r_p = 40\,\text{nm}$, $T_e = 2.0\,\text{eV}$, $n_e = 10^{12}\,\text{cm}^{-3}$, $n_n = 10^{14}\,\text{cm}^{-3}$, $m_i = 40\,\text{amu}$ and different values of the temperature gradient. Solid, dashed, and dotted lines correspond to the gas temperature gradients 50, 0, and $-50\,°\text{C}/\text{cm}$, respectively [105].

the thermophoretic forces dominate, and the total force drives the particle towards the surface. Here we stress that the ion drag and the thermophoretic forces act in the opposite directions; however, the net force on the nanoparticle is directed towards the substrate. Near the wall, a strong electrostatic force creates a potential barrier for the NPs. However, if, despite the strong potential barrier the particle can still reach the area of the significant depletion of the electron density (the charge reversal point), the charge on it can reverse the sign, and the NP may deposit onto the surface.

Therefore, the potential energy

$$U(x) = U(x_0) - \int_{x_0}^{x} F_{\text{tot}}(x)dx \tag{5.7}$$

can be used to describe the nanoparticle motion in the near-substrate area, where $x_0 = -l_{pr}$ is the starting point of the NP motion and $U(-l_{pr}) = 0$. Equation (5.7) allows one to obtain the nanoparticle velocity $v_p(x) = \sqrt{-2U(x)/m_p}$, at any position, where $m_p = 4/3\pi\rho r_p^3$ is the mass of a spherical carbon nanoparticle with the material density ρ and radius r_p. The areas with positive potential energy values are unaccessible for negatively charged particles [269].

Figure 5.7 (left panel) shows the profiles of the potential energy of NPs with different sizes in the sheath area in the case without an external sub-

strate heating. It is clearly seen that larger (40–50 nm) particles have negative values of $U(x)$ in the entire sheath area (presheath is not shown in
the plot) and thus can travel smoothly towards the substrate. However,
smaller particles (10–20 nm) have to overcome a narrow (ca. 150–250 μm
wide) potential barrier to be able to deposit onto the surface. Therefore,
there exists a threshold radius r_p^{thr} of nanoparticles that can be deposited
onto the surface. In this particular example, $r_p^{thr} \approx 24$ nm [105].

The results in Figure 5.7 (right panel) suggest that the value of the near-
substrate gradient of the neutral gas temperature ∇T_n is a very powerful factor that controls the potential energy profiles within the plasma
sheath. One can clearly see that if the temperature gradient exceeds a
certain value, the thermophoretic force can even repel the NPs in the
presheath area thus moving them away from the substrate. For the conditions of the experiment (∇T_n ca. 10–20 °C/cm) [105], the particles with
the radius of approximately 30 nm have the highest chances to reach the
substrate. It was also noted that the near-substrate temperature gradients higher than ca. 20 °C/cm usually exclude most of the nanoparticles
from the deposition process.

Figure 5.8 shows the results of detailed parametric investigations of
the effect of different plasma parameters on the distributions of the NP
potential energy $U(x)$ within the sheath area [152]. To obtain these
plots, a typical set of the plasma and nanoparticle parameters was chosen from previous experiments with high-density inductively coupled
plasmas [50, 53, 93, 94]. These parameters have been called the default
set of parameters and include the plasma density $n_0 = n_{e,i} = 1.1 \times 10^{18}$
m^{-3}, the density of neutrals $n_n = 10^{20}$ m^{-3}, masses of the positive ions
and neutrals $m_i = m_n = 40$ amu, temperatures of positive/negative ions
$T_i^{(\pm)} = 0.2$ eV, population of negative ions not exceeding 20%, temperature of electrons $T_e = 2$ eV, the pre-sheath width $l_{pr} = 2$ cm, and DC bias
$V_s = -80$ V, unless specified otherwise. The default values of the particle
size and mass density are 30 nm and 2×10^3 kg/m^3, respectively. Other
parameters are detailed in each specific case (refer to caption to Figure 5.8
for details). The values of the above parameters control the equilibrium
profiles of the plasma parameters in the pre-sheath and sheath areas. To
elucidate the optimal nanoparticle deposition conditions, each of the basic parameters was independently varied, with all other parameters being held constant.

Briefly summarizing the results of the above detailed optimization
process, we note that a high-density plasma, with low electron temperature, and high number density of a lighter feedstock gas (e.g., He) is
the best environment for the nanoparticle deposition onto the surface.
The minimum radius of the particle which is able to reach the surface

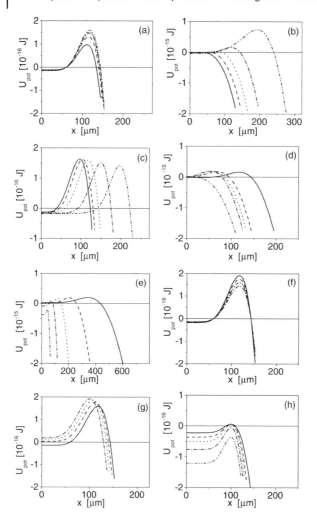

Figure 5.8 Potential energy for different values of (a) ion mass m_i: solid, dashed, dash-dotted, and dash-dot-dotted lines correspond to 4, 15, 30, 27, and 40 amu, respectively; (b) electron temperature T_e: solid, dashed, dash-dotted, and dash-dot-dotted lines correspond to 0.2, 0.5, 1.0, 2.0, and 4.0 eV, respectively; (c) negative ion temperature T_i^- (assuming $n_i^- = 0.2 n_i^+$ in the plasma bulk): solid, dashed, dash-dotted, and dash-dot-dotted lines correspond to 0.06, 0.1, 0.2, 0.5, and 1 eV, respectively; (d) relative density of negative ions $n_i^{(-)}/n_0$ (solid: 20%, dot: 40%, dot: 60%, dash-dot: 80%, dash-

dot-dot: 100%); (e) number densities n_0 of the plasma (solid: 10^{11} cm^{-3}, dot: 3×10^{11} cm^{-3}, dot: 10^{12} cm^{-3}, dash-dot: 3×10^{12} cm^{-3}, dash-dot-dot: 10^{13} cm^{-3}); (f) nanoparticle charge variation (solid: $1.2Z_d$, dash: $1.1Z_d$, dot: Z_d, dash-dot: $0.9Z_d$, dash-dot-dot: $0.8Z_d$); (g) different values of the pre-sheath length (solid: 2 cm, dash: 1 cm, dot: 0.5 cm, dash-dot: 0.25 cm); and (h) nanoparticle radius (solid: 20 nm, dash: 30 nm, dot: 40 nm, dash-dot: 5 nm, dash-dot-dot: 60 nm) [152]. Other parameters are inthe text of this section; see also Section IIIA of the original report [152].

is $a_p^{min} = 30\,$nm under the following set of the discharge parameters: $n_n = 10^{20}\,$m^{-3}, $T_e = 1.0\,$eV, $n_0 = 10^{18}\,$m^{-3}, $T_i = 0.1\,$eV, and without the presence of negative ions. These values are representative of the experiments on PECVD of various carbon nanostructures [53,93,94,105].

It is notable that by changing any one process parameter, one can effectively modify the potential energy profiles. However, variation of some of the parameters, such as the electron temperature or gas temperature gradient, had a much greater affect on U_{pot} than the variations of other parameters. This conclusion can certainly be used for the optimization of the PECVD process. For example, aiming at selective deposition of the nanoparticles in the 20–30 nm range, one should use a lighter gas, such as helium or argon/helium mixture as a hydrocarbon diluent gas. This option follows from the results of Figure 5.8(a). However, due to the relatively high ionization potentials of He, the discharge maintenance would require higher input powers, which, in turn, can lead to overheating of the gas feedstock and deposition surfaces and possibly compromise the film quality. Alternatively, large amounts of easily ionized argon facilitate the discharge maintenance with reasonably low RF powers but increase the threshold size for the nanoparticle fallout [152].

The results of Rutkevych *et al.* [105, 152] suggest that the temperature gradient-controlled deposition conditions and the size-sensitive structural incorporation of nanoparticles in the nanofilm are intimately related through the NP potential energy profile in the near-substrate area. In particular, each value of the temperature gradient (TG) corresponds to the specific distribution functions of the radii of the nanoparticles observable on the nanostructured carbon surfaces, as shown by the experimental points in Figure 5.9. Specifically, the maxima of the NPRDFs increase with the temperature gradient.

On the other hand, numerical results also suggest that the minimum size of the particles that can deposit during the discharge run also increases with the TG (solid curve in Figure 5.9). Thus, both experimental and numerical results prove that higher temperature gradients (and hence stronger counteracting thermophoretic forces) make it more difficult for smaller nanoparticles to deposit on the surface during the discharge run. From Figure 5.9, one can also see that despite similar qualitative tendencies, the discrepancy between the experimental and numerical results is quite large. The main reason for this difference is uncertainty in the plasma and working gas parameters that were not thoroughly measured during the PECVD process. To this end, the numerical results of this subsection can serve as quantitative indicators of the sensitivity of the nanoparticle deposition threshold (e.g., minimum size in Figures 5.8(h) and 5.9) to the plasma and process parameters.

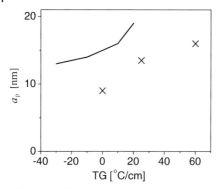

Figure 5.9 Experimentally measured NPRDF maxima for different heating powers ($\times$), and computed minimum size of the NPs able to deposit on the substrate (solid line) suggesting a reasonable correlation between the theory and the experiment [152].

Therefore, the thermophoretic barrier established as a result of the external substrate heating favors the self-organization of the carbon nanofilm to the ordered nanostructured state shown in Figure 5.5(a). In other words, the *nanopowder particle deposition can be thermophoretically controlled*. This result apparently correlates with the earlier report on the threshold values of the near-substrate temperature gradients that enable one to selectively control the PECVD of the microcrystalline and polymorphous (featuring plasma-grown nanoparticles embedded in the amorphous matrix) silicon films [124]. From this point of view, the disordered nanostructured state in Figure 5.5(b), which also shows the elevated amounts of amorphous carbon in the films, can be termed a carbon analog of the Si-based polymorphous films.

One of the most important results stressed in this section is that the momentum gained by the nanoparticles in the presheath may be sufficient to enable particles of a certain size to pass through the sheath and land on the substrate. In the case considered here, the 30–50 nm nanoparticles are the best candidates for this purpose. This is apparently consistent with the scanning electron micrographs of the carbon nanofilms in the disordered nanostructured state which reveal that the size of the major part of the NPs in the film fall within this range. Such a remarkable agreement between the SEM and numerical results further justifies the main assumptions of numerical modeling.

To conclude this section, we emphasize that the above disussion revealed the conditions under which the plasma-generated nanoparticles may pass through the sheath and deposit onto the surface. However, this alone does not answer another important question: 'Where exactly do the nanoparticles land on the surface? The following section aims to shed some light on this puzzling issue.

5.3
Selected-Area Nanoparticle Deposition Onto Microstructured Surfaces

From the previous section, we have learned that nanoparticle motion in the plasma area near the deposition surface is governed by a number of forces unique to a low-temperature plasma environment and is extremely sensitive to the nanoparticle charge, mass and shape. The NPs are usually charged negatively and as such are repelled by the similarly charged surface. However, the negative potential of the surface also repels the plasma electrons, which causes a severe depletion of the electron density within the plasma sheath. Thus, the electron current is significantly reduced. Alternatively, the ion current gets stronger as the particle approaches the surface.

As a result, the nanoparticles can reverse their sign (at the point referred to as the charge reversal point) and become positively charged. In this case the electrostatic repulsion changes to attraction and the NPs can deposit on the substrate surface, as is the case in the synthesis of carbon nanotubes in low-temperature plasmas (see Reference [272]) and precipitation of large amounts of quasi-spherical nanoparticles onto nanostructured carbon surfaces [1, 50, 93, 94, 105, 152] considered in the previous section.

However, a nanoparticle can reach the charge reversal point only in certain cases: namely, when the inertia effects are strong enough to keep the particle moving despite strong electrostatic repulsion [105,152]. Following the original report [153], we will now specify exactly where the nanoparticles land on microstructured surfaces. We stress that this knowledge is a vital yet missing link in the development of robust strategies for nanoparticle contamination management in microelectronics and size-selective nanoparticle deposition in fabrication of nanofilms and nanodevices.

In this section, by means of advanced numerical simulation of the dynamics of variable-charge-nanoparticles in the plasma sheath, the microscopic nanoparticle fluxes onto specific areas on microstructured surfaces will be computed. The effects of variation of the plasma process parameters and micropattern features on selected-area deposition onto microstructured surfaces will also be investigated. We will also discuss specific process conditions that enable site-selective (on top or lateral surfaces of the surface micro-structures (SMSs) or in the inter-SMS valleys) nanoparticle deposition [153]. These results are generic and applicable for a broad range of plasma-assisted nanofabrication, materials synthesis and surface modification processes.

5.3.1
Numerical Model and Simulation Parameters

The simulation geometry (Figure 5.10) presented here is representative of a typical experiment on plasma processing of a microstructured surface in a PECVD reactor [153]. A solid substrate is maintained at a negative electrostatic potential and faces a large-volume plasma as shown in Figure 5.10. The microstructured surface, in turn, represents either a pre-fabricated feature (e.g., a trench) on a microstructured semiconductor wafer or a quasi-two-dimensional morphology element in the plasma-assisted growth of microstructured films. Ions and plasma-grown nanoparticles are simultaneously deposited from the ionized gas phase onto an insulating (e.g., undoped Si) or a conductive (e.g., sp^2-hybridized carbon or highly-doped Si) substrate. The surface is maintained at a variable potential φ_{surf}.

The model in the original report [153] contains two modules: the first module describes the combined sheath-presheath area whereas the second module is related to a specific description of the NP motion and computation of microscopic nanoparticle fluxes. Owing to the extreme sensitivity of NP deposition to momentum accumulated in the process of travel, the presheath area is an integral part of the model. In Figure 5.10, the point of origin ($x = y = 0$) is chosen at the surface microstructure (SMS) base plane, in the middle of the valley that separates the first and the second (from the left) microstructures. The parameters of the plasma process and micropatterns used in these simulations are summarized in Table 5.2.

It is further assumed that the SMSs are maintained at the same negative potential $\varphi_{surf} < 0$. The reference electrostatic potential in the plasma bulk is assumed to be zero. It is convenient to divide the plasma into regions according to the potential values: a collisional planar presheath, a planar upper sheath, and a two-dimensional lower sheath, hereinafter referred to as regions I, II, and III, respectively.

The density of ions in the plasma bulk is n_0 and they enter the presheath with velocity v_0. The potential profile in this area (region I)

(a)

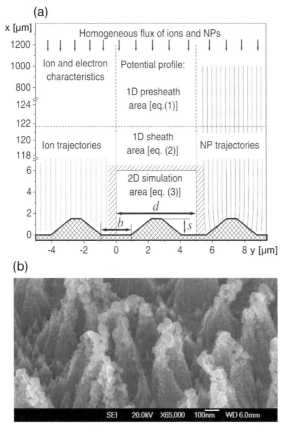

(b)

Figure 5.10 (a) Schematics of the periodic micropattern on the substrate surface used in simulations (panel in the middle). Left and right sections of Fig. 5.10(a) show the focusing effect of the local electric field created by the surface microstructures on the nanoparticle (right) and ion (left) trajectories. This effect is stronger in the nanoparticle case [153]. (b) Scanning Electron Microscopy image of carbon nanoparticles deposited on a Si substrate in a high-density PECVD reactor [50, 93, 94, 153].

can be calculated analytically [153] (note, $\varphi(x) < 0$)

$$x = x_{\text{bulk}} + L_{\text{coll}} \left(\varphi(x)/T_e - \frac{v_B^2}{2v_0^2} \exp\left(2\varphi(x)/T_e\right) + \frac{v_B^2}{2v_0^2} \right)$$

$$\varphi_{\text{pr}} \le \varphi(x) \le 0, \quad (5.8)$$

where $v_B = \sqrt{T_e/m_i}$ is the Bohm velocity, T_e is the electron temperature, m_i is the ion mass, φ_{pr} is the potential at the sheath-presheath boundary, x_{bulk} is the coordinate of the boundary between the pre-sheath and the plasma bulk, and L_{coll} is the collisional mean free path of ions.

Table 5.2 Values of parameters used for numerical simulation [153].

Parameter	Description	Value
n_0	Plasma density	10^{10}–10^{12} cm^{-3}
v_0	Ion velocity in plasma bulk	500 m/s
m_i	Ion mass	40 amu (Ar)
T_e	Electron temperature	2.0 eV
n_n	Density of neutrals	10^{14} cm^{-3}
m_n	Mass of neutrals	40 amu (Ar)
a_p	NP radius	30 nm
φ_{surf}	Surface potential	-13 .. -18 V
s	Height of SMSs	4 μm
b	Inter-SMS distance	2..4 μm
d	Micropattern period	4..10 μm

The upper sheath boundary in region II can be obtained through numerical integration [153]

$$x = x_{\text{pr}} - \int_{\varphi_{\text{pr}}}^{\varphi} \frac{d\varphi}{\sqrt{(\varphi'_{\text{pr}})^2 + 2|e|n_0/\varepsilon_0(J_e + J_i)}} \qquad \varphi_{2D} \le \varphi \le \varphi_{\text{pr}}, \qquad (5.9)$$

where

$$J_e(\varphi) = T_e\left[\exp(\varphi/T_e) - \exp(\varphi_{\text{pr}}/T_e)\right],$$

and

$$J_i(\varphi) = 2\mathcal{E}(\varphi)\left[1 - \varphi/\mathcal{E}(\varphi)\right]^{1/2} - 2E_{\text{pr}}(1 - \varphi_{\text{pr}}/\mathcal{E}_{\text{pr}})^{1/2},$$

and φ_{2D} is the potential at the upper edge of the two-dimensional (2D) grid shown in Figure 5.10. Here $\mathcal{E}(\varphi)$ is the ion energy as a function of potential, and $\mathcal{E}_{\text{pr}}$ is the ion energy at the sheath-presheath boundary x_{pr}, $\varphi'_{\text{pr}} = (d\varphi/dx)|_{x=x_{\text{pr}}}$, and $\varepsilon_0 = 8.85 \times 10^{-12}$ F/m.

In the lower sheath area (near-substrate area $0 < x < x_{2D}$), the following numerical solution of the two-dimensional Poisson equation

$$\frac{\partial^2\varphi}{\partial x^2} + \frac{\partial^2\varphi}{\partial y^2} = -\frac{|e|}{\varepsilon_0}\left[n_i(x,y) - n_e(x,y)\right] \qquad \varphi_{\text{surf}} \le \varphi \le \varphi_{2D} \qquad (5.10)$$

makes it possible to obtain the electrostatic potential in region III.

The boundary potential φ_{2D} between regions II and III determines the coordinate x_{2D}. For the accuracy of the multi-grid method used, the latter should satisfy the inequality $d, s \ll x_{2D}$, where d is the micropattern period and s is the height of the surface microstructures considered (Figure 5.10). In this case, the presheath and upper sheath areas (regions I and II, described by (5.8) and (5.9), respectively) can be safely assumed as one-dimensional. However, the lower sheath area near the substrate (region III) is *essentially two-dimensional* and requires integration of a partial differential equation (5.10).

It is interesting that this condition is very easy to satisfy experimentally. In fact, if the potential difference between the surface and the top edge of the 2D simulation grid is only 4V, the edge between regions II and III is located far above the microstructures at approximately a few hundred microns above the zero level. Suitable boundary coordinates x_{pr} and x_{2D} entering (5.8)–(5.10) can be found from the potential continuity condition between relevant regions [153]. In the lower sheath area ($x < x_{2D}$, region III) a spatially uniform in horizontal (y direction, 47 steps) and variable-step in vertical (x direction, 71 steps) two-dimensional grid has been used. In this area, the solution of the Poisson equation has been obtained numerically.

As usual, the equilibrium nanoparticle charge $Z_p|e|$ was computed by equating the microscopic electron and ion currents [28]. Similar to the previous subsection, the total force acting on a nanoparticle includes the electrostatic $\mathbf{F}_{el}$, ion drag $\mathbf{F}_i^{dr}$, friction $\mathbf{F}_{fr}$ and thermophoretic $\mathbf{F}_{th}$ forces.

The effect of the plasma process parameters and microstructure features has been investigated through variation of the surface potential φ_{surf}, height s of the SMSs, width of the inter-SMS valleys b, and micropattern period d [153]. These and other parameters used in simulations here are summarized in Table 5.2. These parameters are representative of carbon nanoparticle deposition experiments in high-density inductively coupled plasmas discussed in Section 5.2. Through statistical description of a large number of nanoparticles it was possible to compute their microscopic fluxes. Assuming uniform nanoparticle flux j_0 in the presheath, the horizontal distribution $j(y)$ of NPs landed on the microstructured substrate is computed and expressed as a fraction of the initial flux j_0 [153].

5.3.2
Selected-Area Nanoparticle Deposition

As we have stressed in Section 5.2, the substrate bias is perhaps the easiest process parameter to adjust experimentally. The other factors that can significantly affect the NP deposition process are the shape of the microstructures, micropattern features, and plasma parameters. A very important point in this consideration is that the charged microstructured surface in Figure 5.10 creates a horizontal component of the elec-

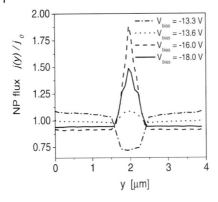 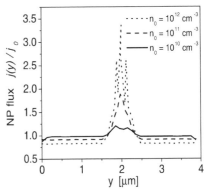

Figure 5.11 *Left panel:* NP fluxes calculated for different values of the surface potential. Here, the SMS and nanopattern dimensions are $s = 4\,\mu m$, $b = 3.2\,\mu m$, and $d = 4\,\mu m$. Density of argon plasma is $n_0 = 10^{11}\,cm^{-3}$. Depending on the surface potential, the nanoparticles can be deposited either between the microstructures, or near the SMS tips. The flux $j(y)$ is expressed in units of the original flux j_0 [153]. *Right panel:* Effect of the bulk plasma density on the nanoparticle deposition flux. Here, the $\varphi_{surf} = -16\,V$, and other parameters of the plasma and surface microstructures are the same as above [153].

tric field $E_\perp$, which deflects plasma-grown nanoparticles from straight trajectories.

The results for nanoparticle deposition on the substrate surface at different surface potentials are depicted in Figure 5.11 (left panel) [153]. It is very clear that the effect of the surface potential is essentially nonlinear. Indeed, when the absolute value of the potential $|\varphi|$ is small, the nanoparticles are predominantly deposited between the SMSs (dash-dot line); further more, the flux deposited onto inter-SMS valleys is more than 50% than that onto the SMS surfaces. However, when the surface potential increases, the NP flux transforms and is increasingly concentrated on the SMS surfaces, with more particles landing closer to the SMS tips (dashed line). A further increase of the potential makes the flux more homogeneous (solid line). One can thus conclude that larger φ_{surf} lead to a more homogeneous nanoparticle deposition over the micropattern. From Figure 5.11 (left panel), one can see that the particles in high-density plasmas tend to deposit near the tips of the SMSs, while in low-density plasma the deposition is homogeneous.

The micropattern period also affects the distribution of the microscopic nanoparticle flux $j(y)$. Firstly, if the SMSs are located further away from each other, each of them draws a larger NP flux compared to that in denser micropatterns. In fact, if the inter-SMS distance is $10\,\mu m$, the nanoparticle flux drawn by individual microstructures is more than

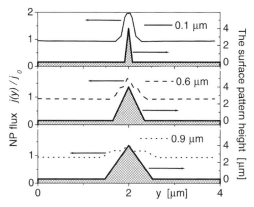

Figure 5.12 Nanoparticle flux over the substrate structures of a different width. The half-width at microstructure base is: $0.9\,\mu\text{m}$ (bottom panel), $0.6\,\mu\text{m}$ (middle panel), and $0.1\,\mu\text{m}$ (top panel). Sharp structures exhibit a stronger nanoparticle focusing effect. Here, $\varphi_{\text{surf}} = -16\,\text{V}$ and $n_0 = 10^{11}\,\text{cm}^{-3}$ [153].

twice as large as j_0, whereas it is only ca. $1.6 j_0$ when the surface density of the SMSs is doubled [152]. Secondly, at the base of the microstructures the nanoparticle flux is depleted and does not exceed $0.8 j_0$, whereas it is almost the same as the initial flux j_0 elsewhere in the valleys and peaks at the mid-point between the SMSs. Indeed, small peaks located at $y = 4, 8, 12$, and $16\,\mu\text{m}$ can be observed. Therefore, the relative probability of the nanoparticle deposition in the middle between the microstructures is higher [152].

Moreover, the microstructure shape changes the electric field above them and strongly affects microscopic nanoparticle fluxes, as can be seen in Figure 5.12. In particular, the focusing effect of the electric field becomes more pronounced as the SMSs sharpen up. In the case of relatively wide microstructures, the NPs are fairly uniformly deposited over the SMS surfaces (bottom panel in Figure 5.12). In other words, the effective nanoparticle deposition area is almost the same as the microstructure width; meanwhile, the NP flux associated with this area only slightly exceeds that in the inter-SMS valleys. However, for very sharp microstructures the effective nanoparticle deposition area is clearly larger than the SMS width (solid line in the top panel in Figure 5.12).

We will now discuss the numerical results presented in Figures 5.11–5.12 and comment on their relevance to other ongoing research efforts in the area and practical implications for the develoment of robust strategies and techniques for nanopowder contamination management and highly-controlled site-selective nanoparticle deposition in complex plasmas [152].

Let us first discuss the effect of the surface potential. If the surface potential is large enough so that the kinetic energy of landing nanoparticles exceeds their cohesive energy, the NPs can break into subnano- and nano-fragments; in this case a build-up of (usually unwanted) amorphous deposits around the area of impact is verylikely [4]. This scenario is illustrated in Figure 2.9 in Chapter 2. Moreover, the microstructures and their ordered patterns can also be substantially damaged. In this case, quite large gains in nanoparticle kinetic energy can be explained as follows [152]. When the DC bias is larger, the sheath width is effectively increased due to the stronger repulsion of electrons towards the plasma bulk. Thus, the electron density depletion is more pronounced and the nanoparticle charge reversal occurs further away from the substrate surface, which in turn means that the NP are able to pass through a larger accelerating potential drop. Note that the nanoparticles slow down as they approach the charge reversal point [152].

Alternatively, when the DC bias is small or the surface is insulating (in the latter case the surface potential equals φ_f, where φ_f is the floating potential [68]), the charge reversal point is nearer the substrate and the nanoparticle kinetic energy gain is smaller. Since typical values of the floating potential in low-temperature, thermally non-equilibrium plasmas are ca. 4–5 T_e, such low potential drops are very unlikely to lead to nanoparticle breaking upon landing [153]. Moreover, in this case the nanoparticle velocities are relatively small and the effects of local (microscopic) electric fields in the vicinity of microstructures are quite strong. This results in a pronounced deflection of nanoparticles from straight downfall trajectories (the latter is usually the case when the substrate is biased), clearly seen in the panel on the right in Figure 5.10.

Interestingly, the electric field-controlled deflection of nanoparticles is much larger than that of the plasma ions, see Figure 5.10. Physically, the nanoparticles are much slower and also carry a much larger electric charge than the ions; however, despite their heavier mass, particle deflection in the horizontal (y) direction turns out larger than that of the ions which move so fast the local microscopic field of the microstructures is unable to significantly deflect them as they cross the lower sheath area. This conclusion is consistent with the reports of a similar effect of plasma-grown nanostructures on the plasma ions. In particular, these results suggest that ion deflection is larger when the surface potential (and, hence, the vertical component of the ion velocities) is lower [159].

As can be seen in Figure 5.11 (left panel), the nanoparticle deposition process turns out to be extremely sensitive to the surface potential. In the original work [153], the range of the surface potentials was chosen to "catch" the moment of nanoparticle charge reversal and maximize

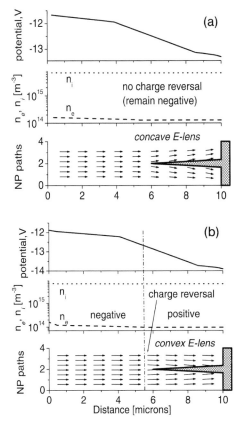

Figure 5.13 Effect of charge reversal on nanoparticle deposition: (a) when the surface potential is −13.3 V, the NPs do not change their sign and are scattered by the *concave* electrostatic lens (E-lens), also charged negatively; (b) a minor change in φ_{surf} to −13.6 V results in the NP charge reversal from negative to positive and attraction to the SMSs which act as *convex* E-lenses. Profiles of the electrostatic potential and electron and ion number densities are also shown. In case (b), the NP charge reversal occurs close to the SMS. Parameters are the same as in Figure 5.11 [153].

the effect of particle deflection by the surface microstructures. This is why the surface potential φ_{surf} is relatively low. In the example shown in Figure 5.11 (left panel), when the surface potential is low ($\varphi_{surf} = -13.3$ V), the nanoparticles remain negatively charged as illustrated in Figure 5.13(a). Nevertheless, due to their inertia, the NPs are still able to land in the valleys between the surface microstructures.

It is worthwhile to remark that the negatively charged SMSs repel and scatter the nanoparticles and operate as *concave* electrostatic lenses as shown in Figure 5.13(a). When the surface potential is only slightly increased to −13.6 V, the electron density is further depleted making the nanoparticle charge reversal possible, yet very close to the surface microstructures. This situation is shown in Figure 5.13(b). In this case the particles uniformly cover the SMS surface as the dotted line in the left panel in Figure 5.11 suggests. A further increase in φ_{surf} leads to a more pronounced focusing of the nanoparticles onto the microstructure tips, see Figure 5.11 (left panel).

This remarkable nanoparticle focusing is a manifestation of the operation of the surface microstructures in which we are interested as effective *convex* electrostatic lenses. It follows from Figure 5.11 that the surface potential of -16 V is the optimum nanoparticle focusing condition. With even higher substrate potentials (as is the case represented by the solid line in the left panel in Figure 5.11) the nanoparticle charge reversal point is located even higher above the substrate. In this case, vertical components of particle velocities increase as they approach the microstructures, which reduces the duration of effective action of $E_\perp$. Hence, the nanoparticle deposition over the micropattern surfaces becomes once again more uniform [153].

However, as we have mentioned above, one should be particularly careful not to overaccelerate the nanoparticles so as to avoid their structural disintegration upon crashing onto the surface. The above mentioned low near-surface potential drops are particularly favorable for minimizing the NP energy upon deposition and thereby of implementing their "soft" landing onto the microstructured surface.

The plasma parameters such as the ion number density and the electron temperature also affect the nanoparticle deposition. In particular, the results in the right panel in Figure 5.11 prove that high-density plasmas should be used for more effective NP focusing at the SMS tips. On the other hand, nanoparticle fluxes from rarefied plasmas are expected to be more uniformly distributed over the nanostructured surfaces. The more uniform deposition of NPs in lower density plasmas shown by the solid curve in Figure 5.11 (right panel) occurs because the nanoparticle charge is lower under the rarefied plasma conditions. Alternatively, nanoparticles in high-density plasmas carry a larger electric charge; thus, the focusing effect due to the electrostatic force appears to be stronger under the dense plasma conditions [152].

Inter-microstructure spacing is another important parameter that affects the redistribution of the nanoparticle fluxes over the surface pattern features. Under conditions when the SMSs draw large and focused NP currents, one clearly sees the zones of depleted particle fluxes around the microstructures. It is noteworthy that quite similar depletion zones are also the case in plasma-assisted growth of discontinuous nano-/micro-islanded films [273]. Meanwhile, when microstructures of the same width and height are arranged in a micropattern of a larger period, the ability of individual SMSs to focus NP fluxes onto their surfaces increases. In this case the microstructures act as effective nanoparticle collectors, similar to pyramid-like submicrometer-sized structures in Figure 5.10(b) [153].

An interesting feature of the nanoparticle fluxes is their excellent uniformity in the inter-SMS valleys. Therefore, "soft" NP deposition from reactive plasmas is an excellent option to fill the space between the microstructures with a porous nanoparticle-made material. This can facilitate the development of low-k nanoparticle composite films for interlevel dielectrics in ULSI technology [274]. Note that for obvious reasons, mechanical strengths of such films can be a matter of concern. Another possiblity is when the nanoparticles break upon crash landing; in this case the filling of the space between the microstructures will most likely be amorphous [152].

Finally, sharper microstructures focus nanoparticles more effectively as demonstrated by the results presented in Figure 5.12. Examining the results in the top panel in Figure 5.12 one can conclude that sharper structures are likely to be coated by porous nanoparticle-made films around their bases, while quite a significant number of particles remain stuck to their tips. Here we recall that quite similar observations have been made in earlier experiments [93, 94] and are discussed in Section 5.2 of this monograph. Alternatively, the NP fluxes onto lateral surfaces of smaller-aspect-ratio microstructures appear to be very uniform. However, this does not necessarily mean that the SMSs should be uniformly coated by the plasma-grown particles – this depends on specific material, structure of the SMSs and other surface conditions [153].

5.3.3
Practical Implementation Framework

Let us now discuss practical aspects of the site-selective nanoparticle deposition from low-temperature plasmas. In large-area plasma deposition experiments the set of parameters is limited to input power, gas feedstock pressure and composition, substrate temperature and bias, and some other parameters. During the first stage, the nanoparticles grow in the ionized gas phase as discussed in Chapter 4. This process involves nucleation, growth, and crystallization in the ionized gas phase, which can be effectively controlled by the parameters of the plasma bulk [4, 28]. The second (nanoparticle size selection) stage, can be implemented by adjusting the temperature gradient in the presheath and the upper sheath [105, 152]. And finally, the manipulation of these nanoparticles in the lower sheath area can be implemented using the results presented in this section.

The effect of the electrostatic focusing can be very useful for controlled synthesis of a broad range of nanomaterials and nanoassemblies using plasma-grown nanoclusters. If the nanocluster charge is positive (this sensitively depends on their size [65]), intense flows of plasma-generated nanoclusters can be created and directed towards the nanoassembly sites where they are needed. This new technique would be a plasma-based equivalent of the cluster beam deposition (CBD), a powerful tool for nanoscale science and technology [79]. Moreover, since the CBD technique mostly uses neutral clusters as building units and complex aerodynamic lenses to focus them, the effective manipulation of charged nanoparticles in the plasma demonstrated in this work can have clear competitive advantages, for example, in precise and size-selective electric field-controlled cluster deposition into specified surface areas with complex surface morphology [153], such as the micropatterns of this section.

A very interesting aspect is to discuss what might happen after the plasma-grown nanoparticles are deposited on the surfaces of the microstructures. Interestingly, the charge on the particles is also crucial in the process of their attachment and/or incorporation into the growing structures. Assuming that positively charged particles land smoothly onto negatively charged surfaces of surface microstructures, electrostatic attraction can be a major factor in nanoparticle attachment to the surface microstructures. This effect has previously been reported for charged nanoclusters [65]. However, the effectiveness of the nanoparticle bonding to the surfaces will depend on the relation between the time of the charge transfer τ_{chtr} and the time required for bond formation τ_{bond}. If the bonding process completes well before the NP charge is neutralized ($\tau_{chtr} \gg \tau_{bond}$), the electrostatic attraction between the particle and the surface serves as a firm "grip" that significantly facilitates the bonding process [153]. It is remarkable that quite a similar principle is used to hold NPs together in binary nanoparticle superlattices (BNSLs). Moreover, electric charges on NPs determine stoichiometry and structural diversity of the BNSLs, which represent a novel class of nanoscale objects [275].

Conversely, if the NP charge is neutralized faster than the bonds form, the particle can move from the point of its initial attachment to the SMS surface. This becomes more obvious in the case of sharp microstructures when the as-attached particle can "roll" down the slope. This conclusion is certainly more accurate when the surface roughness of the

SMSs is much smaller (approaching atomically smooth surfaces) than the nanoparticle size [153].

The ability of NPs to reach the substrate and their specific site and energy of landing are extremely sensitive to a number of factors and parameters. Since these are in most cases process-specific, it would be futile to formulate ready strategies for how to manipulate such particles in a particular experiment. Nonetheless, using the conclusions drawn from the results in Figures 5.10–5.12, one can formulate the following practical suggestions [153]:

- to avoid NP contamination, the best policy is to prevent them from reaching the charge reversal point; for more specific details refer to Section 5.2 and original reports [105,152];

- to enable effective nanoparticle collection near the tips of the microstructures, moderate (ca. φ_f) surface potential conditions should be used (Figure 5.11, left panel);

- alternatively, low-surface-potential (e.g., sub-φ_f) conditions are ideal for smooth filling of the inter-SMS spaces by microporous nanoparticle-made material; preserving the negative charge on NPs can be advantageous as suggested by dash-dotted curve in Figure 5.11, left panel.

These practical hints are certainly not exhaustive and many other interesting features of nanoparticle deposition onto microstructured surfaces will be discovered if research in this direction is continued, both numerically and experimentally [153]. In the following section we will consider an effective experimental arrangement that can be used to prevent undesirable fallout of plasma-grown nanoparticles by using an electrostatic filter.

5.4
Electrostatic Nanoparticle Filter

As we stressed in Chapter 2, reactive plasmas have recently been recognized as a versatile nanofabrication tool [4] and are able to generate various species ranging in size from a few angstroms (atomic units) to several microns and above (large nanoparticles and agglomerates). Large nanoparticle species, a common troublesome problem of plasma-aided semiconductor micromanufacturing [28], should thus be avoided in nanofabrication that relies on subnanometer-sized building units.

From Section 5.2 we have learned that carbon microemitter patterns and other carbon-based nanostructures (CNSs), synthesized in

$Ar + H_2 + CH_4$ low-temperature plasmas suffer from contamination by the plasma-grown nanoparticles (NPs), see also original reports [93, 94, 105, 152]. Such nanoparticles can be partially removed by additional heating of the substrates, which gives rise to a temperature gradient-dependent thermophoretic force that repels the NPs away from the CNSs [105, 152].

However, as was shown in Sections 5.2 and 5.3, particles above a certain size still manage to overcome the potential barrier and incorporate into the nanostructured films. In this section we follow the original report [276] and introduce a simple yet effective technique which makes it possible to significantly reduce the nanoparticle contamination and increase the threshold nanoparticle fallout size in the plasma-enhanced chemical vapor deposition of carbon nanostructures.

To prevent the deposition of the plasma-grown nanoparticles onto carbon-based nanopatterns, Rutkevych *et al.* [276] used a metallic grid placed above the growth pattern. The resulting inhomogeneous electric field repels the plasma-grown nanoparticles (which, as we have stressed several times above, are usually negatively charged [28]) back to the plasma bulk, or otherwise diverts them from the nanofilm being grown. The grid can also collect the nanoparticles before they reach the substrate. Alternatively, by applying a negative (with respect to the plasma bulk) potential to the grid, one can also repel negative ions. Here we recall that according to Section 4.3, negative ions are commonly accepted as nanoparticle growth precursors in a range of reactive plasmas including methane-based plasmas. Therefore, by negatively biasing the filter mesh, one can impede or even potentially completely suppress the nanoparticle growth in the area between the grid and the nanostructures on the surface. This effect can be used in combination with thermophoretic manipulation as discussed in detail in Section 5.2. Using a balanced combination of these two effects, Rutkevych *et al.* [276] managed to achieve a nanoparticle-free deposition.

The carbon nanostructures in this section have been deposited on Ni-catalyzed highly-doped Si(100) substrates by using $Ar(35 sccm)+H_2(30 sccm)+CH_4(35 sccm)$ inductively coupled plasmas (ICPs). A 30 min substrate conditioning in Ar was followed by a 20 min etching of Ni catalyst layer in $Ar + H_2$ plasmas, and, finally, the actual 40 min PECVD in 6.133 Pa (46 mTorr)–6.799 Pa (51 mTorr) $Ar + H_2 + CH_4$ gas mixture. The input power of a 460 kHz RF generator was maintained at the 2 kW level, with the reflected RF power not exceeding 105 W. The surface temperature of negatively biased (-80 V) substrates was maintained between 330 and 375 °C, depending on any additional heating power supplied to the substrates, the latter was varied from 0–110 W [276]. Other details of the plasma reactor used in these series of experiments can be found elsewhere [93,189].

The nanoparticle filter used in the experiments of Rutkevych *et al.* [276] was made of a 2 mm-thick aluminum frame, and two sets of copper wires, each of 0.12 mm diameter. The wires were arranged in two layers separated by an 0.8 mm gap; the distance between the parallel wires within each layer is 1.1 ± 0.2 mm. The bottom surface of the nanoparticle filter is conducting. This arrangement made the filter grid equipotential with the substrate and negatively charged with respect to the plasma bulk. In each PECVD experiment four substrates cut from the same Si(100) wafer uniformly pre-coated with a 20–30 nm Ni catalyst layer were used. Only two of the substrates were protected by the filter, whereas the other two were subject to direct plasma deposition and served as reference samples. The effect of the filter on the film composition was investigated by comparing the SEM images and EDX analysis of the filter-protected and reference samples [276].

It is interesting that the as-deposited filter-protected samples appear to be uniform and feature regular surface morphologies over large surface areas. Moreover, they do not contain any traces of irregularities due to remote wire patterns. In contrast, the reference films appeared less uniform and their surface morphology was more irregular. Another interesting observation is that the films deposited by using the filter are smoother and adhere better to the substrate compared to their unprotected counterparts [276]. Moreover, the reference substrates are almost perfectly black and can easily be scraped off, similar to carbonaceous soot-made nanoparticle films. On the other hand, it appears more difficult to remove the gray-colored films from the filter-protected specimen. In particular, this suggests that adhesion of nanofilms to the substrate can also be improved by using electrostatic nanoparticle filtering. Moreover, the field emission scanning electron microscopy (FESEM) suggests that the presence of small nanoparticles on the nanostructured surface is much higher in the first case, as can be seen in Figure 5.14 [276].

Another important observation is that the reference (unprotected) films are thicker than the filter-protected ones. Since all specimens were subject to the same deposition duration, it is reasonable to conclude that the electrostatic filter reduces the deposition rates. This effect can be attributed to different film building units in the two different cases. Indeed, deposition rates of nanoparticle- and nanocluster-assembled films are usually higher than of those synthesized by using subnanometer-sized species [4]. The results in Figure 5.14 also illustrate that the threshold NP fallout size can be significantly increased by using the electrostatic filter maintained under the same electrostatic potential as the substrate [276].

(a)   (b)

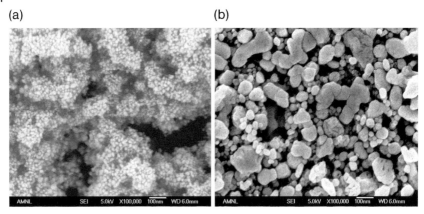

Figure 5.14 FESEM micrographs of disordered nanoparticle fallouts without (a) and with (b) the nanoparticle filter at 22 W substrate heating power [276].

In particular, one can observe that the films deposited at 22 W substrate heating power contain a very large number of disorderly dispersed nanoparticles, with the average particle size in the 10–20 nm range (Figure 5.14(a)). Alternatively, the filter-protected samples in Figure 5.14(b) feature a much larger amount of bigger (typically ca. 50 nm in size) nanoparticles. However, this heating power is below the CNS growth threshold, which is confirmed by the electron micrographs in Figure 5.14 and previous experiments of the same group [105] (see also Section 5.2). Furthermore, when the heating power is increased to 110 W, this additional substrate heating has two major effects [276]. The first effect is stronger catalyst fragmentation which is required for the nanostructure growth. The second effect of the external substrate heating is thermophoretic repulsion of the plasma-grown nanoparticles from the growth specimens.

Under conditions of increased external substrate heating, larger (20–30 nm in size) nanoparticles can reach the reference substrates (Figure 5.15(a)). More importantly, by combining the thermophoretic nanoparticle manipulation with the electrostatic filtering, one can almost completely remove the nanoparticles from the carbon nanostrucures as can be seen in Figure 5.15(b) [276]. Interestingly, the structures in Figure 5.15 are similar to carbon nanotip structures grown by essentially atomic (atomic hydrogen) and radical (e.g., CH_3/CH_3^+) building units [4] (see also Chapters 2 and 3).

The averaged over the surface area EDX results from different experiments are presented in Table 5.3 [276]. In both cases the presence of car-

(a)   (b)

Figure 5.15 Same as in Figure 5.14 for the substrate heating power of 110 W [276].

Table 5.3 Averaged EDX data of elemental composition of the coatings in different heating regimes (in weight %, filter-protected/reference sample) [276].

Element	No heating	22W	50W	110W
Si	55.3/64.3	63.6/61.5	53.8/70.2	63.0/68.7
C	6.0/20.3	7.3/20.3	6.9/12.3	15.2/16.3
Ni	23.6/12.8	21.8/15.8	23.6/14.6	18.7/12.9
Cu	14.2/0.1	6.5/0	14.6/0.2	2.2/0
Al	0/0	0/0	0/0	0/0

bon in the films appears to be high. However, filter-protected samples feature a somewhat lower carbon content, which can be attributed to the loss of carbon material due to elimination of large carbon nanoparticles from the deposition process. Presence of copper in the filter-protected specimen is due to the plasma sputtering of copper wires. When the copper wires were replaced by aluminium wires, the traces of Al were undetectbale by the EDX. This suggests that Al is an ideal wire material for future versions of the electrostatic nanoparticle filter. It is also possible that the Cu contamination problem can be overcome by using stainless steel wires. Meanwhile, one should not exclude the possibility that a Cu intervening layer could be the reason behind the better adhesive strength of the Si-Ni-C interface [276].

It was suggested that the irregular size of the particles deposited through the filter might be an indication of different growth cycle durations of nanoparticles levitated in the areas with different electric field magnitudes [276]. It is also possible that the particles trapped in the

extremely non-uniform electrostatic field may cause a significant contamination in discharge afterglow. Therefore, for better quality nanostructures it is advisable to use pure argon plasma after the deposition
(by switching the inlet of CH_4 and H_2 at the end of the PECVD stage)
for additional conditioning and removal of the particles levitated in the
presheath area.

This work should be continued to study relative contributions of the
gas-phase and surface processes in the PECVD of the carbon nanostructures concerned. To conclude this section, we note that the combinatorial
nanoparticle electrostatic filtering and thermophoretic manipulation approach is generic and can be used in the nanofabrication of a large number of nanofilms and nanostructures in nanoparticle-generating reactive
plasmas.

5.5
Concluding Remarks

In this chapter, we have considered several examples of transporting of
plasma-generated building units towards or away from nano- and microstructured surfaces (depending on the particular application). In Section 5.1, we considered three-dimensional topography of the microscopic
ion current onto the ordered array of vertically aligned carbon nanotips
targeted for electron microemitter array applications. By using Monte
Carlo simulation techniques, individual ion trajectories can be computed
by integrating the ion equations of motion in the electrostatic field created by a biased nanostructured substrate.

There are two distinctive cases of lower- and higher- electrostatic potential drop (U_s) across the plasma sheath. When the substrate is unbiased or floating (which corresponds to the low-U_s case), the nanotip
array focuses the ion flux more effectively than in the higher-U_s case (DC
biased substrate), which is one of the most interesting conclusions of Section 5.1. Quantitatively, under low-U_s conditions, the ion current density
onto the surface of individual nanotips is higher for higher-aspect-ratio
nanotips and can exceed the mean ion current density onto the entire
nanopattern by up to approximately 5 times. However, when the substrate bias voltage increases, this effect becomes less pronounced and the
mean relative enhancement of the ion current density ξ_i is reduced to
ca. 1.7 [51].

Nonetheless, in both cases the relative current density onto the nanotips depends strongly on the plasma density and the electron temperature. The value of ξ_i is higher in denser plasmas and behaves differently

with the electron temperature T_e depending on the substrate bias. When the substrate bias is only 20 V, ξ_i decreases with T_e, with the opposite tendency under higher-U_s (50 V) conditions. Thus, the distribution of the ion current along the nanotip lateral surface is strongly non-uniform and can be controlled by the plasma density and electron temperature.

The nanotip aspect ratio appears to be another important factor, since the ion flux focusing by smaller-radius nanotips appears to be more effective. The results presented in Section 5.1 demonstrate that the plasma parameters and substrate bias are the important factors that enable one to manipulate the microscopic ion fluxes onto the substrate and nanotip surfaces, eventually leading to the possibility of effectively controlling the growth of carbon nanotips.

In Section 5.2 we considered the possibility of manipulating the plasma-grown carbon nanoparticles in the plasma sheath using a combination of the main plasma forces. Here we stress that highly-precise manipulation of a large number of nano-sized particles during the discharge operation still remains a major challenge. Nonetheless, as the results of Section 5.2 suggest, the solution of this problem can be achieved by adjusting the plasma process parameters. In this way the nanoparticle delivery to nanostructured surfaces can be facilitated by dynamically balancing the main plasma forces acting on the particle in the vicinity of the solid surface. More specifically, the size-selective nanoparticle deposition onto the nanostructured surface can be achieved by tailoring the potential energy profiles in the near-substrate areas [152].

By means of direct *in situ* measurements of the near-substrate gradients of the working gas temperature and numerical simulation of nanoparticle motion in the vicinity of the deposition substrate under the action of various plasma forces, a comprehensive investigation of the effect of various factors that affect the nanoparticle deposition process has been undertaken [105,152]. Therefore, these results can be used to select and optimize the two (with and without nanoparticle incorporation into the films) different regimes of PECVD of various carbon nanostructures.

To obtain the optimum conditions for the nanoparticle deposition, one should use a balanced combination of higher operating pressures, lighter gas feedstock, lower temperatures of ions and electrons, as well as longer pre-sheath areas. The numerical model of Section 5.2 allows one to estimate the sensitivity of the actual potential energy profiles to variations of the most important plasma, nanoparticle, and film deposition parameters.

After reading Section 5.2, one can formulate a reasonable practical approach for the nanoparticle manipulation. This approach may be based on the selection of a certain control parameter that satisfies the following criteria:

- has a reasonably, strong affect on the nanoparticle's potential energy U_{pot};

- does not significantly disrupt the plasma discharge; and

- can be varied independently from other process parameters.

Amazingly, the results presented in Section 5.2 show that the near-substrate temperature gradient does satisfy all the above requirements. First, the effect of ∇T_g on U_{pot} is remarkable as can be noted from Figure 5.8(h). Second, this parameter mostly affects the neutral gas through the temperature gradient-controlled thermophoretic force. Third, the value of ∇T_g can be independently controlled by the external heating/cooling of the substrate stage.

Moreover, the nanoparticle size distributions on the nanostructured deposition surface can be controlled by the heating power as suggested by the results in Figure 5.6 (right panel). Interestingly, nanoparticles of a certain size can deposit on the surface even without any external substrate heating. Therefore, additional cooling of the substrate may accelerate the nanoparticle deposition process [152]. This effect might be relevant for the effective deposition of nanocrystallites onto liquid nitrogen-cooled substrates considered in Chapter 3 (see Figure 3.3).

On the other hand, one can also use the thermophoretic force to manipulate the nanoparticles in the applications that treat the plasma-grown fine particles as unwanted contaminants [152]. For example, it is commonly believed that ordered carbon nanotip structures are grown by atomic, molecular, and radical units rather than by using the plasma-grown nanoparticles. In this case, the thermophoretic force can effectively repel the nanoparticles even in the pre-sheath area. An important distinctive feature of this force is its long range which can extend beyond the pre-sheath area in the plasma bulk. However, certain practical limitations can arise due to limited power of the substrate heater and thermal properties of the substrate stage. The dimensions and layout of the chamber and substrate stage can be tailored to confine and eventually divert the plasma-grown nanoparticles to the pump line [152]. However, if these arrangements still do not help, additional nanoparticle filters [276] considered in details in Section 5.4, can be very helpful.

Section 5.3 focused on finding specific strategies to deposit plasma-grown nanoparticles onto specified locations on microstructured surfaces [153]. In Section 5.3 we introduced a two-dimensional simulation technique of a nanoparticle-generating plasma in a combined sheath-presheath area, and applied the model to simulate the NP deposition onto a microstructured solid surface. The discharge parameters and surface morphology strongly affect the nanoparticle deposition flux.

Unlike ions which carry a constant (e.g., positive) electric charge, the nanoparticles feature a dynamically variable charge, which can be used for sensitive control of their site-specific deposition with controlled energy onto the micropattern [153]. The deflection of nanoparticle trajectories in the vicinity of the SMSs is caused by the horizontal component of the electric force. This effect is enhanced due to a large electric charge on the nanoparticles; alternatively, it is suitably reduced at high particle velocities (e.g., when the surface potential is large).

Summarizing the results of Section 5.3, we stress that [153]:

- the localized electrostatic field in the vicinity of the microstructures affects the nanoparticle deposition flux; this effect is most pronounced when the surface potential is reasonably low as is the case for electrically floating insulating substrates;

- depending on the surface potential, the nanoparticles can be deposited either atop the microstructures, or fill up the spaces between the microstructures;

- plasma parameters also affect the nanoparticle fluxes; higher plasma densities are favorable for achieving more effective focusing; and

- sharper microstructures show a better ability to focus microscopic nanoparticle fluxes.

Future work in this direction should be related to more complex situations such as fabrication of binary nanoparticle superlattices using plasma-grown nanoparticles, and should involve various surface processes, broader ranges of plasma and surface parameters, and various nano-sized surface features. The results of the simulations presented in Section 5.3 are generic and are applicable to various processes involving low-temperature thermally non-equilibrium plasmas used for nanomaterials synthesis and surface modification [153].

In Section 5.4, we have reviewed the means of effectively reducing nanoparticle contamination in the reactive $Ar + H_2 + CH_4$ plasmas used for the synthesis of carbon nanostructures highlighted in this chapter. Following the original report [276], it was shown that by complementing the thermophoretic manipulation (discussed in detail in Section 5.2) by additional electrostatic filtering using a mesh of parallel wires, one can significantly reduce the amount of nanoparticles that can reach the substrate and therefore substantially improve the quality of carbon nanopatterns.

It was also observed that a further increase in the substrate heating power leads to an increased size of deposited nanoparticles. By optimizing the wire mesh configuration and adjusting the external heating of the substrate, one can completely eliminate plasma-grown nanoparticles from the surface and eventually achieve nanoparticle-free nanoassemblies. The original combinatorial approach of Rutkevych *et al.* is generic and is applicable to other reactive plasma-aided nanofabrication processes [276].

To conclude this chapter, we stress that the range of options and examples considered here is in no way exhaustive. One of the most important messages delivered in this chapter is that each species, be it a building unit, a working unit, or a deleterious contaminant, requires an individual approach uniquely tailored to specific process conditions. Again, we have arrived to a conclusion that supports the idea that most of nanoscale processes including the BU delivery stage, are indeed process-specific. However, knowledge of effective controls of the delivery processes of the required BUs/WUs is very useful to ensure that the subsequent stages of their interaction (e.g., stacking/incorporation for BUs and etching/sputtering for WUs) with the nanostructured surfaces proceed smoothly.

Indeed, it is extremely important to deliver exactly what is needed to exactly where it is needed, and, moreover, in appropriate amounts!

This "delicate delivery" of plasma-grown building units is one of the most stringent requirements in nanofabrication of arrays of quantum dots considered in detail in the following chapter.

6
Surface Science of Plasma-Exposed Surfaces and Self-Organization Processes

K. Ostrikov and I. Levchenko

In this chapter we discuss the topic of self-organizing processes that involve atomic and nanocluster building units delivered from the plasma
environment to the solid surface. These processes are traditionally covered by surface science, which is commonly understood as a research
field dealing with the structure, arrangements, kinetics, formation and
other properties of a few atomic layers of a solid near its interface with
the environment [277–279]. Atoms in these layers (for simplicity hereafter called surface atoms) are subject to very different conditions compared to their counterparts in the material bulk. Indeed, they face the
environment and so are the first to respond to any action exerted on the
material by external atoms, molecules, radicals, electromagnetic fields,
and so on.

In particular, the collective response of surface atoms to external actions determines whether the solid will accept external atoms and allow them to self-organize on the surface to form ultra-small nanoassemblies [279]. This particular response depends strongly on the properties
of the material itself, its surface, and is also strongly affected by the environment and its specific action. To understand the range of possibilities
arising from the interaction between the surface and the environment,
atomic description of surface processes based on diffusion and dynamic
models is commonly used [277,278].

Moreover, surface science is known as one of the most delicate research fields which deals with the motion of individual adsorbed atoms
(adatoms) along atomic-scale features on the surface and their self-
organization into small clusters, islands, and other nanoscale objects.
For this reason, small (typically submonolayer) amounts of adatoms
are supplied to the surface and then traced by ultra-sensitive tools
such as scanning tunnelling microscopy (STM) [280], atomic force microscopy [281], and several others. Furthermore, because of the delicate nature of atomic-scale surface processes, in most cases ultra-high-

Plasma Nanoscience: Basic Concepts and Applications of Deterministic Nanofabrication
Kostya (Ken) Ostrikov
Copyright © 2008 WILEY-VCH Verlag GmbH & Co. KGaA, Weinheim
ISBN: 978-3-527-40740-8

vacuum and "mild" (e.g. non-reactive) adatom creation and deposition environments are commonly used [279]. Some of the possibilities in this regard have been considered in the previous chapters.

Despite very impressive recent advances in this area, self-organization on solid surfaces exposed to "harsh" environments still remains one of the main challenges of surface science. Low-temperature plasmas are good examples of such environments. However, understanding the way in which plasma-related effects such as electric charge, ion fluxes, plentiful radicals, clusters and other reactive species, electric fields, polarization effects, and so on, may affect the elementary surface processes studied by surface science, and, moreover, the ability to control these processes through the plasma conditions, still remains a major challenge. Moreover, due to the overwhelming complexity of the theoretical description and experimental investigation of the elementary processes involved in plasma-surface interactions, plasma-exposed surfaces are outside the reach of the present-day capabilities of surface science.

As already mentioned in Chapter 1, one of the major aims of plasma nanoscienceis to bridge the three-dimensional (3D) world of the physics of plasmas and gas discharges and the two-dimensional (2D) world of surface science and gainfully use this unique combination to synthesize exotic nanoscale assemblies and nanostructured materials with intricate electronic, optical, structural, thermal and other properties. For an extensive discussion of how exactly these worlds can be bridged, please refer to Section 1.4 and Figures 1.10 and 1.11.

It is interesting that this "microscopic" aspect (involving surface adatoms and related processes) is relatively rarely used in the area of applied plasma research where "more macroscopic" and phenomenological descriptions of plasma-surface interactions are more common. In this chapter, we will introduce the basic approach to surface processes on plasma exposed surfaces, discuss some of the most important elementary processes involved in self-organization of nanoscale objects and highlight unique features of the plasma environment that can be used to control these processes. We will also discus some of the emerging challenges of this multidisciplinary research field.

To be specific, in this chapter we will focus on zero-dimensional nano-assemblies which are commonly referred to as quantum dots (QDs). The QDs are among the smallest nano-sized objects and their synthesis requires atomic precision. Nevertheless, most of the approaches introduced here are applicable to a broader range of nanoscale objects and will be used in subsequent chapters of this monograph.

In Section 6.1 we will introduce the main requirements for the synthesis of arrays of quantum dots, discuss some of the most important elementary (atomic) processes involved in the self-organized growth of such systems, and map a practical plasma-based pathway to overcome some of the acute problems of the existing growth modes/scenarios. Various possibilities in this regard will be considered. This section mostly contains a qualitative discussion and leads to basic understanding of the most important issues, concepts, and approaches. Detailed sets of equations that describe the basic surface processes are case specific and are introduced in Sections 6.1–6.5 of this chapter.

In Section 6.2, the process of deposition of Ge nanocluster seed nuclei using atomic and nanocluster incoming fluxes is considered. These nanoclusters contain up to a few tens of atoms and serve as seed nuclei for further development of arrays of quantum dots with the required nanodot size and surface coverage. Section 6.3 considers an example of a more complex binary Si_xC_{1-x} quantum dot system, and in particular, a practical approach to achieve highly-stoichoimetric SiC QDs and also to tailor the nanodot composition and layering throughout their internal structure.

Section 6.4 is devoted to the modeling of the development of the nanodot array from the initial seed nuclei of Section 6.2. The mechanism of self-organization leading to the higher order of individual QDs within the arrays is also discussed.

Section 6.5 elucidates some of the most important effects the plasma environment exerts on the processes of nanodot self-organization and crystallization on plasma-exposed surfaces. This chapter concludes with Section 6.6, which summarizes some of the most important issues related to the self-organized processes on solid surfaces facing low-temperature plasmas.

6.1
Synthesis of Self-Organizing Arrays of Quantum Dots: Objectives and Approach

We will first introduce the concepts of quantum dots and nanodots and their arrays, which is the main focus of this section. The first two terms are very often used interchangeably and refer to very small (and usually crystalline) objects with all three dimensions ranging from fractions of one nanometer to a few tens of nanometers. When positioned in nanopatterns, they represent clear and distinctive nanostructures. Quite a few real objects may qualify as nanodots. For example, small nanoclusters and nanoparticles, crystallites with different shapes and faceting arrangements, as well as nanoislands, also of different shapes and structures, may well be considered as nanodots, if positioned separately from each other. The nanodots can be arranged in patterns on the surface; in this case they are commonly referred to as surface nanodots. If they are buried in a layer of another material or the same material in a different phase state, they are termed buried nanodots.

However, not every nanodot can be considered a real zero-dimensional quantum dot. The main distinctive feature of quantum dots is their ability to effectively confine the electrons in all three dimensions [282]. To achieve this, the nanodot structure should be crystalline and, moreover, the dimensions should be comparable with the characteristic spatial scale over which electron confinement is effective. For semiconductor materials, this scale is of the order of the exciton's Bohr radius, which varies from one material to another but typically stays within the 1–20 nm range.

Therefore, to achieve effective electron confinement in all three dimensions, one should be able to synthesize crystalline nanodots with sizes of the order of 10 nm as the upper limit. In this case, the electronic, optical, and some other properties of a single QD will be very different from those of related bulk materials. One of the most salient features of electronic spectra of such nanostructures is the presence of discrete energy levels; these are most common with individual atoms. This is why quantum dots are frequently referred to as artificial atoms. However, the response from a single QD might be too weak to be detected by existing electronic and/or optical devices. Therefore, a very large number of individual QDs is usually required. In most current applications, typical densities of surface-bound QDs are 10^{10}–10^{12} nanostructures per square centimeter.

The nanodots are usually arranged on the surface or within the host matrix in a pattern. Depending on the process parameters, these patterns may show quite different degrees of order. Indeed, small nanocrystals of very different sizes may be randomly dispersed within a layer of amorphous material. This is an example of a strongly disordered system. On the other hand, if size-uniform semi-spherical nanoislands are arranged in a regular pattern, one can speak of a high level of order in the system. The latter is often the case in ordered arrays. However, sizes and other parameters of individual nanostructures may be quite different even within ordered arrays.

Let us now examine the most appropriate approach for synthesizing nanopatterns that contain a very large number of nanodots. As we have already discussed in Chapter 1, top-down approaches (lithography, pattern delineation, etc.) traditionally used to fabricate microelectronic devices have inherent limits (e.g. resolution, precision and efficiency) that inhibit their usefulness to fabricate a very large number of tiny QDs. Another possibility is to use a nanomanipulator arm such as the tip of a scanning tunneling microscope and stack individual atoms, one by one, into nanodots. However, taken the huge number of atoms to be manipulated in such a way, this approach would be extremely inefficient. Indeed, if each of 10^{12} QDs in the area $1\,cm \times 1\,cm$ has, in average, 100 atoms in them, the number of runs the nanomanipulator arm needs to make would be ca. 10^{14}; if each run could be performed in $1\,s$, the whole process would take $10^{14}\,s$. To say the least, this is extremely inefficient and perhaps only the universe can wait that long until nanoparticles can be created in this way!

Therefore, referring to Figure 1.3, one can see that the only viable remaining option is self-assembly of individual nanostructures. If the nanodot pattern also needs to feature a certain degree of order, then the growth should proceed via self-organization from a more disordered to a more ordered state. Here it would be prudent to stress that the terms "self-assembly" and "self-organization" are also used interchangeably. In many cases, they indeed refer to the same thing when smaller building units combine and form more complex structures, without any external manipulation. Some of the relevant issues have already been discussed in Chapter 1.

Interestingly, one of the seminal papers on self-assembly of nanostructures via chemical synthesis by Whitesides *et al.* [283] discussed the possibility of utilizing these processes to fabricate delicate nanoscale objects. However, Whitesides *et al.* stated that while self-assembly (or self-organization with no difference drawn between these notions) via chemical synthesis was very promising, this approach was not suited to the production of elements of micro- and nanoelectronic circuits, primarily because of the necessity of regular arrays of nanostructures and their specific positioning (and we would also add interconnections here!) [283].

Each of my brilliant students would immediately argue with this stating that if the system of self-assembled quantum dots can somehow self-organize into ordered and interconnected arrays, then why is this approach not suitable? It would indeed be suitable to have fully self-organized nanodevices made of self-assembled and interconnected nanodots! Then what is the difference between self-assembly and self-organization? Thus, if we follow literally the prediction made by Whitesides *et al.* [283] in 1991, we would come to the inadvertent conclusion that self-organization is unlikely to lead to the degree of order required in the future nanoelectronics.

Let us now compare this with the original meaning of self-organization, which was described as spontaneous ordering in dissipative systems [284].

Moreover, the present-day interpretation of the term "self-organization" is a process through which a system restructures itself and re-orders its constituent elements without any (significant) external "guidance" or direct intervention. Therefore, self-organization is indeed a process that may lead to some sort of better ordering in the system. Then why should it not be suitable for the creation of ordered networks of nanostructures? The answer is simply because we do not completely understand how it works and how it can be controlled! And this chapter is about self-assembly of quantum dots and formation of their ordered self-organized arrays by using plasma-related controls.

To avoid any confusion, more and more researchers try to draw a distinctive line between self-assembly and self-organization. In the examples considered in the following sections, it is considered reasonable to term an essentially unassisted and spontaneous nucleation process of adatoms into nanodots with various (e.g. core-shell, muti-shell, functionally-graded, etc.) internal structures as "self-assembly". On the other hand, repositioning of individual nanoislands to arrange themselves into ordered arrays, also without any external intervention, would perhaps better be termed "self-organization", in order to stress that the system may spontaneously evolve into its more ordered state.

It is very important to emphasize here that because these transitions often happen under non-equilibrium conditions, the new ordered state may be less energetically favorable than the original more disordered state. An excellent example is the transformation of less ordered and energetically more favorable graphite into better ordered and less stable diamond, which commonly takes place under strongly non-equilibrium conditions such as high pressures and high temperatures. From our perspective, it is promising that a low-temperature plasma is a prominent manifestation of a non-equilibrium system. Moreover, low-pressure thermally non-equilibrium plasmas, which are commonly and widely used in microelectronic fabrication are examples of such non-equilibrium systems. If we continue our reasoning by recalling that using non-equilibrium systems may improve the degree of order in the system, we will arrive at the apparent conclusion: *non-equilibrium low-temperature plasma systems are very good candidates to create the as yet elusive self-organized nanodevices!*

Yes, there is very little doubt in that; however, the main question that remains is: How exactly can this be achieved? An answer to this question, which puzzles many researchers around the globe, will not be possible without detailed studies of plasma-assisted and guided self-assembly and self-organization on solid surfaces exposed to the plasma environment. This is one of the main aims of this chapter. Before we proceed with more specific discussions of what happens with the building units when they land onto plasma-exposed solids, let us summarize the most important practical applications and technical requirements for the synthesis of quantum dots and their ordered arrays.

Quantum dots and other low-dimensional semiconducting nanomaterials hold an outstanding promise for advanced optoelectronic, nanoelectronic, biosensor, and third generation photovoltaic applications [128, 285]. The applications of semiconducting QDs span vast fields including biomedical engineering, bionanotechnology, micro- and opto-electronics, quantum computing, data storage, nanoplasmonics, nanophotonics, solid-state lasers, quantum information, quantum computing and several others. In many applications such materials are used as thin films made of a large number of individual nanostructures arranged in patterns/arrays on a solid substrate or otherwise appropriately stacked in a host matrix [286].

For example, low-dimensional nanostructures such as nanoparticles, nanodots or nanowires can be either buried inside a layer of another material or grown on the top surface of a thin solid film. The former is the case, for example for polymorphous silicon (pm-Si) made of ultrasmall Si nanocrystals embedded in an amorphous silicon (a-Si) matrix (discussed in detail in Chapter 3, these quantum dots are termed buried QDs), whereas the latter is common with most epitaxial semiconductor nanomaterials made of single-material (e.g. Si, Ge, In, Ga, etc.), binary (e.g. AlN, SiC, InSb, GaAs, InAs, CdSe, ZnO), ternary (e.g. SiCN, AlInN, AlCN, etc.), and quarternary (e.g. SiCAlN, AlCInN, etc.) nanodots [286,287].

To be suitable for most of the envisaged applications, low-dimensional nanostructures such as QDs should satisfy a number of essential requirements summarized in Figure 6.1. The first group of such requirements is related to individual nanostructures [286]. First of all, they should be made of appropriate material and in most cases have a perfect crystalline structure. Secondly, it is essential to maintain the desired shape (e.g. aspect ratio, facets, etc.) within the size range (which is usually below the exciton's Bohr radius for most semiconductor materials, typically 1–10 nm) where quantum confinement effects are present. For binary, ternary and other multi-element low-dimensional semiconductors, there is one more essential requirement – the elemental composition should be stoichiometric or, alternatively, the elements should be alloyed in any required fractions. There are some more complex requirements relating to the internal structure of individual nanostructures, such as core-shell, heterolayered or compositionally graded structures; these specific

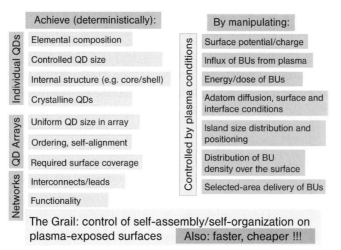

Figure 6.1 Main requirements and some plasma-related controls in the synthesis of ordered quantum dot networks.

attributes can be used to further modify electronic, optoelectronic, bioresponse, and some other properties of the semiconductor nanostructures.

When low-dimensional nanostructures are arranged in microscopic patterns or arrays, a large number of other important requirements arises [286]. The most essential one is the homogeneity of the main attributes of individual nanostructures (size, structure, crystallinity, composition, etc.) over the entire microscopic pattern. In the context of semiconductor nanodots on a solid surface, one can also require the highest possible surface coverage by the largest possible number of individual nanostructures. This is particularly important for applications where an overlap of wavefunctions of individual quantum dots is needed. Moreover, a reasonable degree of nanoparticle ordering and alignment within the array may be expected.

Depending on the nanofabrication technique used, the above requirements are met using quite different approaches. As already mentioned in Chapters 1 and 2, reasonable nanoparticle alignment can be achieved on solid substrates pre-patterned using focused ion beams and various lithographic techniques. In this chapter we explore the potential of nanoparticle self-organization into ordered arrays. For this, the nanodots should be able to self-order and self-align as shown in Figure 6.1. This is one of the most intriguing possibilities in view of a number of the limitations the currently existing pattern delineation techniques experience in the sub-100 nm range [5].

Present-day nanodevice applications further demand the ability to create interconnections between individual nanostructures within ordered arrays. Moreover, these interconnects may also need to be a lot more complex than just one-dimensional links between the nearest neighbors. Indeed, the network may need to be multi-levelled and the directions of interconnects in different levels may be different. Moreover, different levels should be reconnected using vertical interconnects commonly termed vias in microelectronics. There should also be suitable leads connecting the array of interconnected nanodots to external power circuits and so on. Above all, it is extremely important that the nanodevice-grade arrays can function properly in the envisaged applications. Otherwise, what is the point in synthesizing such complex nanoscale systems and networks?

As one can see from Figure 6.1, the Holy Grail of the self-organization-based approach of interest here is the ultimate ability to control the self-assembly and self-organization processes using plasma-related "turning knobs". Some of the most effective controls are also listed in the right column in Figure 6.1. Most of them have already been mentioned elsewhere in this monograph. For instance, low-temperature plasmas can be used to create specific building units in suitable energetic states, deliver them in appropriate amounts to where exactly they are needed, prepare the deposition surface and control/maintain the surface conditions. An example of how these different process conditions (e.g. surface charges, potential, reduced energy barriers, higher surface temperature under conditions of localized energy transfer from impinging ions, etc.) affect self-organized growth of nanodots on plasma-exposed surfaces will be given in Section 6.5. Last but not the least, plasma-based processes of nanofabrication of complex networks of quantum dots needs to be faster and cheaper in order to warrant their widespread industrial applications.

Let us now consider common methods for the self-organized growth of nanodot patterns and discuss the approach we will pursue throughout this chapter. Controlled delivery of building units from the nanofabrication environment and their self-assembly into surface nanopatterns is a commonly accepted and promising pathway to achieving this as yet elusive goal. For example, if the aim is to achieve a high and uniform surface coverage (approaching one monolayer) consisting of ultra-small (< 10 nm), size-uniform quantum dots, one would find it particularly challenging, if indeed possible at all, to direct QD self-assembly. The latter strongly depends on the surface energy and the mismatch between the lattices of the substrate and the nanodot material.

The range of possibilities in epitaxial lattice-matched and mismatched systems is sketched in Figure 6.2(a). If the lattices are matched, the epitaxial films grow via the Frank–van der Merwe (FM), also called the layer-by-layer (LBL), scenario. Quantum dot nanopatterns usually de-

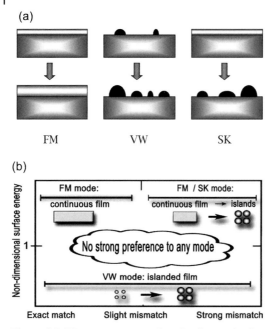

Figure 6.2 Three common modes of self-organized growth of quantum dots.

velop via an island nucleation (Volmer–Weber (VW)) or a strain-driven fragmentation of a few-monolayer thick continuous film (Stranski–Krastanov (SK)) mechanism [176].

Figure 6.2(b) shows the ranges of non-dimensional surface energy and lattice mismatch in epitaxial systems where such growth modes prevail. However, in both modes, the controllability of the QD areal density and uniformity is very limited. In fact, adatom nucleation in a VW mode produces nanopatterns with a broad variation in QD sizes, whereas fragmentation of continuous films into nanoislands in the SK mode is even less predictable owing to its strong dependence on the number of monolayers and other factors [176].

For example, Ge/Si nanodots develop via the Stranski–Krashtanov scenario because of a relatively small (ca. 4%) lattice mismatch in epitaxial Ge/Si systems. In this case 3–4 monolayers develop first and then break into disordered and size-nonuniform islands after the lattice strain (which appears due to the lattice mismatch) exceeds the fragmentation thershold. Above this threshold, the nanoislanded state is more energetically favorable than the strained continuous film. On the other hand, SiC/Si nanodots develop via the Volmer–Weber growth mode since the lattice mismatch between Si and SiC is approximately 20% [129].

Referring again to Figure 6.2(b), we stress that it is still unclear how to grow quantum dot arrays in lattice-matched systems with a high surface energy, which favor the layer-by-layer Frank–van der Merwe (FM) growth scenario (Figure 6.2(a)). Moreover, what should be done under conditions (area in the middle in Figure 6.2(b)) where there is no particular preference for any specific growth mode [176]?

Therefore, in order to be able to control the formation of nanoscale objects in systems with pre-determined lattice matching conditions, one should be able to control the energetics of the growth process. These include, but are certainly not limited to, introduction of lattice-matching interlayers, surface defects, affecting energy barriers of adatom insertion into nanostructures, as well as controlling stress distribution about the surface. This control may lead to better selectivity between the three possible growth modes sketched in Figure 6.2(a).

For a better understanding of the growth mode selectivity issue, let us consider a few basic possibilities in a system when atoms A are deposited onto the surface of a crystalline material B. The two main energies that control the arrangement of the material A on the substrate B are the energies of the bonds between atoms of materials A and B, $\mathcal{E}_{AB}$ and between atoms in material A, $\mathcal{E}_{AA}$. The first energy characterizes the bonding strength between the atoms of the deposited material and the substrate lattice, whereas $\mathcal{E}_{AA}$ quantifies the cohesive ability of the atoms of material A. Depending on the lattice mismatch $\delta l_{AB} = |l_A - l_B|$, where l_A and l_B are the lattice constants of materials A and B, respectively, the following possibilities may arise:

- when $\mathcal{E}_{AB} \gg \mathcal{E}_{AA}$ and lattices match, the film should grow in a layer-by-layer (FM) mode;

- when $\mathcal{E}_{AB} \ll \mathcal{E}_{AA}$ and lattices match, the film should grow in a nanoisland (VW) mode;

- when $\mathcal{E}_{AB} \sim \mathcal{E}_{AA}$ and lattices match, a range of options (FM, VW, SK or mixed modes) is possible;

- when $\mathcal{E}_{AB} \sim \mathcal{E}_{AA}$ and lattices are mismatched, a range of options (FM, VW, SK or mixed modes) is possible;

- when $\mathcal{E}_{AB} \gg \mathcal{E}_{AA}$ and lattices are mismatched, the film should grow in a layer-by-layer (FM) mode, possibly with somewhat corrupted layers;

- when $\mathcal{E}_{AB} \ll \mathcal{E}_{AA}$ and lattices are mismatched, the film should grow in a nanoisland (VW) mode.

The above arguments are quite simple and do not consider a few important factors related to surface and process conditions. However, they give some indication of how interactions between different sorts of atoms affect the arrangement of deposited material A on the solid surface B. To demonstrate the possibility of controlling the self-organized growth process via selective choice among the growth modes, we note that a more rigorous treatment of this problem would ultimately involve finding the most energetically favorable state of the substrate-nanofilm system. For example, if it is a strained layer in a weakly lattice-mismatched system, strain relaxation and a new, more favorable, energetic state is achieved via fragmentation of a continuous film into a nanoislanded film with a broad and non-uniform distribution of nanoisland sizes and their positions over the substrate. As we have discussed above, more ordered patterns of size-uniform islands would be less energetically favorable. Therefore, some non-equilibrium process conditions (such as low-temperature plasmas) are needed to achieve this goal.

The question is what would these non-equilibrium conditions affect and how could they be used to control the development of quantum dot arrays? This problem will be considered in more detail in Section 6.5. Here we only mention that non-equilibrium surface conditions also affect the energetics of the development of nanoislands and their patterns. For example, surface energy/potential can be affected by the deposition rates and surface charging conditions; accordingly, various energy barriers that describe the ability of adatoms to attach, diffuse, insert, detach, or evaporate to/from the growing islands can be affected via adatom polarization and controlled by the strength and configuration of microscopic electric fields.

Thus, our main aim is to learn how to use plasma-related controls (e.g. charging and excited states of building units, electric fields, ion bombardment, surface charges, etc.) to achieve better-ordered arrays of size-uniform quantum dots. In two sections of this chapter we will consider the examples of the Ge/Si epitaxial system which normally develops via the Stranski–Krashtanov scenario. And since this pathway leads to uncontrollable fragmentation of a continuous film into discrete islands, one would immediately pop the question as to whether it is possible to avoid this fragmentation in the first place?

One possible way is to use non-equilibrium plasma conditions to modify the energy barriers and so prevent the initial growth of a few monolayers of germanium on the silicon surface and direct the nanopattern growth straight into the Volmer–Weber mode. Looking at the basic possibilities for the development of epitaxial systems in terms of bond energies itemized above, one can say that to push the moderately lattice-mismatched material system A–B straight into the nanoisland development (VW) mode, one should try to either effectively increase the lattice mismatch or make it more favorable for atoms A to bond to each other rather than with the substrate atomic species B. In other words, the energy barriers for island formation should be reduced whereas those for the layer-by-layer growth should be increased.

Let us now consider these two options separately. An expedient method to increase the effective lattice mismatch would be to intentionally introduce some surface defects or dislocations. Note that this would also effectively reduce the surface energy, which is very desirable for the VW mode. However, introducing defects and dislocations also means intentionally causing quite substantial damage to the surface. This may lead to significant irregularities in surface conditions (e.g. stress distribution) which in turn may adversely affect the development of nanodot arrays at later stages of their growth.

Such considerations further pose more general questions. How does one introduce the defects in an ordered fashion to ensure that the final nanoarray will evolve into the ordered one? What size distribution should the defects have? Should they be of the same size or feature some specific size distribution? Another drawback of this approach is that the defect density should not exceed the maximum acceptable density in epitaxial semiconductor systems. Thus, if the density of quantum dots is to be very high, the defect density should be also increased accordingly. For presently acceptable defect densities in microelectronic manufacturing please refer to the description of the US Patent 5156995 (1992) [288].

A viable way to solve this problem is to create something which would effectively increase the lattice mismatch in the system without introducing surface defects. Thus, defect-free synthesis of a pattern of quantum dot nuclei (QDN) suitable for the subsequent growth of the desired QD array would certainly be an option. This growth stage will hereafter be referred to as stage I and is sketched in Figure 6.3. As will be discussed in Section 6.2, the initial pattern of seed nuclei may be created by intentionally seeding the surface with plasma-grown nanoclusters within the optimized size range [176].

The second possibility to drive the system into the VW mode and bypass the SK mode would be to reduce the energy barriers for islanding/clustering of atoms of material A and increasing the barriers for stronger attachment to substrate B. Interestingly, this is one of the main purposes of using transition metal catalyst layers in the synthesis of carbon nanotubes. As we will consider in Chapter 7, the catalysts may significantly facilitate incorporation of carbon atoms into the nanotube walls. In nanodot synthesis of interest here, there is usually no catalyst involved; therefore, the solid surface should act as a solid whose catalytic activity is enhanced by simply exposing it to the low-temperature

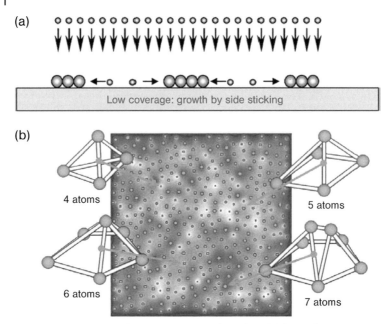

Figure 6.3 Deposition of the initial pattern of seed nuclei (stage I).

plasma! Reduced barriers for adatom clustering on moderately lattice-matched surfaces is a very realistic possibility; however, this assumption should be rigorously tested using advanced *ab initio* atomistic simulations.

However, effective and fast bonding with atoms of the same sort is not the only requirement for the formation of nanoislanded structures. One of the most important considerations in this regard is the ability of adatoms to move about the surfaces of the substrate and the nanostructures being grown. This is particularly important if crystalline nanodots are targeted. In the latter case the adatoms need to find the best attachment/bonding site to incorporate into the structure, after a series of hopping motions between intermediate adsorption sites. Thus, to be able to continue hopping from one site to another, an adatom should not completely lose its energy upon sticking to any intermediate site.

Should adatoms lose their energy, fractal-like structures sketched in Figure 6.4(a, b) are usually formed. In the situation sketched in Figure 6.4(a), an adatom was mobile and had some kinetic energy, which was immediately dissipated upon attachment to the first available atom in the atomic chain. As a result of the assembly of low-mobility adatoms via immediate bonding upon the first collision encountered, fractal-like films such the one shown in Figure 6.4(b), are usually formed.

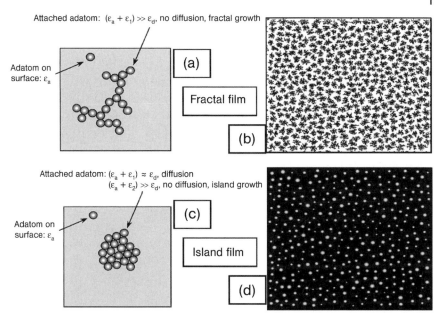

Figure 6.4 Possible arrangements of quantum dot seed nuclei to form fractal (a, b) and islanded (c, d) initial growth patterns. Not drawn to scale.

Figure 6.4(c) shows a quite different situation when an adatom is mobile before and after attachement to the first encountered atom in an island. In this scenario, the adatom sticks for a short while upon collision with the first encountered atom and then quickly detaches and continues its motion along the "path of least resistance", that is heads where it is easier to move to and so overcome any potential barriers for insertion into the nanodot. This path can be along the boundary of the island, a facet with the lowest diffusion energy, or another atomic layer, depending on the prevailing conditions. Furthermore, the more bonding sites an adatom is able to visit, the higher the chance it will find the most stable place in a crystalline lattice. Hence, the probability for the adatom to recrystallize increases with the rate of mobility. Here we recall that one of the main aims set in Figure 6.1 is to achieve crystalline nanodots.

Therefore, we have just reached an important conclusion: poor mobility – fractal films, higher mobility – islanded films, and even higher mobility – possibly highly-crystalline nanodots. As we will see from Section 6.5, exposure of a solid surface to a plasma leads to lower diffusion activation barriers and faster diffusion. As a result, it becomes easier to form nanoislanded and eventually crystalline nanodot films compared with using neutral gas-based processes. If the plasma exposure can also lead to lower bonding barriers, it will further increase the number of sites an adatom can visit during its lifetime on the surface and eventually better crystallization of the nanostructures.

Another important consideration is to deliver appropriate amounts of initial seed nuclei and not to oversupply them. The key factor in determining the exact amount of seed nuclei to produce and arrange on the surface is the desired density of the final nanodot pattern. To work out a suitable criterion, we need to introduce the surface coverage

$$\mu = S_{isl}/S_{tot} \tag{6.1}$$

which is defined as the ratio of the surface area covered by the islands S_{isl} and the total area covered by the film S_{tot}. If the islands are semi-spherical and the total film area has linear dimensions a and b, the surface coverage can be estimated as

$$\mu = \sum_{i=1}^{n} \pi r_i^2 / a \times b,$$

where r_i is the radius of the i-th island and n is the total number of islands in the discontinuous film. The surface coverage is a direct function of the mass deposited onto the surface. Generally speaking, μ increases with the time of deposition, when the mass deposited is transformed into more and more islands that cover larger and larger surface areas. Therefore, if one aims at synthesizing delicate nanoassemblies such as nanodots, it is extremely important to work out the right amount of material to deposit during the process, that is the deposition fluxes.

In the simplest case of infinitely small semi-spherical islands, the dependence of the surface coverage on time can be represented as

$$\mu = \frac{\Psi t}{\lambda^2}, \tag{6.2}$$

where Ψ is the deposition flux to the surface, t is the deposition time, and λ is the lattice constant. The rate of increase of the surface coverage in this case is constant, $\mu' = \Psi/\lambda^2$. From Equation (6.2), one can obtain the process duration

$$t_0 = \lambda/\Psi$$

required to cover the surface fully. Thus, the higher is the flux, the shorter time is required to achieve full (or any required) coverage.

In the limiting case of larger semi-spherical islands distributed over the surface one has

$$\mu = \frac{\pi \lambda^2}{a^2} \left[\frac{3a^2 \Psi t}{2\pi} \right]^{2/3}, \tag{6.3}$$

where a is the spacing between the islands. From Equation (6.3), one can conclude that the surface coverage decreases with the inter-island spacing proportionally to $a^{-2/3}$. Interestingly, for larger islands the temporal dependence of the surface coverage is quite different from the case of infinitely small islands, $\mu \sim t^{2/3}$. Thus, the surface coverage by larger islands increases slower compared to infinitely small islands. Moreover, the rate of surface coverage in this case decreases with time as $\mu' \sim t^{-1/3}$. This conclusion does not consider some more complex effects such as island coalescence. If the islands can coalesce upon contact, the surface coverage may even decrease with time.

Simple surface coverage considerations allow one to make a rough estimate of the appropriate incoming flux. This estimate is particularly useful in helping to work out the ranges of suiatble incoming fluxes of nanocluster seed nuclei during the first stage of fabrication of a QD nanoarray. For example, if one aims at creating a pattern of semi-spherical, 10 nm in diameter nanodots that cover 50% of the surface (thus the final surface coverage is $\mu \approx 0.5$) and assumes that every seed nucleus eventually develops into a 10 nm-sized nanodot, the required surface coverage by the nanocluster seed nuclei would be in the 10^{-3}–10^{-4} range.

If the process of deposition of such nanoclusters is expected to last for 1 s, then the required incoming flux would be approximately $\Psi \approx 10^{-3}$–10^{-4} ML/s. This simple estimate is more appropriate for two-dimensional islands and does not take into account the actual amount of material (measured in monolayers, ML) needed to create semi-spherical islands; therefore, the actual flux would need to be somewhat higher. Nevertheless, these simple estimates have been used in numerical simulations in Sections 6.2–6.4 and can serve as a guide for experimentalists on the appropriate choice of the incoming flux of building units. Quite similar estimates can be made assuming that the incoming flux consists of other (e.g. atomic/radical) building units.

Let us now proceed with the next stage of nanopattern development in moderately lattice-mismatched systems developed into nanoislanded films during the initial stage. As we have seen above, during this stage the initial seed nuclei should be created deterministically to fit the requirements for the targeted final nanopattern. The second stage of this process is shown in Figure 6.5. At this stage, some of the deposited initial seed nuclei (clusters) are still mobile and can migrate as a whole and experience collisions with other species on the surface. However, the rates of cluster mobility are inversely proportional to their masses. Therefore, only the smallest nanoclusters are able to experience any significant motion about the surface as can be seen in the background image in Figure 6.5. Heavier clusters, on the other hand, experience only small dis-

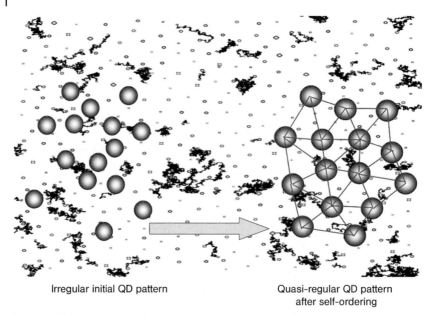

Irregular initial QD pattern   Quasi-regular QD pattern
after self-ordering

Figure 6.5 Possible self-ordering of moveable QD seed nuclei
on plasma-exposed surfaces. Background image shows traces of
cluster motion about the surface (numerical simulation of P. Liu,
I. Levchenko, and K. Ostrikov, unpublished).

placements from their original positions. In principle, this process may
lead to some sort of reorganization of nanocluster seed nuclei on the surface and possibly to better ordering.

As will be discussed in more detail in Section 6.5, unusual and essentially non-equilibrium conditions in plasma-based processes lead to
marked differences in the self-assembly of surface adatoms into QD
nanopatterns. However, simply avoiding nanofilm fragmentation and
seeding the surface with small nanoisland nuclei is not yet sufficient evidence to demonstrate the advantages of plasma-based processes in nanodot fabrication [279]. For this, effective plasma-related means of generating patterns of size-, shape-, and position-uniform quantum dots are
required.

One effective way to generate size-controlled nanoisland patterns on
plasma-exposed surfaces is depicted in Figure 6.6 [279]. Figure 6.6(a)
shows the temporal dependence of the island size for 5 selected islands
from the nanopattern shown in Figure 6.6(c). Figure 6.6(b) shows experimentally generated highly-uniform distributions of SiC nanoislands
on a Si(100) surface in the integrated plasma-aided nanofabrication facility [49,129]. The selected islands are shown by the arrows in Figure 6.6(c).

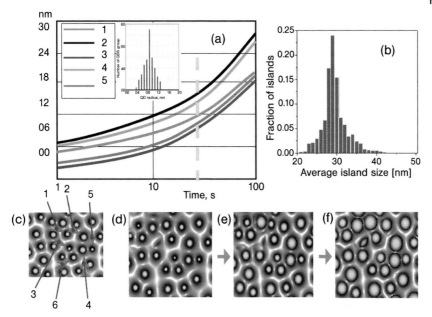

Figure 6.6 Temporal evolution and size distribution of SiC nanoislands on a Si(100) surface in a low-temperature plasma-assisted magnetron sputtering process: (a) nanoisland size versus time into the deposition process; the inset shows a typical island size distribution after approximately 30 s into the process (time shown by a dashed line); (b) representative experimental size distribution [129]; (c) island numbering used in panel (a); (d, e) temporal evolution of the initial island pattern at $t = 0$, 15, and 30 s into the process. Other parameters: surface temperature 450 °C, incoming fluxes Si and C atoms 0.01 and 0.015 ML/s, respectively; ionization degree of Si and C building units is 10^{-3} [279].

The temporal evolution of the selected nanopattern is depicetd in Figures 6.6(d, e). In this case, initial nanoislands nucleate from small nanoclusters (deposited from the plasma during the first stage) as discussed above. Moreover, this deposition also requires the specific optimization of the size and flux distribution of nanoclusters; this process in application to the Ge/Si nanodot system will be considered in Section 6.2. Interestingly, the best results in terms of size uniformity of the developed nanopattern such as the one shown in Figure 6.7 can be achieved from initial seed nuclei with a narrow size distribution [176].

For a better understanding of how seed nuclei eventually develop into nanodots, let us discuss the main processes and various possibilities in the growth of nanoislanded films as sketched in Figure 6.8. The developing system consists of the nanoislands distributed about the surface (in an ordered or disordered way depending on the process requirements

Before self-ordering   After self-ordering

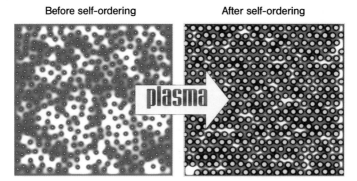

Figure 6.7 The ultimate goal: simulated nanodot arrays before
and after self-ordering in a plasma environment [279].

and conditions) and a two-dimensional field of adatoms from which the
nanoislands grow. The gradients of adatom concentrations determine
their fluxes onto two-dimensional island borders. These gradients and
particle fluxes are calculated with respect to the mean level of adatom
concentration on a "free" (i.e., unprerturbed by the islands) surface. Depending on the specific locations of each individual island with respect to
other islands and the two-dimensional distributions of adatom concentrations around them, different possibilities, sketched in Figures 6.8(a–f),
may appear. The development of individual nanoislands and their patterns also critically depends on the prevailing surface coverage. Let us
now follow Figure 6.8 and discuss various situations that may arise in
the process of nanopattern development.

Figures 6.8(a–c) show basic processes of island growth and dissolution
when the surface coverage is low. In this case, the main factor that determines the nanopattern evolution is the relative concentrations of atom
densities in the growth islands with respect to the background adatom
density. If the atom density within the island is higher than the mean
level, such an island is commonly termed supercritical (Figure 6.8(a)).
On the other hand, if the opposite inequality holds, one deals with undercritical islands as depicted in Figure 6.8(b). For supercritical islands,
adatoms flow away from them, which may lead to eventual dissolution
of such islands as shown in Figure 6.8(a). In constrast, if the island is
undercritical, the adatom flux is directed towards its border and such an
island grows.

Figure 6.8(c) shows two (one undercritical and one supercritical)
nanoislands in the low surface coverage case. Even though each island
may significantly affect the distributions of neighboring adatoms, the
two islands are positioned far enough away from each other so that the

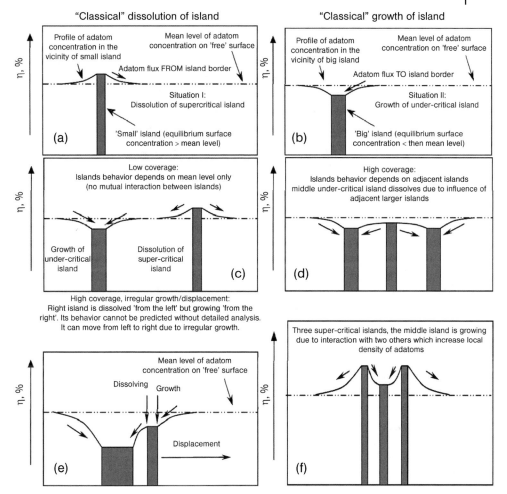

Figure 6.8 Nanoisland growth and dissolution processes in nanopatterns with low (a–c) and high (d–f) surface coverage.

mean adatom density level is not affected. In this case the two islands simply do not feel each other's presence in the pattern.

The situation changes quite substantially when the surface coverage increases (Figures 6.8(d–f)). In this case, nanoislands are not only affected by the mean concentration of adatom density but also by adatom density values and gradients around neighboring islands. The latter two factors may dramatically change the growth conditions compared to the low-coverage cases in Figures 6.8(a-c). For example, a small island in the middle in Figure 6.8(d) is undercritical, yet it dissolves because of the strong adatom outflow towards two adjacent larger islands.

Another situation involving two closely positioned underctitical islands is sketched in Figure 6.8(e). In this case, the left side of the smaller island dissolves because of the presence of a larger island on its right. However, an intense flux from the background two-dimensional adatom field leads to the growth of the island on its right side. If the rates of the growth in the right direction are higher than the rates of dissolution of its left side, one would observe a displacement of the center of mass of the island to the right. Panel (f) in Figure 6.8 shows the possibility of a supercritical island growing in the presence of two adjacent islands with even higher atom densities. In this case the island in the middle grows whereas the other two islands dissolve, until the equilibrium state is reached.

In the growth process the island borders may shift and eventually overlap leading to mass transfer between the islands. In this case, partial or even full coalescence may take place. As a rule of thumb, dissolution is one of the main reasons why small islands disappear, whereas coalescence would be a natural way for larger islands to merge and form even larger nanoscale objects.

The processes sketched in Figure 6.8 are strongly affected by evaporation of adatoms from the surfaces of the islands and the substrate. The two-dimensional nature of the "surface playground" results in quite unusual 2D evaporation processes to and from nanoisland borders. The 2D evaporation processes are complemented by the usual evaporation to the gas phase. These processes strongly depend on the surface temperature and specific activation energies for each particular process. These and several other processes will be rigorously taken into account in the following sections.

Returning to the nanoisland/nanopattern development in Figure 6.6, we stress that to obtain highly-uniform nanodot size distributions, small islands as in Figure 6.6(c) should develop smoothly. This means that they should not dissolve, evaporate, or coalesce and their size should increase gradually to reach the desired sizes and size uniformity as shown in Figure 6.6 [279]. The process duration can be selected to reach any preset nanodot size as shown by a dashed line in Figure 6.6(a); this line corresponds to the final nanoarray shown in Figure 6.6(f). More importantly, direct comparison of nanopatterns of nanoislands of the same average size shows that nanoarrays developed in a plasma-based process form significantly faster and also feature substantially better size uniformity compared to those from a neutral gas-based processes [279].

Here we stress that the simulated patterns of SiC/Si nanodots in Figure 6.6 have been obtained deterministically to achieve the average size and size distribution (shown in the inset in Figure 6.6(a)) corresponding to the nanopattern in Figure 6.6(f). In particular, the number of nanocluster seed nuclei delivered to the surface has been estimated (in the same way as we have done in this section discussing the appropriate choice of the incoming flux) to achieve the final pre-selected (approximately 35% in the case shown in Figure 6.6(f)) surface coverage by size-uniform, in average 10 nm-sized islands. The estimates show that in this case the initial surface coverage by the seed nuclei should not exceed ca. 5×10^{-3}. Moreover, the incoming flux of plasma-generated nanoclusters, ions, and atoms has been adjusted similarly to the Ge/Si case considered in Section 6.2.

The last stage of the nanodot pattern development involves self-ordering and size uniformization of individual nanodots within the array. It is essential that during the growth process the plasma-grown nanodots should be able to self-organize to form quasi-uniform arrays from essentially non-uniform initial patterns of small seed nuclei as shown in Figure 6.7 [279]. Here we stress that the plasma conditions strongly affect two-dimensional distributions of adatom fluxes to the growing quantum dots.

As will be shown in Section 6.4, even reasonably large nanodots (with sizes of ca. 10 nm) may slightly reposition themselves to improve the order in the pattern. All the dots are subject to adatom fluxes from 4 different directions. If these fluxes from different directions are balanced, the nanodots grow without shifting the position of their center of mass about the surface. However, if the fluxes are unbalanced, the dots are shifted predominantly in the direction of the total flux. As will be shown in Section 6.4, the nanopattern in a Ge/Si system self-arranges in such a way that individual dots within the array minimize the differences between their distances to their nearest neighbors [177]. More importantly, this effect becomes more pronounced when more building units arrive to the surface in the ionic form (the ionization degree is higher). Furthermore, fairly uniform stress distribution caused by ion bombardment can improve the outcomes of this process. Indeed, areas of higher stress are expected to be more attractive for nanodot nucleation and exert an effect which is quite similar to the common effect of nanocluster seed nuclei, defects and dislocations. Some of these interesting phenomena will be discussed in Section 6.5.

Finally, the last group of aims stipulated in Figure 6.1 is related to creation of ordered and interconnected networks of nanostructures. This goal requires the development of precise and, more importantly, self-organization-based fabrication techniques to grow interconnects between individual nanostructures in the arrays. These processes are however outside the scope of this chapter. Some examples of interconnected nanostructure networks will be presented elsewhere in this monograph.

In the following section, we will concentrate on the moderately lattice-mismatched Ge/Si system and consider the plasma-based approach for creating initial patterns of seed nuclei. As we have discussed above, these nuclei make it possible to avoid the highly-uncontrollable nanofilm fragmentation and so define the growth pattern for the development of ordered nanodot arrays.

6.2
Initial Stage of Ge/Si Nanodot Formation Using Nanocluster Fluxes

In this section we will follow the original report by Rider *et al.* [176] and simulate the growth of ultra-small Ge/Si quantum dot (QD) nuclei (ca. 1 nm) which are suitable for the synthesis of uniform nanopatterns with high surface coverage. The process considered is based on controlled deposition of atom-only and size non-uniform cluster fluxes onto the growth surface. These simulations predict that initial seed nuclei of more uniform sizes can be formed when clusters of non-uniform size are deposited. This result is counter-intuitive and can be explained via adatom-nanocluster interactions on silicon surfaces. These numerical results are supported by experimental data on the geometric characteristics of QD patterns synthesized by nanocluster deposition. In this section we will also discuss the role of plasmas as non-uniform cluster sources and their impact on surface dynamics. The technique proposed by Rider *et al.* [176] challenges conventional growth modes and is promising for deterministic synthesis of nanodot arrays.

Let us begin our consideration by recalling that in commonly used cluster beam deposition techniques one usually tries to utilize beams of size-uniform nanoclusters generated in the gas phase [78]. However, it is a common observation that in such processes the sizes of seed nuclei on the surface often appear quite non-uniform despite the delivery of uniform cluster fluxes from the gas phase. This provokes an obvious question: Do the gas-borne nanoclusters have to be of uniform size? As already mentioned above, we will introduce a counter-intuitive strategy for *creating size-uniform nanodot patterns by using fluxes of size-non-uniform clusters*. The resulting size-non-uniform seed nuclei then serve to obtain the desired QD nanopatterns via controlled self-assembly. While such cluster distributions may be generated from a range of sources, utilising nanocluster-generating plasmas offers a number of advantages, which include increased control of the clusters production, nanocluster manipulation within the plasma sheath, not to mention the surface energetic considerations such as the impact of plasma-related electric fields on adatom diffusion rates.

6.2.1
Physical Model and Numerical Details

We will consider Ge/Si QDs, one of the most popular QD systems, with a moderate (ca. 4%) lattice mismatch. It is instructive to recall that this system is commonly grown using a variety of neutral gas processes such as chemical vapour deposition and molecular beam epitaxy and typically develops via the SK mechanism, which comes into play after a few monolayers have been epitaxially grown [289]. In the alternative method proposed by Rider *et al.*, size non-uniform nanoclusters are delivered alongside atomic/ionic building units, which then self-assemble and create the desired QDN pattern on a Si(100) surface. Note we use the term QDN to refer to a cluster of atoms on the substrate; this is to avoid any confusion between clusters formed on the surface and the nanoclusters delivered from the gas phase.

To demonstrate this, the initial stage of Ge/Si QD pattern formation on the Si surface exposed to an atom/ion only flux and two different atom/ion/nanocluster fluxes (Figure 6.9(a)) has been simulated and analyzed. These options will be referred to as the atom-only route and non-uniform cluster routes (NUC and NUC2), respectively(see Figure 6.9(b)). The nanocluster fluxes considered include clusters of up to 25 atoms. For the atom-only route, a flux absent of any nanoclusters was used. For the NUC and NUC2 cases, size non-uniform nanocluster fluxes as shown in Figure 6.9(b) were used which are typical for the low-temperature plasmas [262,290].

The species included in the model for the NUC and NUC2 fluxes were restricted to atoms, monatomic Ge ions and neutral clusters consisting of between one and twenty atoms. The QDN pattern formation was simulated separately for the atom-only, NUC and NUC2 fluxes. It was assumed that the surface is stress- and defect-free. This assumption well justified for the moderate lattice mismatch and very low (ca. 10^{-4} ML) surface densities considered.

The QDN densities examined here (e.g. a rough estimate for a QD of approx. 10 nm in diameter with a spacing of 20 nm between QDN centres, on a substrate with lattice density 4×10^{18} m^{-2}, implies a nanodot density of 2.5×10^{15} m^{-2}) are significantly higher than the highest surface defect densities acceptable in microelectronics [288] which as we have stated previously is ca. 10^{10} m^{-2}. Given that the number of islands formed equals the number of defects, this implies that QD fabrication

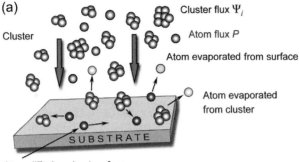

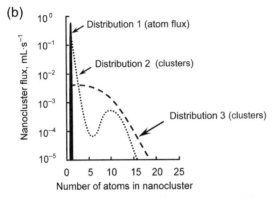

Figure 6.9 Main processes on substrate surface during quantum dot deposition from atom and nanocluster-containing fluxes (a); and atom-only flux (Distribution 1), nanocluster-containing flux for NUC2 process (Distribution 2) and nanocluster-containing flux for NUC process (Distribution 3) used in simulations (b) [176].

techniques based on intentional surface defects will not be capable of producing the dense QD patterns in which we are interested [176].

Our model takes into account the main processes of BU delivery and consumption on the surface. Building units are delivered to the seed formation sites by diffusion about the surface and are consumed by adatom attachment to the growing seeds. The surface density η_i [m^{-2}] of QDN consisting of (i) atoms (where $i > 1$) can be obtained from:

$$\frac{\partial \eta_i}{\partial t} = \Psi_i + \zeta_{i,C} + \zeta_{i,2D} + \zeta_{i,3D}, \tag{6.4}$$

where $\partial \eta_i / \partial t$ is the change in density (m^{-2}s^{-1}) of (i)-atom QDN and Ψ_i is the deposition rate of clusters (m^{-2}s^{-1}) consisting of (i) atoms ($i > 1$)

$$\zeta_{i,C} = 2v_d\eta(r_{i-1}\eta_{i-1} - r_i\eta_i) \tag{6.5}$$

is the rate of density variation of nanoclusters $[\mathrm{m^{-2}s^{-1}}]$ consisting of (i) atoms due to adatom collisions with nanoclusters consisting of (i) and ($i-1$) atoms

$$\zeta_{i,2D} = \eta_{i+1}n_{i+1}v_{i+1} - \eta_i n_i v_i \tag{6.6}$$

is the rate of nanocluster density variation due to atom evaporation to the 2D (surface) vapor from nanoclusters consisting of (i) and ($i+1$) atoms, and

$$\zeta_{i,3D} = \pi\eta_k(r_{i+1}^2\eta_{i+1}\mu_{i+1} - r_i^2\eta_i\mu_i) \tag{6.7}$$

is the rate of nanocluster density variation due to the atom evaporation to the 3D (external) vapor from nanoclusters consisting of (i) and ($i+1$) atoms. Other variables used in Equations (6.5)–(6.7) are as follows: η is the surface density of adatoms,

$$v_d = \lambda_{\mathrm{lat}}v_0\exp(-\varepsilon_d/k_BT_s)$$

is the adatom surface diffusion rate (m/s), λ_{lat} is the lattice parameter (m) for the silicon substrate, $v_0 = 2k_BT_s/h$ is the lattice atom oscillation frequency $(\mathrm{s^{-1}})$, k_B is Boltzmann's constant, h is Planck's constant, T_s is the surface temperature (K), ε_d is the surface diffusion activation energy (eV), r_i is the radius (m) of QDN consisting of (i) atoms, n_i is the number of atoms at the border of QDN consisting of (i) atoms, $\eta_k = 1/\lambda_{\mathrm{lat}}$ is the surface density of Si atoms on the substrate surface,

$$v_i = v_0\exp(-\varepsilon_{b,i}/k_BT_s)$$

is the rate of atom evaporation $[\mathrm{s^{-1}}]$ to the 2D (surface) vapor from borders of QDN consisting of (i) atoms, $\varepsilon_{b,i}$ is the energy of atom evaporation to the 2D (surface) vapour from borders of QDN consisting of (i) atoms,

$$\mu_i = v_0\exp(-\varepsilon_{a,i}/k_BT_s)$$

is the rate of atom evaporation to the 3D vapor from surface of QDN consisting of (i) atoms, and $\varepsilon_{a,i}$ is the energy of atom evaporation to the 3D vapor from surface of QDN consisting of (i) atoms [176].

The balance of adatom density on the solid surface is described by [176]

$$\frac{\partial\eta}{\partial t} = P + P_{ne} - P_e - P_{na}, \tag{6.8}$$

where P is the external flux of atoms $(\mathrm{m^{-2}s^{-1}})$ to the substrate surface,

$$P_{ne} = \sum_2^{\infty} n_i\eta_i v_i \tag{6.9}$$

is the flux of atoms evaporating from the QDN borders to the 2D vapor on the substrate surface,

$$P_e = \eta v_0 \exp(-\varepsilon_a / k_B T_s) \tag{6.10}$$

is the flux of adatoms evaporating from the substrate surface, ε_a is the energy of Ge adatom evaporation from the substrate (Si) surface, and

$$P_{na} = 2\eta v_d \sum_2^\infty r_i \eta_i \tag{6.11}$$

is the flux of adatoms attaching to QDN [176].

The evaporation energies $\varepsilon_{a.i}$, $\varepsilon_{b.i}$, and QDN radius r_i depend on the QDN size. The quantum dots are treated as hemispherical, which is close to the shapes observed in experiments [291]. The characteristic energies can be calculated by taking into account the number of bonds between the atoms constituting the QDN. Thus, for Ge atom evaporation to the 2D vapor one has: for $i = 2$, $\varepsilon_{b.2} = \varepsilon_b$ (1 bond); for $i = 3$, $\varepsilon_{b.3} = 2\varepsilon_b$ (2 bonds); for $i = 4$, and so on approaching $\varepsilon_{b.i} = (1/2)\varepsilon_{a.Ge}$ is the energy of atom evaporation from the surface of bulk germanium.

On the other hand, for evaporation from a 2-atom nucleus to the 3D vapor, a Ge atom should spend the energy ε_b for breaking one bond with a Ge atom in addition to the energy ε_a for breaking the bond with the surface. Here we recall that the model considers quantum dot nuclei that consist of a discrete number of atoms and thus exhibit properties that depend discretely on their size. Therefore, the energy of Ge atom evaporation to the 3D vapor from the surface of QDN consisting of (i) atoms is $\varepsilon_{a.2} = \varepsilon_b + \varepsilon_a$ for $i = 2$ and approaches $\varepsilon_{a.i} = \varepsilon_{a.Ge}$ for large i, where $\varepsilon_{a.Ge}$ is the energy of Ge atom evaporation from the surface. The number of atoms at the QDN border n_i was also determined by analyzing the QDN geometrical shape; this number approaches $n_i = 4.83i^{1/3}$ for large i.

In the simulations, Rider *et al.* [176] used the surface diffusion activation energy $\varepsilon_d = 1.072 \times 10^{-19}$ J (0.67 eV), evaporation energy $\varepsilon_a = 4.304 \times 10^{-19}$ J (2.69 eV), and bonding energy $\varepsilon_b = 2.400 \times 10^{-19}$ J (1.5 eV), as well as the lattice parameter $\lambda_{lat} = 5.4 \times 10^{-10}$ m, representative of the Ge/Si(100) system. The substrate temperature was $T_s = 600$ K, the total QDN surface density $\rho = 0$–2×10^{-3} ML, total external flux $P = 1$–10 ML/s, and the mass of nanoclusters $m_{clust} = 72$–1800 amu.

Despite its apparent simplicity, this model of the initial stage of development of nanodot patterns, is comprehensive and accurately describes the size distributions and surface coverage by small Ge seed nuclei with the number of atoms in them not exceeding 25–30. The quantum dot seed distribution function was calculated at the end of the initial stage of the simulation process. To explicitly show the advantage offered by the nanocluster route over the atom-only route in generating size-uniform patterns of nanosized quantum dots, after obtaining the distribution of the quantum dot nuclei, a further numerical experiment was conducted.

In these computations, the nanodot growth was examined using a standard diffusion model, with the sole aim being to demonstrate how the difference between the two QDN patterns formed from these fluxes (atom only and NUC) impacts on the further evolution of the quantum dot pattern. To do this, a diffusion model within the ranges of QD surface coverage up to 0.25 and quantum dot radii 5–10 nm was used. This model is based on a standard diffusion equation of adatoms on the substrate surface and will be considered in details in Section 6.4. For further details of numerical procedures and assumptions made please refer to the original report [176].

6.2.2
Physical Interpretation and Relevant Experimental Data

Figures 6.10(a) and (c) show the temporal evolution of the surface densities of adatoms and QDN consisting of 2–10 atoms computed for the atom-only and the NUC case. For the atom-only process, as shown in Figure 6.10(a), all densities increase for the first 10 s, and then tend to saturate. During the initial seconds smaller QDN have higher densities, but between 15 and 25 s the ordering of densities gradually changes and becomes inverted; namely, the QDN consisting of 10 atoms now have the highest densities, whereas the adatoms and 2-atom QDN have the lowest [176].

The two most striking observations in the NUC case (Figure 6.10(c)) are the very strong fall of adatom density and the similar behaviour exhibited by the densities of all QDN during the first 10 s. Between 0 and 25 s, the difference between the densities of the QDN and adatoms reaches approximately 2 orders of magnitude. The temporal evolution of surface densities for the NUC2 case (Figure 5(a) in the original report [176]) shows that QDN densities rapidly increase during the first 5 s, and then tend to saturate at levels that depend on the number of atoms involved in the QDN. Moreover, at the early stage (ca. 5 s) the density of adatoms exceeds that of i-atom QDN ($i > 1$). Thereafter, it decreases rapidly and becomes lower than any QDN density. In the NUC2 case, it was also observed that during the entire period of simulation there was an inverse relationship between the QDN densities and the number of constituent atoms, that is, the densities of smaller QDN always remain higher.

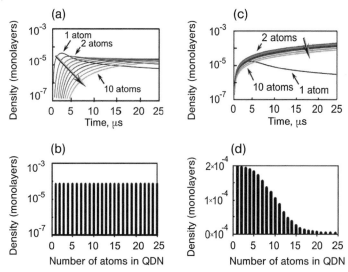

Figure 6.10 Temporal evolution of QDN surface density in the atom-only case (Distribution 1 in Figure 6.9(b)) (a); final QDN distribution function (b). Panels (c) and (d): same as above for the NUC case (Distribution 3 in Figure 6.9(b)). Large arrows in (a) and (c) show the direction of the QDN mass increase [176].

The corresponding equilibrium quantum dot nuclei distributions (taken at 1 ms) are presented in Figure 6.10(b) and (d) [176]. The QDN size distribution taken at equilibrium for the atom-only case is a uniform function in which the numbers of the smallest and largest QDN are approximately equal (Figure 6.10(b)). Figure 6.10(d) shows that the equilibrium distribution function of QDN deposited from the NUC flux exhibits a very strong decrease in the density of QDN consisting of 15–20 atoms. The density of QDN of 20 atoms is approximately 10 times lower than that of QDN consisting of 2 and 3 atoms, whereas the density of quantum dot nuclei consisting of 25 atoms and more approaches zero as can be seen in Figure 6.10(d).

The NUC2 QDN distribution function (not shown here) yields a similar result, albeit with a significantly less pronounced decrease. It also exhibits a clear descending shape; the decline is rather strong and covers approximately one order of magnitude. In fact, the density of QDNs that consist of 2 and 3 atoms is approximately twice that of 25 atom-nuclei. The decrease is observed to be almost a linear function. An important conclusion is that the surface coverage in the atom-only process is lower than both the NUC and NUC2 cases [176].

Comparison of the distributions of the QDN obtained in the atom-only (Figure 6.11(a, b)) and NUC (Figure 6.11(c, d)) processes demonstrates a

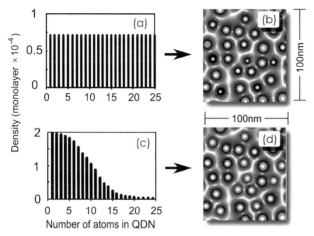

Figure 6.11 Size distributions and simulated patterns of quantum dot nuclei: (a, b) deposition from atom only flux; (c, d) deposition from NUC flux [176].

major advantage of the NUC-based synthesis of a dense QDN pattern suitable for further growth of the uniform QD array with high surface coverage. Indeed, by using this process, one can deposit a high-density seed pattern of small (< 15 atoms) QDN (Figure 6.11(c, d)). Further growth of these QDN results in the formation of a high-surface-coverage pattern of QDs of approximately the same size (Figure 6.11(d)).

Here we reiterate that Figures 6.11(b,d) were both produced using the diffusion model, which will be described in detail in Section 6.4. In contrast, the atom-only flux grown QDN patterns feature a large number of QDN consisting of 25 or more atoms. The accelerated growth of such large seed nuclei results in a suppression of the growth of smaller QDN, which are eventually dissolved via 2D evaporation. Thus, the density of QD patterns formed from an atom-only flux is in fact lower than that formed via the NUC process [176].

From the nanofabrication perspective, a dense nanopattern of same-size QDNs is an important prerequisite for obtaining dense arrays of nanodots of the same size; hence our aim of minimizing the width of the QDN size distribution. The result obtained in the atom-only process is apparently the worst from this point of view. On the other hand, when using non-uniform cluster distributions, one can expect a much more uniform QD growth, which will develop from seed nuclei of more-or-less similar size (< 15 atoms). The abrupt cut-off in the number densities of large QDN reduces the chances of the growth of over-sized QDs, thus ensuring a substantial improvement in the quality of the entire nanopattern.

One of the main conclusions made by Rider *et al.* [176] is that the *QDN distributions on the surface do not mirror the building unit distributions in the gas phase*. Indeed, in the atom-only process, a narrow building unit distribution results in a broad spectrum of seed nuclei sizes. Conversely, broader BU size distributions in the NUC and NUC2 sources lead to much narrower QDN size distributions on the surface. One thus arrives at a counter-intuitive conclusion, namely, that *the synthesis of size-uniform QDN nanopatterns may not necessarily require an influx of size-uniform BUs as is commonly believed*.

Thus, the density and sizes of the QD seed nuclei can be controlled by the building unit composition and size distribution in the NUC flux. A specific method is to use a reduced atom influx to the substrate, thus reducing the adatom density and hence avoiding uncontrollable formation of the QDN pattern from diffusing adatoms, and so to have a controllable influx of size-non-uniform nanoclusters. In this way one can create an initial pattern for the deposition of a quantum dot array with the required surface coverage.

Let us reiterate that in a coventional Stranski–Krashtanov growth mode, an atom flux is deposited on the substrate, leading to the formation of a few monolayers. The strain-induced fragmentation of these monolayers (a result of the lattice mismatch of the system) is what causes the formation of large islands, with further growth via adatom diffusion. The method examined in this section differs in that clusters as well as atoms/ions are delivered to the deposition surface. The clusters act as "already separated film fragments", effectively frustrating the strain-induced fragmentation favored by the SK growth.

To estimate the final surface coverage by the quantum dots grown from the computed QDN pattern, one can assume a specific value for the final QD radius (e.g. $R_{end} = 6\,nm$) and use the computed QDN distribution functions (e.g. in Figure 6.11(c)) and the total surface coverage by all QDNs (which is approximately 10^{-3} in this case). This estimate shows that the mean step between the QDN on the surface is 15 nm. Hence, the surface coverage for the final QD pattern in Figure 6.11(d) may reach 0.5. Therefore, if the quantum dots continue to develop smoothly (without coalescence), the required nanopattern density and QD size distribution can be achieved deterministically [176].

Strong support for the model and numerical simulations can be obtained by analyzing available experimental data on the dependence of is-landed film morphology on the distribution of nanoclusters in the incoming flux. The experiments convincingly support the counter-intuitive fact that the use of a NUC flux incorporating larger nanoclusters leads to formation of denser patterns of quantum dots of smaller size. The depo-

Table 6.1 Comparison of calculated and experimental values of difference in nanoislands size and surface coverage for atom and nanocluster flux, equivalent film thickness 5 nm [176].

Parameter	Flux	Value	Reference
Difference in nanodot size, nm, $d_{max} - d_{min}$	Atomic/molecular	70	Experiment, Figure 2(e) [292]
		56	Computed [176]
	Nanocluster	5	Experiment, Figure 2(f) [292]
		4	Computed [176]
Surface coverage	Atomic/molecular	0.15	Experiment, Figure 2(e) [292]
		0.2	Computed [176]
	Nanocluster	0.5	Experiment, Figure 2(f) [292]
		0.47	Computed [176]

sition of antimony films with a thickness of 1 and 5 nm [292], as well as 10 nm [293] has revealed a striking advantage of nanocluster deposition over atom flux deposition in terms of uniformity of the fabricated nanopatterns. More importantly, the numerical results of Rider *et al.* appear to be in fair agreement with the experimental findings. The numerical and experimental results are compared in Table 6.1.

However, experimental verification of the very initial stage of QDN pattern development is extremely challenging. It is worth emphasizing that despite impressive recent advances in nanofabrication and analytical surface science/materials science techniques, at this stage an experimental investigation of the very short transition processes (up to $\approx 25\,\mu s$) of our interest here seems to be quite difficult, if indeed possible at all. The main reason is the ultra-small (subnanometer) nanocluster sizes studied and their very short time scales of nucleation on the surface. Presently available atomic-resolution analytical tools of surface science and materials science [294] still cannot meet the strict requirements (e.g. adequate time resolution to scan sufficiently large surface areas with atomic (ca. 0.2 nm) precision and vacuum compatibility with plasma-based and other vacuum processes) for time-resolved *in situ* measurements of the nanocluster size distributions. Therefore, at this stage numerical experiments still remain the only viable way to investigate the initial (core structure-determining) stage of self-assembly of Ge/Si quantum dots on silicon surfaces [176].

Another significant problem that arises at the initial stage of nanodot growth is how to tailor their elemental composition and chemical structure from the very beginning of the process. This problem becomes even more complicated for material systems which contain atoms of more than one element. Binary and ternary quantum dots are good examples of such systems. The following section deals with an important issue of controlling the elemental composition and internal (e.g. core-shell) structure of binary SiC quantum dots.

6.3
Binary Si$_x$C$_{1-x}$ Quantum Dot Systems: Initial Growth Stage

Self-assembly of SiC quantum dots with precisely controllable elemental composition and internal structure still remains a major challenge for the gas/plasma-based nanodot synthesis. In this section we will consider an example of a binary nanodot system, namely Si$_x$C$_{1-x}$, wherein the proportions of both elements can vary depending on the value x. This value can be precisely tailored by adjusting the incoming fluxes of Si and C atoms. Interestingly, the ratio of the above fluxes is not necessarily the same as the required elemental ratio $x/1 - x$.

Here we will consider two different problems related to the two important requirements for nanodot synthesis set in Section 6.1 (see also Figure 6.1), namely the elemental composition and internal structure. We will first consider, Section 6.3.1, how to adjust incoming fluxes of Si and C atoms to obtain the required elemental ratio, for example, the stoichiometric one with $x = 0.5$. As will be seen from Section 6.3.2, the internal structure of the SiC nanodots can be engineered using time-variable incoming fluxes. We will further explore how different layers of Si and C can be arranged to create core-shell and even multi-shell structures.

6.3.1
Adatom Fluxes at Initial Growth Stages of Si$_x$C$_{1-x}$ Quantum Dots

As usual, we will begin with a brief outline of what will be discussed in this subsection. Here we will again follow the original report of Rider *et al.* [174], and discuss some of the main applications of SiC and related issues and show that at the initial stage (0.1–2.5 s into the process) of deposition of SiC/Si(100) quantum dot nuclei, equal Si and C atom deposition fluxes are likely to result in strongly non-stoichiometric nanodot composition due to the very different surface fluxes of Si and C adatoms to the quantum dots.

Nevertheless, the surface fluxes of Si and C adatoms to SiC nanodots can be effectively controlled by manipulating the Si/C atom influx ratio and the Si(100) surface temperature. In particular, when the surface temperature is 800 K, the surface fluxes can be equalized after only 0.05 s; however, it takes more than 1 second at a surface temperature of 600 K [174]. This study can be used to develop effective strategies to maintain a stoichiometric ([Si]/[C]=1:1) elemental ratio during the initial stages of deposition of SiC/Si(100) quantum dot nuclei in both neutral gas and plasma-based processes.

Let us begin with a discussion on the importance of nanostructured silicon carbide for a variety of advanced applications and the range of associated issues. This material system is particularly promising on account of its many favorable characteristics such as resistance to corrosion and thermal shock, robust chemical, thermal and mechanical properties, high elastic modulus, specific stiffness, and fracture toughness [174,295–303]. These and several other properties make SiC an ideal candidate for use in high-temperature, high-voltage and chemically reactive environments [301]. Such strength and stability also lead to application of *a*-SiC in molecular sieves for high-temperature gas separation [304]. An interesting feature of nanostructured SiC is its wide bandgap and associated unusual photoemission properties [302, 305–307]. This gap lends itself to blue light emission, which has recently been the subject of extensive research endeavors [174]. Enhanced field emission from SiC-capped Si nanotip arrays has also been noted as very promising for field emission displays [308].

Furthermore, SiC is not only of interest in optoelectronics but is also an important material in the field of bionanotechnology. Utilization of SiC quantum dots as nanostructured labels of biological material is an example of one of the applications of the biocompatibility of SiC [174, 306]. Moreover, both surface structure and composition play a crucial role in determining the stability and optical properties of SiC quantum dot arrays [306].

Different approaches to fabrication of nanostructured SiC films have been put forward (see, e.g. [303, 309] and references therein). However, reliable and robust methods for fabrication of stoichiometric and hydrogen-free SiC quantum dots and associated nanopatterns are still in their infancy [174]. While some methods do make it possible to achieve fairly stoichiometric SiC thin films [303,310], such films are in most cases hydrogenated. Achieving a stoichiometric (1:1) ratio of Si to C without any hydrogen, in the case of SiC/Si QDs, is inherently more difficult. The main reason is that the silicon surface shows quite different affinity to Si and C atoms; thus, carbon atoms are subject to very different conditions on a silicon surface as compared to Si atoms. Moreover, this elemental ratio is often required throughout the entire quantum dot structure, from the internal core to the outer shell. Unfortunately, it presently appears quite challenging, if possible at all, to control and characterize the elemental composition of ultra-small nano-objects, which develop within very short time periods.

The difficulty in obtaining stoichiometric SiC QDs (or any desired elemental ratio $x/1 - x$) is ultimately due to the fact that it is a complex binary system and thus more difficult to control than Si/SiC [311], Ge/Si(100) [312], or Ge/SiO$_2$ [313]. Given that Si and C atoms are subject to different conditions on the surface (different characteristic energies, migration rates, etc.), great care must be taken, not only to deliver the right amount of each type of atom to the surface but also to use an appropriate substrate temperature as it strongly affects surface reaction rates. Moreover, because of their different behavior on the substrate (Si, for example is more likely to epitaxially recrystallize on an Si surface than C), it is difficult to separately control the delivery and incorporation of Si and C atoms into SiC QDs [174].

Ideally, the stoichiometry of a binary quantum dot should be controlled from the earliest possible growth stage, when the initial composition and structure of the QD nucleus effectively dictates its future evolution. In the case considered here, controlling the balance between delivery and consumption of Si and C atomic building units is clearly important when attempting to fabricate stoichiometric quantum dots. The delivery and relative amounts of Si and C atoms on the substrate are determined by the precursor influx ratio $k_p = P_{Si}/P_C$, where P_{Si} and P_C are the incoming fluxes of Si and C atoms, respectively. On the other hand, the consumption of Si and C (e.g. by the quantum dot nuclei) depends on the number of building units on the solid surface and surface conditions, such as the surface temperature, morphology, electric charge, and so on.

The surface temperature is particularly important, in part because it largely determines the rates of adatom migration, evaporation to/from the surface, rates of adatom insertion into crystalline lattice, and so on. This complex set of surface conditions determines how long the adatoms may remain on the surface, that is, adatom lifetime. From a practical perspective, the precursor influx ratio and the substrate temperature appear to be the most effective controls to achieve a stoichiometric ([Si]/[C]=1:1) or any required ([Si]/[C]=$x/1 - x$) elemental ratio of Si and C on smooth silicon surfaces [174]. It is worthwhile to stress that existing numerical efforts on modeling quantum dot growth, whilst accounting for the effects of incoming fluxes and surface temperature, in most cases sidestep the important issue of the elemental composition and stoichiometry, which are essential for binary, ternary, and so on quantum dot systems [314].

Rider *et al.* [174] have determined the optimal conditions to obtain stoichiometric SiC/Si(100) QDs at the initial stage of growth, by investigating the surface fluxes of Si and C adatoms to SiC QD nuclei (QDNs) consisting of 20 atoms or less. An important conclusion is that the Si/C ele-

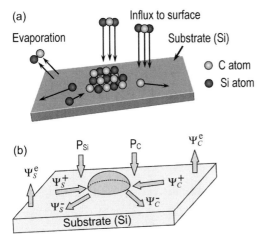

Figure 6.12 (a) Schematics of SiC QD growth on silicon surface; (b) atom/ion and adatom fluxes to/from the quantum dot [174].

mental ratio on the surface is time-dependent at the initial growth stage and usually levels off at a steady-state value. Moreover, this ratio can be precisely controlled by the precursor influx ratio, deposition rate, and the substrate temperature.

Intuitively, one would expect that by setting $k_p = 1$, it might be possible to obtain stoichiometric SiC QDNs. However, this is not the case at the early stages of nanodot self-assembly when the core is formed [174]. Below, we will consider various strategies aimed at achieving stoichiometric elemental ratios of Si and C in SiC quantum dot nuclei at the early stages of self-assembly. More importantly, these approaches are generic and can also be applied to other binary nanodot systems such as GaAs, InP, InAs, GaSb, InSb, and so on.

For the convenience of the reader, we will now describe the original model and numerical details of Rider *et al.* [174]. Figure 6.12(a) is a schematic representation of the simulation geometry of SiC quantum dot nuclei deposition in a neutral/ionized gas-based process. Silicon and carbon atoms/ions are delivered from the neutral/ionized gas phase and after becoming (charge neutral) adatoms on the substrate are redistributed about the Si(100) surface via surface diffusion processes. Depending on the surface temperature, the adatoms can also re-evaporate from the QD surfaces to the two-dimensional adatom field on the surface and/or back to the gas phase as shown in Figure 6.12(a).

The QDN growth model of Rider *et al.* [174] is based on a set of species balance equations on the surface, which take into account the incoming fluxes from the gas phase. Such an approach has been widely used in surface science and is very suitable for describing the surface growth processes examined here [179,315]. In this case, an extended two-component (Si and C) system covering all the most important surface processes [273], as sketched in Figure 6.12(b), was used. The surface processes taken into account include atom attachment to and detachment from QDN surface borders, as well as atom evaporation from the substrate surface (Figure 6.12). Clusters of two atoms and more are treated as immobile for the range of surface temperatures simulated. As can be seen in Figure 6.12(b), the balance of adsorbed atoms (adatoms) takes into account the Si and C atom influx from the neutral/ionized gas phase

$$P^+ = P_{Si} + P_C,$$

flux of Si and C atom evaporation from the substrate surface

$$\Psi^e = \Psi^e_S + \Psi^e_C,$$

flux of Si and C adatoms to the QDN

$$\Psi^+ = \Psi^+_S + \Psi^+_C,$$

and flux of Si and C adatoms from the QDN

$$\Psi^- = \Psi^-_S + \Psi^-_C.$$

Here, P_{Si} and P_C are the incoming Si and C fluxes to the solid surface, and

$$\Psi^e_S = n_k v_0 \exp[-\epsilon_{a.S}/kT],$$

$$\Psi^e_C = n_k v_0 \exp[-\epsilon_{a.C}/kT]$$

are the fluxes of silicon and carbon evaporation, respectively; v_0 is the lattice atom oscillation frequency, $\epsilon_{a.S}$ and $\epsilon_{a.C}$ are the silicon and carbon adatom evaporation energies, n_k is the adatom surface density, T is the substrate temperature, k is Boltzmann's constant, Ψ^+_S and Ψ^-_S are the Si surface adatom fluxes to and from the QDN, respectively, Ψ^+_C and Ψ^-_C are defined similarly for carbon [174]. The balance equation for adatom density on substrate η_j is

$$\frac{\partial \eta_j}{\partial t} = P^+_j - \Psi^e_j - \Psi^+_j + \Psi^-_j, \tag{6.12}$$

where the subscript j represents either Si or C.

The state of the substrate surface influences QD growth mainly via the adatom diffusion activation energy ϵ_d. In this model, there was no simplifying assumption of a defect-free surface. Instead, an experimental value for the silicon atom surface diffusion activation energy $\epsilon_{d.S}$ was used; this measured value accurately accounts for the energy of adatom detachment from the surface defects, steps, and other important factors [316]. Reliable experimental data for the carbon atom surface diffusion activation energy $\epsilon_{d.C}$ is not available; therefore, it was assumed that $\epsilon_{d.C} = 1/3\epsilon_{b.C}$, where $\epsilon_{b.C}$ is the energy of a carbon-silicon bond [174]. A complete set of the rate equations for SiC QD formation incorporates two balance equations describing Si and C atoms on the substrate surface, as well as equations for the density of SiC quantum dots consisting of 2, 3, ... i atoms.

Thus, the balance equations for the C and Si atoms are [174]

$$\frac{\partial \eta_S}{\partial t} = P_{Si} - \Psi_S^e + \sum_{i=2}^{\infty} n_{S.i}\eta_i\mu_{S.i}$$

$$- \eta_S \sum_{i=2}^{\infty} \sigma_i\eta_i v_{d.S} - 2\sigma_1\eta_S^2 v_{d.S} - \sigma_1\eta_S\eta_C v_{d.S}, \tag{6.13}$$

$$\frac{\partial \eta_C}{\partial t} = P_C - \Psi_C^e + \sum_{i=2}^{\infty} n_{C.i}\eta_i\mu_{C.i}$$

$$- \eta_C \sum_{i=2}^{\infty} \sigma_i\eta_i v_{d.C} - 2\sigma_1\eta_C^2 v_{d.C} - \sigma_1\eta_S\eta_C v_{d.C}, \tag{6.14}$$

where $n_{S.i}$ and $n_{C.i}$ are the numbers of Si and C atoms at the borders of the QDN consisting of i atoms; η_i is the surface density of (i)-atom QDN; $\mu_{S.i}$ and $\mu_{C.i}$ are the frequencies of Si and C atom re-evaporation from borders of (i)-atom QDN; σ_i is the diameter of i-atom nanodot nucleus. Similar to Section 6.2

$$v_{d.S(C)} = \lambda v_0 \exp[-\epsilon_{d.S(C)}/k_B T_s]$$

are the linear velocities of Si and C adatom movement about the substrate surface, where λ is the lattice constant of the Si(100) substrate and $\epsilon_{d.S(C)}$ is the Si or C adatom surface diffusion activation energies. The first sum in Equations (6.13) and (6.14) represents the surface influx of Si and C adatoms to the substrate surface due to adatom re-evaporation; the second sum represents the adatom flux from the surface due to the attachment to the QDN consisting of two and more atoms. The remaining two terms denote the adatom outflux from the surface due to the adatom-adatom collisions which lead to the formation of 2-atom QDN [174]. The rate equations for the formation of i-atom QDN are [174]

$$\frac{\partial \eta_i}{\partial t} = \sigma_{i-1}\eta_{i-1}[\eta_S v_{d.S} + \eta_C v_{d.C}] - \sigma_i\eta_i[\eta_S v_{d.S} + \eta_C v_{d.C}]$$

$$+ \eta_{i+1}n_{i+1}[\mu_{S.(i+1)} + \mu_{C.(i+1)}] - \eta_i n_i[\mu_{S.i} + \mu_{C.i}] \tag{6.15}$$

Table 6.2 Parameters and representative values of computations [174].

Parameter	Value	Reference
Physical constants		
Lattice atom oscillation frequency, ν_0, s^{-1}	1 x 10^{13}	[317]
Lattice parameter, λ, m	5 x 10^{-10}	[318]
Si atom diffusion activation energy, $\epsilon_{d.S}$, eV	1.35	[316]
Si atom evaporation activation energy, $\epsilon_{a.S}$, eV	3.04	[319]
Si atom bonding energy, $\epsilon_{b.S}$, eV	2.3	[319]
C atom diffusion activation energy, $\epsilon_{d.C}$, eV	1.1	[318]
C atom evaporation activation energy, $\epsilon_{a.C}$, eV	4.8	[319]
C atom bonding energy, $\epsilon_{b.C}$, eV	3.6	[319]
Si-C bonding energy, $\epsilon_{b.S-C}$, eV	3.3	[320]
Parameters of simulation		
Influx to surface, P^+, monolayers $\cdot$ s^{-1}	0.001–0.1	
Surface coverage, η, monolayers	0–0.5	
Number of quantum dot nuclei in pattern, K	1000	
Maximum number of atoms in QDN, N	2–25	
Substrate surface temperature, T, K	500–850	
Time of deposition, t, s	0–3	
Influx ratio, $k_p = P_{Si}/P_c$	0–3	
Adatom balance factor, $k_{st} = \Psi_S^+/\Psi_C^+$	0–4	

where the first term describes the Si and C adatom collisions with $(i-1)$-atom QDN; the second term stands for the Si and C adatom collisions with the (i)-atom QDN; the third term is related to the Si and C adatom detachment from the QDN consisting of $(i+1)$ atoms; and the fourth term represents Si and C adatom detachment from (i)-atom QDN. Evaporation and diffusion activation energies for the Si and C atoms, as well as the energies of the Si–Si and Si–C bonds have been calculated using standard bond enthalpies (except where noted otherwise). The specific values for all the parameters used are listed in Table 6.2.

The ratio of silicon and carbon adatom fluxes to the borders of quantum dots (the adatom balance factor) is defined as $k_{st} = \Psi_S^+/\Psi_C^+$. Using Equation (6.15), one obtains

$$k_{st}(t) = \sum_{i=1}^{N} \zeta_{S.i}/\zeta_{C.i},\qquad(6.16)$$

where

$$\zeta_{S.i} = \eta_S \nu_{d.S}\sigma_i \eta_i - \eta_i n_i \mu_{S.i},$$
$$\zeta_{C.i} = \eta_C \nu_{d.C}\sigma_i \eta_i - \eta_i n_i \mu_{C.i}$$

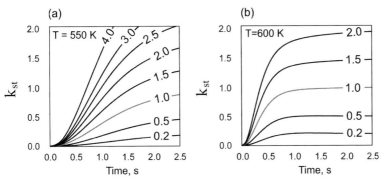

Figure 6.13 Dependence of adatom balance factor k_{st} on time with influx ratio k_p as a parameter, for the surface temperature (a) $T = 550$; and (b) $600\,\text{K}$ [174].

and N is the maximum number of atoms constituting the quantum dot nuclei. We emphasize that k_{st} is a differential ratio, that is, it is computed at each instant and thus provides a time-dependent measure of the evolution of the QD composition and internal structure [174].

Numerical simulations [174] have been conducted for different substrate surface temperatures T_s, deposition times t, and values of the precursor influx ratio k_p. Here we stress that obtaining time-dependent behavior of the adatom balance factor k_{st} is the main objective. The dependence of k_{st} on deposition time and the influx ratio factor with the surface temperature and deposition time is a key to achieving any desired elemental composition and internal structure of Si$_x$C$_{1-x}$ nanodots.

Let us now consider the results of numerical simulation of silicon carbide quantum dot nuclei formation on a Si(100) surface [174]. The dependence of the adatom balance factor k_{st} on time with the surface temperature and precursor influx ratio k_p as parameters is shown in Figure 6.13. The first important observation is that k_{st} starts from the zero point for all k_p values within the range of 0.2–4. Therefore, at the beginning of the deposition process the surface flux of carbon adatoms is much larger than the Si adatom flux. The adatom balance factor k_{st} increases with time and eventually saturates at k_p ($k_{st} \rightarrow k_p$) for longer deposition times. It is also notable that at higher k_p, k_{st} levels off more slowly. For example, at the same surface temperature $T_s = 700\,\text{K}$ (not shown here, see Figure 3(a) of the original report [174]), the saturation level $k_{st} = k_p = 0.2$, is reached in 0.15 s whereas more than 1 s is required to reach $k_{st} = k_p = 2.0$.

When the surface temperature increases, the k_{st} factor increases rapidly; it takes 2.5 s to reach $k_{st} = 0.2$ at $k_p = 0.2$ and $T_s = 550\,\text{K}$; however, the same process requires only 0.05 s at $T_s = 800\,\text{K}$. Therefore, the general behavior of the adatom balance factor clearly also depends on the

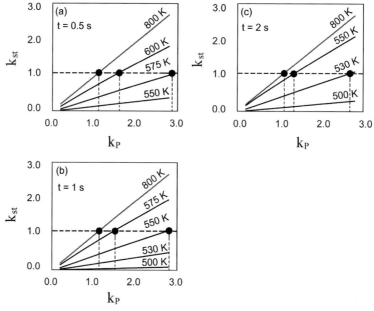

Figure 6.14 Dependence of adatom balance factor k_{st} on influx ratio k_p with surface temperature T_s as a parameter for a deposition time of 0.5 (a), 1.0 (b), and 2.0 s (c) [174].

surface temperature: k_{st} tends to the corresponding k_p value at lower temperatures (up to 700 K) but shows an oscillating behavior (exceeding a saturation level after the first rise followed by a decrease to the equilibrium level) at a temperature of 800 K as can be seen in Figure 3(b) of the original report [174].

Let us now consider the dependence of the adatom balance factor k_{st} on the influx ratio k_p with surface temperature and deposition time as parameters (Figure 6.14). The general tendency is that k_{st} increases with k_p. As can be seen from Figure 6.14, when surface temperature decreases, k_{st} becomes smaller; this means there is a decreased influx of silicon adatoms to the QDN, compared to the carbon adatom influx. It is important to note that at surface temperatures equal to or exceeding 800 K, the saturation level $k_{st} = k_p$ is reached quickly (in less than 0.5 s); thus the upper curve in Figure 6.14(a–c) is also the same for $T > 800$ K. On the other hand, at lower surface temperatures (the range 500–600 K is shown in Figure 6.14) the slope of the k_{st} *versus* k_p curve gradually increases with time; it takes longer to reach the saturation level at lower substrate temperatures [174].

Here we note that the "stoichiometric" precursor influx ratio $k_p = 1$ is most commonly used in many gas-based deposition processes because of the commonsense expectation of achieving a stoichiometric elemental ratio of the elements constituting the quantum dots. An interesting observation that follows from numerical experiments of Rider *et al.* [174] is that if the precursor influx ratio $k_p = 1$ does not vary with time, the k_{st} factor starts from the zero point at all temperatures and then gradually increases with time to reach $k_{st} = 1$. Therefore, at the beginning of the process, the carbon flux is much larger than that of silicon. This transition process typically lasts from 0.1–0.4 s at higher temperatures (650–800 K) and 1–5 s at temperatures below 600 K. It should be noted that the above dependencies have been calculated using a total atom influx P^+ of 0.1 monolayer/s (ML/s) [174].

The total incoming flux also appears to be a very important factor that determines the surface fluxes of Si and C adatoms. Let us consider an example where the deposition time is 1.0 s and the surface temperature is 600 K. With the total influx equal to 0.1 ML/s, k_{st} is approximately equal to k_p for any k_p value. On the other hand, when the total influx is less than 0.1 ML/s (0.01 and 0.001 ML/s in this simulation), the k_{st} factor is much smaller than k_p. One can thus conclude that under lower deposition rate conditions the carbon adatom flux well exceeds the silicon adatom flux.

As we have seen from the above consideration, the compositional structure of the quantum dots (crystalline structure, polytype, etc.) develops continuously, from the moment the precursor species are released until the desired surface coverage (up to ca. 0.5 in this work) is achieved. The duration of this stage can be estimated by $t = \mu / P^+$, where μ is the required surface coverage. One of the main aims of this study is to obtain process conditions that provide stoichiometric QD composition, and avoid unwanted QD formation from unbalanced ("non-stoichiometric") silicon and carbon fluxes. The most important fact discovered is that, if one keeps the precursor influx ratio k_p constant at very early process stages, the ratio of surface fluxes of Si and C adatoms is time-dependent, and, moreover, "non-stoichiometric" [174]. The equilibrium level of the balance factor k_{st} can be reached after a time lag, ranging from 0.05 s to several seconds; this time lag depends on the surface temperature and the precursor influx ratio k_p [174].

Therefore, at early deposition stages it is not possible to synthesize stoichiometric SiC nanodot cores on a Si(100) surface by using time-invariable and equal fluxes of Si and C atoms or ions and so special arrangements should be made to ensure that the surface fluxes of Si and C adatoms are properly balanced at every growth stage. In brief, this can be achieved by transiently increasing the surface temperature or the silicon atom influx with respect to that of carbon. One further possibility would be to appropriately control the deposition rate P^+. Some of these options will be considered in Section 6.3.2.

Let us now discuss specific strategies to mediate the unwanted self-assembly of non-stoichiometric SiC quantum dot nuclei at initial growth stages. Examining Figure 6.13, one can conclude that at lower surface temperatures (550 and 600 K) and a "stoichiometric" ratio of Si and C atom fluxes to the surface (curves labelled "1.0" in Figures 6.13(a) and (b)), the C adatom flux to the SiC quantum dots is almost twice as large as that of silicon adatoms during the first 0.5–1.0 s into the process. With a total influx of 0.1 mL/s, the total amount of deposited material in this time interval is ca. 0.05–0.1 mL. Under such conditions, QDN consisting of 10–20 atoms are formed by highly-unbalanced fluxes of Si and C adatoms. As a result, the quantum dot nuclei will be rich in carbon with a strongly non-uniform elemental composition. This conclusion will be confirmed in Section 6.3.2.

It is important to stress that the effects of unbalanced surface adatom fluxes and time lags before the stoichiometric level can be reached are stronger at lower substrate temperatures. This can be seen in Figure 6.13. Therefore, in low-temperature-demanding processes, such as polymer processing, one could expect highly non-stoichiometric (carbon-rich) SiC nanodot cores. At higher surface temperatures, this effect is less important, but still plays a role at $T = 700$ K. However, when $T \geq 800$ K, the time required to reach the saturation level $k_{st} = k_p = 1$ is less than 0.05 s. During this time, the total number of atoms deposited is ca. $2 \times 10^{16}\,\mathrm{m}^{-2}$.

The time dependence of k_{st} at very early growth stages can be attributed to different Si and C adatom mobilities on the surface, which depend strongly on the substrate temperature. Hence, the undesired growth of non-stoichiometric QDs is unavoidable when adatoms of different elements (with very different surface diffusion activation energies) are involved and their incoming fluxes are not properly balanced [174]. One possible way to mediate this unwanted effect is to increase the surface temperature during the initial stage of deposition. Indeed, at higher surface temperatures the difference in the Si and C adatom mobilities decreases. The surface temperature can be transiently increased during the first 0.01–0.1 s of the process by using energetic ion bombardment in a pulsed substrate bias regime. After the pulse is turned off, the excess energy accumulated in the superficial layer of the substrate will then be dissipated in the substrate material without any significant heating of the substrate. It is clear that this method requires the use of a plasma-based process and cannot be implemented in a neutral gas-based environment [174].

Another effective way to achieve "stoichiometric" adatom fluxes at the initial stage of SiC nanodot self-assembly is to use an increased influx ratio k_p during the first 0.01–0.1 s of the process. In this case, as can be seen in Figure 6.13, the equilibrium value of the balance factor $k_{st} = 1$ can be achieved much quicker. Thereafter, the influx ratio $k_p = 1$ should be used to maintain the stoichiometrical composition of the SiC QDs.

Thus, two-step strategies can be used to fabricate stoichiometric SiC/Si(100) quantum dots. Specifically, the deposition can first be conducted at a higher surface temperature or increased precursor influx ratio k_p. Upon achieving the adatom balance factor $k_{st} = 1$, the process should be continued by using equal fluxes of silicon and carbon atoms. At this stage, T can also be lowered to the most suitable surface temperature, which can be determined from the conditions of QD crystallization and/or thermal stability of the substrate [174].

Figure 6.14 illustrates that the two major control parameters, the surface temperature and the precursor influx ratio k_p are highly interdependent. It becomes apparent that at lower temperatures, fluxes with a higher k_p should be used, and vice versa. In practice, there is no need to increase both parameters simultaneously to the limit, as the results can best be achieved by an optimal combination of slightly increased T and k_p.

A number of suitable process "working points" are mapped in Figure 6.14 and can be chosen depending on specific requirements. For example, our target is to obtain "stoichiometric" Si and C adatom fluxes after 1 s. As follows from Figure 6.14(b), one can use, for example the following options: (i) at $T= 800$ K, a "stoichiometric" precursor influx ($k_p = 1$) can be used; (ii) at $T= 575$ K, k_p should be ca. 1.3; and (iii) at $T= 550$ K, k_p should be ca. 2.7. These combinations are different at other time moments and can easily be worked out by using the results in Figure 6.14 [174].

As discussed earlier, there is a requirement that the adatom balance factor levels off at $k_{st} = 1$ as early as possible in order to ensure that the SiC quantum dots have equal amounts of Si and C atoms from the innermost volume of the dot to its outermost layers, and at all stages of the self-assembly. Thus, one more convenient way to speed up this process would naturally be to synthesize the nanodots faster, at higher deposition rates.

As the results of Rider *et al.* [174] suggest, higher deposition rates at a constant precursor influx ratio do indeed result in faster "stoichiometrization" of the Si and C adatom fluxes. However, this important process parameter cannot be increased indefinitely as very high deposition rates may lead to undesired formation of continuous films rather than nanodot arrays.

Therefore, at a given deposition rate, a *balanced combination of a transiently elevated surface temperature and increased delivery of silicon atoms (with respect to carbon) to the surface* can be the best way to obtain stoichiometric composition of SiC quantum dot nuclei at the initial stage of their self-assembly. However, it should be noted that higher substrate temperatures whilst improving not only the stoichiometry but also the crystallinity of SiC QDs may the same time dramatically restrict the range of substrate materials that can be used. For example, polyethylene terephtalate (PET) and PET/silica nanocomposites, widely used in biomedical applications, have very low melting points ca. 540 K [321]. For these and similar easily fusible materials the best way to maintain the proper elemental composition of the SiC QDs is to use larger ratios (P_{Si}/P_C) of incoming fluxes of silicon and carbon atomic building units [174].

Therefore, determining the k_p factor for which the ratio of Si and C adatom fluxes to the growing quantum dots tends to 1 at all deposition stages, and then establishing the minimum substrate temperature necessary to achieve acceptably stoichiometric dots throughout the entire structure is a viable approach to active dynamical control of the nanoassembly process. To conclude this subsection, it is worthwhile to reiterate that such a process is generic and is applicable to other binary QD systems such as GaAs/Si(100) and a variety of building unit delivery methods such as molecular beam epitaxy (MBE), chemical vapour deposition (CVD), atomic deposition and plasma-based methods (PECVD, magnetron sputtering, pulsed laser deposition, etc.) [174].

6.3.2
Control of Core-Shell Structure and Elemental Composition of Si$_x$C$_{1-x}$ Quantum Dots

Let us now follow the original report [175] and consider the possibility of initial stage control of the elemental composition and internal (e.g. core-shell) structure of binary SiC quantum dots briefly mentioned in Section 6.3.1. This can be achieved by optimizing the temporal variation of Si and C incoming fluxes and surface temperatures [175]. At higher substrate temperatures and stronger incoming fluxes a stoichiometric SiC outer shell is formed over a small carbon-enriched core. Conversely, lower process temperatures result in a larger carbon-enriched core, Si-enriched under-shell and then a stoichiometric SiC outer shell. A quite similar treatment can be applied to a broad range of semiconductor materials and nanofabrication techniques.

Let us begin our consideration by noting that single-element and binary self-assembled quantum dots feature highly-unusual electron confinement properties, which make them particularly appealing for a large number of potential applications and extensive investigations in various fields [322,323]. In addition to size-dependent electronic and optical properties intrinsic to single-element QDs , binary quantum dot systems (e.g. InSb, GaSb, CdSe, CdTe, GaAs, InAs, AlN, SiC, GaN and others) offer an additional way to tune these properties via controlling the relative amounts of elements A and B (x and $1 - x$ in a binary system A$_x$B$_{1-x}$) or arranging them in a layered stack commonly referred to as a core-shell structure.

As stressed in Section 6.1 (see Figure 6.1), one of the requirements along the road to deterministic nanofabrication is the ability to synthesize highly-stoichiometric (elemental ratio [A]/[B]$= 1$ throughout the entire structure), selected-composition ([A]/[B]$= x/(1 - x)$ throughout the entire structure), compositionally-graded (with x varying) [324], and core-shell ($x = 0$ in the core and $x = 1$ in the outer shell or vice versa) quantum dot structures and that appropriate, for example atomic/molecular/cluster beam epitaxy [325] or ion and plasma-based nanofabrication techniques are developed. As QDs continuously shrink in size, controlling their elemental composition starting from the initial stage of deposition is a particularly challenging task.

The epitaxial semiconductor QDs of interest here are usually formed via self-assembly on solid surfaces. As we have seen from the previous subsection, precisely controlling the delivery of elements A and B is therefore crucial in obtaining QDs with the desired elemental composition. The rates of delivery of both elements to the growth surface are also crucial for precise tailoring of the internal nanodot structure.

From previous sections of this chapter we have learned that the surface diffusion rates of elements A and B are generally quite different; thus, their incoming fluxes must be properly balanced and the surface conditions (e.g. substrate temperature) appropriately adjusted [174,175]. However, the outcome of this commonly accepted strategy for ultrasmall quantum dots nuclei and developed QDs appear to be quite different due to transition processes that originate as a result of different surface conditions and which typically last from fractions of to a few seconds depending on the materials and process parameters involved (see Section 6.3.1 for more details). More importantly, this time lag may be sufficient for the development of a QD core with an elemental composition which is quite different from that in other areas within the quantum dot. The structure formed at very early growth stages in turn can affect the later stages of the QD evolution [175].

Let us now discuss how the elemental composition and internal structure of the binary SiC quantum dots can be controlled during the initial growth stage by changing the time dependence of silicon and carbon incoming fluxes, and adjusting the temperature of the growth surface. Depending on the surface temperature T_s and the temporal variation of the Si/C incoming fluxes, the nanodots can self-assemble either as a heterolayered stack consisting of a carbon-enriched core, silicon-enriched under-shell and stoichiometric SiC outer shell or as near-stoichiometric SiC QDs with a small carbon-enriched core. Moreover, high-influx and higher-T_s conditions (common to plasma-assisted epitaxial growth) result in a relatively quick equalization of surface fluxes of Si and C adatoms. In fact, this may eventually result in highly-stoichiometric SiC quantum dots throughout their internal structure [175].

The initial stage of SiC quantum dot formation involves a very small number of atoms, typically up to several tens of atoms. In this size range, the nanodot properties depend discretely on the number of atoms. Nucleation rate equations are used to compute the QD size distribution, and the Monte Carlo technique is employed to determine the QD parameters [171]. A hybrid approach, such as this, is accurate in calculation of kinetic QD parameters (size, collision cross-sections, etc.) which are used in the rate equations module, which is quite similar to what was considered in the previous subsection.

The main processes that determine the silicon and carbon adatom balance on the substrate surface are [175]: fluxes of Si and C atoms to the surface $\Psi^+ = \Psi_S^+ + \Psi_C^+$; Si/C adatom diffusion about the surface, collisions, and attachment to the quantum dots. These can be quantified by calculating the total and separate fluxes of Si and C adatoms to the QDs $\Gamma^{to} = \Gamma_S^{to} + \Gamma_C^{to}$; total and separate fluxes of Si and C adatoms from the QDs due to the 2D evaporation $\Gamma^{from} = \Gamma_S^{from} + \Gamma_C^{from}$; as well as adatom evaporation from the surface $\Psi^- = \Psi_S^- + \Psi_C^-$. Here, Ψ_S^+ and Ψ_C^+ are the incoming fluxes of Si and C atoms, respectively; Ψ_S^- and Ψ_C^- are the Si/C atom evaporation fluxes from the substrate, respectively.

Furthermore, Γ_S^{to} and Γ_C^{to} are fluxes of silicon and carbon adatoms to the QDs, whereas Γ_S^{from} and Γ_C^{from} are the Si/C adatom fluxes from QDs. In the case considered here, the surface coverage is very small; thus, the probability of atom attachment from the gas/plasma directly to the QD is low. One can thus assume that the QDs are formed mainly from the fluxes of Si and C adatoms supplied to their borders by surface diffusion. The two balance equations for Si and C adatom densities η_S and η_C on the surface are [175]

$$\frac{\partial \eta_S}{\partial t} = \Psi_S^+ - \Psi_S^- + \sum_{i=1}^{N} \eta_i (\Gamma_{S,i}^{to} - \Gamma_{S,i}^{from}) \qquad (6.17)$$

$$\frac{\partial \eta_C}{\partial t} = \Psi_C^+ - \Psi_C^- + \sum_{i=1}^{N} \eta_i (\Gamma_{C,i}^{to} - \Gamma_{C,i}^{from}), \qquad (6.18)$$

where N is the number of atoms that make up a quantum dot. The density η_i of SiC quantum dots consisting of i atoms is calculated by a set of nucleation rate equations similar to what has been used in Section 6.3.1 [174, 175]. The balance equations for C and Si adatom fluxes to a quantum dot consisting of i atoms are

$$\Gamma_{S(C),i}^{to} - \Gamma_{S(C),i}^{from} = v_{d,S(C)} \sigma_i \eta_{S(C)} - n_i \mu_{S(C),i}, \qquad (6.19)$$

where σ_i is the cross-section of a QD of i atoms; n_i is the number of Si or C atoms on the border of a QD of i atoms; $\mu_{S(C),i}$ is the frequency of atom re-evaporation from QDs to the two-dimensional vapor; $v_{d,S(C)} = \lambda_{Si} v_0 \exp(-\varepsilon_{S(C)}/k_B T_s)$ is the linear velocity of Si/C adatom movement about the substrate surface, where λ_{Si} is the Si lattice constant, v_0 is the frequency of lattice atom oscillations, T_s is the substrate temperature, k_B is Boltzmann's constant, and $\varepsilon_{S(C)}$ is the Si/C adatom surface diffusion activation energies. A more detailed description of the model can be found elsewhere [171, 174].

Similar to in the previous subsection, the Si(100) surface was not treated as defect-free; instead, an experimental value was used for the silicon atom surface diffusion activation energy $\varepsilon_S = 2.160 \times 10^{-19}$ J (1.35 eV), which takes into account the energy of adatom detachment from the surface defects, steps, and so on. The carbon atom surface diffusion activation energy on Si is assumed to be $\varepsilon_C = 1/3\varepsilon_{b,C} = 1.760 \times 10^{-19}$ J (1.1 eV), where $\varepsilon_{b,C}$ is the carbon-silicon bond energy equal to 5.280×10^{-19} J (3.3 eV). The Si-Si bonding energy was assumed $\varepsilon_{b,S} = 3.680 \times 10^{-19}$ J (2.3 eV) and the silicon lattice constant $\lambda_{Si} = 5 \times 10^{10}$ m. The surface temperature was varied from 600 to 800K, the range of the influx ratio factor $\bar{\Psi}_{S-C} = \Psi_S^+/\Psi_C^+$ was 0–40, and the total influx of Si and C atoms $\Psi^+ = \Psi_S^+ + \Psi_C^+$ was varied from 0.01 to 10 ML/s [175].

The elemental composition of quantum dots has been quantified by an averaged elemental ratio factor

$$\wp_{S-C}(t) = \int_0^t \bar{\Gamma}_{S-C}(\tau) d\tau \qquad (6.20)$$

which was calculated by integrating the adatom flux ratio factor

$$\bar{\Gamma}_{S-C}(\tau) = (\Gamma_S^{to} - \Gamma_S^{from})/(\Gamma_C^{to} - \Gamma_C^{from})$$

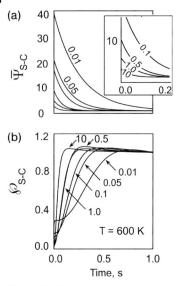

Figure 6.15 Temporal dependence of the influx ratio factor (a) and averaged elemental ratio factor (b) with total influx (marked on curves) as a parameter; $T_s = 600$ K [175].

from 0 to t. Thus, $\wp_{S-C}(t)$ is a measure of the ratio of Si and C atoms that form the QDs at a given time t.

We recall from the previous section that when one uses a "stoichio-metric" ($\bar{\Psi}_{S-C} = 1$) influx, a carbon-enriched core is formed at the initial growth stage [174, 175]. The size of this core, however, depends on the total incoming flux and the surface temperature. Rider et al. [175] simu-lated the QD formation using a time-variable influx ratio factor $\bar{\Psi}_{S-C}(t)$ in order to provide the fastest equalization of Si and C adatom fluxes to the QDs and therefore the most uniform elemental composition (elemen-tal ratio factor $\wp_{S-C}$ close to 1) throughout the QD internal structure.

Not surprisingly, this approach results in the controlled formation of QDs of the core-shell/core-undershell-shell structure. Taking into ac-count that the carbon-enriched core is formed from stoichiometric Si/C fluxes, the incoming flux should be initially silicon-rich and then needs to be gradually changed to include equal amounts of Si and C atoms. In this way, one can obtain the optimum temporal variation (quantified by the rate and curvature of $\bar{\Psi}_{S-C}(t)$) of the influx ratio factor that results in $\wp_{S-C}$ leveling off at unity in the shortest possible time [175].

Figure 6.15 displays the graphs of the optimum temporal variation $\bar{\Psi}_{S-C}(t)$ and the corresponding elemental ratio factor $\wp_{S-C}$ with total in-flux and temperature as parameters. One can see that at the initial time moment $\wp_{S-C}$ is close to zero and does not exceed 0.4 for any influx ratio

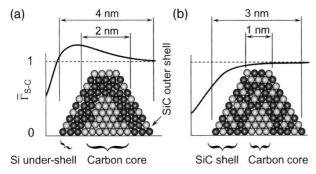

Figure 6.16 Formation of carbon core silicon under-shell – outer SiC shell (a) and near-stoichiometric with a smaller carbon core (b) SiC quantum dot structures (drawn not to scale) for $T_s = 600$ and 800 K, respectively. Approximate dimensions are shown for total influx 0.1 ML/s [175].

factor $\bar{\Psi}_{S-C}(t)$. Then, at $T_s = 600$ K (see in Figure 6.15(b)), $\wp_{S-C}$ increases rapidly and reaches unity within 0.3–0.8 s. At higher temperatures ($T_s = 800$ K), however, the numbers of C and Si atoms in a quantum dot can be balanced much faster, typically within 0.04–0.08 s; this time interval is approximately 10 times shorter than at $T_s = 600$ K.

After this initial increase, the behavior of the elemental ratio factor depends on the surface temperature. In a lower-temperature process, $\wp_{S-C}$ intersects the $\wp_{S-C} = 1$ level and then approaches it from above. In contrast, at higher surface temperatures, the elemental ratio factor does not intersect the $\wp_{S-C} = 1$ level at any point; instead, it always approaches from below. In this way one can determine the optimum temporal variation of the influx ratio factor $\bar{\Psi}_{S-C}(t)$ required to obtain the fastest possible equalization of the Si and C surface fluxes. Different temporal dependencies of $\bar{\Psi}_{S-C}(t)$ result in larger deflections of the $\wp_{S-C}$ factor from unity, and hence, longer times are required to achieve stoichiometric composition of SiC QDs [175].

Interestingly, larger total incoming fluxes lead to faster equalization of the Si and C atom fluxes but do not change the general trends shown in Figure 6.15. For example, at the maximum flux $\Psi^+ = 10$ ML/s the $\wp_{S-C} = 1$ level can be intersected in a $T_s = 600$ K process. However, this does not happen at $T_s = 800$ K.

The results of Figure 6.15 can be understood better if one considers the temporal variation of the ratio of equilibrium fluxes of Si/C adatoms shown in Figure 6.16. In the first case (Figure 6.16(a) is related to the results in Figure 6.15), the adatom flux to the quantum dots is clearly carbon-enriched. In this case, a carbon-enriched core is formed, with the averaged elemental ratio factor $\wp_{S-C}$ less then 1 as shown in Figure 6.16(a).

Thereafter, the adatom flux ratio factor $\bar{\Gamma}_{S-C}$ overshoots the stoichiometric level. In this case the surface flux to the QDs is rich in silicon; as a result, a Si-enriched shell is formed with the elemental ratio factor $\wp_{S-C} > 1$. Then, the adatom flux ratio factor $\bar{\Gamma}_{S-C}$ tends to unity causing the averaged elemental ratio factor to gradually level off at the $\wp_{S-C} = 1$ level. As a result, the quantum dot structure shown in Figure 6.16(a), is formed. This structure has a carbon core, silicon shell and near-stoichiometric SiC outer shell [175].

As can be seen in Figure 6.16(b), at higher surface temperatures ($T_s = 800\,\text{K}$), a smaller carbon-rich core is formed which smoothly transits to a stoichiometric SiC shell, without any Si-enriched under-shell. This behavior can best be understood by noting that at the initial stage of deposition the more movable adatoms, which have lower surface diffusion activation energy, move faster about the surface and thus provide higher flux to QD according to Equation 6.19. This explains the excess of carbon in the QD core found in both cases. With time, the surface density of the Si adatoms rises and eventually provides an adatom flux large enough for an effective drain of Si adatoms from the substrate surface to QDs. Eventually, the fluxes of both Si and C adatoms to the quantum dots become equal, and the stoichiometric shell is formed [175].

To conclude this section, we stress that binary semiconductor systems introduce an additional level of complexity for quantum dot fabrication. In this case there is an additional requirement for specific elemental ratios throughout the nanodot internal structure. The simplest case is that of the near-stoichiometric composition throughout. Owing to very different surface conditions (energies, mobilities, etc.) for the two atomic species involved, their incoming fluxes may not necesarily need to be equal during the process. Another complication is that delicate objects such as quantum dots require very small amounts of material to be delivered. In some instances, a dot of a couple of tens of atoms can self-assemble within only fractions of a second. The time required to form their core or any specific layers is even shorter. The surface fluxes of the two adatom species may be significantly unbalanced, and moreover, time-varying during this crucial growth stage. The results presented in this section suggest an effective approach to adjust the process conditions to tailor the elemental composition and internal structure of binary semiconductor QDs in any way desired.

The following section will be devoted to later stages of development of nanodot arrays when the number of atoms composing the QDs is substantially larger. In this case the nanodots grow via attachment of various species (similar to the surface growth mechanism of nanoparticles discussed in Chapter 4) to the surfaces of QDs rather than clustering/nucleation of adatoms, which was the main mechanism of self-assembly at initial growth stages considered in this section. Nevertheless, even at the later stages, self-organization plays a significant role, in particular, in the development of ordered arrays of size-uniform Ge/Si nanodots as considered in the following section.

6.4
Self-Organization in Ge/Si Nanodot Arrays at Advanced Growth Stages

As already mentioned in previous chapters, self-assembly of size-uniform and spatially ordered quantum dot arrays is one of the major challenges in the development of the new generation of semiconducting nanoelectronic and photonic devices. In this section, we will follow the original report [177] and consider an advanced stage of assembly of Ge QD (in the ca. 5–20 nm size range) arrays from randomly generated position- and size-non-uniform nanodot patterns on plasma-exposed Si(100) surfaces.

It will be shown that by properly manipulating both the incoming ion/neutral flux from the plasma and the surface temperature, the uniformity of the nanodot size within the array can be improved by 34–53%, with the best improvement achieved at low surface temperatures and high external incoming fluxes, which are intrinsic to plasma-aided processes. Moreover, using a plasma-based process also leads to an improvement (ca. 22% at 700 K surface temperature and 0.1 ML/s incoming flux from the plasma) of the spatial order of a randomly sampled nanodot ensemble, which self-organizes to position the dots equidistantly to their neighbours within the array.

Remarkable improvements in quantum dot ordering and size uniformity can be achieved at high growth rates (a few nm/s) and surface temperatures as low as 600 K, which broadens the range of suitable substrates to temperature-sensitive ultra-thin nanofilms and polymers. The results of Ho *et al.* [177] are generic and can also be applied to non-plasma-based techniques and a range of material systems. This effort is particularly useful for the development of deterministic strategies of nanoassembly of self-ordered arrays of size-uniform quantum dots, in the size range where nanodot ordering cannot be achieved by presently available pattern delineation techniques.

We have already learned form Section 6.1 that the ability to arrange QDs with the same size, shape, structure, and so on in regular, uniform spatial arrays remains, despite a decade or so of intense research efforts, one of the main issues in the devlopment of quantum dot-based nanodevices. Such an arrangement greatly enhances the collective optoelectronic properties of the quantum dots through coupling interactions within the ensemble. In particular, these interactions substantially improve the efficiency and intensity of photoemissions in numerous applications including quantum dot laser and biomedical tagging devices [326–330]. Moreover, ultra-fine tuning of the emission frequency can be achieved by controlling the size and positioning of quantum dots within an array. Therefore, nanodevice applications utilizing quantum dot systems will ultimately require a high level of quantum dot size uniformity and positional ordering [177,331].

Many presently available pattern delineation techniques provide a reasonable level of control over quantum dot position yet cannot meet the requirements for the size and shape uniformity of individual QDs [332]. As already mentioned in Chapter 1 introducing the self-organized nano-world, these pre-patterning techniques have a number of intrinsic resolution-related limitations, and as such are not suitable for nanofabrication of ordered QD arrays compatible with the emerging sub-40 nm semiconductor technology, which will essentially rely on *bottom-up* self-assembly approaches [162]. Another major difficulty of top-down approaches is in achieving uniform distributions in quantum dot sizes and shapes [177].

It becomes clear that controlled self-organization is the most promising (and moreover, the only practical) way to overcome the above difficulties and so deterministically fabricate nanodot arrays with attributes suitable for their eventual nanodevice applications. The effectiveness of this approach critically depends on the nanofabrication environment used and the ability to create and manipulate nanoassembly building units [4]. Ho *et al.* [177] used advanced multi-scale hybrid numerical simulations [171] to show that plasma-assisted nanofabrication can be used to achieve deterministic control of the positional order and size uniformity of Ge/Si quantum dots. As we have already learned from Section 6.2, this is one of the most commonly used semiconducting systems possessing unique optoelectronic properties, a lattice mismatch of ca. 4% and widespread industrial applications [333–335].

Here we also reiterate that self-organized growth of Ge/Si quantum dots usually proceeds via strain-induced fragmentation of a continuous Ge film into nanoislands, commonly referred to as the Stranski–Krashtanov (SK) growth mode (see also Sections 6.1 and 6.2). As a result, very non-uniform patterns of size-non-uniform nanoislands are formed. As suggested by the results of Ho *et al.* [177], these irregular nanoisland patterns can be improved using a plasma-based process and can be eventually brought to the level required for nanodevice applications.

More specifically, by exposing the growing pattern to ion fluxes from the plasma, one can significantly improve the size uniformity of individual Ge quantum dots and their positional uniformity in originally non-uniform nanopatterns on plasma-exposed Si(100) surfaces. The reason for this original non-uniformity does not actually matter; it might have arisen from non-uniform nuclei developed in the process discussed in Section 6.2 or is a consequence of uncontrollable fragmentation of a continuous Ge film into size-nonuniform islands.

Similar to Section 6.3, the approach considered here is based on precise manipulation of both the incoming fluxes of plasma-generated building units and the surface temperature. This leads to a rearrangement of the two-dimensional surface adatom fluxes, which in turn results in faster growth of smaller dots and eventually better QD size uniformity. Additionally, quasi-displacement of quantum dots due to unbalanced surface adatom fluxes from different directions results in a more equidistant positioning of the dots with respect to their neighbours within the array [177].

6.4.1
Model of Nanopattern Development

Let us consider the growth and displacement of Ge/Si QDs arranged in a non-uniform nanopattern subjected to incoming fluxes of germanium atoms and ions from the plasma environment. A rectangular $1\mu m \times 1\mu m$ nanopattern with 400 dome-shaped nanodots arranged randomly was used to reproduce typical distributions of Ge nanoislands on a silicon surface [177, 336–339]. The initial QD sizes range from 4–13 nm and are assumed to possess a Gaussian distribution. Similar to the assumptions made in Section 6.2, the exact method by which the plasma is generated is not important. For simplicity, it is assumed that only germanium adatoms act as building units of Ge quantum dots on a Si surface. A controlled flux of Ge^+ ions can be achieved via the argon plasma-facilitated ionized physical vapor deposition (i-PVD); in this case the argon plasma serves the purpose of ionizing Ge atoms and also activating silicon surface [5].

Upon deposition onto a silicon surface, Ge ions are neutralized and become adatoms, the main contributors to the Ge/Si quantum dot growth. The values of the incoming fluxes of Ge^+ ions and Ge atoms are consistent with relevant experiments on synthesis of Ge/Si and related quantum dot systems in low-pressure plasmas [49, 50]. The pressure range in the simulations was 0.667 Pa (5 mTorr)–13.332 Pa (100 mTorr). Meanwhile, the ionization degree was varied from ca. 10^{-3} representative of high-density, low-pressure RF plasmas commonly used for microelectronic fabrication to ca. 0.5, which is typical for advanced pulsed i-PVD systems [177]. Other parameters of the simulations are summarized, together with the main physical constants used, in Table 6.3.

Table 6.3 Physical constants and simulation parameters used in computations [177].

Physical Constants	Value
Lattice atom oscillation frequency, ν_a, s^{-1}	1×10^{13}
Lattice parameter, λ_{lat}, m	5×10^{-10}
Ge diffusion activation energy, $\mathcal{E}_d$, eV	1.3
Energy of Ge evaporation to 2D vapor, $\mathcal{E}_{\text{evap}}^{2D}$, eV	0.70
Energy of Ge evaporation to 3D vapor, $\mathcal{E}_{\text{evap}}^{3D}$, eV	0.75
Energy of Ge evaporation from Si, $\mathcal{E}_{\text{evap}}^{S}$, eV	0.65
Ge atom bonding energy, $\mathcal{E}_b$, eV	2.6
Simulation Parameters	
Total influx Ψ_{pl}, ML/s	0.01–0.15
Influx of neutrals Ψ_{pl}^{n}, ML/s	0.006–0.1
Ion influx $\Psi_{\text{pl}}^{\text{ion}}$, ML/s	0.004–0.05
Initial number of QDs	400
Surface temperature, T_s, K	600–900
Surface coverage, ζ	0.1–0.46
QD mean radius, ϕ, nm	8–19
Process duration, t, s	0–0.5
Pressure range, mTorr	5–100
Ionization degree	10^{-3}–0.5

As before, the primary variable simulation parameters are the Si substrate surface temperature and the incoming external flux of Ge building units from the plasma. The optimum combination of these parameters may lead to a high level of deterministic control over the spatial and/or dimensional order within a Ge/Si QD array of a reasonably high surface coverage. Interestingly, the effectiveness of each control parameter is quite different in every case [177].

During the Ge/Si QD deposition germanium atoms are delivered from the ionized gas environment either directly to the nanodot or via the wafer surface similar to as shown in Figure 1.2. Upon contact with the Si(100) substrate, Ge adatoms migrate about the surface via surface diffusion. Depending on the surface temperature, adatoms may re-evaporate from the QD either to the two-dimensional adatom field or back into the plasma bulk.

The hybrid QD growth model of this section is based on a set of adatom flux balance equations, which takes into consideration incoming fluxes from the plasma, fluxes due to surface diffusion and a range of adatom evaporation processes. This method has been commonly employed in surface science as a means for modelling a broad range of growth processes [179]. Previous efforts [340, 341] at modelling QD nanoassembly have been attempted but this model is generic in that it does not specify the method of preparation of the initial patterns of seed nuclei [177].

The growth process of the i-th quantum dot is described by [177,178]

$$(\partial V_i / \partial r_i) dr_i = \lambda_{\text{lat}}^3 \Psi_i dt, \tag{6.21}$$

where λ_{lat} is the lattice constant, V_i and r_i are the i-th nanodot's volume and radius, respectively, and t is the time into the growth process. Here

$$\Psi_i = \Psi_{\text{pl},i} + \Psi_{\text{surf},i} - \Psi_{\text{vap},i} - \Psi_{\text{svap},i} \tag{6.22}$$

is the total flux of species to the i-th QD, where $\Psi_{\text{pl},i}$ is the flux of incoming building units from the plasma, $\Psi_{\text{surf},i}$ is the two-dimensional surface flux of adatoms onto the border of the i-th nanodot, and $\Psi_{\text{vap},i}$ and $\Psi_{\text{svap},i}$ are the bulk and surface evaporation fluxes from the i-th quantum dot, respectively.

The incoming flux from the plasma [177]

$$\Psi_{\text{pl},i} = \Psi_{\text{pl},i}^{\text{ion}} + \Psi_{\text{pl},i}^{n} \tag{6.23}$$

involves contributions from the ion and neutral fluxes represented by the first and the second terms in this equation, respectively. Here, $\Psi_{\text{pl},i}^{\text{ion}} = \int_{S_i} j_{S_i} ds_i$ and $\Psi_{\text{pl},i}^{n} = \bar{S} \psi_n$, where j_{S_i} is the density of the ion flux onto the nanodot surface with the area S_i, and ψ_n is the density of the neutral flux; the latter is assumed to be uniform over the entire substrate surface.

The two-dimensional surface flux of adatoms is [177]

$$\Psi_{\text{surf},i} = -D_S l_i \left[\nabla \eta(x, y, t) \right]_{l_i}, \tag{6.24}$$

where $[\nabla \eta(x, y, t)]_{l_i}$ denotes the gradient of the adatom surface density $\eta(x, y, t)$ at the border of the i-th QD with the perimeter l_i, and

$$D_S = [v_a \lambda_{\text{lat}}^2 / 4] \exp(-\mathcal{E}_d / k_B T_s)$$

is the surface diffusion coefficient. Here, v_a is the characteristic frequency of atom oscillations in a lattice, $\mathcal{E}_d$ is the surface diffusion activation energy. As before, T_s is the temperature of the silicon surface, and k_B is Boltzmann's constant.

The time-varying density of Ge adatoms on the Si(100) surface can be obtained numerically from the two-dimensional diffusion equation

$$\frac{\partial \eta(x, y, t)}{\partial t} = D_S \nabla_{2D}^2 [\eta(x, y, t)] + \psi_{\text{in}} - \psi_{\text{evap}}, \tag{6.25}$$

where $\nabla^2_{2D} = \partial^2/\partial x^2 + \partial^2/\partial y^2$, and $\psi_{in} = \psi_{in}^{ion} + \psi_n$ is the incoming flux of the ionic (first term) and neutral (second term) building units from the plasma onto open surface areas uncovered by the quantum dots. Here, $\psi_{in}^{ion} = [1/\bar{S}] \int_{\bar{S}} \bar{j}_S d\bar{s}$, where $\bar{j}_S$ is the density of the ion flux onto the substrate surface between the QDs (area $\bar{S}$). On the right hand side of Equation (6.25)

$$\psi_{evap} = [\nu_a/\lambda^2_{lat}] \exp(-\mathcal{E}^S_{evap}/k_B T_s)$$

is the flux of adatom evaporation (with the characteristic energy $\mathcal{E}^S_{evap}$) from the substrate surface between the QDs [177].

The remaining two terms in Equation (6.22), $\Psi_{vap,i}$ and $\Psi_{svap,i}$ represent evaporative losses from the i-th QD to the gas phase and the 2D surface gas vapor, respectively. More specifically,

$$\Psi_{vap,i} = [\nu_a S_i/\lambda^2_{lat}] \exp(-\mathcal{E}^{3D}_{evap}/k_B T_s)$$

and

$$\Psi_{svap,i} = [\nu_a l_i/\lambda_{lat}] \exp(-\mathcal{E}^{2D}_{evap}/k_B T_s),$$

where $\mathcal{E}^{3D}_{evap}$ and $\mathcal{E}^{2D}_{evap}$ are the energies of adatom evaporation to the 3D vapor in the plasma bulk and the 2D vapor on the surface, respectively [177].

The quantum dot growth is determined by the two-dimensional field of microscopic adatom diffusion across its circular border. If adatom fluxes from different directions are balanced the dots grow at the same position, otherwise one can observe nanodot quasi-displacement as sketched in Figure 6.17. To quantify this displacement, the surface flux term entering Equation (6.22) is decomposed into four components Ψ_N, Ψ_S, Ψ_E, and Ψ_W directed to the North, South, East and West, respectively (Figure 6.17(a)). In this case, the quasi-displacement of the dots within the ensemble is described as their asymmetric growth, and the net shift of the QD centers resulting from differential influxes in the two-dimensional adatom field as shown in Figure 6.17(b) [177].

Referring to Figure 6.17, one can define the two-dimensional positional shift (dX, dY) of a nanodot center subjected to two-dimensional surface fluxes as

$$dX = \frac{\bar{\xi}_x}{\sqrt{\bar{\xi}_x^2 + \bar{\xi}_y^2}} \frac{dr}{2}$$

and

$$dY = \frac{\bar{\xi}_y}{\sqrt{\bar{\xi}_x^2 + \bar{\xi}_y^2}} \frac{dr}{2},$$

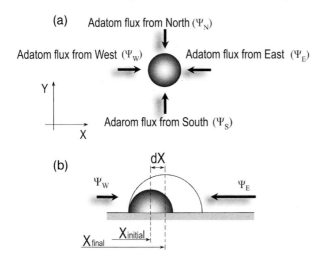

Figure 6.17 Schematic of (a) the two-dimensional flux on a QD and (b) the QD center shift as a result of a non-uniform two-dimensional adatom flux [177].

where

$$\tilde{\zeta}_x = (\Psi_W - \Psi_E)/(\Psi_W + \Psi_E)$$

and

$$\tilde{\zeta}_y = (\Psi_N - \Psi_S)/(\Psi_N + \Psi_S)$$

are the flux differential terms, and r is the (time-varying) quantum dot radius; subscript i has been omitted for simplicity.

6.4.2
Ge/Si QD Size and Positional Uniformity

Let us now consider the dynamics of the quantum dot self-organization in a randomly generated starting pattern. Firstly, the effect of the surface temperature and incoming flux on nanodot size uniformity is studied; and secondly, the dynamics of QD displacement will be presented demonstrating the possibility of their positional self-ordering [177].

Figure 6.18 shows the simulated quantum dot nanopatterns at different incoming fluxes at a surface temperature of 600 K. When the substrate temperature is changed, the levels of size uniformity within the pattern also change. For instance, the process at $T_s = 900$ K produces broader size distributions. The temporal dynamics of the dot growth displayed in Figure 6.18 shows an initial wide distribution in the size of QDs, which narrows at surface coverage $\zeta = 0.255$ and then broadens again at $\zeta =$

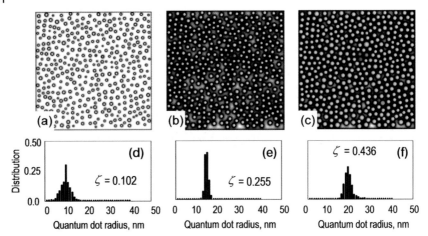

Figure 6.18 QD pattern (a–c) and QD size distribution (d–f) for a surface temperature of $600\,\mathrm{K}$ and flux $\Psi_{\mathrm{pl}} = 0.1\,\mathrm{ML/s}$ ($\Psi_{\mathrm{pl}}^{\mathrm{ion}} = 0.3\Psi_{\mathrm{pl}}$) for surface coverage (left to right) of 0.102 (a, d), 0.255 (b, e), and 0.436 (c, f) [177].

0.436. This narrowing illustrates the improvement from the initial pattern with surface coverage $\zeta = 0.102$ and subsequent deterioration in the size uniformity. The uniformity in the nanodot size initially improves from the starting seed nuclei pattern but is followed by a steady deterioration in uniformity when a certain surface coverage is achieved.

Interestingly, the mean nanodot radius and the level of size uniformity strongly depend on the surface temperature. Specifically, when the temperature is increased, there is a decrease in the mean dot radius at the points of the optimum size uniformity. We should also stress that the improvement (as compared to the initial pattern) in the QD size uniformity is less for higher surface temperatures [177]. Most amazingly, smaller dots exhibit accelerated growth while the growth of larger dots is retarded. Thus, *the nanopattern tends to self-organize to reduce the nonuniformity of QD size distribution.*

We now consider the results related to the positional order of QDs within the array and the dynamics of nanodot movement to achieve such an order [177]. It has been repeatedly observed that dot movement follows predictable routes from highly occupied (by neighboring QDs) towards less occupied (e.g. vacant) regions of the wafer. On the other hand, dots in close proximity often appear to move in opposite directions. To elucidate the motion dynamics of 5 selected QDs (labelled 1–5), in Figure 6.19, plots of dot-specific radii, displacements from initial positions, and velocities associated with the quasi-displacements are pre-

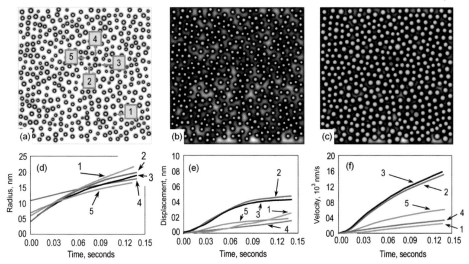

Figure 6.19 QD pattern and adatom field at (a) 0; (b) 0.035; and (c) 0.141 s. Temporal evolution of (a) radii; (b) displacement; and (c) velocity values for selected QDs of the initial size: 1) 6 nm (green); 2) 4 nm (blue); 3) 5 nm (grey); 4) 11 nm (magenta); and 5) 7 nm (red) [177].

sented alongside 3 snapshots of the same nanopattern with these dots at three different time moments.

Figure 6.19 corresponds to a typical surface temperature of 700 K with 0.1 ML/s incoming flux. In particular, Figure 6.19(f) indicates that smaller QDs appear to move faster. In fact, upon closer evaluation of Figures 6.19(e) and (f), the QD labeled (1) moves with a considerably lower velocity (and is displaced by a smaller distance) than its more mobile counterparts (2) and (3) despite their similar size as shown by Figure 6.19(d). Physically, this is caused by the difference in the relative location of the dots with respect to their neighbors. Examining the QD patterns in Figure 6.19 it is seen that the QD labeled (1) is positioned centrally in that it is approximately equidistant from its adjacent neighbors. In contrast, the dots labelled (2) and (3) are positioned assymmetrically with respect to their neighbors [177].

A further illustration of the improvement of the positional uniformity within the nanopattern is shown in Figure 6.20. From Figure 6.20 one sees that a local site with a QD in the middle self-organizes to equalize the distances between the central dot and its neighbors. In this example, Figure 6.20(a) shows that the central dot is initially positioned closer to its neighbors on the left. After only 0.14 s at $T_s = 700$ K and $\Psi_{pl} = 0.1$ ML/s the same dot has positioned itself equally between its right- and left-side

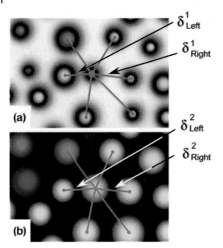

Figure 6.20 QD pattern at (a) start ($t = 0$ s) and (b) end
($t = 0.14$ s) of the growth simulation showing the distance of a
dot (gray) to its neighbours for surface temperature 700 K and
external flux $0.1 \, \mathrm{ML/s}$. The ion flux is the same as in Figure 6.18.
The distances are seen to become more uniform [177].

neighbors following quite minor quasi-displacements of all the nanodots
in the ensemble, as shown by Figure 6.20(b). In fact, the ratio of the most-
unequal distances (labeled in the figure) changed from 0.7 for the initial
state to 0.96 for the final pattern [177].

One can thus conclude that the growth of individual quantum dots is
affected by their relative size and their vicinity to neighboring dots. In
fact, smaller dots located in an area of a low local surface coverage usually experience accelerated growth as compared to larger QDs located in
close proximity to nearby nanodots [177].

6.4.3
Self-Organization in Ge/Si QD Patterns: Driving Forces and Features

Before we proceed any further, it should be stressed that despite a large
number of reports on self-organization phenomena in complex physical
systems, it is still unclear what the common driving force of such amazing phenomena is. It is understood however, that the physical mechanisms responsible for the self-organization are numerous and case-
specific. The key to self-organization in the plasma-surface environment
of our interest is the collective behavior of neutral and ionized building
units, BU transport via the plasma sheath followed by their surface migration, collisions, clustering and self-assembly into specific nanostructures on the solid surface discussed in detail in Chapter 2. If properly un-

derstood and controlled through manipulation of the plasma and surface parameters and initial conditions, this self-organization phenomenon may be harnessed to enable deterministic nanofabrication of nanodevice-grade quantum dot arrays and ultimately their interconnected functional networks. In this section, we have explored some plausible ways to create ordered Ge/Si QD arrays in a plasma-assisted process by manipulating the incoming fluxes from the plasma and the surface temperature [177].

One of the most striking observations is that QDs grow faster in areas with a lower local surface coverage. Moreover, QDs displace themselves to establish uniform size and distance in relation to their neighbors. Individual QDs appear to move into spaces with the lower surface coverage ζ but higher adatom number density. We emphasize that the quantum dot itself does not actually move on the substrate surface (we examined the QDs of 10 to 20 nm which cannot move about the surface like adatoms or very small nanoclusters), and thus we consider the displacement of the QDs centers as being due to the irregular QDs growth. The apparent cause for this displacement is the unbalanced two-dimensional adatom fluxes (Figure 6.17). Examining Figure 6.18 one can see that the adatom density shown as a gray field is higher in surface zones between the distant QDs. As a result, the increased adatom fluxes from the surface zones of higher density cause an increased irregular growth of the QDs and finally displacement of their centers, usually towards higher adatom densities. Finally, the QDs arrange themselves within the two-dimensional adatom field in order to maintain or achieve uniform adatom fluxes about their perimeter [177].

This behavior presents an undeniable picture of the self-organization phenomena on plasma exposed surfaces. Once the mechanisms behind this self-organization are established, the next stage is to harness them in practical applications that require self-assembled, highly-ordered arrays of size-uniform quantum dots. It is worth noting that the numerically simulated nanodot arrays qualitatively resemble those Ge/Si(100) QD patterns synthesized experimentally [313, 339]. The growth and movement of QDs have been clearly influenced by the two-dimensional non-uniform field of the surface adatom fluxes.

Therefore, one can state that *the plasma-aided approach considered here is an effective self-organization route whereby BUs from the plasma are deposited onto a solid surface and then contribute to self-assembly of ordered arrays of size-uniform quantum dots.* We emphasize that the self-assembly of individual QDs and their self-ordering into self-organized nanoarrays makes it possible to fabricate such nanoassemblies without the need of a continuous external control, which is a common feature of most of self-organized complex systems [177].

However, as the results of the numerical experiments suggest, the plasma-based QD growth process is very fast and requires only a fraction of a second to substantially increase the nanodot size as can be seen in Figure 6.19. Therefore, the process should be terminated at the point where the best size uniformity and/or positional ordering are achieved, for example, as shown in Figure 6.6.

The above consideration suggests that the Ge/Si nanodot size uniformity can indeed be effectively controlled and in most cases improved by appropriately manipulating the incoming fluxes and the surface temperature. However, extreme caution should be taken not to overheat the surface since the mean size deviation tends to rise with T_s. Deterioration of nanodot size uniformity becomes even more apparent for larger QDs.

Table 6.4 Mean square deviation of distances of randomly selected QD to adjacent QDs for $T_s = 700\,\text{K}$ at $\Psi_{pl} = 0.1\,\text{ML/s}$ [177].

	Quantum Dot	Mean Square Deviation	
Number	Starting coordinate of central QD (pixels)	Initial	Final
1	(611, 421)	105.46	86.96
2	(481, 809)	145.86	103.94
3	(903, 553)	133.87	64.00
4	(215, 413)	35.16	37.75
5	(165, 215)	75.71	60.69

Table 6.4 quantifies the improvement of the spatial order in localized regions of the wafer such as that seen in Figure 6.20. The mean square deviation reflects on the level of uniformity in the distances of the selected central dot to adjacent QDs. One notices that positional uniformity improves significantly in the areas of the initially high non-uniformity. It is also notable that selected regions containing numerous smaller dots exhibit a greater net gain in order. Due to the greater ability of smaller QDs to move (illustrated by larger values of their quasi-displacements relative to their original size), this is likely to result in a greater improvement in the spatial order. The displacement of these dots, therefore, has a better chance of reaching a maximum before a coalescent contact can be reached.

Thus, in order to improve the spatial order during this growth process the initial nanodot pattern should contain seed nuclei with the minimum possible size. This was one of the main conclusions of Section 6.2! One more important thing to stress is that the nanodot growth was initiated from randomly generated initial size- and position-non-uniform nanopatterns. Nevertheless, the numerical experiments revealed pronounced "self-uniformization" of QD sizes as well as self-ordering of the nanodots within the developing array [177].

We now turn our attention to discussing some of the salient features of plasma-assisted nanoassembly [177]. In one respect, it has been already discussed as the preferred route in enabling deterministic control in the growth and shape of nanostructures. Another important feature is that high rates of material delivery from the plasma allow for faster QD growth compared to most of the conventional thermal deposition methods.

The simulation results also indicate that uniform nanopatterns of size-uniform Ge/Si nanodots can be synthesized at surface temperatures as low as 600 K. This growth temperature is notably lower than those commonly used in thermal chemical vapor deposition (CVD). This can substantially expand the range of substrate materials that can host semiconductor quantum dots to include temperature-sensitive polymers, plastics, and ultra-thin nanofilms.

Another advantage is the possibility using very short deposition processes (e.g. based on pulsed iPVD) not exceeding 1 s (see Table 6.3); this is much shorter than the previously reported deposition times of CVD typically ranging anywhere between 0.5 to 40 minutes. We emphasize that plasma-based nanoassembly routes offer the possibility of controlling the dosing of specific (e.g. more complex) BUs [4].

For example, the Si(100) surface can be covered with a monolayer of atomic hydrogen in less than 0.5 s using an ion flux extracted from a 1.333 Pa (10 mTorr) H_2 plasma at room temperature and an ionization degree of ca. 10^{-4} [342]. As we have seen from Section 6.3, plasma-based processes can also offer great precision in controlling the elemental composition of binary quantum dots, such as SiC/Si [174, 175]. One of the most important advantages of the plasma-controlled self-organization approach is the ability of QDs to self-organize into ordered arrays without any lithographic pre-patterning or strain-driven site allocation [177].

To conclude this section, we stress that by using plasma-based nanoassembly, it may eventually be possible to deliver controlled combinations of different plasma-generated species and arrange them selectively to fabricate intricate nanoassemblies, interconnected networks and even elements of nanodevices. In the following section, we will consider more plasma-related controls and features of self-organized nanostructures and their arrays on plasma-exposed surfaces.

6.5

Self-Organized Nanodot Arrays: Plasma-Specific Effects

To elucidate the effects of the ionized gas environment on the self-organization phenomena on plasma exposed solid surfaces, let us now consider once again the environment where the elementary self-organization processes occur and re-examine Figure 1.2 which shows a solid surface (denoted as a substrate) exposed to a low-temperature plasma. As we already know, the size of the plasma bulk is of the order of fractions of a meter, which is a typical size of plasma processing reactors. The plasma bulk is separated from the solid surface by an area of uncompensated surface charge called a plasma sheath with a typical width ranging from microns to millimeters. The electric field in the sheath accelerates positively charged ions towards the surface as shown in Figure 1.2. Surface processes involve a few of the topmost atomic layers as well as adatoms migrating about the surface. As a result of assembly of these adatoms, nanostructures of a reasonably small number of atoms are formed. Thus, the spatial scales of the processes involved differ by up to 9 or even 10 orders of magnitude [279].

Those nanodot structures considered in this chapter emerge in most cases as nanoislands with size and shape determined by prevailing surface conditions such as energies of activation of adatom diffusion and desorption/re-evaporation to the gas phase, chemical potential of the surface, distribution of surface stress, as well as available surface features such as stepped monoatomic terraces, dislocations, and surface defects [278]. The surface processes encircled by a dash-dotted line in Figure 1.2 also involve small nanostructures being created since their presence substantially affects the distribution, motion, and self-arrangements of adatoms on the surface [279]. The line also encircles a few of the topmost atomic layers since interactions of adatoms with them determine the ability of adatoms to move and self-assemble.

Further, interactions of the nanostructures with the substrate atoms underneath determine the nanostructure growth processes and also set a two-dimensional distribution of surface stress, which in turn affects adatom migration and assembly into nanostructures. In fact, this can already be considered as a simple self-organization process where the nanostructure growth affects the topmost atomic layers in the solid and the resulting stress affects surface adatom diffusion and evaporation, which in turn affects the nanoassembly process [279].

The above picture of self-organization is common to "mild" nanoassembly environments such as those in thermal chemical vapor deposition (CVD) systems. However, what about the plasma environment of our interest in this monograph? What changes to this common scenario can one expect? This chapter will aim to shed some light on these and other related issues.

6.5.1
Matching Balance and Supply of BUs: a Requirement for Deterministic Nanoassembly

Before we move on with this discussion, it will be instructive to recall that present-day nanotechnology aims at achieving a deterministic (highly-controllable and predictable) level in the synthesis of nanostructures and their assemblies. One of the main requirements for achieving the *deterministic* nanoscale synthesis of delicate nanoscale assemblies is to identify the most appropriate building units [4] and properly balance their demand and supply. For example, if a square centimeter array of 10 billion 1 nm-sized, crystalline, cubic silicon quantum dots is to be grown in a one second-long process, a minimum of approximately 3×10^{11} atoms would need to be delivered within 1 s. The required growth flux is thus $\Psi_{growth} = 3 \times 10^{11}$ atoms/cm^{2}s.

Hovever, this amount should be increased to account for those adatoms that are unable to attach to the surface or desorb/evaporate shortly after adsorption. These processes may take a substantial fraction $\varsigma (0 < \varsigma < 1)$ of adatoms, especially at higher temperatures. Realistically assuming $\varsigma = 0.5$, we obtain that the flux of adatoms lost form the surface Ψ_{loss} is also 3×10^{11} atoms/cm^{2}s. Thus, the total incoming flux under conditions of perfect match of building unit balance and supply

$$\Psi_{tot} = \Psi_{growth} + \Psi_{loss}$$

is 3×10^{11} atoms/cm^{2}s. Taken that $1 \, cm^2$ of silicon surface contains approximately 5×10^{14} (1 ML) silicon atoms, the "deterministically" required total incoming flux will be of the order of $10^{-3} \, ML/s$. In real experiments, the total flux can differ from $\Psi_{growth} + \Psi_{loss}$. If $\Psi_{tot} < \Psi_{growth} + \Psi_{loss}$, the nanostructures would not grow or under-develop during the adatom lifetime on the surface. On the other hand, if $\Psi_{tot} > \Psi_{growth} + \Psi_{loss}$, excessive delivery of building material results in a growth of undesired formations such as shapeless islands or amorphous films instead of the expected crystalline quantum dots [279].

The latter situation is quite common to plasma processing tools. To show this, let us now estimate the flux of SiH$_3$ radicals from a typical silane-based plasma considered in Chapters 3 and 4. Assuming the radical density $n_{rad} \sim 4 \times 10^{13} \, cm^{-3}$ and their thermal velocity $V_{Tr} \sim 500 \, m/s$, we obtain $\Psi_{tot} = (1/4) n_{rad} V_{Tr} \sim 5 \times 10^{17}$ radicals/cm^{2}s, which is approximately 1000 ML/s, which is 6 orders of magnitude higher than required! This enormous oversupply of building units (although certainly good for the growth of bulk films) is one of the main reasons why plasma tools have not been particularly successful in the synthesis of very delicate nanostructures such as quantum dots.

This is in stark contrast with the very impressive results achieved by using plasma-based techniques and approaches in the area of synthesis of larger nanostructures that contain 10^5–10^6 atoms such as various vertically aligned nanostructures (nanotubes, nanowires, nanocones, etc.). The latter nanostructures, although surface-bound, usually have the third (normal to the surface) dimension much larger than the other two dimensions and as such extend far beyond the traditional area of surface science shown in Figure 1.2. Therefore, proper balancing of the demand and supply of the required building units is one of the main challenges of self-assembly processes on plasma-exposed surfaces [279].

Therefore, one sees that high-density reactive silane-based plasmas produce fluxes which exceed in almost six orders of magnitude what is actually required to synthesize the array of 1 nm-sized quantum dots. Let us now consider how to match the demand and supply of the building units in a plasma-based nanoassembly environment.

Fortunately, the number of possible options is quite large [279]. First of all, it is possible to reduce densities of BUs by running plasma discharges under lower input power and gas temperature conditions, depositing radicals from a discharge afterglow, using various configurations of remote plasma deposition systems, or releasing the needed building units in much smaller amounts from solid targets using, for example magnetron sputtering.

Another option is to reduce the duration of the building unit delivery τ_{deliv} to the growth surface, for example, by shortening the pulse duration during which the BUs are released. Plasma-assisted pulsed magnetron sputtering is a good example of a suitable plasma-based system. We emphasize that regardless of the system used, the nanostructures need to be given sufficient time to develop.

In a pulsed plasma facility, this can be done by using plasma-on τ_{ON} and plasma-off τ_{OFF} sequences of appropriately chosen durations. In this case the duration of the plasma-on sequence can be estimated as

$$\tau_{ON} = \xi N_a / \Psi_{tot},$$

where ξ is the number of individual QDs per cm^2 and N_a is the average number of atoms in each nanostructure. Likewise, the duration of the plasma-off sequence should exceed the time needed for the quantum dots to nucleate and reach the required size. For simplicity, the nanodot growth time τ_{growth} has been assumed to be 1s. In reality, τ_{growth} strongly depends on prevailing surface conditions, such as surface temperature, diffusion activation and evaporation energies, surface stress, presence of seed nuclei and several others.

Under conditions of heavy incoming fluxes and relatively long growth times, the plasma duty cycle $\tau_{ON}/(\tau_{ON} + \tau_{OFF})$ can be quite low, which is undesirable in practical applications. Therefore, to improve the efficiency of plasma-based pulsed processes or ultimately achieve a continuous process in which the rates of nanodot growth are perfectly matched with the building unit delivery rates, one has to significantly shorten the growth time, in this example from 1s to small fractions of a second [279].

6.5.2
Other General Considerations

It is amazing that the rates of nanocrystal growth in a plasma environment can be substantially higher than in neutral gas environments with very similar parameters except for the presence of an ionized gas component. As we will see from the following consideration, the increase of the growth rates is primarily due to two factors [279, 343]:

- higher surface temperatures due to ion bombardment (if building units arrive to the surface as ions, they transfer their energy locally upon deposition);

- reduced surface diffusion activation barriers due to electric field-related polarization effects.

Moreover, the estimates made later in this section will show that under typical conditions of suitable i-PVD systems the nanodot growth rates can be increased by up to a couple of thousand times!

In this regard, using ions instead of neutral radicals can bring a range of indisputable advantages [279]. Indeed, in weakly-ionized plasmas the ion densities are typically 3–5 orders of magnitude lower compared to neutrals. The ion flux can be precisely controlled by electric fields either existing within the plasma sheath or applied externally. More importantly, since ions usually move much faster than neutrals, very competitive fluxes (with much lower particle number densities) can be obtained. Therefore, ion-based processes are indeed very promising to satisfy the requirement of delivery of smaller amounts of building material at faster rates. As mentioned above, under conditions of a local energy transfer, ionic building units also increase surface temperature and can assemble into nanostructures at much faster rates than their low-energetic adatom counterparts. Faster mobility on the surface also enables adions (which can neutralize upon landing but have a higher energy compared to adatoms) to move faster about the facets of growing crystals and find their places in a crystalline lattice shortly after deposition. This effect is quite similar to ion-enhanced crystallization common to nanostructured films.

Let us now provide some more quantitative information on the effect of the plasma environment on self-organization on the surfaces of solid materials immersed in the plasma. To do this, we will follow the original reports [286, 343] and mainly concentrate on the self-organization phenomena during nanostructure formation in low-temperature plasma environments. The effect of the plasma appears to be quite different at different stages of nanopattern development.

For this reason, we will split this complex process into three main stages: the initial growth, the growth of individual nanostructures, and the development and self-arrangement of nanopatterns/arrays. The relevant experimental data, complemented with the results of hybrid numerical simulations, convincingly demonstrate that the growth process, self-assembly and self-organization in large arrays of nanostructures can be effectively controlled in plasma-based processes. More specifically, the initial seed pattern formation on the surface, nanostructure crystallization and nanostructure array re-organization involve processes which can be effectively influenced by the plasma parameters.

Thus, the whole process of the nanoarray formation may be conditionally divided into the three main growth stages [343]; namely, the initial (sub-monolayer) stage which consists in the growth of ultra-nano (up to 1 nm) nanoclusters; separate growth stage which involves the growth of nano-objects of several nm size without mutual influence; and, finally, the nanopattern self-development stage. During the latter stage, the nanoscale objects that make the array continue their growth and while growing, interact with other nanostructures. In this case, individual nano-objects interact with each other. As we have seen from Section 6.4, quantum dots in dense arrays interact through two-dimensional adatom fields, which are in turn strongly affected by the nanostructures. Thus, we have a clear example of a complex self-organized system. Interestingly, the above three-stage picture of nanoarray development is also applicable for some other nanoassemblies such as dense forests of carbon nanotubes.

In this section we will not examine the various aspects of the nanostructure growth at different stages; instead, we focus on the self-organization aspects at each stage. More precisely, our task is to point out the specific physical processes that (i) can be effectively controlled via the plasma parameters; and (ii) strongly influence the surface processes involved in the nano-structuring/nano-array formation.

Before we proceed with consideration of specific plasma-related effects on self-organized processes on solid surfaces, we should recall that methods involving self-organization are commonly used along with "top-down" techniques that involve natural or artificially made templates or pre-delineated patterns. We have already discussed this issue in Chapter 1. What we would like to stress is that contrary to the above "top-down" techniques which cannot be used below a certain size thershold, approaches based on self-organization have virtually no restriction on the sizes of objects and their patterns that can be created. However, this is not strictly true for large nanostructures which may develop in a non-interacting, uncorrelated manner.

A general examination of the physics involved in nano-array formation shows that there are actually two main parameters that appear to be both plasma-controllable and simultaneously important for the self-organization of the nano-arrays. These factors are the surface temperature and surface diffusion activation energy. In fact, in previous sections of this chapter we have already stressed the importance of the above two factors in controlling elementary surface processes that involve adatoms. Thus, let us now try to discuss how and to what extent the temperature and surface diffusion activation energy, being controlled via plasma-related parameters, may influence the kinetics of surface processes that in turn control self-organization on plasma exposed surfaces.

6.5.3
Plasma-Related Effects at Initial Growth Stages

Let us start from the initial, or sub-monolayer, growth stage of the nano-array formation, with the characteristic size of nano-objects up to 1 nm. Similar to Section 6.2, we consider very small nano-objects, or nanoclusters, consisting of several atoms, typically up to 20–25. At this stage, the kinetics of the process is mainly determined by the kinetics of an individual adsorbed atom (adatom) interaction with the substrate surface and nano-objects. The influence of plasma-related parameters on the self-organization at the initial stage can be analyzed by the rate equation (6.4) of Section 6.2. In the absence of lattice/surface defects and impurities, the rates

$$v_d = \lambda_{\text{lat}} v_0 \exp(-\varepsilon_d / k_B T_s)$$

of adatom diffusion,

$$v_i = v_0 \exp(-\varepsilon_{b.i} / k_B T_s)$$

adatom evaporation from the borders of quantum dot nuclei to the two-dimensional vapor, and

$$\mu_i = \nu_0 \exp(-\varepsilon_{a.i}/k_B T_s)$$

adatom evaporation to the bulk, appear to be extremely sensitive to the values of the surface temperature T_s and the characteristic energy barriers (activation energies) for surface diffusion ε_d and two- and three-dimensional evaporation, $\varepsilon_{b.i}$ and $\varepsilon_{a.i}$, respectively.

In a plasma environment, the surface temperature T_s is determined by the energy balance on the surface and is strongly influenced by the ion flux and the electric field within the plasma sheath. Thus, ions transfer their kinetic energy upon deposition; as a result, the temperature of surfaces subjected to ion fluxes is usually higher than that of the same surfaces in a neutral gas environment. In some cases this difference can reach 100 K or even higher [1,5,94]. This temperature increase is mostly controlled by the electric field within the plasma sheath.

Moreover, the electric field in the vicinity of nanostructured surfaces is quite different from that in the bulk of the plasma sheath (some examples will be considered in Chapter 7) and is primarily responsible for the changes in the surface diffusion activation and evaporation energies. The plasma-related change in the surface diffusion activation energy is usually negative and can be estimated as [343]

$$\delta\varepsilon_d = -|\nabla E|\lambda_{\text{lat}}\tilde{p},$$

where $\tilde{p}$ is the dipole moment of the adatom, and E and ∇E are the microscopic electric field and its gradient in the vicinity of the nanostructured surface, respectively.

Let us now briefly discuss the results of the numerical simulations of the effect of plasma-related variation of the surface diffusion activation energy on the assembly of Ge nanocluster seed nuclei on a Si(100) surface (this process has been considered in Section 6.2 without accounting for the effect of the plasma environment on the surface diffusion activation energy). The value of ε_d unaffected by the plasma is 1.072×10^{-19} J (0.67 eV) and the histogram of the nanocluster size distribution is shown in Figure 6.21(c).

Even small reductions in ε_d (not exceeding 10%) cause a dramatic change in the distribution of the nanocluster seed nuclei in terms of the number of atoms they contain. In fact, by reducing the diffusion activation energy to 0.960×10^{-19} J (0.6 eV) (Figure 6.21(b)) and then further to 0.880×10^{-19} J (0.55 eV) (Figure 6.21(a)), the population of small nanoclusters can be substantially reduced. However, the total nanocluster

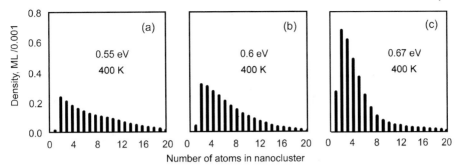

Figure 6.21 Nanocluster distribution function with surface diffusion activation energy as a parameter for (a) 0.880×10^{-19} J $(0.55\,\text{eV})$; (b) 0.960×10^{-19} J $(0.6\,\text{eV})$; and (c) 1.072×10^{-19} J $(0.67\,\text{eV})$. Surface temperature $400\,\text{K}$ [343].

density is also reduced. This reduction is primarily due to the effect of the electric field on the adatom evaporation rates.

Even though more detailed investigations into this effect are required, Figure 6.21 indicates the possibility of controlling the elementary surface processes that involve adatoms by adjusting the characteristic energies of their surface diffusion and two- and three-dimensional evaporation. This adjustment can be implemented by varying the electric field within the near-surface sheath, in particular, its strongly non-uniform microscopic component in the vicinity of the nanostructures being grown.

6.5.4
Separate Growth of Individual Nanostructures

We will now consider the next stage of nanopattern formation, namely the stage of separate growth of the nanostructures, when the latter are large enough (include at least several hundreds of atoms) for crystallization, but small enough that they do not interact with neighboring structures (in particular, due to the relative large distances between them). At this stage, the processes that determine the shaping and crystallization play the main role.

In this case, the use of the plasma environment becomes particularly important due to the increased energy of ions extracted from the plasma, which facilitates the nanostructure crystallization during their growth. In fact, experimental results on the synthesis of ZnO nanodots and CdS nanostructures using RF magnetron sputtering suggest that additional exposure to in low-temperature inductively coupled plasmas of inert

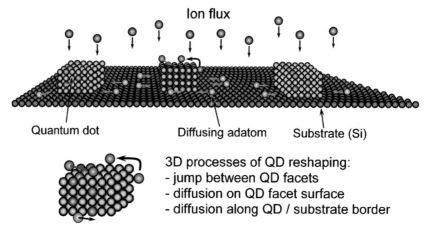

Figure 6.22 Basic processes involved in nanoparticle growth and crystallization on a plasma-exposed surface [286].

gases leads to significantly improved nanofilm quality and, in particular, crystallinity [286].

We will now concentrate on the above mentioned common observations as related to the improvement of growth rates, nanostructure quality and crystallinity under conditions when the inductively coupled plasma discharge was in operation. To qualitatively explain the observations, let us consider a simple model of plasma-assisted growth and crystallization of a nanoparticle on a solid surface exposed to a plasma discharge proposed by Ostrikov *et al.* [286].

Figure 6.22 shows a schematic representation of nanocrystal growth and crystallization on a plasma-exposed surface and includes the most important processes involved in the growth: incominig flux of ionic species, (green), migration of adsorbed species (shown as blue spheres) in the areas between the nanoparticles (shown as ordered cubic stacks of golden spheres) and over the surfaces of different crystalline facets of the nanoparticles (red spheres). The surface of the silicon substrate is represented by pink spheres. This graphic representation is extensively used in complex multi-scale kinetic Monte Carlo (KMC) modeling of the growth of various nanoparticles (e.g. nanodots).

The KMC approach ensures an accurate and detailed modeling of all the major processes (sketched in Figure 6.22) including (i) ion motion through the plasma sheath (first scale level ca. 10^4 nm); (ii) adatom diffusion about the surface (second scale level ca. 10^2 nm, top panel in Figure 6.22); (iii) adatom diffusion about nanoparticle surfaces, including jumps between facets (third scale level ca. 1 nm, lower panel in Figure 6.22) (see also Section 3.3 of this monograph).

Following the original report [286], we will now quantify the relevant electric field-related effects and show the influence of the electric field and plasma ions on the formation kinetics and crystallinity of small crystalline nanoparticles (e.g. zero-dimensional nanodots of this chapter or other nanostructures considered elsewhere in this book). Referring to Figure 6.22, the characteristic adatom diffusion time (time of residence in a single lattice site) is [95]

$$\tau = \tau_0 \exp(e\varepsilon_d/k_B T_s), \tag{6.26}$$

where $\tau_0 = 1/\nu_0 = h/2k_B T_s$ is the lattice oscillation time, $\nu_0 = 2k_B T_s/h$ is the lattice oscillation frequency, and h is Planck's constant. As before, ε_d is the surface diffusion activation energy, k_B is Boltzmann's constant, and T_s is the surface temperature.

When an adatom diffuses about a nanocrystal surface under conditions of local heating by incoming plasma ions, the ratio of the characteristic surface diffusion time at the equilibrium substrate temperature T_s to the characteristic surface diffusion time at the increased nanoparticle temperature (crystallization temperature) T_{Cr} is [286]:

$$\nu_\tau = (\tau/\tau_e) = \exp\left[(e\varepsilon_d/k_B T_s)(\delta T_s/T_{Cr})\right], \tag{6.27}$$

where $\delta T_s = e\mathcal{E}_i/N_{ns}k_B$ is is the local temperature change due to local ion energy transfer. Here, $\mathcal{E}_i$ is the ion energy, N_{ns} is the total number of atoms in the nanostructure, and $T_{Cr} = T_s + \delta T_s$ is the effective temperature of a nanoparticle surface subject to incoming ion fluxes. Under conditions of a low-temperature (600–700 K) deposition process [286] when local heating is particularly important, one can assume that nanoparticle crystallization occurs only at the effective temperature T_{Cr} (numerical estimations show that it can reach 1500 K for ion energies of 50–100 eV) and does not occur when the equilibrium surface temperature T_s is low. It is evident that the coefficient ν_τ quantifies an increase in the nanoparticle crystallization rate and thus can be named a characteristic crystallization time coefficient.

Using Equation (6.26) and expressions for δT_s and T_{Cr}, Equation (6.27) can be represented in the form [286]

$$\nu_\tau = \exp\left[\frac{e\varepsilon_d}{k_B T_s}\left(\frac{1}{k_B T_s N_{ns}/e\mathcal{E}_i + 1}\right)\right] \tag{6.28}$$

which explicitly incorporates the ion energy. Using Equation (6.28), one can demonstrate a possibility of increasing the incoming flux of the impinging species to the growth surface due to a shorter time of nanoparticle crystallization under conditions of ion irradiation.

It is natural to assume that the best crystallinity of the nanoparticles can be achieved when the adatoms are supplied to the NP border at the rate $\zeta_{in} = (\nu_\tau \tau_{Cr})^{-1}$, where τ_{Cr} is the total time of adatom incorporation into a crystalline nanoparticles structure (shown in Figure 6.22 (bottom panel))

$$\tau_{Cr} = \sum_{i=1}^{n_C} \tau_i, \tag{6.29}$$

where n_C is the number of sites on the nanocrystal facets as a function of a total number of atoms. For example, for a crystalline nanoparticle with a pyramidal shape we have

$$n_C = 2\sqrt{2}(3N_{nc})^{2/3} \tag{6.30}$$

and one can see that $n_C \propto N_{nc}^{2/3}$ [286].

Equations (6.26)–(6.30) can now be used to estimate characteristic times required for adatoms to diffuse about the substrate surface between the nanoparticles and on their crystalline facets. Although the results vary quite significantly from one set of parameters to another (owing to exponential dependence in Equations (6.26) and (6.28) on the surface temperature and diffusion actication energy), the main conclusion is that when the building units of the nanostructures are deposited on the solid surface as ions (and then locally dissipate their energy, which is in the range 50–100 eV in Figure 6.23), the rates of their surface diffusion (which determines how fast the adatoms can reach the nanoparticles) and migration between surface sites on nanocrystalline facets (which eventually determines the crytallization rates) are much higher (up to 10^3 times) than for the case of neutral atom/radical deposition.

Figure 6.23 shows the dependence of the characteristic crystallization time coefficient on the number of atoms in nanoparticles under different values of the diffusion activation energy (which is 0.5–0.7 eV for the materials of our interest here such as Ge/Si in Figure 6.22); the top and bottom panels correspond to the ion energies 50–100 eV, respectively. First of all, the enhancement of the crystallization process is much more pronounced for smaller nanocrystals (ca. 500 atoms). Secondly, this effect strengthens when the ion energy is increased from 50–100 eV. Furthermore, at lower diffusion activation energies the enhancement of crystallization is even more significant. Thus, the graphs in Figure 6.23 show that the rates of surface diffusion of adatoms on plasma-exposed surfaces (including surfaces of nanoparticle) can be $10 - 4 \times 10^4$ times higher compared with purely neutral gas routes.

This has two very important implications which, explain the common experimenal observations [286] briefly mentioned above. In fact, us-

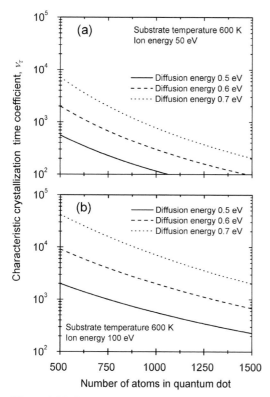

Figure 6.23 Dependence of the characteristic crystallization time coefficient on the number of atoms in a quantum dot (QD size) with the surface diffusion activation energy and the ion energy as parameters [286].

ing more intense ion fluxes (which is achieved by running inductively coupled plasmas external to the plasmas produced in the vicinity of the heads of magnetron sputtering assemblies), one can

- substantially increase the growth rates while keeping nanoparticle crystallinity at the same level;

- achieve much better crystallinity of the nanoparticles at growth rates comparable with the inductively coupled plasma-unassisted process.

Moreover, through a proper choice of the plasma process parameters, with the most important ones being the RF power input from the external inductive coil into the plasma and partial pressures of working gases, one can achieve significant improvements in both growth rates and crystallinity of nanocrystals. It would also be prudent to note that these con-

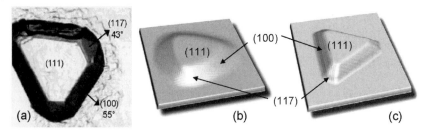

Figure 6.24 Experiment (a) [344]: a typical shape of a crystalline Ge nanodot on Si(111) surface showing the faceted structure. Numerical simulations (b, c) under the neutral Ge atom flux (a) and Ge$^+$ ion flux (b) conditions (I. Levchenko and K. Ostrikov, unpublished).

clusions are also applicable to other low-dimensional semiconducting nanostructures such as one-dimensional nanorods and nanowires [286].

An example of a plasma-related improvement of crystallinity of Ge/Si quantum dots is shown in Figure 6.24. This figure shows a representative germanium nanoisland on a Si(111) surface observed in experiments of Motta *et al.* [344]. This island shows a clear crystalline structure with a top (111) facet that basically reproduces the structure of the silicon surface and two other clearly resolved (100) and (117) facets. These preferential growth directions were used to simulate the effect of the neutral gas and plasma environments on the development of this nanoisland. The nanoisland shape in Figure 6.24(b) is computed under conditions of excessive deposition rate of Ge atoms in a neutral atom beam process. Even though the main crystallographic facets are visible, the whole structure is covered with smooth amorphous deposits. In thermal CVD processes this can be avoided in two ways: (i) either decreasing the deposition rate to allow adatoms to incorporate into the most energetically preferable site; or (ii) increasing the substrate temperature to accelerate adatom migration and insertion into the crystalline structure.

However, in many applications the above two options may be impractical because of low deposition rates (hence, poor process throughput) and strict temperature limits for the substrate material. Therefore, one can use a plasma to enable a faster and more effective process of nanodot synthesis. When an ion is deposited onto a nanoisland, its energy is transferred to the nanostructure which causes a local temperature rise without significantly affecting the equilibrium substrate temperature in the areas unaffected by this ion impact. As a result, the adatom diffuses about the nanoisland surfaces much faster and eventually finds the optimum location to recrystallize on a specific crystal facet. This eventually leads to much better shaped nanoislands such as the one depicted in Figure 6.24(c). In this case all the (111), (100), and (117) facets are well shaped and flat.

6.5.5
Self-Organization in Large Nanostructure Arrays

Let us now consider the specifics of self-organization on the plasma-exposed surfaces at the third stage of the nano-array formation, when a strong interaction (via adatom field and also via the electric field) between the nanostructures that form a large array is present [343]. At this stage, the adatom diffusion about the substrate surface plays an important role. In this chapter we have concentrated mainly on quasi-zero-dimensional nanostructures such as small clusters, nanoislands, and nanodots. However, other nanostructures can also demonstrate self-organized behavior, for example, carbon nanowalls [345].

Let us first consider the experimental results of the SiC/Si quantum dot array formation on plasma-exposed substrates [129, 346]. The three consecutive scanning electron micrographs are shown in Figure 6.25. The analysis of temporal dependence of the mean radius of quantum dots shows that at early growth stages the experimental dependence nicely fits the cubic-root dependence $d = \kappa t^{1/3}$, which is common to the situation of independent, uncorrelated growth of individual nanoislands. However, at more advanced stages the experimental curves overshoot the thoretically predicted dependence $\sim \kappa t^{1/3}$. For example, the theoretically predicted mean diameter of nanoislands in Figure 6.25(c) is approximately 24–25 nm, which is 4–5 nm less than in the experiments of Cheng *et al.* [129, 346].

After careful examination of the experimental data and the growth scenario, one can make the following conclusions. Firstly, the total number of quantum dots changes during the growth. Moreover, the mean radius gradually increases above the value determined by the requirement of mass conservation. This means that the number of quantum dots decreases (in 1.7–2.5 times in the example shown in Figure 6.25). Hence, dissolution and other surface processes that lead to the decrease of the nanodot number occur. Then, at the final stages the size of quantum dots does not change (or if so just decreases slightly). This means that the number of quantum dots begins to increase.

To summarize, the development of the quantum dot ensemble can be split into three stages. During the first stage, the number of QDs increases. At the second stage, the total number of quantum dots decreases, mostly due to coalescence and dissolution. Finally, during the third stage the total number of quantum dots increases. The latter usu-

14 nm   20 nm   29 nm

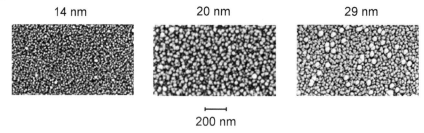

200 nm

Figure 6.25 Development of SiC quantum dot array in a plasma-based process. The mean size of the quantum dots is indicated above the SEM micrographs [343].

ally happens due to re-enhanced nucleation of new nanoislands on the substrate surface [343].

Now, let us consider the effect of the plasma environment on surface diffusion in the presence of an electric field (for example, on biased surfaces) and under conditions of interacting nanostructures at advanced stages of their development [343]. In the process of deposition from a plasma, the quantum dots acquire electric charge and hence produce the electric field. The total electric field appears to be a sum of the plasma-surface (sheath) component $\mathbf{E}_\lambda$ directed from the surface to the plasma, and the microscopic component $\mathbf{E}_S$ which is present in the vicinity of a single quantum dot and directed towards it. The plasma effect on the QD formation is caused by these two electric field components, which affect the surface diffusion rate.

Let us estimate the change in the surface diffusion activation energy caused by the microscopic electric field component $\mathbf{E}_S$, which originates due to the presence of the nanostructures on the surface. In general, an adsorbed particle has a dipole moment $\tilde{p}$ and polarizability α. The total dipole moment in the electric field $E(r)$ is $\tilde{P} = \tilde{p} + \alpha E$, and the energy acquired by an adatom in one jump along the lattice spacing λ_{lat} is [343]

$$W_e = \frac{\partial E}{\partial r} \left[\tilde{p} + \alpha E(r) \right] \lambda_{\text{lat}} \tag{6.31}$$

which can be conveniently normalized (for simplicity of notations, only a radial component of the electric field gradient is included)

$$\varepsilon_e = -\frac{\lambda_{\text{lat}}}{k_B T_s} \frac{\partial^2 \phi}{\partial r^2} \left[\tilde{p} + \alpha E(r) \right] \tag{6.32}$$

and expressed in terms of the electrostatic potential ϕ. Thus, to determine ε_e, one needs to know the distribution of the microscopic electric field in the vicinity of nanostructures.

An example of such distribution is shown in Figure 2.13. Generally, this electric field can be determined using the following relationship between the electrostatic potential and the density of surface charges σ

$$\phi = \iint_S d\phi = \iint_S \frac{\sigma dS}{4\pi\varepsilon_0 r}, \tag{6.33}$$

where S is the surface of a nanodot, r is the distance between the surface and point of potential determination. Surface integral (6.33) depends on the quantum dot growth shape. There are three main cases: flat quantum dot (the quantum dot height is less than the radius $h \ll r_0$); quantum dot in the form of a spherical segment; and a cylindrical quantum dot.

For the flat quantum dot, the calculations lead to the dependence

$$\phi(r) = \frac{r\sigma}{\pi\varepsilon_0} \left[K_2(\bar{r}_0) - (1 - \bar{r}_0^2) K_1(\bar{r}_0) \right], \tag{6.34}$$

where $\bar{r}_0 = r_0/r$, and $K_{1,2}$ are the elliptical integrals of the first and the secon kind, respectively. When $r \to \infty$, $\bar{r}_0 \to 0$, and Equation (6.34) reproduces the potential of a point charge.

If a nanodot has the shape of a semispherical segment, one obtains

$$\phi(r) = \frac{\sigma}{4\pi\varepsilon_0} \int_0^{\pi/2} \int_0^{2\pi} \frac{r \cdot \bar{r}_0^2 d\alpha d\beta}{\sqrt{\bar{r}_0^2 - 2\bar{r}_0 \cos(\alpha)\cos(\beta) + 1}} \tag{6.35}$$

for the electrostatic potential. Equation (6.35) can be integrated numerically and used in simulations.

If the nanodot has a cylindrical shape, both lateral and flat surfaces make a contribution to the total electric field. The contribution of the flat surface is calculated according to Equation (6.34), whereas the lateral cylindrical surface produces the following potential distribution

$$\phi(r) = \frac{\sigma}{4\pi\varepsilon_0} \int_0^H \int_0^{2\pi} \frac{r\bar{r}_0 d\alpha d\bar{h}}{\sqrt{\bar{h}^2 + \bar{r}_0^2 + 1 - 2\bar{r}_0 \cos(\alpha)}}, \tag{6.36}$$

where $\bar{h} = h/r$ and h is the height of the cylindrical QD.

After the appropriate electrostatic potential is determined using Equations (6.34)–(6.36), the electric field-related variation of the diffusion activation energy can be found through Equation (6.31). In this case the surface diffusion coefficient is modified

$$D_S = \lambda_{\text{lat}}^2 \nu_0 \exp\left[-\frac{\varepsilon_d - W_e}{k_B T_e} \right], \tag{6.37}$$

where ε_d is the surface diffusion activation energy in the absence of the plasma-related effects.

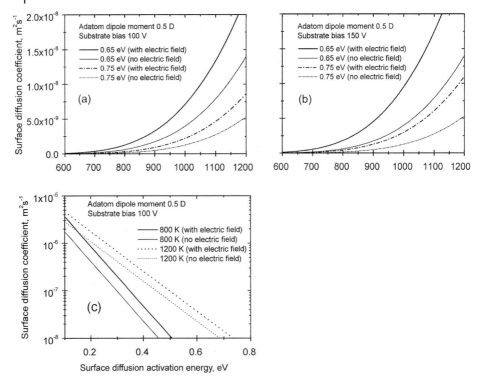

Figure 6.26 Dependence of the surface diffusion coefficient on the surface temperature with surface diffusion activation energy in a range of 0.65 to $0.75\,\mathrm{eV}$ as a parameter, with and without the plasma-related electric field, for the substrate bias $-100\,\mathrm{V}$ (a) and $-150\,\mathrm{V}$ (b). Panel (c) shows the dependence of D_S on the surface diffusion activation energy with the surface temperature as a parameter. The adatom dipole moment is 0.5 D $(1.6 \times 10^{-30}\,\mathrm{C \cdot m})$ and the substrate bias is $-100\,\mathrm{V}$ [343].

The above dependencies can potentially provide a significant decrease in surface diffusion activation energy and hence result in a higher rate of adatom supply to quantum dots. In addition, according to the detailed analysis in the previous subsection, this factor also contributes to more effective formation of quantum dots with a better crystalline structure. The calculated dependencies of the surface diffusion coefficients on the substrate temperature for various substrate biases and surface diffusion activation energies shown in Figures 6.26(a) and (b) demonstrate a strong increase with the surface temperature and a significant reduction with the surface diffusion activation energy.

Figure 6.26(c) illustrates the dependence of the surface diffusion coefficient in the electric field on the surface diffusion activation energy. This dependence spans over nearly three orders of magnitude when the sur-

Quantum dots interacting via fields of adatoms

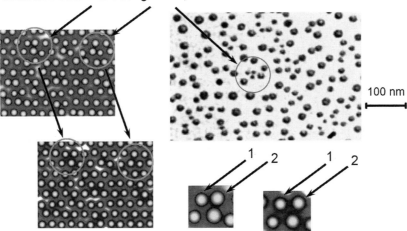

100 nm

Figure 6.27 Scanning electron micrograph and numerically simulated self-organizing patterns of SiC quantum dots [343].

face diffusion activation energy changes from 0.1 to 0.7 eV, thus demonstrating a great potential for controlling the diffusion and hence the self-organization processes on plasma-exposed biased surfaces covered with the growing quantum dot arrays.

As we have seen in Section 6.4, the value of the surface diffusion coefficient (6.37) is crucial for the self-organized development of nanodot arrays. It is instructive to recall that quantum dots in the array grow mainly by the mass supply from the two-dimensional field of adatoms with the adatom density η_a described by the following diffusion equation

$$\frac{\partial \eta_a}{\partial t} = D_S \left(\frac{\partial^2}{\partial x^2} + \frac{\partial^2}{\partial y^2} \right) \eta_a + \Psi_+ - \Psi_-, \tag{6.38}$$

where Ψ_+ and Ψ_- are the influx and evaporation flux from the substrate surface.

The two-dimensional adatom field in which the quantum dots grow causes the spatial self-ordering of the array through re-distribution of the adatoms between quantum dots, in a way we have seen in Section 6.4. These processes are illustrated in Figure 6.27, where the fragment of an SEM micrograph (the experimental details are described elsewhere [50]) is shown along with the simulated pattern, on which the quantum dots numbered '1' and '2' displace themselves in a way which eventually results in a better (more uniform) nanopattern [343].

The examples given in this section certainly do not cover the whole range of possible effects the plasma environment may exert on the self-organized growth of nanodots and the development of their ordered arrays. Instead, we have only discussed some of the most important issues such as deterministic matching of the incoming flux of BUs and the actual demand of building material on the surface. Among numerous possible effects of the plasma environment, we have highlighted the faster surface diffusion with reduced activation barriers, substantial improvement of nanodot crystallinity, as well as higher surface temperatures caused by ion bombardment. These changes alone lead to quite significant improvements in nanopattern quality such as in the example shown in Figure 6.27.

6.6
Concluding Remarks

The main aim of this chapter was to introduce specific and unusual features of processes that occur on solid surfaces exposed to low-temperature non-equilibrium plasmas and elucidate plasma-based approaches to nanoassembly of ordered nanodot arrays. In Section 6.1 we introduced the main requirements and targets for the synthesis of ordered arrays of quantum confinement nanostructures and their interconnected networks that need to be achieved before they can be used in the envisaged applications. The fabrication processes involved need to be engineered to meet these stringent requirements. As we know from surface science, the growth of very small quasi-zero-dimensional nanostructures proceeds via self-organization of adatoms and atomic clusters on two-dimensional solid surfaces.

The main issue we have raised in this chapter was what happens to the elementary surface processes and self-organized growth of nanostructures if the surfaces are exposed to low-temperature plasmas. Thus, we have tried to remain within the classical domain of surface science while demonstrating that the above processes may be quite different in a new, "harsh" environment. Indeed, we have identified a few things that are particularly peculiar to surfaces of solids immersed in a plasma. The most obvious features of the environment we have considered in this chapter are surface charges, non-uniform microscopic electric fields in the vicinity of nanostructures, and ion bombardment.

To be more specific, we started from considering one of the most common semiconductor nanoscale systems, namely Ge/Si and followed what difference the plasma environment can bring into the self-organized growth of Ge/Si nanodot arrays. We have also proposed a way to overcome the major difficulties that arise in neutral gas-based techniques due to uncontrollable fragmentation of few monolayer-thick continuous Ge films into nanoislands commonly referred to as the Stranski–Krashtanov growth mode. Using plasma-grown nanoclusters with typical size distributions, one can control the densities and sizes of initial seed nuclei, which further develop, via self-organization, into the nanodot arrays of our interest. In Section 6.1, we further discussed what happens with the above seed nuclei at later growth stages and which elementary processes are involved. One of the most intriguing possibilities is that of developing nanodot arrays with narrow size distributions and ordered positioning of individual dots, in a self-organized way, without any pre-patterning of the growth surface.

In Section 6.2, we have discussed the results of numerical simulations of Ge/Si quantum dot seed pattern formation from atom-only and non-uniform cluster fluxes. The results of Rider *et al.* demonstrate that the NUC flux (nanocluster flux with a depleted influx of atoms) provides a very narrow size distribution function of the quantum dots seed pattern, with a sharp decrease in the number of QDN consisting of 15 (or more) atoms. However, if the atoms are deposited only, the seed size distribution becomes much wider and the numbers of QDNs consisting of 2–3 atoms are almost the same as the numbers of nuclei consisting of 25 atoms. Moreover, the adatom density on the Si(100) surface in the NUC process can be very low. In particular, this suppresses the unwanted formation of new quantum dots during the growth process. The calculated parameters of nanodot patterns are consistent with relevant experimental data.

It transpires that it is possible to form a very dense pattern, $(2 \times 10^{-4}$ ML for smallest QDNs) with the final QDs coverage of 0.5 from nanocluster-dominating influx, compared to the density of 2×10^{-5} ML and final coverage of 0.2 for the atom-only case. Therefore, the nanocluster-based technique can be considered as a viable alternative QD nanopattern formation method which does not involve the commonly accepted Stranski–Krastanov route and, in addition, provides much greater process controllability.

Therefore, plasma processing turns out to be a promising and competitive fabrication environment for quantum dot arrays; in the example considered in Section 6.2 the plasma served both as a source for non-uniform clusters as well as a factor that influences surface diffusion rates. Future research should be focused on the refinement of this model for later growth stages and increased focus on the role of surface charges due to the plasma, not only in the initial formation of nanodots but also in the effect of ions on the deterministic placement of QDs in the uniform arrays required for technological applications.

Interestingly, the sources of size non-uniform clusters which have been shown advantageous at initial stages of Ge/Si nanodot formation in Section 6.2, are numerous. Indeed, there are a number of environments conducive for preparing such clusters – from neutral gases to complex, reactive plasmas. The most important element is the ability to precisely tailor the cluster size distribution and the most promising pathway from this point of view is the plasma route [176].

As we have already discussed in Chapters 1–4, the complex chemistry occuring in a typical plasma discharge results in the formation of a wide range of species that can act as potential building units - from atoms, ions, radicals, and molecules to nanoclusters [4]. The amount of each species produced, however, largely depends on how the plasma parameters such as the working pressure, temperature, degree of ionization, power, and composition of the precursor gas feedstocks can be manipulated to favor certain reactions taking place. Plasmas hold a great deal of promise in that specifically tailoring cluster distributions for a wide variety of deposition scenarios is not their sole function, they may also be used in surface preparation (energetic ions such as Ar^+, frequently included in plasma mixtures are commonly used to activate a deposition surface, similarly atomic hydrogen, also a common constituent is employed to terminate surface dangling bonds) in addition to controlling the transport of particles and clusters via the plasma sheath to the substrate [4,176].

Now the importance of low-temperature plasmas in the generation and transport of BUs becomes clear. However, for the purpose of nanodot array fabrication it is particularly important to understand the influence plasmas exert on surface reactions and processes. What must be particularly considered in plasmas, as opposed to neutral gas cluster sources, is the role of ions. As we have seen from Section 6.5, the plasma ions significantly influence surface dynamics due to their effect on the surface activation energy. Therefore, the ion flux used may be chosen such that the surface activation energy may be lowered; this makes the surface diffusion and associated surface processes increasingly energetically favourable and therefore more likely to occur [347].

Thus, the plasma ions can substantially increase surface reaction rates [5]. Given that surface diffusion is the dominant formation mechanism at the initial growth stage, this is very important indeed. In addition, it has been noted by Wegner *et al.* [78] that a requirement for clusters in nanostructured films is that they should possess sufficient impact energy to dislodge a surface lattice atom in order to anchor the incident cluster to the surface. Furthermore, Roca i Cabarrocas *et al.* [348] noted that the impact energy of positively charged ultra-small Si nanoparticles may be controlled by applying a DC potential drop between the plasma and substrate.

For many technological applications, QDs must be able to be fabricated in a uniform, regular array [349]; therefore, correct placement of QDs is also a concern. An article by Krinke *et al.* [350] reveals that the presence of ionized species (as typically found in a plasma environment) results in less agglomeration on the substrate. Instead, a more randomized distribution is obtained – this means less clumping in particular areas on the substrate [176].

As we have seen from the previous chapters, in the fabrication of other nanostructures, carbon nano-tips for example, the use of plasmas via PECVD has resulted in significantly better size and positional uniformity than that recorded for the neutral gas - thermal CVD route. Similarly, it has been noted that the growth process of carbon nanotubes depends on the residence time of the plasma-generated nanotube seed particles in the preferential growth region. The use of ion fluxes has also been noted as a way to reduce hydrogenation – leading to higher purity films and nanostructures. Several authors have also noted that additional external substrate heating was not required in many plasma-based processes and in some cases even proved detrimental. It is commonly accepted nowadays that due to the greater dissociation of the feedstock gas by plasmas and thus the greater variety of species for carbon nanotube growth, the growth temperatures for PECVD are ultimately lower than those required for thermal CVD (which employs a neutral gas route). The substrate temperature also affects the elementary surface reactions, carbon dissolution and diffusion into metal particles, as well as playing a role in surface preparation. Some of these processes will be considered in more details in Chapter 7 of this monograph.

It also becomes clear that (sometimes substantially) lower growth temperatures employed when using non-thermal equilibrium plasmas broaden the range of substrates that may be processed – including temperature sensitive materials such as polymer substrates. As discussed in Section 6.5, lowered surface diffusion activation energies and higher substrate temperatures result in higher diffusion rates. The use of plasmas, therefore, is not restricted to the ability to precisely tailor the cluster size distribution – they are also particularly important in increasing the rate of surface diffusion, and by extension, the speed of the nanoassembly process. The nanocluster-dominated distributions of Section 6.2 are very similar to the nanocluster size distribution representative of reactive silane plasmas. Clustering and nucleation in silane plasmas have already been discussed in Chapter 4. Other plasma cluster sources, besides reactive plasmas (i.e. silane), include magnetron sputtering, laser vaporization cluster source (LVCS), pulsed microplasma cluster source (PMCS) and pulsed arc cluster ion source (PACIS), amongst others [78]. The wide range of choices reinforces the observation that plasmas are effective tools for the whole spectrum of nanofabrication processes [176].

We reiterate that an in-depth discussion of species production mechanisms via the of myriad chemical reactions occurring in complex plasma discharges and their subsequent modification via manipulation of plasma parameters represents a significant research effort by itself. The detailed discussion in Section 6.2 is intended to expose the advantages inherent in using a nonuniform cluster flux instead of an atom-only flux in the initial QD growth stage. An extensive technical description of quantum dot fabrication at all stages from species generation, to surface preparation to the final QD product is clearly outside the scope of Section 6.2. Nevertheless, Section 6.2 provided plenty of convincing evidence that partially-ionized low-temperature non-equilibrium plasmas can generate suitable non-uniform cluster distributions and, as such, can be used to effectively control the nuclei nanopattern development, ultimately giving rise to size-uniform and dense patterns of quantum dots.

In Section 6.3, we have discussed one of the most important issues at the early stages of nanofabrication of binary quantum dot systems, namely, the ability to control the elemental composition and the internal structure. This can be achieved by properly balancing the incoming fluxes of the two elements involved. In the example of SiC/Si QDs considered, the Si and C adatom fluxes to the quantum dots are not equal even in the case when the Si and C influxes from the neutral/ionized gas to the substrate surface are equal. This interesting fact owes to quite different surface mobilities of silicon and carbon adatoms on a Si substrate. Thus, the deposition process conducted at the equal Si and C influxes to the surface may result in the undesired formation of non-stoichiometric SiC nanodots. Based on the extended discussion in Section 6.3.1, one can draw the following conclusions which may help to improve the predictability and controllability of nanofabrication of highly-stoichiometric SiC quantum dots on Si(100) surface [174]:

- the surface temperature and the ratio of Si and C atom/ion influxes to the substrate surface are the main factors that determine the time dependence of the ratios of Si and C adatom fluxes k_{st};

- the use of a transiently increased surface temperature results in a faster "stoichiometrization" of the Si and C adatom fluxes but may

be restricted by the substrate and quantum dot material characteristics;

- correction of the Si and C influx ratio by increasing Si atom flux to
  the substrate for 0.1–1 s is very effective in maintaining the Si and
  C adatom fluxes ratio close to unity;

- a balanced combination of the elevated substrate surface temperature and increased precursor influx ratio k_p at the initial stage of
  deposition should be considered as the most promising way to sustain self-assembly of highly-stoichiometric SiC quantum dot nuclei.

Based on the above conclusions, the follow-up work [175] reported
on the possibility of using a time-variable ratio of incoming Si and C
atom fluxes to control the Si/C adatom flux ratio at all stages of the
nanoassembly process. This approach may eventually lead to the synthesis of SiC/Si quantum dot arrays with the required elemental composition and internal sturcture. In particular, as we have seen from Section 6.3.2, binary SiC quantum dots can grow by one of the two scenarios
depending on the surface temperature, forming a carbon core and then
either a Si-enriched under-shell followed by a stoichiometric shell, or just
a stoichiometric shell. An increase in total influx results in the stoichiometric composition being reached faster and hence a smaller core and a
thinner shell are obtained.

Both scenarios are of interest, given that deterministic control of composition and internal structure is a viable way in which the electronic and
optoelectronic properties of binary quantum dot systems can be finely
engineered [175]. This approach can also be used to find specific process
conditions to synthesize compositionally graded, heterolayered, and hybrid binary quantum dot systems for a large variety of semiconducting
materials and nanofabrication processes.

In Section 6.4, we introduced an effective low-temperature plasma-
based technique to control self-organization within an array of Ge/Si
quantum dots for establishing or improving spatial and dimensional order. In particular, it was demonstrated that the self-organization phenomena can be explained by the ability of the system to maintain equal or
unequal fluxes of adatoms about the perimeters of individual quantum
dots. The dynamics of dot growth and displacement act accordingly to
establish this self-organization scenario. Given the above results, several
specific conclusions may assist in improving control and predictability
of plasma-aided nanofabrication of Ge quantum dot arrays on a Si(100)
surface [177]:

- Surface temperature is a factor that determines the mean size of quantum dots, with decreased temperatures producing larger dots while increased temperatures are shown to deteriorate the size uniformity within the nanopattern.

- Plasma influx is a factor in the surface coverage of nanodots across the surface. Lower fluxes result in an increased final coverage (prior to coalescence) while for higher incoming fluxes the surface coverage is larger at the point of the maximum dot size uniformity. An apparent disadvantage of the lower influx is the extended growth time.

- The quality of the initial nanopatterns is an important factor in obtaining a high-level spatial order. However, some spatial and size variation is permitted as the self-organized growth process does provide a substantial improvement in the spatial alignment and size uniformity.

- Spatial order may be further improved by decreasing the QD size in the initial nanopatterns.

- Spatial uniformity may be improved at temperatures as low as ca. 600 K, which is difficult to achieve via thermal deposition methods such as CVD.

However, more research efforts should be aimed at establishing ultimate control in the quality of the nanopatterns of Ge/Si and other nanodot systems by determining the precise combination of the main process parameters and expanding the numerical simulations to account for the most essential features of QD internal structure, explore the effect of time-varying process parameters, and, more importantly, appropriately account for the plasma-related effects on the main elementary surface processes. Some relevant efforts were reviewed in Section 6.5.

In Section 6.5, we discussed some of the major issues and challenges on the way to successful and widespread application of low-temperature thermally non-equilibrium plasmas in the deterministic fabrication of delicate nanoassemblies (such as arrays of size-uniform quantum dots on soild surfaces) via precise balancing and manipulating motion and self-assembly of adatoms. This research area is traditionally covered by surface science.

One of the major issues we have identified was how to properly balance the fluxes of plasma-generated building units, the required nanopattern parameters (e.g. size distribution and number of QDs in the pattern), and the duration of the growth process. It turns out that a large variety of plentiful atomic, molecular, radical, and so on species common to low-temperature plasmas and very high rates of their deposition on solid surfaces might be another hurdle in this direction.

Several viable ways to achieve the required balance between the demand and supply of the building units and to eventually link macroscopic and microscopic processes that occur at spatial scales differing by up to 9–10 orders of magnitude have been discussed. We have also identified several important changes the plasma environment introduces to surface conditions and how these changed conditions can be used to synthesize better nanodot patterns compared to neutral gas-based routes. Some of the plasma-related effects include faster growth rates, better crystallization, better control of the average size and size uniformity in lattice-mismatched systems and also self-ordering of QDs in nanoarrays. The above conclusions make us very optimistic with regard to the possibility of successful bridging the 9–10 order of magnitude spatial gap between plasma physics and surface science.

One of the greatest challenges in this direction is the ability to predict the energetic characteristics of elemetary surface processes involving plasma-related electric fields, polarization, surface charges, and so on. This missing information is expected to be provided by research on the computational materials science area, focusing particularly on "harsh" nanofabrication environments.

In Section 6.5, we also addressed some of the most important benefits of using plasma-assisted techniques for the growth of quasi-zero-dimensional nanostructures. In particular, local energy transfer from impinging ions upon deposition may result in a substantially enhanced surface diffusion of adatoms. In this case, the deposited material is removed to the growth sites of individual nanoparticles much faster, with the rates exceeding up to three orders of magnitude the rates of corresponding plasma-unassisted processes. This, in turn, allows one to significantly improve the crystallinity of the nanostructures and also increase the deposition/growth rates compared with the neutral gas-based routes. These conclusions are generic and are applicable to a much broader range of low-dimensional semiconducting nanomaterials and plasma-based nanofabrication techniques and approaches.

We have also thoroughly discussed the importance of the plasma-related parameters (surface bias and the presence of the electric field) and surface conditions for the self-assembly, self-organization and self-ordering processes on plasma-exposed surfaces. In Section 6.5, it was shown that the presence of the microscopic electric field strongly affects the kinetics of surface diffusion and leads to better crystallization of quantum dots and also to better self-organization of large quantum dot arrays. The use of plasma-extracted ions with increased kinetic energy also results in better crystallization by local heating of the nanostructures, and more effective self-organization of the quantum dot arrays. Based on these fascinating results, one can conclude that self-organized nanostructure and nanoarray synthesis can be effectively controlled in a low-temperature plasma environment.

From the material presented in this chapter, it becomes clear that low-temperature thermally non-equilibrium plasmas do indeed offer many competitive advantages for nanofabrication. Most notably, they may be used in every step of the nanofabrication process, even for such delicate nanoassemblies as self-organized ordered arrays of quasi-zero-dimensional quantum dots. The approaches and conclusions of this chapter will be further refined and applied for plasma-assisted fabrication of a range of other nanoscale assemblies. Finally, the area of surface science of plasma-exposed surfaces is new and full of exciting, rapidly emerging and developing opportunities for everyone involved either from plasma physics or the surface science side!

7
Ion-Focusing Nanoscale Objects

To proceed further from the quasi-zero-dimensional nanodots of Chapter 6, it would be logical to increase the nanostructure dimensionality by one and then consider one-dimensional nanostructures such as nanotubes, nanowires, and nanorods. However, as we have stressed in Chapters 1–3, these and even some two-dimensional (e.g., nanowalls) nanostructures align vertically in the direction of the electric field in the plasma sheath. This direction is also the path along which the plasma ions move from the plasma bulk towards the nanostructures. Thus, both the one- and two-dimensional nanostructures may focus ion fluxes in a plasma. Interestingly, tiny pores in the surface also significantly affect the ion fluxes and often draw them into their interior.

As we will see from the material presented in this chapter, the presence of intense ion fluxes makes a remarkable difference in terms of improving the quality of individual nanoscale objects (including nanopores) and their nanoarrays. For this reason we have decided to combine all these nanostructures to form one common category of *ion focusing nanoscale objects*. However, most common quasi-one-dimensional vertically aligned nanostructures such as single- and multiwalled nanotubes, conical nanotips, and high-aspect-ratio nanorods will be given a higher priority. Nevertheless, some of the most important effects of the plasma environment on quasi-two-dimensional nanowall-like structures and nanopores will be discussed elsewhere in this chapter and Chapter 8. This chapter will predominontly focus on the growth aspects of the ion-focusing nanoscale objects and particular features the plasma environment contributes to the relevant growth processes. As one can surmise from the title of this chapter, the effects induced by focused ion fluxes will be in the spotlight of our consideration.

This chapter will commence with a discussion of electric field-related effects on the growth of single- and multiwalled nanotubes in two common cases of the growth of gas-phase-borne and surface-bound carbon nanotubes (CNTs, Section 7.1). In Section 7.1, we will also consider possible reasons for so pronounced vertical alignment of one-dimensional

Plasma Nanoscience: Basic Concepts and Applications of Deterministic Nanofabrication
Kostya (Ken) Ostrikov
Copyright © 2008 WILEY-VCH Verlag GmbH & Co. KGaA, Weinheim
ISBN: 978-3-527-40740-8

nanostructures in a plasma and the contribution of surface adatom diffusion (the main source of building material of quantum dots from the previous chapter) to the nanotube growth. Interestingly, electric field-related effects make it possible to explain very high growth rates of carbon nanotubes in dense plasmas with relatively strong ion fluxes.

Building on this semi-quantitative analysis, in Section 7.2 we will introduce a more rigorous treatment of the effects of ion fluxes, dissociation and radical creation, and adatom/adradical surface diffusion on the growth of single-walled carbon nanotubes and carbon nanofibers in low-temperature plasmas. In particular, it will be shown that ion-induced hydrocarbon dissociation on nanotube surfaces may be the main process through which carbon atoms are supplied to the catalyst particles. It will also be explained why the activation energy of carbon nanofiber growth is lower in a plasma environment.

Section 7.3 considers the possibilities for controlling the shape of carbon nanostructures using plasma-based processes. In many cases this growth shows strong self-organization features. For example, platelet-structured carbon nanocones sharpen up during the final stages of their growth as a result of a complex interplay between adatom diffusion, surface termination, and ion focusing effects.

In Section 7.4 we will consider a striking example of how using a plasma process environment can lead to a significant improvement in the overall quality of large microscopic patterns of conical carbon nanotips. This improvement is characterized by a better size uniformity and higher-aspect-ratios of individual nanostructures throughout the entire nanoarray.

As suggested in Section 7.5, self-organization on plasma-exposed surfaces may be even more complicated; namely, it may involve metal catalyst nanoislands and may also develop in all three dimensions. Moreover, this complex self-organization may even lead to the development of position- and size-uniform carbon nanocone arrays from essentially nonuniform Ni catalyst nanoislands on an Si(100) surface.

In Section 7.6, examples of other ion-focusing nanostructures are considered. These examples of relevant nanosclae objects include quasi-two-dimensional nanowall-like structures and nanopores of various shapes in solid substrates. And finally, the main points and results discussed in this chapter are summarized in the concluding section (Section 7.7).

7.1
General Considerations and Elementary Processes

Modern nanotechnology requires a reasonably high degree of determinism in fabricating nanostructures of various dimensionality such as zero-dimensional (0D) quantum dots, one-dimensional (1D) nanorods, nanotips and nanowires [351–355], two-dimensional (2D) nanowalls [156, 356] and nanocombs [357, 358], and also numerous three-dimensional (3D) nanostructures (NSs) of complex shapes and nanocrystalline films.

Despite a great variety of shapes, densities, sizes, materials and geometrical characteristics, all nanostructures should satisfy the two main groups of practical requirements: *geometrical* and *structural* [173]. These requirements are intimately interrelated and strongly influence each other. Nevertheless, proper geometrical characteristics (e.g., the width, height, height-to-width (aspect) ratio, and size distribution function) may be regarded as somewhat "primary", as the most essential parameters. Moreover, as we have already stressed elsewhere in this book, when the sizes of nanoscale objects fall below the exciton Bohr radius (ca. 10 nm), their size and shape control their electronic, optical and other parameters. This is why the development of the methods for shaping surface nanostructures is one of the most important aims of present-day research in nanoscience.

At present, there are several methods that ensure, in principle, a proper shaping of the growing nanostructures [173]; however, controllability of such methods still requires substantial improvement. Generally speaking, it is often said that the degree of determinism is strongly insufficient in the modern nanotechnology, meaning a complete control of the interrelation between the input, process, and resultant parameters. Analyzing this problem systematically and revisiting our discussions in Chapters 1–3, one can see that neutral gas processes such as Chemical Vapor Deposition (CVD) are not the best candidates for deterministic nanofabrication due to the intrinsically random nature of such processes. On the other hand, plasma environments offer long-range forces and extended, long-range interactions between numerous plasma-generated species. One can then expect that plasma-based methods, such as plasma-enhanced chemical vapor deposition (PECVD), can provide, in principle, a much greater degree of determinism compared to neutral gas routes, owing to the long-range electric field-based interactions. Here we will discuss how electric field-related effects and ion fluxes can affect the growth of ion-focusing nanostructures of different dimensionality and morphology of their nanopatterns.

In this chapter we consider plasma-based growth of nanostructures on biased conductive substrates such as those sketched in Figure 7.1. Our

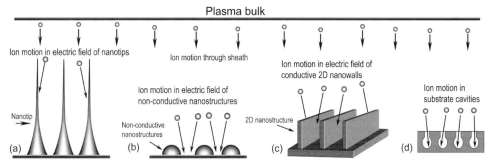

Figure 7.1 Schematics of growth of various nanostructures in a plasma environment: (a) conductive nanotips; (b) non-conductive nanostructures; (c) conductive nanowalls; and (d) nanopores/nanocavities [173].

main aim here is to present the most essential general features and salient trends of the plasma-based growth of nanostructures. Figure 7.1 displays a representative set of four nanostructures with different geometrical and physical characteristics, namely, conductive nanotips/nanocones (1D or 3D depending on size), non-conductive (insulating) nanoislands (0D or 3D), conductive nanowalls (2D), and also porous structures and cavities in the substrate surface. Here, the term "conductive" is used for materials that do not show any significant voltage drop when a current flows onto them. On the other hand, "insulating" nanostructures show a larger electric resistance which prevents or substantially slows down electric charge leak from their surfaces. The range of nanostructures considered in this chapter also includes gas-borne (floating) and surface-bound carbon nanotubes. The main focus here will be on numerical simulations, which are usually done by splitting the whole complex model into three main sub-models that describe the microscopic topology of ion currents in nanopatterns, surface process, and nanostructure growth and nanopattern development [173].

In most cases, we will consider a system consisting of a biased substrate with the growing nanostructures immersed in the plasma (Figure 7.1). However, we usually do not specify the method of the plasma generation and thus do not restrict ourselves by the plasma parameters; accordingly, the plasma can be produced, e.g., by inductive and capacitive RF discharges, cathodic and anodic vacuum arcs, magnetron and microwave discharges, pulsed laser deposition or some other configurations of ionized physical or chemical vapor deposition. The plasma density usually varies in the range 10^{16}–10^{19} m^{-3}, whereas the electron temperature T_e varies from 2 to 10 eV.

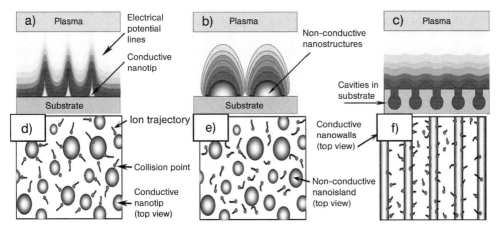

Figure 7.2 Equipotential lines of the electric field in the nano-patterns of: (a) conducting quasi-1D nanostructures; (b) non-conducting nanoislands; (c) nanopores/nanocavities. View from the top of ion trajectories in nanopatterns of (d) conducting quasi-1D nanostructures; (e) non-conducting nanoislands; and (f) conducting nanowalls [173].

When a nanoassembly process is conducted in an ionized gas environ-ment, an ion flux from the plasma to the nanoassembly site is controlled by the electric field $\mathbf{E}(\mathbf{r})$ created by the entire population of NSs on the surface, for example, nanotips or nanoislands as shown in Figures 7.2(a) and (b). Thus, the microscopic ion fluxes to the various areas of the sub-strate and NSs are affected by the nanostructures themselves, bias and the plasma-nanostructure interaction. In the case of conductive nanotip patterns, the electric potential of the nanotips is the same as that of the surface. On the other hand, insulating nanoislands can acquire both pos-itive and negative electric charges, depending on the deposition condi-tions.

If the NSs are composed of a conductive material (e.g., metal or graphitic carbon, Figure 7.2(a)), the electric field increases strongly near conductive peaks (this is the case for nanotips and nanowalls). By com-parison, in the case of non-conductive or highly-resistive nanostructures (e.g., hydroxyapatite (HA) islanded films [359], Figure 7.2(b)) the islands are exposed to the ion flux to the negatively biased substrate and acquire a positive charge due to the retarded drain of positive charges through the substrate.

From Figure 7.2, one can see that these two cases feature a very differ-ent pattern of the electric field. Indeed, in the case of conductive nanos-tructures, the electric field is "patterned" to attract the ion flux to sharp structures. On the other hand, the electric field in nanopatterns of non-conductive islands features a dipole-like structure due to mirror electric charges induced in the substrate. In both cases, the electric field appears to be strongly patterned in the above-surface layer to a thickness several times larger than the nanostructure sizes [173].

Let us now discuss in more detail the electric field configurations computed for different nanopatterns. Figure 7.2 shows the topology of the microscopic electric field for (a) conductive nanotips, (b) non-conductive nanoislands and (c) cavities in a conductive substrate. From Figure 7.2(a), one can see that conductive nanotips generate very sharp potential peaks, which protrude well outside the nanotip height domain. Nanoarrays of carbon nanotubes or conducting high-aspect-ratio nanorods have quite similar electric field distributions. One can thus expect a reasonably strong focusing of the ion current by the nanotips. In the case of non-conductive nanoislands (Figure 7.2(b)), the electric field has a dipole-like shape with the force lines entering the substrate surface; therefore it is apparent that the maximum current density should be expected on open surface areas between the nanoislands. In the third case (Figure 7.2(c)), when cavities/pores in a conductive substrate are considered, the electric field is shaped so as to focus the ion flux into the interior.

Typical patterns of the ion trajectories for the conductive nanotip and non-conductive nanoisland patterns are shown in Figures 7.2(d) and (e), respectively. From Figures 7.2(d), it is seen that the ions are strongly attracted by the nanotips; as a result, their trajectories are directed to the nanotips and the ion current increases at the surfaces of conductive nanostructures. Figure 7.2(e) depicts the opposite situation; namely, ions are deflected in such a way that they avoid the nanoislands and precipitate on the uncovered surface of the substrate. In this case the curvature of the ion trajectories is much larger. This is the consequence of the dipole-like shape of the electric field in non-conductive patterns, which generate a stronger field capable of deflecting fast ions in the vicinity of the substrate surface. Indeed, when an ion starts moving through the sheath over the nanoisland, it deflects strongly from the nanoisland centre and moves to the uncovered surface of the substrate. However, when an ion starts moving between the nanoislands, it does not show any strong deflection of the trajectory.

Let us now use our knowledge of distributions of electric fields and microscopic ion fluxes in the vicinity of nanostructures to model the growth of carbon nanotubes grown in the gas phase and on the surfaces in high-density plasmas [351]. In the following, we will analyze the conditions of nanotube growth on solid surfaces and in the gas phase of dense low-temperature plasmas produced, for example, by an arc discharge. These conditions will be related to the ion flux intensity and distribution. In particular, Levchenko *et al.* [351] have shown that the densities and microscopic topology of ion flux distribution appear to be very different for the plasma and surface-grown nanotubes. This difference ultimately results in very different growth rates for the surface-bound and plasma-borne nanotubes.

Before we proceed any further, let us introduce the growth model and recall the most important basic growth modes of carbon nanotubes. Some of the basic ideas have already been introduced in Chapters 2 and 3. Nevertheless, despite the apparent simplicity, there is still no consensus on whether or not metal catalyst nanoparticles are always required to synthesize carbon nanotubes in these situations. Interestingly, the reports on metal-catalyzed nanotube growth significantly outnumber those on catalyst-free synthesis. Indeed, in many cases nanotubes can grow without specifically prepared transition metal nanoparticles. However, under certain conditions some surface features, such as areas with a high density of defects and ultra-thin deposits segregated into small catalyst islands, may trigger nanotube growth. When nanotubes are produced in the ionized gas phase of high-density plasmas, products of erosion of liquid and solid fragments from the cathode may also serve as effective catalysts.

Therefore, in the following we will only consider the catalyst-based growth of the surface-bound and gas-borne nanotubes. In a plasma environment, multi-walled carbon nanotubes (MWCNTs) very often follow the tip growth mode, in which the catalyst particle is located on the top of the nanotube as shown in Figure 7.3(a). On the other hand, single-walled carbon nanotubes (SWCNTs) can only grow via the base growth (also called root growth) scenario (Figure 7.3(b)). For the arc discharge-grown SWCNTs shown in Figure 7.3(c), the above terms are not applicable. However, it is frequently observed that a bunch of nanotubes may grow on a single nanoparticle. As a substrate for the nanotube growth Si or SiO_2 are most commonly used. It is also important to stress that the catalyst particle on which the nanotubes grow in arc discharge plasmas are usually much larger (up to several μm) than surface-bound catalyst nanoislands.

Analyzing the nanotube growth from the plasma or neutral gas environment and using the knowledge we already have from Chapter 6, we can state that the main processes involved are the same for all the three cases considered in Figure 7.3: carbon particle (atoms, ions, radicals such as CH_2 etc. – for simplicity these will be further referred to as "carbon") accumulation on the substrate or catalyst particle surface; carbon diffusion about the substrate, catalyst or nanotube surfaces (surface dif-

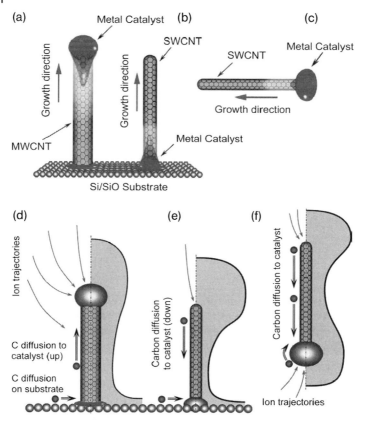

Figure 7.3 Schematic of the three main scenarios of carbon nanotube growth in a plasma.(a) Tip-mode growth of MWCNT on surface; (b) base-mode growth of SWCNT on surface; (c) and catalytic growth of SWCNT in dense plasmas. The catalyst is being drawn partially into the nanotube in the (a) and (b) scenarios and probably in the (c) scenario. Distribution of ion flux and direction of diffusion processes for nanotubes on the surface in case of (d) MWCNT tip growth mode; (e) SWCNT base growth mode; and (f) growth on catalyst particle in arc plasma. Nanotube parameters: MWCNT height 2 μm and diameter 100 nm; SWCNT height 1 μm and diameter 2 nm; length and diameter of SWCNT in arc plasma (f) are 2 μm and 2 nm, with the diameter of metal catalyst particle 10 nm [351].

fusion); carbon diffusion through the catalyst (bulk diffusion); and carbon re-evaporation from the substrate, catalyst and nanotube surfaces. The nanotube nucleation is often described using the vapor-liquid-solid (VLS) model [360]. Thus, the processes that determine the carbon balance at the nanotube growth surfaces (Figure 7.3(a–c)) and eventually determine the nanotube growth rates and possibly the transitions between the growth modes (e.g., involving multiplication of the number of graphene walls) are common for all the above three CNT growth cases [351].

We reiterate that in this section we will only analyze the ion flux-related effects that are determined by the electric field, and do not consider the flux of neutral carbon particles to the nanotubes and substrate surface. In general, the neutral particle flux to the growing nanotube array is quite large and may be comparable or even larger than the mean ion flux, especially when the ionization degree of the plasma is low. However, the neutral component of the plasma is not affected by the electric field and thus it provides only a uniform background flux. In comparison, the ion flux follows non-uniformities of the nano-structured electric field and leads to many features of the nanostructures which cannot be obtained through neutral gas-based processes.

The above features appear quite different for nanotubes arranged in microarrays of different densities. One of the reasons is the number of different options [351] for carbon delivery to catalyst particles at the bases or tops of individual nanotubes. In those arrays of a relatively low density, both neutral and ion fluxes can reach the substrate surface and thus participate in the nanotube growth through the base growth mechanism. This growth mode requires the carbon delivery to the lower part of the nanotubes where the metal catalyst particle is located. In this case, a significant part of the total carbon flux is delivered directly to the surface and then reaches the catalyst via surface diffusion about the substrate surface and those sections of the nanotube walls that are closest to the substrate.

If the density of nanotubes in the array is higher than a certain threshold, direct delivery of neutral species to the substrate becomes problematic or even impossible. The main reason for this is a random orientation of their velocity vectors which results in a predominant deposition of neutral species onto side surfaces of the nanotubes. Even worse, the denser the array, the higher the density of neutral flux deposited onto the uppermost section of the CNT walls. In this case, the neutral carbon is being deposited on nanotube surfaces but the ion flux can still reach the substrate surface and the lower parts of the nanotube side surface and possibly the catalyst particles anchored to the substrate.

The situation changes even further when the nanotubes form a closely packed array or a dense forest. In this case, neither neutral nor ion fluxes can be delivered to the substrate or the catalyst particle (in the root growth scenario) directly from the plasma. Therefore, carbon material can only be supplied to the surface-bound catalyst particles via surface diffusion about the nanotube walls.

In Figure 7.4(a) the calculated ion flux distribution on substrate in a rarefied square nanotube array is shown [351]. It can be seen that the ion flux is mainly deposited around the base of the nanotube, exactly where

(a)

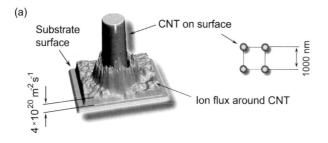

(b)

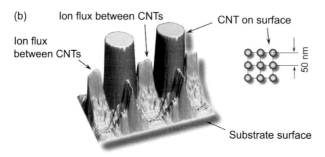

(c)

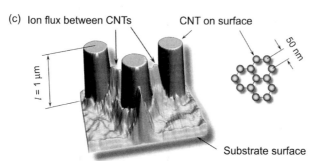

Figure 7.4 Distribution of ion flux around CNTs in regular patterns. (a) Rarefied pattern; (b) dense square pattern; and (c) dense hexagonal pattern. In dense patterns, ion flux demonstrates peaks between nanotubes die to ions drawing into inter-CNT gaps. In rarefied patterns, ion flux is distributed with angular symmetry around each nanotube. Nanotubes height 1 μm, diameter 2 nm, step between the nanotubes 50 nm (dense patterns) and 1000 nm (rarefied patterns). Distributions obtained by MC simulation of ion trajectories in electric field of nanotubes [351].

the metal catalyst nanoparticle is located. The mutual influence of the neighboring nanotubes is low, thus the ion flux distribution is symmetrical around each nanotube's circumference. In the case of dense square and hexagonal patterns (Figures 7.4(b) and (c)) the ion flux on the surface is mainly distributed between the adjacent nanotubes, thus forming well pronounced current peaks. The configuration of these peaks reflects the nanotube arrangement on the surface: in the square pattern, each nanotube is surrounded by four radially extended peaks, and in the hexagonal pattern by the three similar peaks. The peaks are rather strong and exceed the averaged ion current density between the nanotubes by approximately one order of magnitude [351].

We will now consider the ion flux distribution over the surfaces of surface-bound and arc discharge-produced nanotubes [351]. These distributions were computed using the above growth models and assuming that the SWCNT in arc discharge plasmas developed on metal catalyst particles of 10 nm in diameter. The surface-bound nanotubes were split into two groups: MWCNTs of 100 nm in diameter with a length of 2 μm (Figure 7.3(a), top growth mode) and 1 μm-long SWCNTs of 2 nm in diameter (Figure 7.3(b), base growth mode). The arc discharge-grown nanotubes had the same thickness as the surface-bound counterparts, yet were twice as long.

The results of the calculations and the main diffusion processes that deliver carbon to the catalyst are shown in Figures 7.3(d–f). The case of MWCNT is illustrated in Figure 7.3(d), together with typical trajectories of carbon ions being deposited on the nanotube and substrate surface. From this figure it is clearly seen that the carbon deposition profile is strongly non-uniform, with the current density much stronger at the MWCNT top where the catalyst nanoparticle is located. In this area the ion current density is 4 times larger than at the nanotube base.

Likewise, a quite significant proportion of the total ion flux is deposited on the substrate surface as can be seen in Figure 7.3(d). Since the MWCNT growth proceeds through the metal catalyst particle on its top, all carbon-bearing species deposited on the nanotube walls and on the substrate should migrate, mainly by surface diffusion, to the nanotube top as it is shown in Figure 7.3(d). As a result, the surface of the MWCNT is exposed to very strong diffusional fluxes. Hence, one achieves a high surface density of carbon adatoms, which can be a plausible reason for the genesis of new graphene walls.

The ion current distribution on the SWCNT shown in Figure 7.3(e) is similar to that of the MWCNT, with stronger non-uniformity of the distribution and higher flux density at the nanotube top (without a metal catalyst particle in this case). The degree of non-uniformity reaches 1:10 for the top and bottom parts of the nanotube, respectively; that is, the current density is lower at the substrate surface. As already mentioned, we assumed that the SWCNT grow by the base mode, with the catalyst located at the substrate; thus, the diffusion fluxes differ from those on MWCNTs and are directed to the SWCNT along the substrate and also downwards from the nanotube top. As a result, only carbon deposited directly on SWCNT diffuses along the nanotube [351].

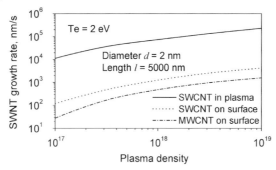

Figure 7.5 Dependence of carbon nanotube growth rates on the plasma density, for plasma-grown SWCNT and SWCNT/MWCNT grown on the surface from the ion flux. Electron temperature 2 eV, nanotube diameter 2 nm, nanotube length 5000 nm, nanotube array density factor 0.05 [351].

The third case of the nanotube growing in arc discharge plasmas is illustrated in Figure 7.3(c). In this case the two distribution peaks are observed (on the nanotube top and on the catalyst particle), with the "top" peak stronger (due to a larger curvature of the nanotube top and higher electric field strength). The carbon fluxes in this case are similar to the case of surface-bound SWCNTs; namely, carbon species deposited on the nanotube diffuse toward the catalyst particle. On the other hand, carbon deposited directly on the catalyst diffuses partially along the catalyst particle and partially through the catalyst [351].

A rigorous analysis of the above fluxes of carbon-bearing species to catalyst nanoparticles made it possible to calculate the total flux of adsorbed carbon species to the nanotube nucleus Γ_s [351, 361], which was then used to calculate the nanotube growth rate

$$\eta_{growth} = \frac{\Gamma_s}{\pi N_w \rho_s d}, \tag{7.1}$$

where ρ_s is the surface density of carbon atoms in graphene sheets, N_w is the number of nanotube walls ($N_w = 1$ for SWCNT and $N_w > 1$ for MWCNT), and d is the nanotube's diameter. The results of calculations of the nanotube growth rates as a function of the plasma density are shown in Figure 7.5. It is seen that the growth rate of SWCNT in arc discharge plasmas is 2×10^5 nm/s for the plasma density 10^{18} m^{-3}, and well exceeds the growth rate of the SWCNTs on the surface which is ca. 3000 nm/s and the growth rate of the MWCNT on the surface of ca. 400 nm/s for the same plasma density. From Figure 7.5 one can also see that the dependence of the growth rate on the plasma density is rather strong. The calculations were made for nanotubes of length/height 5 μm and diameters 2 and 100 nm for the SWCNT and MWCNT, respectively [351].

The results of Figure 7.5 can be compared with experimental information on nanotube growth on the surface and in the bulk of arc discharge plasmas. The growth rates of SWCNT/MWCNT on the surface can be directly compared with the rates measured in experiments; furthermore, the experiments demonstrate a good agreement with the calculations of Levchenko *et al.* [351]. Growth rates for surface-bound SWCNTs of up to 2000 nm/s have already been reported [362]; these are in good agreement with the value 3000 nm/s calculated above. For the MWCNT, the reported rates from 20 to 300 and even 1000 nm/s [363–365] are also in good agreement with the above rates ranging from 400 to 1500 nm/s obtained for plasma densities of 10^{18} and 10^{19} m^{-3}, respectively.

For the SWCNT grown in the bulk of dense plasmas, the growth rate can be obtained from the typical velocity of nanotubes in the arc plasma discharge and the width of a typical inter-electrode gap [272]. For the nanotube velocity 0.05 m/s and inter-electrode gap of 1 mm, one obtains the growth rate of ca. 250 μm/s, which is also in a good agreement with the value of 200 μm/s calculated by Levchenko *et al.* [351] for the plasma density of ca. 10^{18} m^{-3}.

Let us now discuss some important features of the nanotube growth stipulated by the electric field effects. In arc discharge plasmas, mainly SWCNTs grow. This mode is characterized by a very large and very nonuniform (Figure 7.3(d–f)) specific flux to the nanotube surface, which leads to very high growth rates (Figure 7.5). Note that higher fluxes can be deposited onto both the nanotube top and the catalyst particle. The latter feature is particularly useful to enhance the nanotube growth in either the top or the base growth mode, as applicable.

Another important feature of nanotube growth in a plasma is the state of the metal catalyst, which may be completely or partially molten. Indeed, metal catalysts, can be considered molten, due to the additional energy of the surface tension and high curvature [366]. Further, in arc discharge plasmas the metal catalysts can melt during emission from the cathode. In some cases carbon diffusion through the molten catalyst may be much faster than the surface diffusion [170]. This may lead to insufficient surface densities of those carbon adatoms on nanotube walls near the catalyst particle. As a result, formation of new walls becomes likely and in this case rapid growth of SWCNTs is usually observed.

Catalyst poisoning, which is usually associated with covering a catalyst surface by a carbon layer [367, 368], is also unlikely in this case due to the liquid surface of the catalyst and the high density of the ion flux which will provide an intense sputtering or etching of the deposited amorphous carbon and so restore the catalyst activity [364]. Besides, the size of catalyst particles used for SWCNT growth in a plasma are rather large (as compared with the nanotube diameter), and so the conditions are optimal for accumulation of a large amount of carbon in the catalyst. This is possible due to the fact that carbon solubility in Fe and Ni is very large and carbon saturates in Fe at 20 at.% if the particle size does not exceed 100 nm, and even at 50 at.% in particles of 3 nm in diameter.

In the above, we have always assumed that the nanotubes and other related one-dimensional nanostructures of our interest are aligned vertically, that is, in the direction of the electric field within the plasma sheath. However, what is the reason for this remarkable and indisputably electric field-related effect? Although this fact is acknowledged by everyone who works in the area, it still remains essentially unclear why exactly it occurs.

One of the first attempts to explain this phenomenon was undertaken by Merkulov *et al.* [160] who demonstrated that PECVD-synthesized carbon nanofibers grown in the tip mode align vertically irrespective of their density on the surface. However, base-grown nanofibers were frequently observed to grow in random orientations. The alignment cannot be explained using either the crowding or van der Waals force arguments. In the top growth mode the catalyst particle follows the direction of the plasma electric field as the nanofiber develops. Merkulov *et al.* [160] proposed a mechanism of vertical alignment based on self-adjustable precipitation of carbon material to the nanofiber walls through the catalyst particle.

Put simply, if the rates of carbon atom delivery to the left and to the right are equal, the nanofiber grows straight. These rates depend on the distribution of mechanical stress around the particle. If the nanofiber with the catalyst on its tip is bent, the above mechanical stress is mediated through the effect of the electric field, which effectively pulls the particle and the entire fiber up. This results in the restoration of uniform and balanced carbon precipitation around the nanofiber circumference and eventually in its perfectly vertical growth. On the other hand, if the catalyst particle remains at the nanofiber's base, the electric field only further strengthens the stress distribution and the nanofiber may bend even further.

The above conclusions are further supported by numerical simulations by Bao *et al.* [369], which suggested that in reasonably dense arrays the vertical alignment of multi-walled carbon nanotubes may be due to (or at least improved via) direct electrostatic interactions of individual MWCNTs within the array. Their numerical results, in particular, suggest that: (i) the electrostatic repulsive force induced by the similar charges of neighboring nanotubes is strong enough to become a major factor in nanotube alignment; (ii) the electrostatic attractive force along the the field direction acting on the catalyst particle at the MWCNT tip is much larger than that on the tube; this effectively pulls the "head" of the nanotube to the plasma bulk and eventually leads to the vertical growth; (iii) the magnitude of the attractive electrostatic force acting on the nanotube and the catalyst particle is related to the height and diameter of the MWCNTs as well as to their density in the array [369]. These numerical results also confirm that vertical alignment further improves (and in some cases quite substantially) when the DC substrate bias is increased. This is frequently observed in experiments [1].

Now the question is why *single-walled* carbon nanotubes also show so strong alignment along the direction of the electric field in the plasma sheath? This phenomen cannot be explained using the arguments of uniform and non-uniform stress and carbon material precipitation as we have done in the case of carbon nanofibers with the catalyst particle anchored to their base. Indeed, since the SWCNTs usually grow via the base-growth scenario, then why do not they bend just like nanofibers in the same growth mode?

This puzzle remained essentially open until Kato *et al.* [370] suggested that the SWCNTs can be treated as dipoles which align in the electric field within the plasma sheath. In this case, if a single-walled carbon nanotube is treated as an electric dipole with polarizability $\mathbf{P} = \alpha\mathbf{E}$, the potential energy of the nanotube dipole U_E in the electric field created in the plasma sheath $\mathbf{E}$ is

$$U_E = -\alpha E^2 \cos\phi,$$

where α is the principal term of the polarizability tensor and ϕ is the angle between the nanotube and the electric field. Here we recall that depending on their chirality, SWCNTs can be either semiconducting or metallic (see Section 3.2).

Since such nanotubes require quite significant thermal activation (e.g., growth temperatures at ca. 700–800 °C), a substantial amount of free carriers should exist even in semiconducting nanotubes [371]. Therefore, Kato *et al.* [370] assumed that all freestanding nanotubes in their experiments have a metallic structure. They have also approximated the static polarizability of metallic SWCNTs using a classical electrostatic model for a continuous metallic cylinder in an electric field

$$\alpha = 4\pi\varepsilon_0 \frac{l^3}{24[\ln(l/r) - 1]}\left[1 + \frac{4/3 - \ln 2}{\ln(l/r) - 1}\right],$$

where l and r are the cylinder's length and radius, respectively [372].

The absolute value of the maximum potential energy in this case turns out to be approximately 4.3 eV. In contrast, the thermal excitation energy U_T of the nanotubes grown at T_s ca. 750 °C, is ca. 0.088 eV. Hence, one has $|U_E| \gg U_T$, which means that the magnitude of the electric field within the plasma sheath is sufficient to align individual SWCNTs vertically [370].

To complete this section, we stress that the range of electric field- and ion-related effects is not limited to what has been discussed in this section, which was mostly devoted to carbon nanotubes and related structures. We will continue this theme in the following section and focus on carbon nanofibers and single-walled carbon nanotubes.

7.2
Plasma-Specific Effects on the Growth of Carbon Nanotubes and Related Nanostructures

As we have seen in the previous section, the electric field in the plasma and the associated ion fluxes and charging effects cause many interesting phenomena such as a remarkable vertical alignment and high growth rates of one-dimensional nanostructures. These growth rates are partly controlled by the state of the transition metal catalyst particles and are substantially higher than the rates of development of low-dimensional epitaxial nanostructures considered in Chapter 6.

In this section we will concentrate on low-temperature growth conditions and show that the plasma conditions affect many aspects of the growth of carbon nanofibers (CNFs) and single-walled carbon nanotubes (SWCNTs). As the nanostructures develop, their surface areas increase, and elementary processes on these surfaces such as adatom/adradical creation and diffusion become increasingly important.

The plasma-related effects considered in this section include ion-enhanced production of the required building units on top and lateral surfaces of the nanostructures followed by their diffusion from the point of creation to the point of insertion through the catalyst particle and control of the presence of reactive atomic hydrogen. This is a typical example of a reactive plasma environment where numerous transformations of plasma-created species take place not only in the gas phase but also on the surfaces of individual nanostructures!

7.2.1
Plasma-Related Effects on Carbon Nanofibers

We will now follow the structure of the original report [167] and consider the growth kinetics of carbon nanofibers in hydrocarbon plasmas. In addition to gas-phase and surface processes common to chemical vapor deposition, the model considered also includes several important processes that are unique to plasma-exposed catalyst surfaces such as ion-induced dissociation of hydrocarbon molecules and radicals, interaction of adsorbed species with incoming hydrogen atoms, and also dissociation of hydrocarbon ions.

This elegant model made it possible to demonstrate that at low, nanodevice-friendly process temperatures, carbon nanofibers develop through surface diffusion rather than bulk diffusion mechanisms. For an extended discussion of these two channels of incorporation of carbon species into the growing nanostructures please refer to Section 7.1 of this monograph.

Atomic carbon building units, which are delivered to the growth sites through surface diffusion, are generated on the (tip-attached) surface of the catalyst metal nanoparticle via ion-induced dissociation of hydrocarbon precursor species. These results explain the lower activation energy of nanofiber growth in a plasma and can also be used for the plasma-assisted synthesis of other one-dimensional nanoassemblies [167].

Similar to carbon nanotubes, carbon nanofibers are also one-dimensional nanostructures but with a different arrangement of graphene sheets; namely, wrapped in cones and stacked in one another. The top cone supports a metal catalyst nanoparticle as shown in Figure 7.6. This kind of carbon nanofiber is commonly referred to as bamboo-like nanofibers which develop in the tip growth mode. For an extended discussion of other arrangements, properties and applications of CNFs please refer to the dedicated review [22].

What is more important for the purpose of this monograph is that using plasma-enhanced CVD (PECVD), it is possible to grow the CNFs with better vertical alignment in addition to improved size and positional uniformity, at higher deposition rates and at substrate temperatures remarkably lower than in most neutral gas-based processes [21,22, 170,373]. As we have already discussed in Chapter 3 and Section 7.1, elementary processes on the surface and within an apical metal catalyst nanoparticle determine the subsequent growth and structure of vertically aligned carbon nanostructures.

However, how the plasma environment exactly (for instance, through ion bombardment and/or reactive chemical etching) affects these processes and leads to frequently reported in experiments

- higher growth rates;

- lower activation energies for CNF growth; and ultimately

- lower growth temperatures,

remains unclear despite extensive efforts to explain the growth kinetics or invoke modelling of neutral gas-based CVD, atomistic structure of related nanoassemblies or a limited number of plasma-related effects (ion/radical composition, surface heating, and so on) [167,231,374–377]. This issue remains one of the major obstacles on the way to deterministic plasma-aided synthesis of carbon nanofibers and related nanostructures of nanodevice quality [4,167].

Denysenko and Ostrikov [167] accounted for carbon diffusion over the catalyst particle surface and through the bulk of the catalyst as well as for a range of ion- and radical-assisted processes on the catalyst surface that are unique to plasma environments yet are frequently sidestepped by the existing models of carbon nanofiber/nanotube growth [378–380]. This approach makes it possible to conclude that at low surface temperatures T_s, which are insufficient for effective catalytic precursor decomposition, the plasma ions play a key role in the production of carbon atoms on the catalyst surface. The effect of the ion bombardment of the catalyst surface has been quantified and related to a remarkably lower CNF growth activation energy in the plasma-based process [167]. This has been one of the most-debated yet intractable issues in the last decade.

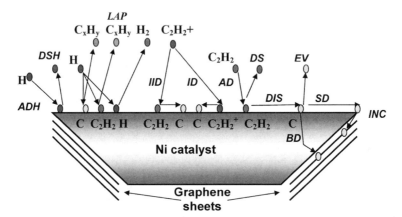

Figure 7.6 Processes that are accounted for in the PECVD. AD = adsorption of C_2H_2; DS = desorption of C_2H_2 (activation energy E_{aCH}); DIS = dissociation (δE_i); EV = evaporation (E_{ev}); SD = surface diffusion (E_s); INC = incorporation into a graphene sheet (δE_{inc}); BD = bulk diffusion(E_b); ADH = adsorption of H; DSH = desorption of H (activation energy E_{aH}); LAP = loss of adsorbed particles at interaction with atomic hydrogen; IID = ion-induced dissociation of C_2H_2; ID = $C_2H_2^+$ ion decomposition [167].

Let us now consider the plasma-assisted growth of a CNF with a metal catalyst particle on top, as shown in Figure 7.6 [167]. It is assumed that carbon atoms, the primary building units of the nanofibers, are created on the flat, circular top surface of the particle via a number of elementary processes (as sketched in Figure 7.6) and then incorporated into the growing graphene sheets (shown as stacked cones in Figure 7.6) via surface or bulk diffusion. This makes it possible to calculate the CNF growth rate H_t, single out specific contributions of the two competing diffusion processes (H_s and H_v for the surface and bulk diffusion, respectively), and apply these rates to explain and quantify the relevant experimental results [170,373,381,382].

The total CNF growth rate

$$H_t = H_s + H_v$$

can be split into two components originating from the surface

$$H_s = m_C J_s / (\pi r_p^2 \rho)$$

and bulk

$$H_v = m_C J_v / (\pi r_p^2 \rho)$$

diffusion, where J_s and J_v are the fluxes of carbon atoms to the graphene sheets over the catalyst particle's surface and bulk, respectively [167].

Here, r_p is the radius of the catalyst particle, ρ is the CNF material density, and m_C is the mass of a carbon atom. The flux of C atoms throuth the catalyst bulk is

$$J_v = \int_0^{r_p} (n_C D_b / r_p^2) 2\pi r dr,$$

where n_C is the surface density of carbon atoms, and

$$D_b = D_{b0} \exp(-E_b / k_B T_s)$$

is the bulk diffusion coefficient with D_{b0} a constant and $E_b \approx 2.560 \times 10^{-19}$ J (1.6 eV) [373] and, as usual, k_B is Boltzmann's constant.

To calculate the surface diffusion flux

$$J_s = -D_s \frac{dn_C}{dr} \Big|_{r=r_p} \times 2\pi r_p$$

one can assume that diffusing carbon atoms incorporate into the graphene sheet at the border of the catalyst particle ($r = r_p$), with the rate determined from $-D_s dn_C / dr = k n_C$, where

$$D_s = D_{s0} \exp(-E_s / k_B T_s)$$

is the surface diffusion coefficient (please note a similarity to adatom diffusion processes considered in Chapter 6), D_{s0} is a constant, E_s is the energy barrier for carbon diffusion on the catalyst surface

$$k = A_k \exp(-\delta E_{inc}/k_B T_s)$$

is the incorporation constant and A_k is the constant that depends on the carbon nanostructure size [383]. Here, δE_{inc} is the barrier for C diffusion along the graphene-catalyst interface, which is approximately 0.640×10^{-19} J (0.4 eV)–0.800×10^{-19} J (0.5 eV) for different nickel facets [165,373,383,384].

To calculate the surface density of carbon atoms n_C it was assumed that the top surface of the catalyst nanoparticle (here, Ni nanoparticle) is affected by fluxes of hydrocarbon neutrals (here, C_2H_2), an etching gas (here, H) and hydrocarbon ions (here, $C_2H_2^+$). Similar to the established CVD growth models of carbon nanotubes and related structures this model accounts for adsorption and desorption of C_2H_2 and H as well as thermal dissociation of C_2H_2 on the catalyst surface (Figure 7.6). It is also assumed that C_2H_2 and H are adsorbed only on the uncovered part of the catalyst surface.

Similar to the approach introduced in Chapter 6, this model also rigorously accounts for evaporation of carbon atoms from the catalyst surface and for the following processes on the catalyst surfaces, unique to the plasma environments yet not accounted for in the existing models: ion-induced dissociation of C_2H_2, interaction of all the adsorbed species with incoming hydrogen atoms, and dissociation of hydrocarbon ions (Figure 7.6) [167].

The model of Denysenko and Ostrikov [167] also includes mass balance equations for C_2H_2 and H species on the catalyst surface [385] and the following equation for atomic carbon on the surface

$$J_C + \mathrm{div}(D_s \mathrm{grad} n_C) - O_C = 0, \tag{7.2}$$

where

$$J_C = 2n_{CH}v \exp(-\delta E_i/k_B T_s) + 2(n_{CH}/v_0)j_i y_d + 2j_i$$

is the carbon source term describing the generation of C on the catalyst due to thermal and ion-induced dissociation of C_2H_2, and decomposition of $C_2H_2^+$, respectively.

The second term in Equation (7.2) describes the carbon loss due to surface diffusion. Likewise,

$$O_C = n_C v \exp(-E_{ev}/k_B T_s) + n_C \sigma_{ads} j_H + n_C D_b/r_p^2$$

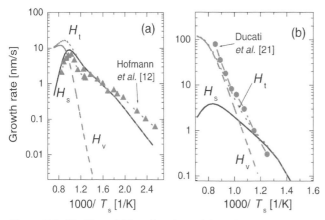

Figure 7.7 H_s, H_v and H_t as functions of the substrate temperature for (a) PECVD and (b) CVD. The triangles and circles represent the experimental points of Hofmann *et al.* [170] and Ducati *et al.* [381] for PECVD and CVD, respectively [167].

accounts for the loss of carbon atoms due to evaporation (with the energy barrier E_{ev}), interaction with atomic hydrogen from the plasma, and bulk diffusion. Here, n_{CH} is the surface concentration of C_2H_2; v_0 is the number of adsorption sites per unit area [385]; v is the thermal vibrational frequency; and σ_{ads} is the cross-section for the reactions of atomic hydrogen with adsorbed particles [385].

The flux of the impinging species α is given by $j_\alpha = \tilde{n}_\alpha v_{th\alpha}/4$, where subscript $\alpha = $(CH, H) stands for C_2H_2 and H species, respectively; $\tilde{n}_\alpha$ and $v_{th\alpha}$ are the volume density and thermal velocity, respectively. Furthermore, δE_i is the energy barrier for thermal dissociation of acetylene, $j_i \approx n_i(k_B T_e/m_i)^{1/2}$ is the ion flux, n_i is the ion density in the plasma, $T_e(\approx 1.6\,\text{eV})$ is the electron temperature, m_i is the ion mass, $y_d \approx 2.49\cdot 10^{-2} + 3.29\cdot 10^{-2} \times E_i$ [385], where E_i is the ion energy in eV. In the above, it was assumed that the C_2H_2 loss due to ion-induced dissociation is the same as that in the growth of diamond-like films [167,385].

We will now discuss the numerical results that follow from the above model. To elucidate the relative roles of the surface and bulk diffusion channels under typical experimental conditions of carbon nanofiber growth in a direct current PECVD system [170,373], Denysenko and Ostrikov [167] studied the dependence of H_t, H_s, and H_v on the surface temperature. Figure 7.7(a) shows the comparison of the computed nanofiber growth rates (using $\tilde{n}_{CH} = 5 \times 10^{14}\,\text{cm}^{-3}$, $\tilde{n}_H = 10^{-3}\tilde{n}_{CH}$, $E_i = 500\,\text{eV}$, $n_i = 3 \times 10^{10}\,\text{cm}^{-3}$, $r_p = 25\,\text{nm}$, $E_{aCH} = 4.640 \times 10^{-19}\,\text{J}$ (2.9 eV), $E_{aH} = 2.880 \times 10^{-19}\,\text{J}$ (1.8 eV), $\delta E_i = 2.080 \times 10^{-19}\,\text{J}$ (1.3 eV) ($E_{a\alpha}$ is the desorption activation energy for species α), $E_s = 0.480 \times 10^{-19}\,\text{J}$ (0.3 eV), and

$\delta E_{inc} = 0.640 \times 10^{-19}$ J (0.4 eV)) and measured experimentally by Hofmann *et al.* [170].

It is seen that the calculated H_t reproduces the experimental trend in the CNF growth rate. One can only observe minor deviations at low ($\beta_T = 1000K/T_s > 1.9$) and large substrate temperatures ($\beta_T < 1.1$). The minor difference at low T_s may be attributed to heating of Ni catalyst particles by intense ion fluxes from the plasma [231].

The most striking observation from Figure 7.7 is that the surface diffusion curve fits best to the experimental curve in the broad range of temperatures ($\beta_T = 1000K/T_s > 0.9$). This confirms and quantifies the earlier conclusion [373] that the CNF synthesis in the experiments of Hofmann *et al.* [170] may indeed be due to surface diffusion of carbon atoms over the catalyst particle surface [167].

More importantly, at low substrate temperatures, the temperature dependence of the CNF growth rate due to surface diffusion

$$H_s \sim \exp(-\delta E_{inc}/k_B T_s)$$

appears to be the same as that of the constant k of carbon incorporation into graphene sheets. One can thus conclude that the activation energy in PECVD is about the same as the energy barrier for carbon diffusion along the graphene-catalyst interface. Given that δE_{inc} is only ca. 0.640×10^{-19} J (0.4 eV) [373], this very low activation energy of the plasma-based growth of CNF in fact explains the higher growth rates in plasma-aided processes compared to CVD and some other neutral gas-based processes [170].

Letting $j_H = j_i = 0$, the growth rates for the CVD case (Figure 7.7(b)) have been also calculated [167]. From Figure 7.7(b) one notices that the computed total growth rate H_t is very close to the experimental results of Ducati *et al.* [381]. Moreover, it is clearly seen that at lower temperatures ($\beta_T > 1.2$), surface diffusion controls the growth whereas at higher temperatures ($\beta_T < 1.0$), CNF growth is due to the bulk diffusion. In the intermediate range $1.0 < \beta_T < 1.2$, both growth channels make comparable contributions [167].

It is also interesting to elucidate how the ion and atomic hydrogen fluxes from the plasma affect the CNF growth rate H_t. The growth rates H_t as functions of T_s are presented in Figures 7.8(a) and (b) for different ion and hydrogen atom densities in the plasma, respectively. One can see from Figure 7.8(a) that at low substrate temperatures the growth rate increases with j_i. This increase is mostly due to the enhanced ion-induced dissociation of C_2H_2 on the catalyst nanoparticles. On the other hand, Figure 7.8(b) suggests that H_t decreases with j_H because of the larger

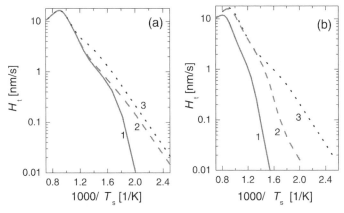

Figure 7.8 Variation of H_t with T_s for different densities of ions n_i (a) and atomic hydrogen $\tilde{n}_H$ (b) in the plasma bulk. Curves 1, 2 and 3 in panel (a) correspond to $n_i = 10^8, 10^{10}$ and 10^{11} cm^{-3}, respectively. Curves 1, 2 and 3 in panel (b) are for $\tilde{n}_H = \tilde{n}_{CH}, 0.05\tilde{n}_{CH}$ and $5 \times 10^{-4}\tilde{n}_{CH}$, respectively, where $\tilde{n}_{CH}$ is the density of C_2H_2 species in the plasma bulk. Here, $\tilde{n}_H = 10^{-3}\tilde{n}_{CH}$ for panel (a) and $n_i = 3 \times 10^{10}$ cm^{-3} for panel (b). Other parameters are the same as in Figure 7.7(a) [167].

loss of C_2H_2 and C species in reactions with impinging hydrogen atoms (see Figure 7.6).

Thus, the numerical modelling of the plasma-assisted growth of carbon nanofibers accounts for a number of processes on the catalyst surfaces which are unique to plasma environments. The results of Denysenko and Ostrikov [167] suggest that at low substrate temperatures ion-assisted precursor dissociation and surface diffusion may be the main processes enabling rapid, low-temperature growth of carbon nanofibers in low-temperature plasmas. This is consistent with the experimental results of Tanemura *et al.* [382] and Woo *et al.* [187] suggesting that carbon nanofibers do not grow when an ion-repelling positive potential is applied to the substrate or when the ion-attracting negative potential of the substrate is small. At low T_s, the loss of carbon material on the catalyst surface in a plasma-based process is mainly due to deposition of etching gas on the catalyst. On the other hand, at relatively high substrate temperatures ($T_s > 800$ K), both surface and bulk diffusion may be important in the CNF growth [167].

In the following subsection we will consider a quite similar approach to describing and quantifying plasma-related effects in the growth of single-walled carbon nanotubes on plasma-exposed surfaces. The main difference with the above case of carbon nanofibers will be that the processes of ion-assisted precursor dissociation, etching by reactive hydrogen species and associated diffusion of the as-created building units will take place on the lateral surface of high-aspect-ratio SWCNTs, which effectively makes this problem quasi-one-dimensional.

7.2.2
Effects of Ions and Atomic Hydrogen on the Growth of SWCNTs

Let us now extend the diffusion models of Chapter 6 and of the previous subsection to describe the growth of single-walled carbon nanotubes in plasma-enhanced chemical vapor deposition [386]. Similar to the results in the previous subsection, atomic hydrogen and ion fluxes from the plasma can strongly affect nanotube growth at low substrate temperatures ($\leq 1000\,$K). Likewise, plasma ion-assisted hydrocarbon dissociation appears to be the main process that supplies carbon atoms for SWCNT growth and is also responsible for the frequently reported higher (compared to thermal chemical vapor deposition) nanotube growth rates in plasma-based processes.

However, excessive deposition of plasma ions and atomic hydrogen can hamper the ability of carbon-bearing species to diffuse freely about the nanotube lateral surface. Thus, their diffusion length and the lifetime on the surface may be reduced. The results of Denysenko *et al.* [386] are in good agreement with the available experimental data and can be used for optimizing SWCNT growth in a variety of plasma-assisted processes.

Plasma-grown single-walled carbon nanotubes (SWCNTs) are often produced using microwave or other specific (e.g., remote plasmas) low-temperature plasmas with or without moderate biasing of the substrate to avoid ion-induced damage [387–390]. In many cases, SWCNTs appear as spaghetti-like networks. Recently, growth of vertically aligned single- and double-walled CNTs at substrate temperature $T_s \approx 700\,$K with moderate substrate voltages ($\leq 100\,$V) was reported [391]. Nozaki *et al.* synthesized vertically aligned SWCNTs using atmospheric-pressure PECVD [145]. Interestingly, in atmospheric pressure plasmas, near-surface potential drops are very small which makes it possible to avoid the ion damage of SWCNTs frequently reported in other kinds of plasmas. Here we stress that despite a substantial number of publications on SWCNTs, the controlled plasma-aided synthesis of vertically aligned SWCNTs at low processing temperatures still remains a major challenge.

As we have stressed several times above, the surface-bound SWCNT follow the base (root) growth scenario. In this case, the catalyst particles remain anchored to the substrate [144, 392]. The size of the catalyst particles and the etching gas can substantially affect the SWCNT synthesis [393]. Zhang *et al.* [394] noted that reactive hydrogen species may be unfavorable to SWCNT formation and can etch the latter. However, the nanotube wall etching is quite unlikely owing to the excellent structural stability of the SWCNTs. Nonetheless, as will be shown below, hydrogen atoms can effectively etch carbon material as it is deposited onto the surfaces of the nanotubes or between them. Furthermore, Gohier *et al.* synthesized SWCNTs under conditions of heavy dilution of hydrocarbon precursor in hydrogen gas [391].

However, the conditions for SWCNT synthesis by PECVD have not been intensively studied, and knowledge on how the radical and ion fluxes from the plasma affect the growth is rather scarce. Moreover, it is still not clearly understood how the reactive species produced in the plasma reach the catalyst particles anchored to the SWCNTs when the latter are long and crowded. A recent report [386] filled this gap by reporting the results of an in-depth investigation into the growth mechanisms of SWCNTs in low-temperature plasma-assisted processes. Denysenko *et al.* [386] investigated the PECVD growth of SWCNTs by extending the microenergetic surface diffusion model of Louchev, Sato, and Kanda [395]. This approach was used earlier to explain the growth of carbon nanotube forests by the ball-milling and CVD techniques [379, 396]. Simple estimates show that carbon atoms can migrate about the nanotube surface at micron scales [397].

Below, the SWCNT base growth will be considered; in particular, deposition of hydrocarbon neutrals and ions, as well as of the particles of the etching gas, on and between the SWCNTs will be accounted for. As in the previous subsection, in addition to the processes common to SWCNT growth in CVD, the original model [386] also takes into account processes unique to PECVD, such as the interaction of species adsorbed on and between the SWCNTs with the incoming etching gas, the decomposition of the adsorbed hydrocarbon species from ion bombardment, the decomposition of hydrocarbon ions on the SWCNT surface, etching and sputtering of carbon films between the SWCNTs, and some other elementary processes.

Let us consider close-ended growth of SWCNTs. The catalyst nanoparticles are anchored to the base (at $x = L_{NT}$, where x is along a SWCNT axis and L_{NT} is the length) of a CNT. The plasma (for example, created in a CH_4/H_2 gas discharge) is located above the SWCNTs and the main particles that interact with the surfaces of the SWCNTs are hydrocarbon neutrals (here CH_3), hydrocarbon ions (here CH_3^+) and the atoms or molecules of an etching gas (here atomic hydrogen H). Please note that the species participating in the growth of the SWCNTs are different from those considered in the previous subsection. The CH_3 and H radicals are adsorbed and desorbed on the SWCNT surfaces as well as on the substrate surface between the SWCNTs. The adsorption and desorption fluxes of the radicals are [385,386]

$$j_{\alpha ads} = j_\alpha(1 - \theta_t) \tag{7.3}$$

$$j_{\alpha des} = \theta_\alpha v_0 \nu \exp(-E_a/k_B T_s), \tag{7.4}$$

where α = CH and H denote CH_3 and H neutrals, respectively. Here, $j_\alpha = \tilde{n}_\alpha v_{th\alpha}/4$ is the flux of impinging neutral species, $v_{th} = \sqrt{8k_B T_s/\pi m_\alpha}$ is the thermal velocity, k_B is the Boltzmann constant, $E_a \approx 2.880 \times 10^{-19}$ J (1.8 eV) [383] is the adsorption energy, $\tilde{n}_\alpha$, θ_α and m_α are the plasma bulk density, surface coverage and mass of species α, respectively. It is assumed that the SWCNT surfaces and the surface between the SWCNTs are covered by CH_3 and H radicals and carbon atoms C, and that the total surface coverage by the particles is $\theta_t = \theta_{CH} + \theta_H + \theta_C$, where θ_C is the surface coverage by C. In Equation (7.4), $v_0 \approx 1.3 \times 10^{15}$ cm^{-2} [385] is the number of adsorption sites per unit area, $\nu \approx 10^{13}$ Hz is the thermal vibration frequency, and T_s is the SWCNT surface temperature. It is also assumed that T_s is constant along a SWCNT.

The adsorbed species can react (for example, $CH_{3(ads)} + H_{(plasma)} \rightarrow CH_{4(plasma)}$) with the atomic hydrogen from the plasma, yielding gas-phase products. The consumption flux of an adsorbed neutral participating in an adsorbed-layer reaction is [385,386]

$$j_{reac} = \theta_\alpha v_0 \sigma_{ads} j_H,$$

where $\sigma_{ads} \approx 6.8 \times 10^{-16}$ cm^2 is the cross section of the adsorbed-layer reaction and j_H is the incident flux of atomic hydrogen.

Carbon atoms can be generated on the SWCNT surfaces by the following reactions: (i) thermal dissociation [398]; (ii) ion bombardment of adsorbed CH_3 radicals; and (iii) decomposition of CH_3^+ ions. The carbon yield due to thermal dissociation is

$$\theta_{CH} v_0 \nu \exp(-\delta E_i/k_B T_s),$$

where $\delta E_i \approx 2.1$ eV is the activation energy of thermal dissociation [398].

It is also assumed that ions impinging from the plasma have sufficient energy E_i (≥ 2.1 eV) to decompose the CH_3 radicals on the SWCNT surfaces, and the carbon yield from the ion bombardment is $\theta_{CH} j_i y_d$, where $y_d = E_i/\delta E_i$, $j_i \sim n_i \sqrt{T_e/m_i}$ is the ion flux, $T_e \approx 1.5$ eV is the plasma electron temperature, and n_i and m_i are the ion density and mass, respectively. The ions bombarding a SWCNT decompose ($CH_3^+ \rightarrow C_{(ads)} + H_{(ads)} + H_{2(plasma)}$) on the surface, resulting in deposition of both carbon and hydrogen on the SWCNTs [386].

The carbon atoms generated on the SWCNT surface by thermal dissociation and ion bombardment diffuse to the catalyst located at $x = L_{NT}$ in the tube base and are incorporated into the growing SWCNTs [399]. This scenario is based on several atomistic microenergetic studies [397, 400–404] on the activation energy for surface diffusion, concluding that the carbon atoms can migrate over micron-scale distances along the carbon SWCNT surface.

Another important assumption is that the characteristic time of surface geometry variation is much larger than that of surface diffusion (i.e., $L_{NT}V_{NT}/D_s \ll 1$, where $V_{NT} = d_t L_{NT}$ and D_s is the surface diffusion coefficient), and one can describe the SWCNT growth using a quasi-steady model [383]. Accordingly, the surface diffusion coefficient is given by

$$D_s \cong a_0^2 v \exp(-\delta E_d / k_B T_s),$$

where $a_0 \cong 0.14 \, \text{nm}$ is the interatomic distance, and $\delta E_d \approx 0.8 \, \text{eV}$ is the activation energy of surface diffusion for carbon on a SWCNT surface [400, 401]. A full list of the processes that take place on the SWCNT surface included in the model are summarized in Table I of the original report [386].

The model equations include the mass balance equations for CH_3, H, and C species on a SWCNT surface. Density of carbon adatoms is determined using a one-dimensional diffusion equation. From this set of equations, one can determine the surface coverages of the species involved as well as the distribution of densities of carbon adatoms over the entire length of the nanotube. The deposition flux in the area between the SWCNTs is accounted for separately; this allows one to describe the growth of thin carbon films in those areas. The above mentioned model equations are complemented with the appropriate boundary conditions.

One can obtain the following differential equation for the SWCNT length [383, 386, 399]

$$V_{NT} = d_t L_{NT} = -\Omega D_s d_x n_C \big|_{x=L_{NT}} = \frac{k \Omega Q_C \tau_a \sinh(\zeta)}{\sinh(\zeta) + (k \lambda_D / D_s) \cosh(\zeta)}, \quad (7.5)$$

where Ω is the area per unit carbon atom in a SWCNT wall, Q_C is the effective carbon flux to the SWCNT surface, and τ_a is the characteristic time of residence of carbon atoms on the SWCNT surface, $\zeta = L_{NT}/\lambda_D$, and $\lambda_D = \sqrt{D_s \tau_a}$ is the surface diffusion length [386].

The growth rate of the carbon film between the nanotubes is

$$V_{dep} = d_t L_{dep} = j_{dep} M_{dep} / \rho N_A, \quad (7.6)$$

where j_{dep} is the effective deposition flux (which accounts for thermal insertion of neutrals, direct incorporation of ions, ion-induced incorporation of CH_3 neutral radicals, etching and sputtering processes), L_{dep}

is the film deposition width, N_A is the Avogadro number, ρ is the film density (e.g., $\rho = 1.2\,\mathrm{g cm}^{-3}$), and $M_{dep} \approx 12\,\mathrm{g mol}^{-1}$ is the mole mass of the growing film material. To obtain the SWCNT length L_{NT} and the width of the deposited film L_{dep} as functions of time t, Equations (7.5) and (7.6) were integrated numerically using a 4th-order Runge–Kutta scheme [386].

First we will discuss how the ion and hydrogen atom fluxes affect the SWCNT growth. Using the above model, Denysenko *et al.* [386] calculated the SWCNT length L_{NT}, the thickness of the film between the SWCNTs L_{dep}, the diffusion length λ_D, the time characterizing carbon loss from the surface of SWCNTs, and the surface coverage θ_C, θ_{CH}, and θ_H for different substrate temperatures T_s, the ion fluxes j_i, the ion energy E_i, and the hydrogen fluxes j_H, that are typical for experiments on CNT and SWCNT synthesis by PECVD. Here we stress that the growth of the SWCNTs is possible if at all times no film between the SWCNTs covers the anchored catalyst particles. This implies that any amorphous carbon film between the CNTs has to be continuously etched or/and sputtered away ($L_{dep}(t_0) \rightarrow 0$ at any t_0), so that they do not block the access of carbon adatoms to the catalyst particles.

The role of the plasma environment in nanotube growth was quantified through investigation of the effect of ion flux density, ion energy, and the ratio of fluxes of atomic hydrogen and hydrocarbon precursor [386]. Let us first consider the dependence of the SWCNT length on the substrate temperature for different ion densities. The increase of the SWCNT length ΔL_{NT} ($= L_{NT} - L_0$, where L_0 is the SWCNT length at $t = 0$) for a growth time of $t = 1\,\mathrm{s}$ as a function of T_s is shown in Figure 2(a). For comparison, the results for CVD ($j_i = j_H = 0$) are also shown. In the evaluation, the following parameters have been used: $\tilde{n}_{CH} = 10^{15}\,\mathrm{cm}^{-3}$, $j_H = 10^{-3} \times j_{CH}$, $E_i = 2.1\,\mathrm{eV}$ ($y_d = 1$), $L_0 = 1\,\mathrm{nm}$, and $n_i = 10^9$, 10^{10}, and $10^{11}\,\mathrm{cm}^{-3}$.

One can see from Figure 7.9(a) that at low substrate temperatures the SWCNT length increases with j_i. This increase is due to enhanced ion-induced dissociation of CH_3 and direct decomposition of CH_3^+ on the SWCNTs. These processes increase the effective carbon flux Q_C, as can be seen in Figure 7.9(b). The latter can be much larger in PECVD than in CVD, where $j_i = j_H = 0$. Figure 7.9(c) shows that the film thickness L_{dep} between the SWCNTs also depends on the ion flux density. At low T_s ($< 1000\,\mathrm{K}$) the film thickness increases with increase of j_i due to ion-induced incorporation of carbon atoms and direct incorporation of CH_3^+ [385].

Moreover, similar to the growth of diamond-like films [385], L_{dep} decreases with increase of T_s. As indicated by Equation (7.4), increase of the

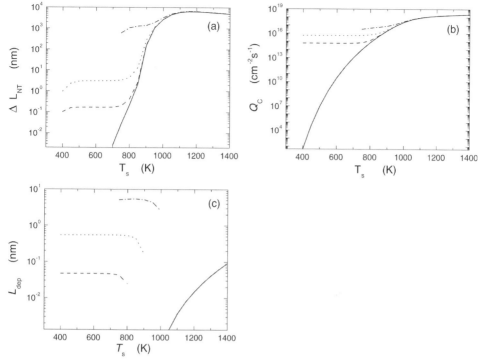

Figure 7.9 The SWCNT length increase (a), effective carbon flux
(b) and film thickness (c) for $L_0 = 1$ nm and different ion densities:
$n_i = 10^9$ cm^{-3} (dashed curve), 10^{10} cm^{-3} (dotted curve) and
10^{11} cm^{-3} (dash-dotted curve). The solid curve corresponds to
CVD ($j_i = j_H = 0$) [386].

substrate temperature at $T_s > 900°$K is accompanied by increase of the
desorption flux. This results in a much lower total surface coverage, as
well as enhancement of film etching and sputtering. Consequently, L_{dep}
drops as T_s increases until the film growth between the CNTs stops at a
relatively large ($T_s > 1000$ K) surface temperature [386]. We recall that in
CVD ($j_i = 0, j_H = 0$), L_{dep} grows with increase of T_s, which is due to the
exponential enhancement of the carbon flux $n_C(L_{NT})v \exp(-\delta E_f/k_B T_s)$
to the solid surface [386].

Next, we consider how the SWCNT growth parameters depend on
the energy E_i at which the plasma ions impinge on the CNT surface.
In Figures 7.10(a) and (b) the variations of the SWCNT length $\Delta L_{NT} =
L_{NT} - L_0$ and the film width L_{dep} for the growth time $t = 1$ s, $L_0 = 1$ nm
and different substrate temperatures are shown. One can see that both
ΔL_{NT} and L_{dep} increase with increase of E_i. The former increases because
of enhancement of ion-induced CH$_3$ dissociation and CH$_3^+$ decomposi-

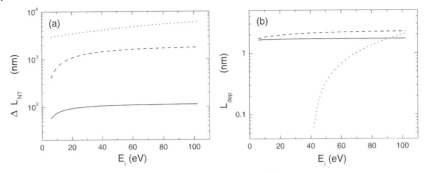

Figure 7.10 The SWCNT length (a) and film thickness (b) as functions of E_i for $\tilde{n}_{CH} = 10^{15}\,\text{cm}^{-3}$, $n_i = 3 \times 10^{10}\,\text{cm}^{-3}$, $j_H = 10^{-3}j_{CH}$ and different substrate temperatures: 800 K (dashed curve), 1000 K (dotted curve) and 600 K (solid curve) [386].

tion on the SWCNTs, which effectively increases the surface coverage by carbon. The film width L_{dep} grows with increasing E_i (Figure 7.10(b)) because of intensification of ion stitching and direct incorporation of CH_3^+ into the growing film. Furthermore, the influence of E_i on L_{dep} at $T_s = 600$ and 800 K is smaller than that at 1000 K, probably because of a relatively large uncovered area at $T_s = 1000$ K, which is subject to etching and sputtering [386].

The influence of atomic hydrogen on the SWCNT growth was also studied by computing the SWCNT length as a function of the atomic hydrogen flux for different ion densities n_i. At relatively high hydrogen fluxes ($k_H = j_H/j_{CH} > 0.1$) the SWCNT length decreases with increase of j_H for all the ion densities considered. At small hydrogen flux ($k_H < 10^{-2}$ for $n_i = 3 \times 10^{10}\,\text{cm}^{-3}$) ΔL_{NT} increases with increase of j_H. The increase is due to the fact that at low j_H the surface coverage by atomic hydrogen is large, and θ_H decreases with increase of j_H because of interaction of atomic hydrogen on the SWCNTs with H incoming from the plasma. The decrease of θ_H at $k_H < 10^{-2}$ is accompanied by increase of θ_{CH} as well as the area $(1 - \theta_t)$ uncovered by particles [386].

Let us now discuss the main features of the SWCNT growth model, as well as some of the results presented in this section in more detail. First, it was assumed that the surface temperature is constant along the SWCNTs. If there is a difference between the temperatures of the gas and the substrate, the SWCNT surface temperature may not be uniform. For example, the tip of SWCNTs can be additionally heated by neutral and ion fluxes from the plasma [231]. However, for relatively short CNTs ($\leq 5\,\mu\text{m}$) the temperature variation along SWCNTs can be expected to be small [404].

In their model, Denysenko *et al.* [386] have accounted for the deposition of the most important species: hydrocarbon radicals (CH_3), etching gas atoms (H) and ions (CH_3^+). Generally, in PECVD growth experiments the species number of ions and radicals that deposit on SWCNTs is much larger. Moreover, the flux densities of CH_3, H and CH_3^+ and the ion energy are input parameters here. In reality, the ion fluxes and substrate bias can affect the densities of neutral particles in the plasma. Consequently, in order to achieve a better insight of the PECVD of SWCNTs the model here should be complemented by an appropriate plasma discharge and chemistry model. Ion-induced physical sputtering, which depends on the substrate bias [391] and can affect the plasma parameters at the surface, was also neglected. However, the model presented here should nevertheless be useful for qualitative analysis of SWCNT growth in PECVD, as well as a basis for further theoretical investigations [386].

Further, this model substantially advances the existing approaches in that it accounts for the most important plasma-related processes on the SWCNT surfaces. These processes, such as ion-induced dissociation of hydrocarbons, interaction of adsorbed species with incoming hydrogen atoms, and dissociation of hydrocarbon ions, are unique to the plasma environment but are usually not accounted for in many existing models. Therefore it allows one to estimate how the fluxes of the ions and the etching gas affect the nanotube growth.

We note that the ions and electrons in the plasma bulk produce a variety of hydrocarbon radicals and excited species which can be easily adsorbed onto the catalyst and SWCNT surfaces. Moreover, ions bombarding adsorbed hydrocarbons can also produce carbon atoms on the latter surfaces. Denysenko *et al.* [386] suggested that the effect of ion-induced C production can be as significant as the generation of carbon atoms in the plasma bulk and thermal dissociation of hydrocarbons on the catalyst and SWCNT surfaces. This in fact answers the question *"is there a specific role for ions related to the growth (of CNTs) and [...] how does it depend on ion energy?"* posed by Meyyappan *et al.* [21], and sheds some light on the roles of the atomic hydrogen, radicals, species responsible for CNT growth, contamination by amorphous carbon, and factors determining the CNT growth rates. Most of these conclusions can be rather straightforwardly verified experimentally.

It is also interesting that for typical experimental conditions the surface diffusion length is of the order of 100 nm. This may well explain the observation in the experiments of Gohier *et al.* [391], where SWCNTs grew only up to a few hundred nanometers. On the other hand, the model in which we are interested in this subsection does not explain how carbon-bearing species reach the anchored catalyst particles in the experiments on the synthesis of 20 μm-long SWCNTs [145].

One explanation could be that under the experimental conditions the activation energy for the carbon surface diffusion is actually smaller than $\delta E_d \approx 0.8 \, eV$. At small δE_d (for example, $\delta E_d = 0.13 \, eV$ [397]) carbon atoms on SWCNT surfaces can migrate micron-range distances. Enhancement of surface diffusion can also occur because of the (positive) effective charge of the species chemisorbed on the CNT lateral surfaces, enhancing their acceleration to the substrate bottom in the sheath [386].

Another possible reason that carbon atoms are still able to reach catalyst nanoparticles at the base of very-high-aspect-ratio CNTs is that the plasma ions deposit non-uniformly onto their surface (see Section 5.1 for more details). As we have seen in Section 5.1, the location of the peak of the ion flux deposition on the nanotube surface can in fact be controlled by varying the DC bias of the substrate. If the substrate bias is low or absent, the ions are predominantly deposited in the upper section of the SWCNTs, with the maximum density near their sharp tips.

In this case the carbon adatoms form as a result of ion impact dissociation, and travel significant distances until they reach the catalyst particle and contribute to the CNT growth. However, when the substrate bias becomes larger, more ions are deposited closer to the base of the nanostructures. Under certain conditions, most of the ions land on the lowest section of the CNT and thus find themselves within a short walk from the catalyst nanoparticles [386]. We have discussed this possibility for surface-bound nanotubes of Section 7.1.

In general, the results of Denysenko *et al.* [386] are in a good agreement with that of the existing studies on carbon nanofiber growth by PECVD. In particular, experiments prove that the SWCNT growth rate attains its maximum at a certain temperature (about 1000 K), and that surface coverage by atomic hydrogen at low T_s can be large [406–409].

In Sections 7.1 and 7.2, the shape of the nanostructures considered did not change during the growth process. Well, why should and how can the shape of carbon nanotubes change if their growth is largely determined by the size of the catalyst nanoparticle either at the CNTs tip or at its base? The only possibility we have discussed above was the origin of new walls. There could be other options, such as closing of originally open-ended multiwalled nanotubes. Nevertheless, CNTs always remain in a tubular shape and only change their aspect ratio as they elongate. Interestingly, other nanostructures may change their shape significantly. Some examples of significant nanostructure reshaping in plasma-based processes are described in the following section.

7.3
Plasma-Controlled Reshaping of Carbon Nanostructures

In this section we consider nanoscale objects that are different from the carbon nanotubes and related structures of Section 7.2. The nanostructures of interest here are continuous (e.g., crystalline) rather than hollow as is the case for the nanotubes; therefore, the structure and the growth processes are very different. More importantly, their behavior under plasma exposure is also quite different. One of the most amazing manifestations of plasma-related effects is the possibility of dynamic self-reshaping of carbon nanocones into even sharper nanotips with a higher aspect ratio. We will first present the relevant experimental results on self-sharpening of single-crystalline conical carbon nanostructures and discuss their growth kinetics (Section 7.3.1). Further possibilities to re-shape such nanoscale objects to fit the requirements for electron field microemitter applications are explored via advanced numerical simulation in Section 7.3.2.

7.3.1
Self-Sharpening of Platelet-Structured Nanocones

In this section we will follow the original report [258] and discuss the mechanism and experimental verification of the model for the vertical growth of platelet-structured vertically aligned single-crystalline carbon nanostructures by the formation of graphene layers on a flat top surface. More importantly, plasma-related effects lead to self-sharpening of tapered nanocones to form needle-like nanostructures. This observation is in remarkable agreement with the theoretically predicted dependence of the radius of a nanocone's flat top on the incoming ion flux and surface temperature. This growth mechanism is also relevant to a broad class of nanostructures including nanotips, nanoneedles, and nanowires and can be used to improve the predictability of plasma-aided nanofabrication.

As usual, we start our consideration by stressing the significance of the nanostructures of interest. Arrays of one-dimensional (1D) vertically aligned nanostructures such as nanocones, nanotubes, nanotips, nanowires and nanofibers are currently of enormous interest owing to their unique optical, electronic, mechanical, chemical and other properties that serve to pave their way into applications as diverse as gene/drug delivery systems, electronic interconnects in nanoelectronics, structural scaffolds for composite material reinforcement, photovoltaic devices, and electron field microemitters [258, 410–412]. Amongst a plethora of possible nanofabrication techniques, methods based on low-temperature plasmas have shown an outstanding promise in creating arrays of vertically aligned nanostructures (VANs) of various materials such as carbon, ZnO, InN, GaN, and several other technologically important material systems [413–415].

Single-crystalline platelet-structured nanostructures is a special class of nanoscale assemblies made of cylindrical platelets oriented normal to the growth direction [258]. Depending on the growth shape characterized by the aspect (height to width) ratio and tapering, they can develop into nanorods, nanowires, nanofibers, and nanocones [416]. As we have already stressed elsewhere in this monograph, the shape selection is primarily controlled by the conditions of the catalyst nanoparticle such as composition, size and location (e.g., on top or at the base of the VAN) ans so on [416, 417]. However, it is not always clear exactly what role the nanofabrication environment plays in this amazing shape selection. Moreover, it is very difficult to create appropriately tapered nanostructures, mostly because of relatively poor understanding of the basic processes involved.

Levchenko *et al.* [258] conducted a dedicated experiment on the plasma-based growth of single-crystalline platelet-structured carbon nanocones and have shown that by altering the process parameters one can effectively create tapered nanocones with size-controlled or very sharp tops. They also proposed a growth model which convincingly related the radius of the flat top to the surface temperature and the incoming ion fluxes. More importantly, the experimental and modeling results are in remarkable agreement and support the proposed growth mechanism [258].

Successful synthesis and applications of carbon-based platelet-structured VANs had been reported earlier [93, 94, 140] (see also Chapter 4 of the relevant monograph [1]). Interestingly, the Ni-based catalyst remains at the base of single-crystalline carbon nanocones, which often develop into nanoneedle-like structures with an aspect ratio of up to 100, with or without a flat top [50]. Quite similar structures with flat tops have been reported for other materials such as ZnO [418]. The experiment of Levchenko *et al.* [258] elucidated the role of the plasma-based environment in the formation of either tapered or very sharp carbon nanocones. The nanostructures were synthesized on Ni-catalyzed lightly-doped Si(100) substrates in low-frequency (460 kHz) inductively coupled plasmas of $CH_4 + Ar + H_2$ gas mixtures under low-pressure (~ 6.666 Pa (50 mTorr)) conditions. The surface temperature was maintained in the range 400–550 °C, while the DC bias was fixed at -100 V. Other experimental details are the same as in earlier reports of the same group [50, 140].

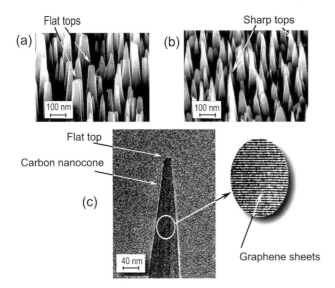

Figure 7.11 (a) Array of flat-top carbon nanocones grown on a Si substrate in Ar + H$_2$ + CH$_4$ plasma at a substrate bias of −100 V and surface temperature 500 °C, for 22 min into the process; (b) array of sharp carbon nanocones grown under the same conditions for 25 min; (c) and HRTEM image showing a typical shape of platelet carbon nanocones, and structure and direction of graphene layers (inset). The tapered nanocone shown on the HRTEM micrograph is ∼ 1 μm in height and features a flat top with the radius 15 nm [258].

The most striking experimental observations were [258]:

- the radius of the flat top of the nanocones decreases as they grow and the transition to an ultra-sharp needle-like shape occurs at the final growth stage;

- growth still continues after a complete coverage of Ni catalyst nanoparticles by nanocone bases and even in cases when the bases of some nanostructures come into contact.

Figure 7.11 shows two SEM micrographs of tapered flat-top (a) and needle-like (b) nanocones developed after 22 and 25 min into the growth process. The nanocone array is rather dense, with the substrate surface coverage reaching approximately 0.5. The observations are consistent with the results of earlier experiments where the radii of flat tops of tapered nanocones varied in the ca. 8–20 nm range [53]. Additionally, high-resolution transmission electron microscopy (HRTEM) shows clear horizontal stacks of carbon sheets normal to the growth direction (Figure 7.11(c)), which is typical for platelet-structured single-crystalline nanocones.

These observations were interpreted by noting that the "bottom-up" growth mode implies formation of new graphene sheets at the surface of the catalyst particle. Hence, vertical growth of the nanocone should stop when the catalyst particle is completely covered by the nanocone base, and therefore their vertical growth at later stages cannot be explained by the "bottom-up" model. Based on these observations, it was proposed that the growth of a single-crystalline platelet-structured carbon nanocone proceeds via three stages [258].

In the first stage, the nanocones develop via formation of new graphene sheets on a Ni catalyst particle anchored to the substrate surface. As a result, nanostructures of a near-cylindrical shape are formed, with the base and top radii being approximately the same. During the second stage, when the lateral surface area increases, the nanocones grow by attachment of carbon atoms to the borders of hydrogen-terminated graphene sheets. At this stage, the nanocones increase their height via new layer formation at the catalyzed surface, and widen due to carbon atom attachment to the borders of the parallel carbon platelets [258].

Here we note that termination of dangling carbon bonds at the periphery of graphene sheets is an essential prerequisite of structural stability of nanocones, as confirmed by *ab initio* density functional theory computations of structural stability of tapered structures that resemble real nanostrtuctures [50] (see also Section 4.4 of related monograph [1]). The dangling bonds are activated through bombardment of the nanocone's lateral surface by the plasma ions and are passivated by attachment of hydrogen atoms. The dynamic balance of these processes enables a certain number of dangling bonds, at any time, for bonding with carbon adatoms migrated from the flat top and substrate surfaces.

After the nanocone radius exceeds the radius of the catalyst particle, formation of new carbon sheets at the nanostructure base becomes impossible and the third stage comes into play. In this case, the nanocone radius increases in the same way as before, while the only way for the height to increase at this stage is through the formation of new graphene sheets on the flat top.

The growth model assumes that the graphene sheets are hydrogen-terminated at the edges, as shown in Figure 7.12; thus, the entire lateral surface of the nanocone is hydrogen-terminated. The particle fluxes to the nanocones and substrate surface are shown in Figure 7.13. The flat top surface collects a flux of carbon atoms and positive ions. This means that hydrocarbon radicals are stripped of hydrogen atoms upon landing on the flat top surface of the nanocones. As was discussed in Section 7.2, ion bombardment can significantly enhance this process.

Referring to the results discussed in Section 5.1, we can confidently state that in a plasma the ion flux is strongly focused by the non-uniform electric field present in the vicinity of the nanostructures and can sub-

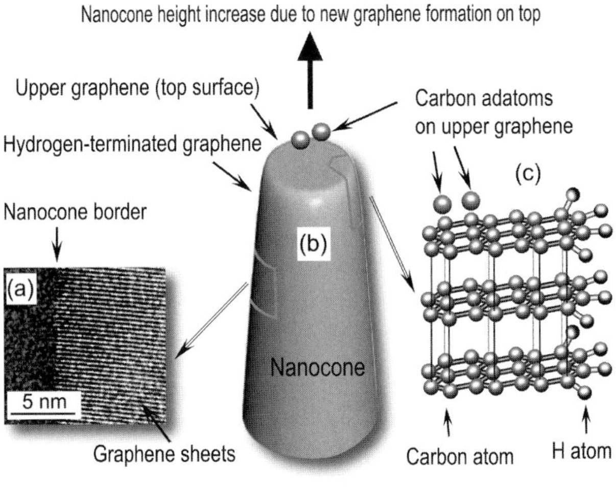

Nanocone height increase due to new graphene formation on top

Upper graphene (top surface)

Carbon adatoms
on upper graphene

Hydrogen-terminated graphene

(c)

Nanocone border

(b)

(a)

Nanocone

5 nm

Graphene sheets

Carbon atom   H atom

Figure 7.12 Schematics of nanocone structure and growth mechanism. (a) HRTEM micrograph of nanocone structure that shows the arrangement of graphene layers; (b) schematic of new graphene sheets formation on the nanocone top; (c) schematic of nanocone structure and hydrogen-terminated graphene layers [258].

stantially increase the total particle flux to the nanocone, as compared with neutral gas-based processes. Furthermore, carbon adatoms diffuse on the top surface and can pass to the lateral surface (Figure 7.13), Ψ_d of the nanocone. If the adatom surface density and their lifetime on the surface (which are controlled by the rates of adatom escape to the lateral surface or re-evaporation to the gas phase represented by the flux Ψ_e in Figure 7.13) are sufficient, the nucleus of a new graphene sheet can be formed. Moreover, adatoms from the flat top and substrate surfaces may diffuse about the lateral surface and attach to the borders of

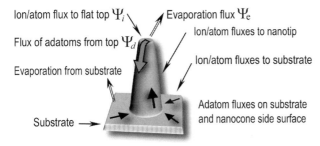

Ion/atom flux to flat top Ψ_i Evaporation flux Ψ_e

Ion/atom fluxes to nanotip

Flux of adatoms from top Ψ_d

Ion/atom fluxes to substrate

Evaporation from substrate

Substrate

Adatom fluxes on substrate
and nanocone side surface

Figure 7.13 Schematic of ion and adatom fluxes contributing to the nanocone growth. A new graphene sheet on the flat top surface is formed by direct carbon influx from a plasma and carbon outflow by evaporation and diffusional escape to the lateral surface of the nanocone [258].

carbon sheets, eventually increasing the nanocone's radius. Therefore, formation of new graphene sheets on the nanocone flat top appears to be the main process which is responsible for nanocone vertical growth at advanced growth stages [258].

Levchenko *et al.* [258] obtained the dependence of the radius of the nanocone flat top on the main process parameters, and related the numerical results to the experimental observations. They assumed that the surface diffusion flux from the flat top Ψ_d is determined by the adatom migration over the edges of graphene sheets. The flux balance on the flat surface of the tapered nanocone's top is described by

$$\frac{dn}{dt} = \Psi_i - \Psi_d - \Psi_e - \Psi_c,$$ (7.7)

where n is the adatom density on the surface, Ψ_i is the total incoming flux, and Ψ_c is the flux from the surface due to adatom collisions with the plasma ions, and all the fluxes are in units of $m^{-2}s-1$. The surface diffusion flux from the flat nanocone top is

$$\Psi_d = nVL/S = 2n\lambda v/r,$$

where $L = 2\pi r$ is the perimeter, $S = \pi r^2$ is the surface area, $v = v_0 \exp(-\varepsilon_d/ k_B T_s)$ is the rate of adatom jumps on a graphene sheet, $\varepsilon_d = 1.6\,eV$ is the surface diffusion activation energy [397], k_B is Boltzmann's constant, T_s is the surface temperature, and v_0 is the lattice atom oscillation frequency.

The evaporation flux from the flat surface is $\Psi_e = nv_0 \exp(-\varepsilon_a/k_B T_s)$, where ε_a is the surface evaporation energy. Furthermore, $\Psi_c = n\lambda^2\Psi_i$. The initial flux balance equation (7.7) can now be rewritten as

$$\frac{dn}{dt} = \Psi_i(1 - n\lambda^2) - \frac{2nk_B T_s}{h}\left(\frac{2\lambda}{r}\exp(-\varepsilon_d/k_B T_s) - \exp(-\varepsilon_a/k_B T_s)\right)$$
(7.8)

using the assumptions made above [258].

An assumption regarding the minimum adatom density necessary for the graphene sheet formation on the flat surface can be explicitly derived from the requirement that there should be at least two adatoms at a time on the nanocone flat surface. In this case, from Equation (7.8) one obtains

$$\Psi_i\left(1 - \frac{\lambda^2}{r_{cr}^2}\right) - \frac{2k_B T_s}{hr_{cr}^2}\left(\frac{2\lambda}{r_{cr}}\exp(-\varepsilon_d/k_B T_s) - \exp(-\varepsilon_a/k_B T_s)\right) = 0,$$
(7.9)

where r_{cr} is the critical radius of the nanocone.

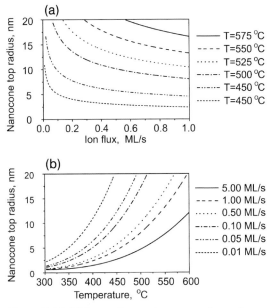

Figure 7.14 (a) Dependence of nanocone top radius on the ion flux with surface temperature as a parameter and (b) on the surface temperature with the ion flux as a parameter [258].

If the top radius of the nanocone is less than that described by Equation (7.9), the nanocone widens since no new graphene sheets can be formed on the top. Indeed, in this case all adatoms deposited onto the flat top may escape to the lateral surface and eventually attach to the hydrogen-terminated graphene borders. And vice versa, the nanocone shows a fast vertical growth when the top radius exceeds the value determined by Equation (7.9), which provides the formation of new layers.

Let us now consider the dependencies of the critical nanocone top radius on the process parameters. Figure 7.14 shows the results of the numerical solution of Equation (7.9), which quantify the dependence of the critical radius on the incoming ion flux (Figure 7.14(a)) and the substrate temperature (Figure 7.14(b)). From these graphs one can see that the critical radius decreases, and hence the nanocone becomes sharper, with increasing incoming flux and decreasing substrate temperature.

This is consistent with the experiments that show higher-aspect-ratio (better sharpness) in plasma processes that involve biased substrates and hence an increased ion influx to nanocone top surfaces [93, 140]. One should also point out the very good agreement of the calculated nanocone radii with the experimental observations presented in Figure 7.11 and previous reports [50, 140].

What is even more interesting is that the radii of top surfaces of various platelet-structured nanostructures such as ZnO nanocantilever (ca. 11 nm [418]) and carbon nanotips (ca. 10 nm) reported by other authors are well within the range predicted by the results in Figure 7.14. The temperature dependence of r_{cr} in Figure 7.14(b) is also consistent with low-temperature growth of thin nanofibers on biased substrates [419].

The results in Figure 7.14 can be used to explain the observed self-sharpening of tapered carbon nanocones to form nanoneedle-like structures. Indeed, as the VANs grow in height, their aspect ratio increases; leading to a local increase and stronger focusing of the electric field in the vicinity of the nanostructures. This in turn causes larger incoming fluxes of the plasma ions and leads to smaller nanocone top radii according to Figure 7.14(a). This process is self-organized and eventually results in the very sharp nanocones shown in Figure 7.11(b).

Furthermore, the temperature at the nanocone's flat top surface is very likely to increase due to ion bombardment and associated energy transfer. The effect of this local temperature increase is a larger surface area of the top platelet, as the results of Figure 7.14(b) suggest. In fact, a competition of the above two effects plays a major role in determining the equilibrium shape of single-crystalline platelet-structured carbon nanocones. It is also relevant to mention that the flat top surfaces of those nanocones in which we are interested are not primarily attributed to etching/sputtering effects caused by focused ion fluxes. Indeed, quite similar VANs can be formed in neutral gas-based processes with no ion- and electric field-related effects [418].

To conclude, we stress that the proposed growth mechanism [258] is applicable to a variety of other growth routes and platelet-structured single-crystalline vertical nanostructures such as nanorods and nanowires. In the following subsection we will present the results of numerical simulations which suggest that by steering the ion flux over the lateral surface of the nanotip, one can effectively reshape and eventually make it fit the requirements for electron microemitter array applications.

7.3.2
Plasma-Based Deterministic Shape Control in Nanotip Assembly

We will follow the original report [100] and describe the possibility of deterministic plasma-assisted reshaping of capped cylindrical seed nanotips by manipulating the plasma parameter-dependent sheath width. In turn, this makes it possible to steer the ion flux about the lateral surfaces of the nanotips and relate the microscopic distributions of the ion current density and the nanostructure growth process. More specifically, under the wide sheath conditions the nanotips widen at the base and when the sheath is narrow, they sharpen up. More importantly, by combining the wide- and narrow-sheath stages in a single process, it is possible to synthesize wide-base nanotips with long and narrow-apex spikes, ideal for electron microemitter applications.

Before we proceed, it should be stressed that the nanostructures of interest here will be quite similar to what was considered in the previous section. However, the ultimate aim will be to create very sharp nanoneedle-like structures, yet wide enough at their bases to be stable on the substrate. For simplicity, we will refer to such objects as nanotips, to be consistent with the original report [100]. It is important to mention that such nanostructures and their arrays have a number of unique and tunable structural and electronic properties and possess outstanding flexibility with regard to functionalization and eventual nanodevice integration [420]. Ordered arrays of C, Si, W, WO_3, GaAs, GaP, and Al nanotips with different shapes and capping/functional overcoats have been successfully syntheisized and tested in various applications [421–425]. In particular, nanotips can be used in non-volatile data storage elements, interconnects in nanoelectronic integrated circuits, electron emitting and lasing optoelectronic functionalities, nanoplasmonic and photonic devices, biosensors, bioscaffolds, protein and cell immobilization arrays and some others [4,420,426]. The major issues that still await their solutions are related to deterministic (highly-controllable and predictable) nanotip synthesis and nanodevice integration [100].

As we have already stressed several times in the introductory chapters, it is crucial both to select the most suitable nanofabrication process and to optimize the synthesis parameters to achieve the desired *size* and *shape* (which in turn determine the electronic and some other properties) and also position individual nanotips in the specified device locations. The success of this endeavor critically depends on the nanoassembly technique. Currently, neutral gas routes (NGRs), such as various modifications of the chemical vapor deposition (CVD), molecular beam epitaxy, cluster beam deposition are among the preferred fabrication methods. However, the degree of shape tunability still remains below the expectations of the as yet elusive deterministic nanofabrication. For example, CVD-synthesized conical or pyramid-like carbon nanotips frequently appear short and wide, and also lack vertical alignment. This compromises their applications as electron microemitters, which, in particular, demand vertically aligned, sharp and high-aspect-ratio nanostructures [253].

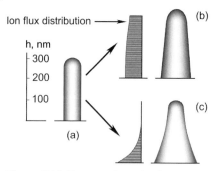

Figure 7.15 Two nanotip reshaping processes: (a) original (seed) nanotip, (b) $T_e = 2.0\,eV$, $U_s = 20\,V$, $n_p = 1.5 \times 10^{18}\,m^{-3}$, (c) $T_e = 2.0\,eV$, $U_s = 50\,V$, $n_p = 4.5 \times 10^{17}\,m^{-3}$ [100].

The existing reports suggest that by using a higher-complexity, plasma-enhanced CVD (PECVD) one can dramatically improve nanotip vertical alignment in a quite similar way to carbon nanotubes and nanofibers [90] (see also Section 7.1). Tam *et al.* [100] conducted a numerical experiment which involved a multiscale hybrid Monte Carlo (gas phase) and self-organization of adatoms on the surface. In particular, it was shown that by manipulating the width of the plasma sheath one can effectively tune the shape, size and features (such as the apex angle, capping, and base radii) of the nanotips. Furthermore, by applying a sequence of unipolar bias voltages to the growth substrate, it appears possible to synthesize a conical convex-shaped nanoassembly with a base width of ca. 150 nm, height of ca. 1 µm, and an apex angle of only 2–3°, a perfectly shaped nanotip that fits the requirements for the optimized electron field emission [100,253].

Multiscale numerical simulations incorporated three physical/numerical models [100] (see also Section 3.3 and Table 3.1 for more details of the spatial scales involved and numerical modules used): (i) microscopic ion flux topography in the immediate vicinity of the substrate surface and nanotip lateral surfaces (ion motion model) based on the MC technique; (ii) self-organization of adsorbed atoms (adatoms) on the substrate surface not covered by the nanotips (surface conditions-controlled adatom diffusion); and (iii) model of the nanotip growth.

The calculated shapes of the nanotips for the two different process conditions (electron temperature $T_e = 2.0\,eV$, substrate bias $U_s = 20\,V$, plasma density $n_p = 1.5 \times 10^{18}\,m^{-3}$ and $T_e = 2.0\,eV$, $U_s = 50\,V$, and $n_p = 4.5 \times 10^{17}\,m^{-3}$) are shown in Figure 7.15. It is seen that the denser plasma case features a quasi-uniform distribution of the ion flux along the nanotip height (Figure 7.15(a)). Complemented by the diffusion influx of adatoms to the nanotip border, this leads to the wide tip formation with a rounded cap and a large apex angle. The second numerical experiment, conducted in the plasma of lower density, provides an increased influx of the ions to the nanotip base and thus leads to the formation of the capped tip with a very wide base as can be seen in Figure 7.15(b).

We will now discuss how the sheath width can be controlled and affects the nanotip shape. The width of the plasma sheath depends on the bias voltage, plasma temperature and plasma density and is the main factor in determining the microscopic topology of the ion flux (see Section 5.1). Ions entering the sheath with a finite velocity (which can be calculated taking into account that the potential drop in the presheath is $\sim T_e/2$) begin to accelerate towards the substrate in the direction normal to the substrate.

If the sheath is large compared with the mean nanotip height, the ions acquire the energy corresponding to the sheath potential drop in the upper layer of the sheath where the influence of the electric field produced by the individual nanotips is weak. As a result, the ions will acquire an almost total sheath energy ($\sim U_s$) when entering the irregular electrical field above the nanotip pattern, and will not deflect in the local fields of the individual nanotips.

On the other hand, under the narrow sheath conditions, the ions will have a lower energy when approaching the nanotip surface, and the electric field created by the nanopattern will deflect low-energetic ions. As a result, in the wide sheath case the ion trajectories are mostly straight lines, and the ions land on the top of nanostructures or hit the substrate without colliding with the lateral nanotip surfaces. In the narrow sheath case the ion trajectories are curved, and a significant amount of ions incorporate into the growing nanostructures via lateral surfaces. This difference in ion energies and trajectories is decisive in the nanotip shape control [100].

Thus, by appropriately manipulating the plasma parameters, one can effectively control the nanotip shape. However, as can be seen from Figure 7.15, none of the growth processes (with either a wide or a narrow sheath) produces the optimum nanotip shape (with a wide base and a sharp tip) for microemitter applications. Indeed, either the base is not wide enough (Figure 7.15(b)) or the tip is not sharpened (Figure 7.15(c)). A possible solution for this problem is to combine the above two processes: the first one with a wider sheath to form a wide base first, and then, by narrowing the sheath, shape up a thin and sharp top (Figure 7.16).

The process depicted in Figure 7.15 is an example of sophisticated deterministic nanotip shape control in plasma-aided nanofabrication.

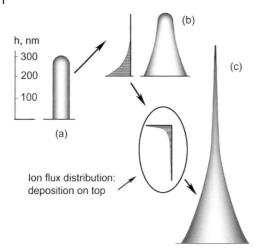

Figure 7.16 Reshaping nanotips in a two-stage process: (a) original nanotip; (b) formation of the nanotip base ($T_e = 2.0\,eV$, $U_s = 50\,V$, $n_p = 4.5 \times 10^{17}\,m^{-3}$); (c) formation of the emissive spike ($T_e = 2.0\,eV$, $U_s = 20\,V$, $n_p = 10^{17}\,m^{-3}$) [100].

Combining two stages that form first a wide base, and then a thin low-apex top, one can obtain an optimal nanotip microemitter structure [253] with a low electrical resistance, high mechanical strength, rigidity, and a very high electron emission current from the emission spike (Figure 7.15(c)).

During the first stage, the process is carried on at a substrate bias voltage of 50 V, electron temperature of 2.0 eV, and the plasma density $4.5 \times 10^{17}\,m^{-3}$. In this case, the focussing of the ions by the nanopattern is weak, and the nanotips mainly grow due to the diffusion fluxes over the substrate surface and by direct ion incorporation into the nanotip base.

At the second stage ($T_e = 2.0\,eV$, $U_s = 20\,V$, $n_p = 10^{17}\,m^{-3}$), narrow-sheath conditions cause a strong focusing of the ions to the upper part of the nanotips close to the top. As a result, a high and narrow spike grows atop of the wide-based nanotip shown in Figure 7.15(c).

Therefore, by appropriately manipulating the plasma process conditions (such as the electron temperature, plasma density and the substrate bias which control the density and energy of ion fluxes), one can effectively control the shape of nanotips and tailor their shape according to the specific requirements in a particular application. However, one should always keep in mind the requirements not only for individual nanostructures but also for their arrays and networks. Such requirements and the ways to achieve them by using plasma-based techniques will be discussed in the following section.

7.4
Self-Organization of Large Nanotip Arrays:
Why Using Plasmas is Better Than Neutral Gas Routes?

In this section we will show that, owing to selective delivery of ionic and neutral building blocks directly from the ionized gas phase and via surface migration (see Figure 3.13), plasma environments offer a higher degree of deterministic synthesis of ordered nanoassemblies compared to thermal chemical vapor deposition. The results of hybrid Monte Carlo (gas phase) and adatom self-organization (surface) simulation by Levchenko *et al.* [52] suggest that higher-aspect-ratios and better size and pattern uniformity of carbon nanotip microemitters can be achieved via the plasma route.

This time we begin by recalling the essence of the concept of deterministic nanoassembly which is both a current demand and the ultimate crux of modern nanoscience and nanotechnology (see Chapter 1). At the macroscopic level, this implies the ability to adequately select and adjust the process parameters to achieve the desired properties of individual nanoassemblies (NAs), such as their positioning, alignment, shape, elemental composition, crystallinity, and so on [427–429].

At the microscopic level, determinism implies a certain degree of control over the building units that self-assemble into the required nanoassemblies and optimization of elementary processes in the nanofabrication environment [4]. Therefore, the choice of the most favorable environment, which should be dictated by the desired parameters of the nanoassembly, turns out to be a critical factor to reduce process costs and achieve the long-held but as yet elusive goal of deterministic nanofabrication [52].

In their article [52], Levchenko *et al.* used the microscopic-level viewpoint and argued that partially ionized environments of the plasma-enhanced chemical vapor deposition (PECVD) can offer a better deal of controlling the size, shape, and pattern uniformity in deterministic synthesis of selected nanoassemblies, compared to charge-neutral thermal CVD. The main focus of their work was on arrays of conical carbon nanotip microemitter structures (representative scanning electron micrograph images of these nanostructures are shown in Figures 3.9(b) and 7.11) which ideally should have the highest possible aspect (height to width) ratio for higher electron emission yield [253].

A hybrid Monte Carlo (gas phase) and adatom self-organization (surface) simulation were used to demonstrate that the ionized gas environment is decisive in sustaining the growth of tall and sharp nanotip structures as opposed to the short and wide nanotips grown by the CVD under the same process conditions [52]. These multi-scale numerical simulations include the motion of neutral and ionic building units in the partially-ionized gas phase and the growth of the nanotips by adatom and adion insertion via surface migration and directly from the gas phase. This process is sketched in Figure 3.13. In this scenario, developing conical nanostructures selectively focus ionic building units onto their lateral surfaces effectively excluding them from migration over open substrate areas. Ultimately, this results in faster growth rates and eventually in the sharper and longer nanotips grown by the PECVD [52].

The model and numerical approach in this case uses the same modules as in Section 7.3.2, with the only major difference in the nanostructure growth module, in which all nanotips within the array were allocated unique numbers and were treated separately, accounting for incoming ion fluxes and adatom fluxes over the surface. In this model of the nanotip growth on nickel-catalyzed Si(100) surface, the growth simulation starts from a small nanotip nucleus.

It is also assumed that the outer carbon layers of the growing nanotips are able to accommodate insertion of adatoms arriving to the nanotip base from the open surface areas and adions landing directly onto the lateral surfaces from the ionized gas phase. This simple model is an adequate representation of the dynamic growth of various carbon nanofilms and nanostructures that involve hydrogen-terminated surfaces. For more datails of hydrogen termination of these carbon nanocone-like structures, please refer to Section 7.3.1 and also to Sections 4.3 and 4.4 of the related monograph [1].

The surface self-organization module incorporates the following processes: surface diffusion of adsorbed species to the nanotips, evaporation of adatoms from nanotips to the gas phase and to the two-dimensional vapor, and attachment of adatoms to the nanotip borders. This module also includes a dynamic growth and reshaping of the nanotips, which is described by

$$(\partial V_n / \partial r_{0n}) dr_{0n} = J_{sn} dt \tag{7.10}$$
$$(\partial V_n / \partial h_n) dh_n = J_{en} dt, \tag{7.11}$$

where V_n, r_{0n}, and h_n are the volume, base radius, and height of n-th nanotip, respectively. Here, J_{en} is the combined flux of ions and neutrals from the plasma bulk to the n-th nanotip lateral surface, J_{sn} is the total surface flux of adatoms to the nanotip's border, and

$$\partial V_n / \partial r_{on} = (2/3)\pi r_{0n} h_n$$

and

$$\partial V_n / \partial h_n = (1/3)\pi r^2$$

are the shape- and size-dependent nanotip growth functions in the radial and vertical directions, respectively.

It was thus assumed that the influx of adsorbed species to the border of each individual nanotip causes an increase in its radius, whereas the direct influx from the ionized gas phase leads to an increase in the nanotip's height. The model also implicitly involves adatoms that diffuse through the catalyst bulk by assuming the rate of their delivery and insertion into the nanotips is the same as that of the surface-migrating adatoms; the corresponding fluxes are included into the adatom fluxes onto nanotips from open surface areas J_{sn}. It is further assumed that the surface fluxes J_{sn} are stationary and are sustained under conditions when equilibrium between the species deposition and their removal due to the reactive chemical etching and physical sputtering (not considered in detail here) is established [52].

The simulation starts from the pre-set pattern of 400 nanotip nuclei covering the simulation area of $S = 1\,\mu m \times 1\,\mu m$. As the height and radius of the nuclei increase, they reshape to the conical nanotips. In computations, the following set of parameters was used: plasma density $n_p = 10^{17}$–$3 \times 10^{18}\,m^{-3}$, electron energy 2.0–5.0 eV, bias voltage $U_S = 20$–50 V, surface temperature $T_S = 750\,K$, gas temperature $T_G = 1000\,K$, gas pressure $P_G = 1\,Pa$. This set of parameters is representative of PECVD of carbon nanotip structures in RF plasmas [50,93,94,140].

The microscopic topography of the ion flux on open surface areas and nanotip lateral surfaces was simulated by a Monte Carlo method similar to Section 5.1. An initial surface coverage μ_0 was 0.1, a typical value in the low surface coverage case. The simulations were terminated at higher surface coverages when nanotip coalescence becomes unavoidable. An initial distribution function of the nanotip base radii was chosen Gaussian. Thus, the results obtained from the multi-scale simulation of the building unit dynamics in the gas phase and on the surface served as the input conditions required for detailed simulation of the nanotip growth kinetics by integrating Equations (7.10) and (7.11) for each individual (n-th) nanotip from the entire nanopattern.

A representative three-dimensional distribution of the ion current density on the nanostructured substrate surface (however, from a different array than published in Figure 2 of the original report [52]) is shown

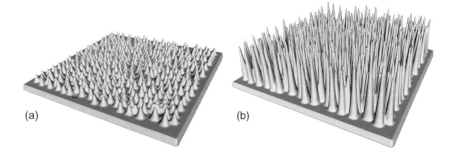

Figure 7.17 Developed carbon nanotip patterns (a) grown by CVD and (b) by PECVD in a plasma with density $3.0 \times 10^{18}\,\text{m}^{-3}$. In both cases the numerical simulation was terminated when the surface coverage reached 70% [52].

in Figure 2.11. The positions of individual carbon nanotips can be easily identified as well-pronounced sharp ion current peaks, surrounded by the significantly reduced background ion flux onto open surface areas. Evidently, the simulated 3D ion current distribution suggests an enhanced influx of the building units directly to the nanotip lateral surfaces, without allowing them to deposit on open surface areas and migrate to nanotip borders over the surface. Furthermore, such a strongly focused microscopic pattern of the ion deposition enables one to control the growth rates and aspect ratios of the nanostructures [52].

A striking observation of numerical experiments of Levchenko *et al.* [52] is depicted in Figure 7.17, which suggests that *the nanotips grown on plasma-exposed surfaces (Figure 7.17(b)) are much taller and sharper than those grown by the CVD process (Figure 7.17(a)) under the same deposition conditions.* To quantify this main conclusion, the dependence of the mean nanotip apex angle α and mean nanotip height h_m on the mean nanotip radius (Figure 7.18) with the plasma density as a parameter was studied.

Another important feature of the development of the nanotip array is the essentially different behavior of the apex angle at the initial and developed growth stages. Specifically, an initial increase of the apex angle is followed by its gradual decrease resulting in nanotip sharpening with time. This self-sharpening phenomenon is quite similar to what was discussed in Section 7.3.1 and may be explained as follows [52].

When the nanotip height is small, the non-uniformity of the electric field is too weak to focus the ion current onto the nanotip lateral surfaces, and so the ions are predominantly deposited to open surface areas where they are neutralized and become adatoms. In this case the nanotip

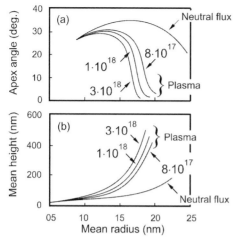

Figure 7.18 Dependence of the nanotip apex angle (a) and mean height (b) on their mean radius for the neutral and plasma-aided processes with the plasma density (m^{-3}) as a parameter [52].

growth is predominantly maintained by the adatom BUs that migrate over the surface to nanotip borders.

As a result, the nanotip base radius increases rapidly, with the height increasing slowly. When the nanotips become taller, the ion current is focused by non-uniform electric fields and thus is increasingly diverted to their lateral surfaces. This causes, in turn, a noticeable decrease in the apex angle. This effect turns out to be more pronounced in denser plasmas, and the sharpest nanotips have been observed for $n_p = 3 \times 10^{18}$ m^{-3} [52].

Indeed, higher plasma densities result in better precipitation of the ionic BUs onto the nanotip surfaces. This also results in smaller sheath widths and more focused ion deposition onto the upper sections of the nanostructures, closer to their crests. Furthermore, one can achieve aspect ratios of more than 30 by using the PECVD route. In comparison, the neutral gas-based process can offer nanotip aspect ratios below 20 under the same conditions. Moreover, dynamic changes of the nanotip aspect ratios turn out to be more pronounced in the plasma-aided process. In other words, the chances of an initial nanotip nucleus evolving as a sharp, high-aspect-ratio nanocone is much higher on plasma-exposed surfaces. These results further support the conclusions of Section 7.3.1 obtained from a quite different viewpoint.

Another very important conclusion is that the *plasma-based process makes the nanotip array more uniform in the substrate plane by significantly improving the uniformity of nanotip diameters*. Figure 7.19 shows the square

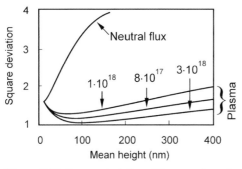

Figure 7.19 Square deviation of the nanotip radius distribution function on their mean height for PECVD and CVD processes with the plasma density (m^{-3}) as a parameter [52].

deviation $\sigma(R)$ of nanotip radii from the mean radius as a function of the nanotip mean height and reflects the main features of the nanotip radii distribution function (NRDF).

It can clearly be seen that the distribution of radii of the nanotips synthesized by the plasma-aided process is much more uniform compared to the CVD process. This difference is most pronounced for mean nanotip heights exceeding 100 nm. In this case $\sigma_{PECVD}(R)$ is almost 10 times smaller than $\sigma_{CVD}(R)$. For more discussions on the reasons for the improved size uniformity of nanotips in the surface plane please refer to the second report of the same group [173].

Self-Organization of Large Nanotip Arrays It is noteworthy that the above "macroscopic" model does not include atomic forces or interactions of individual atoms and is based on physical evaporation of adions and adatoms in and out of the growing conical nanotip structures. The model builds on the established, commonly used, well proven and justified principles and approaches of surface science to surface diffusion phenomena and island nucleation and growth [277,278,430]. These models have been advanced by involving individual treatments of the growth process of 400 "macroscopic" nanotips (each containing approximately 1.5–2 million atoms) arranged in an array on the plasma-exposed surface.

Here we stress that these hybrid multiscale simulations bridged processes occurring at length scales different by several orders of magnitude and involved a very large number of atoms and ions, and so are far beyond the capabilities of the present-day *ab initio* atomic-level numerical techniques, such as the Molecular Dynamics (MD) of Density Functional Theory (DFT) approaches. Therefore, atomistic models would only offer a better deal of accuracy when considering the growth of individual nanostructures with a substantially reduced number (typically not exceeding a few hundred) of atoms.

We emphasize that a number of recent experimental and computational results corroborate the fidelity of the chosen nanotip growth model. These include:

- SEM analysis of the nanotip shapes at different growth stages (see, e.g., Figure 7.11 and References [50,140]);

- experimental evidence of the nanotip sharpening when a DC bias is applied to the substrate (see Section 7.3.1 and original report [258] for more details);

- Transmission Electron Microscopy (TEM) of carbon nanotips composed of parallel graphite layers and terminated by hydrogen on lateral surfaces and made of stacked conical sheets (Figure 7.11). These show that the nanotips are crystalline and fully filled by carbon atoms and have apex angles (ca. 5–9 degrees) and geometrical sizes very similar to the results of the above numerical simulations;

- Results of *ab initio* DFT computations showing that (substantially downscaled) nanotips made of parallel graphite layers and terminated by hydrogen on lateral surfaces represent a stable atomic configuration (Reference [1], Section 4.4).

In the following section we will discuss an even more amazing bit of evidence that plasma environments can even contribute to the creation of uniform arrays of carbon nanocones from non-uniformly fragmented catalyst films. Even more interesting is that some of the processes considered in Sections 7.1–7.4 will take part in a more complex three-dimensional self-organized nanoarray development on plasma-exposed surfaces. This self-organization is powered by the higher complexity and unique features of the low-temperature plasma process environment.

7.5
From Non-Uniform Catalyst Islands to Uniform Nanoarrays: Plasma-Directed Three-Dimensional Self-Organization in Large Nanotip Arrays

In this section we will follow the original report [95] and consider the self-organized growth of uniform carbon nanocone arrays using low-temperature non-equilibrium $Ar + H_2 + CH_4$ plasma-enhanced chemical vapor deposition (PECVD). The experiment of Tsakadze *et al.* [95] shows that size-, shape-, and position-uniform carbon nanocone arrays can develop even from non-uniformly fragmented discontinuous nickel catalyst films.

Here we will consider the three-stage scenario by which the primary nanocones grow on large catalyst particles during the first stage and the secondary nanocones are formed between the primary ones at the second stage. During the last stage, plasma-related effects lead to preferential growth of the secondary nanocones and eventually a uniform nanopattern is formed. More importantly, this does not happen in a neutral gas-based CVD process with the same gas feedstock and surface temperature. The proposed three-stage growth scenario is supported by the numerical experiment which generates nanocone arrays very similar to the experimentally synthesized nanopatterns [95].

This self-organization process can be explained in terms of re-distribution of surface and volumetric fluxes of plasma-generated species in a developing nanocone array. One of the main conclusions of the original report [95] is that *plasma-related self-organization effects can significantly reduce the non-uniformity of carbon nanostructure arrays which commonly arises from imperfections in fragmented Ni-based catalyst films*.

Before we proceed with a description of the relevant experiments and theoretical interpretations, we will highlight the significance of the carbon nanostructures of our interest in various applications. The unique electrical, magnetic, optical and mechanical properties of nano-carbons including microporous carbons and vertically aligned carbon nanostuctures such as nanocones, nanorods, nanotubes, nanotips, and nanofibers make them fascinating and attractive structures for a variety of potential applications such as electron-emitting panels, polymer-carbon reinforced composites, sensors, supercapacitors and various nanoelectronic devices [53, 431–445]. Here we recall that one of the ultimate goals of plasma-aided nanofabrication is to achieve a reasonable level of control and predictability in the ordering, size, and architecture of such nanostructures [4,95] (see also Chapters 1–3 of this monograph).

As already mentioned above, catalyzed chemical vapor deposition is a popular method for assembling vertically aligned carbon nanostuctures (VACNs). Nevertheless, a substantial lack of controllability in the size, position and shape of the VACNs assembled by the CVD method has stimulated research into alternative methods of fabrication such as PECVD and other plasma-based techniques. As we have seen in Chapters 2 and 3 and previous sections of this chapter, plasma-enhanced CVD (PECVD) has an outstanding ability to produce high-quality vertically-aligned nanostructures [93,446,447].

In addition to precursor dissociation on the surfaces of catalyst particles, common to CVD, the plasma-based approach also involves dissociation of carbon-bearing precursors into multiple reactive species that incorporate into the developing nanostructures directly from the plasma. Furthermore, by applying a negative bias to the substrate, one can achieve an excellent ordering of the carbon nanostructures in the preferential growth direction (see Section 7.1 for more details). Another amazing plasma-related possibility is that self-assembled carbon nanostructure arrays can be grown without any external heating of the substrates, as they are heated internally by intense substrate bias-controlled ion fluxes [94] (see also Chapter 4 of the related monograph [1]).

The growth of vertically aligned nanostructures on a catalyzed solid (e.g., Ni-catalyzed highly-doped n-Si(100)) surface usually exhibits very complex behavior, which is still far from being completely understood despite the large number of experimental and theoretical works published on this topic. Apart from the deposition conditions such as deposition rate and degree of ionization, the surface conditions such as surface temperature, catalyst fragmentation and surface roughness also strongly influence the parameters of the final nanostructure array. In particular, it is commonly believed that uniform fragmentation of an initially continuous Ni catalyst layer into a discontinuous islanded film is an essential prerequisite in the fabrication of highly-uniform arrays of vertically aligned carbon nanostructures [95].

However, this is not always the case. In many cases catalyst layers turn into very non-uniformly fragmented nanoislanded films, quite similar to the case of stress-driven Stranski–Krashtanov fragmentation discussed in Section 6.1. This immediately prompts a question: If the surface pattern of catalyst islands is non-uniform (both in sizes and positions on the substrate), would it still be possible to grow size- and position-uniform nanotip arrays?

The main conclusion of the experimental and computational work of Tsakadze *et al.* [95] is a definite yes: if a low-temperature, thermally non-equilibrium plasma is used. Tsakadze *et al.* [95] also proposed a complex mechanism for the self-organized growth of uniform carbon nanocone patterns on a nickel-catalyzed n-Si(100) substrate. Moreover, it was shown that plasma-related effects can substantially improve the uniformity of the nanostructure arrays grown on essentially non-uniformly-fragmented Ni catalyst film.

7.5.1
Experiment and Film Characterization

Let us briefly summarize the details of the experiment of Tsakadze *et al.* [95]. The nanostructures were grown in a low-frequency inductively coupled plasma (ICP) reactor with an external coil configuration described in detail elsewhere [189] (see also Chapter 2 of the related monograph [1]). A schematic diagram of the experimental setup used for the plasma-assisted synthesis of carbon nanocones is shown in Figure 4.1. A lightly-doped n-Si(100) wafer coated with a pre-deposited 30-nm-thick Ni-based catalyst layer was used as a substrate. The wafer was placed on the top surface of a DC-biased substrate stage positioned in the area of the maximal electron/ion density in the plasma reactor. Initially, the plasma chamber was evacuated to a base pressure of $\sim$ 0.0013 Pa (10^{-5} Torr). The base pressure was monitored by Pirani and Penning gauges. The gas pressure inside the chamber was controlled by a capacitance manometer which was connected to a power supply digital readout. A capacitance L-type network was used to optimize the RF power transfer to the plasma source.

Reactive gases Ar, H_2, and CH_4 were introduced into the chamber sequentially. At first, argon was introduced to ignite the plasma and condition the substrate surfaces. After a 30 min-long substrate conditioning in argon, hydrogen was added for 20 min catalyst activation in a $Ar + H_2$ mixture followed by the introduction of high-purity methane into the chamber. In the growth regime of interest here, termed the floating temperature growth (FTG) regime [94], the substrates were heated internally by the hot working gas and intense ion fluxes controlled by the DC bias applied to the substrate. The total pressure of the gas mixture was maintained at 5.999 Pa (45 mTorr) and the substrate temperature was 500 °C [95]. The high-density ($n_{e,i}$ ca. 10^{12} cm^{-3}, where $n_{e,i}$ is the electron/ion density) plasma was sustained with RF power densities of 0.09–0.11 W/cm^2, while $V = -300$ V DC bias was applied to the substrate stage. Under such conditions, the high-density, vertically aligned nanocone-like structures were developed.

The crystal structure of the films was analyzed using an X-ray diffractometer (XRD) operated in a 2Θ mode, wherein the incident x-ray wavelength is 1.5405 Å (Cu K$_\alpha$ line). Raman spectra of nanostructured carbon films were acquired at room temperature. The excitation wavelength of the Raman spectrograph was 514 nm, with the spot size being approximately 2 square micrometers. High-resolution scanning electron imaging was conducted with the aid of a field-emission scanning electron microscope. More details on the experimental facility, process parameters, and material characterization can be found in earlier reports of the same group [93–95,189].

We will now consider the self-organized growth of the carbon nanocone arrays [95]. To do this, we will first describe the experimental observations on how such arrays develop and then elaborate on their growth kinetics and compare the latter to the non-uniform nickel catalyst fragmentation at the preliminary catalyst activation stage. The main accent

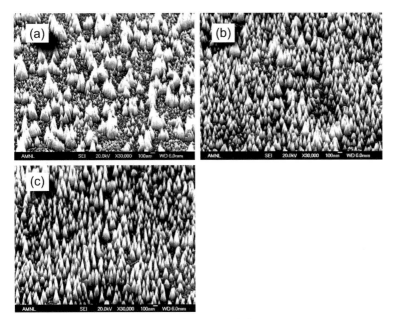

Figure 7.20 FE SEM images for different growth times:
(a) 5 min; (b) 10 min; (c) 20 min [95].

will be on the self-organization-related features of the formation of uniform arrays of the carbon nanocones on non-uniform catalyst pattern.

Scanning Electron Microscopy images of nanocone-like structures
grown under floating substrate temperature conditions are shown in Figure 7.20 for three consecutive times: 5 (initial stage of growth), 10, and
20 min (final structure) [95]. The main feature noticeable in this figure is
the presence of two very different types of carbon nanocones. The first
type, which are large structures called primary nanocones, have heights
up to $h_P = 800$ nm and base radii up to $r_P = 300$ nm (Figure 7.20(a)). One
can see that the primary nanocone array is very non-uniform, with height
dispersion reaching 300%. In this case the dispersion of the nanocone
base radii may reach up to 500%.

In addition, some primary nanocones have a specific shape resembling
aggregates of several closely-located (contacting) nanocones. The surface
area free of primary nanocones is much larger than the covered area;
thus, the surface coverage with primary nanocones μ_P (hereinafter referred to as the primary surface coverage) is approximately 0.25 [95].

From Figure 7.20 one can also notice that between the large nanocones
there is a dense pattern of very small secondary nanocones which are
typically ten times shorter than the primary ones. The small secondary
nanocones have height h_S and base radius r_S of approximately 8–12 nm

and cover nearly the entire surface of the substrate not yet covered with the primary nanostructures. The surface coverage by the secondary nanostructures (secondary coverage) is thus $\mu_S \approx 1 - \mu_P$. The difference between the two patterns is well pronounced, one can thus confidently state the presence of two nanocone patterns with distinctive shapes.

As can be seen in Figure 7.20(b), increasing the growth time to 10 min changes the nanocone array completely. It is seen that the two (primary and secondary) arrays equalize and transform into a pattern which inherits some features of both the primary and secondary arrays: the nanocone height reaches 800 nm and the final surface coverage μ_F approaches unity.

A further increase of the deposition time to 20 min results in the development of the final pattern shown in Figure 7.20(c). Interestingly, in the final array the nanocone height h_F does not change significantly compared with the 5 min array. However, the size- and shape uniformity of carbon nanocones improve substantially as can be seen in Figure 7.20.

One can also observe that the base radii of the nanostructures reduce to ≈ 40 nm; moreover, the aggregates of several contacting nanocones are absent in the developed nanopattern. Figure 7.20 also suggests that the nanocones are uniformly distributed over the substrate. In this case the distributions of the nanocone heights and base radii appear to be very uniform, with the dispersion not exceeding 25%.

Even more important is that the pattern exhibits a high order and a high density over the surface. It is also worth noting that, quite similar structures of carbon nanocones and nanowires can be obtained using a combination of DC plasma deposition and lithographically prepatterned catalyst [95].

Raman spectra of the samples synthesized in processes of different durations feature the D band at ca. 1351 cm^{-1} which is due to the presence of amorphous carbon or defects, and the G band at ca. 1592 cm^{-1} which is a characteristic of longitudinal vibrations along the graphite lattice [184]. It is commonly accepted [448] that the ratio of the intensities of G and D bands quantifies the relative disorder of the graphitic material in the specimen. Interestingly, after 5 min of deposition the intensity of the D peak was higher than that of the G peak, indicating that the specimen contained a substantial amount of defects. However, at a longer nanostructure growth time of 20 min, the ratio of the G and D bands (for simplicity, referred to as the G:D ratio) increased [95]. This demonstrates an increase in the graphite crystallite size and a decrease in the amount of unorganized carbon in the samples [449].

From the XRD spectra (Figure 7.21) one can observe the evolution of

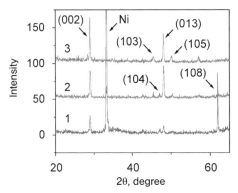

Figure 7.21 XRD spectra for different growth times:
1) 5 min; 2) 10 min; 3) 20 min [95].

the crystalline structure of the nanostructured carbon surface. In particular, one can notice the presence of the amorphous carbon phase after 5 min of deposition (curve 1). With increasing deposition time, the amount of the amorphous phase decreases, giving rise to a crystalline structure. Moreover, some new peaks appear in the spectra. From the XRD spectra, we can conclude that the intensity of the peak at $2\Theta = 29°$ corresponding to (002) increases, reaching its maximum value in specimens synthesized in a 20 min deposition process.

As suggested by Tsakadze *et al.* [95], this can be attributed to the fact that the sample grown in a 20 min process exhibits the highest crystallinity and uniformity of the vertically aligned nanostructures. Here we note that the peak at $2\Theta = 34°$ is attributed to the Ni catalyst. Other important peaks (103), (104), (105), (108), and (013) are also marked on Figure 7.21. It can be noted that (013) peak is weaker than (002) after 5 minutes of deposition, after 10 minutes these peaks are approximately equal, and then the (002) peak becomes the highest carbon-related peak. This is a good indication of the development of vertically oriented crystalline structures in the final pattern deposited for 20 minutes.

It also needs to be stressed that Tsakadze *et al.* [95] observed a striking similarity of the islanded catalyst film (see inset in Figure 7.22) to the primary nanocone array shown in Figure 7.20(a). In this case, the surface coverage of the catalyst islands (approximately 0.25) also corresponds to that of the primary nanocone array.

7.5.2
Growth Model and Numerical Simulations

Let us now recall the most striking observation of Tsakadze *et al.* [95]; namely, nanocone arrays in a low-temperature plasma exhibit very com-

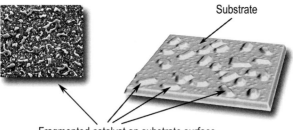

Substrate

Fragmented catalyst on substrate surface

Figure 7.22 Simulated pattern of fragmented Ni catalyst film on Si substrate surface. The inset shows the corresponding FE SEM micrograph of the fragmented Ni catalyst film obtained in the experiment after 20 min-long treatment in Ar + H_2 plasmas [95].

plex behavior, including the formation of two very different patterns of large (primary) and small (secondary) nanostructures. This is followed by the pronounced equalization and transformation into a single array of narrower and more uniformly distributed nanocones. Indeed, their results suggest a strong similarity of the islanded catalyst film with the primary nanocone pattern and a major difference from the final pattern topography. More importantly, *uniform nanocone arrays can develop from an essentially non-uniformly fragmented Ni catalyst*! What is more, a similar phenomenon was not observed in a CVD process at very similar process conditions [95].

We will now discuss the growth scenario of the nanocone array proposed by Tsakadze *et al.* [95]. It was assumed that the nanocones grow on the Ni catalyst particles and their growth shape is formed due to the carbon supply both from the substrate surface and directly from the plasma to nanocone surface, which is quite similar to what happens in the case of carbon nanotubes of Section 7.1 (see also Figure 7.13). The growth scenario of interest here includes one preliminary and three main stages. During the preliminary stage, a Ni-based catalyst film is fragmented into separated islands. Then, during stage 1 (the primary nanocone growth stage), an array of primary nanocones is formed on the large catalyst islands, which grow due to the carbon adatom supply from the substrate surface to the nanocones through the metal catalyst, as well as through the deposition of ions and atoms delivered from the plasma to the nanocone surfaces, similar to Section 7.4.

It is important to mention that in a plasma-based process the solubility of carbon in a Ni catalyst is very high [255], in particular, due to surface heating and activation via intense ion bombardment. This is yet another benefit of plasma-based synthesis of nanostructures via transition metal-catalyzed growth. Therefore, carbon adatoms penetrate relatively easily into the catalyst, diffuse through it and finally incorporate into growing nanocones. At this stage, the density of carbon adatoms on the substrate surface is very low due to the very intense outflow of adatoms to the large catalyst islands not yet covered by nanostructures. Moreover, small catalyst islands are depleted with the carbon due to the high equilibrium density of adatoms at their borders, and thus they cannot sustain the growth of nanocones on their surface [95].

Subsequently, the primary nanocones grow, their height and base radius increase and finally the large catalyst islands become completely covered with nanocones. At this moment, the whole situation on the substrate changes completely, and the second stage (the direct growth stage) comes into play [95]. At this stage, the nanocone walls completely cover the edges of large catalyst islands and contact with the substrate surface. Therefore, carbon adatoms can no longer penetrate into large metal catalyst islands and are thus unable to incorporate into the primary nanocone structure through the catalyst. Instead, carbon adatoms are forced to diffuse about the nanocone surfaces as sketched in Figure 7.13. Hence, incorporation of carbon material into the nanocone structure is retarded and adatoms are set to migrate back from the nanocone walls to the silicon substrate surface.

This leads to a rapid increase in the carbon adatom density on the surface. As a result, the small catalyst islands (visible in Figure 7.22 (picture on the left) between larger ones) saturate with the carbon and give rise to the growth of secondary nanocones on their surfaces. Thus, a secondary array of small nanocones is created. We stress that the growth rate of the small nanocones is higher compared to the primary nanocones. Therefore, the two nanocone arrays (primary and secondary) tend to equalize. Finally, the primary nanocones that were still growing during Stage 2 reach the height at which ion focusing by microscopic electric fields becomes significant. Under the surface biasing conditions of our experiment, ion sputtering effects are quite strong [450]. For more ion-related effects on the growth of carbon-based nanostructures, please refer to Sections 7.1–7.4.

As a result of the above processes, the height of the primary nanocones decreases. In fact, this is one of the most important features of stage 3, the equalization stage. In this stage, two important processes determine the development of the nanocone arrays. First, sputtering of large primary nanocones decreases their height and base radii due to a very strong electric field at their top (for more details see Section 5.1). Second, the secondary nanocones grow faster. These two concurrent processes, namely the growth and the sputtering by ion bombardment, determine the nanocone shape formation at this stage. At the end of stage 3, the two (primary and secondary) nanocone arrays equalize and form a final nanopattern. Moreover, the dispersion of their height and base radii tends to minimize due to the concurrent ion sputtering and nanostructure growth. A quite similar conclusion has been made in Section 7.4. These processes tend to equalize the distribution functions of the geometrical characteristics of the nanostructures, and the final pattern depicted in Figure 7.20(c) is eventually formed.

To simulate the development of uniform patterns of carbon nanocones from size- and position-non-uniform nickel catalyst nanoparticles, all modules described in Section 3.3 were used. The first step on this direction is to appropriately model the fragmentation of the nickel catalyst film. The simulated final catalyst pattern is depicted in Figure 7.22 and is fairly similar to the SEM micrograph in the inset of the same figure. Subsequently, the simulations proceeded using the atom/ion delivery, carbon adatom nucleation and nanostructure growth models. The results of these numerical experiments are shown in Figure 7.23.

Firstly, one can observe the growth of primary nanocones on large catalyst particles, consistent with the predictions made using the empirical growth model described above. The numerical experiments of Tsakadze *et al.* [95] also suggest that at this stage the growth rates of the primary nanocones grown on large catalyst islands are much higher compared with the rates of growth of the secondary nanocones. This results in the formation of the primary nanocone array (Figure 7.23(a)) with the characteristic height and base radius (500 and 300 nm, respectively), which is quite consistent with the experimental nanocone size distributions deduced from Figure 7.20(a).

During the next growth stage, the density of carbon adatoms on the substrate surface increases. Consequently, the growth rate of the secondary nanocones (which develop between the primary nanostructures) also increases. As a result, an initial secondary pattern of small nanocones forms between the primary nanocones as can be seen in Figure 7.23(b). The results of numerical experiments also suggest that the growth of the primary nanocones is retarded, whereas the secondary nanocones grow faster, tending to equalize the two arrays [95].

As the nanostructures elongate, ion sputtering significantly reduces the height increase of the primary nanocones. This leads to a significant decrease in the difference between the primary and secondary nanocone arrays. Finally, due to the simultaneous decrease in the primary nanocone height and an intense growth of the secondary pattern, the two arrays equalize completely so one can no longer distinguish between them; this is clearly seen in Figure 7.23(c).

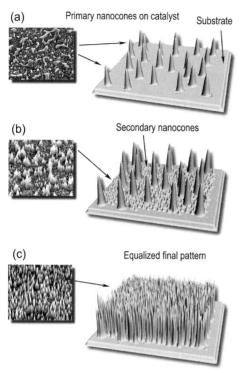

Figure 7.23 Simulated nanocone arrays: (a) the primary nanocone array on an islanded catalyst film; (b) the primary nanocone array and secondary self-assembled nanocones between the primary nanocones; (c) the final equalized pattern. Insets show the corresponding experimental patterns [95].

Thereafter, the entire pattern grows in a uniform fashion [95]. This can be explained by noting that in dense arrays the ion currents to the nanocones are reduced significantly. This is consistent with reports which suggest a strong dependence of the ion current in dense arrays or forests of vertically aligned nanostructures on the surface density of these objects [34].

To conclude this section, we stress that *by using a plasma process one can substantially improve the uniformity of a carbon nanocone array grown from a non-uniformly fragmented catalyst layer*. More importantly, this process occurs via three-dimensional self-organization where the systems involved interact in a way to improve the uniformity of the final nanoarrays. The plasma plays a prominent role at all stages of this complex self-organizing process.

7.6
Other Ion-Focusing Nanostructures

As we have seen from the previous sections of this chapter, ions contribute to the nanoassembly processes quite differently compared to their neutral counterparts. The main focus so far has been on quasi-one-dimensionl nanostructures such as nanotubes, nanorods, nanoneedles, and so on. However, there are some other nanoscale objects that are capable of focusing or significantly affect microscopic ion fluxes in their vicinity. These objects may have a different dimensionality as is the case for nanowall-like structures or represent a sort of "inverted" nanostructures. The latter is the case for various voids, pores, and channels which may be created in solid substrates either through bottom-up self-assembly or top-down nanofabrication (e.g., using lithography).

Similar to other nanostructures already mentioned in this book, the specific effect of nanowalls and nanopores on the ion flux depends on the material from which they are made. As we have seen in Chapter 5, conducting and dielectric nanostructures affect the ion fluxes very differently. As a rule of thumb, if a conducting object is able to concentrate the electric field lines and consequently focus ion fluxes, an identical object made of an insulating material would be expected to exert an opposite action. In this section we will only briefly recap some of the most interesting results related to the above two types of nanoscale objects. In the following, we will discuss some of the features of ion-assisted growth of carbon nanowall-like structures followed by a brief summary of possible ion-related effects on the development of nanoporous/voided structures.

Let us now compare the growth kinetics of carbon nanowall-like nanostructures in the plasma and neutral gas synthesis processes predicted via multi-scale hybrid numerical simulations [345]. Interestingly, the low-temperature plasma-based process shows a significant advantage over the purely neutral flux deposition. Similar to the case of conical carbon nanotips, a plasma-assisted process provides a notably better uniformity in respect of the size distribution within the developed nanoarray.

More specifically, Levchenko *et al.* [345] have demonstrated that the uniformity of nanowall widths is expected to be the best (square deviations not exceeding 1.05) in high-density plasmas of $3.0 \times 10^{18} \, m^{-3}$, worsens in lower-density plasmas (up to 1.5 in $1.0 \times 10^{17} \, m^{-3}$ plasmas) and is the worst (up to 1.9) in a neutral gas-based process with the same parameters. Similar to quasi-one-dimensional nanostructures of previous sections, this significant improvement can be attributed to the focusing of ion fluxes by irregular electric fields in the vicinity of plasma-

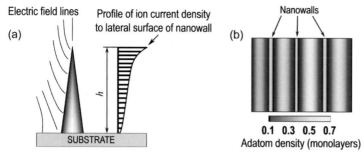

Figure 7.24 (a) Scheme of electric field lines and representative profile of calculated ion current distribution on lateral surface of nanowall; (b) adatom density field on the open surface between nanowalls for surface coverage $\mu = 0.12$. Nanowalls (top view) are shown as thin golden strips (not to scale) [345].

grown nanostructures, and also to the differences in the two-dimensional adatom diffusion fluxes in the plasma and neutral gas-based processes.

Figure 7.24(a) shows a representative distribution of the ion current on the lateral surface of a typical nanowall [345]. It is seen that the ion current is effectively re-distributed in the electric field created by the nanostructures. Furthermore, the density of the ion current in the upper part of the nanostructure is significantly increased and becomes larger than its mean value on the open surface areas uncovered by nanowalls. A profile of the computed adatom density in the trenches between the quasi-two-dimensional nanostructures is shown in Figure 7.24(b), for the surface coverage $\mu = 0.12$. The golden strips correspond to the nanostructure bases. It is seen that the surface density peaks in the areas between the nanowalls and tends to the equilibrium density near the nanostructure borders.

Interestingly, the nanowall-like structures develop quite differently in neutral gas-based processes. The most notable observation from the numerical experiments [345] is that the distribution function of nanowall widths broadens in a neutral gas-based processes. Moreover, the maximum of the distribution function decreases with time. As a result, the final distribution function appears to be significantly wider than the initial one, thus reflecting a larger and continuously increasing dispersion of the nanostructure widths. A comparison of the width distributions from the plasma- and neutral gas-based processes shows that (i) the maximum is higher in the plasma-based process; and (ii) the width of the nanostructure size distribution is significantly higher in a neutral gas-based process.

Moreover, the quasi-two-dimensional nanostructures grown from the low-density plasma have a smaller mean width. Figure 7.25 illustrates

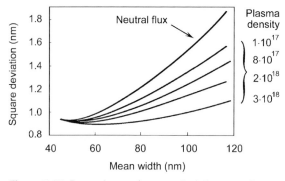

Figure 7.25 Dependence of square deviation σ on the mean nanowall width with the plasma density as a parameter [345].

the dependencies of the square deviation σ of the nanostructure width distribution function on their mean width and deposition time, for a plasma with electron temperature $T_e = 5.0\,\mathrm{eV}$ and working gas pressure of $2\,\mathrm{Pa}$. One can see that the difference in σ is quite strong: at a mean width of $120\,\mathrm{nm}$, the deviation reaches $\sigma = 1.9$ in the neutral gas process but remains almost twice as low ($\sigma = 1.05$) in the plasma with the density $3 \times 10^{18}\,\mathrm{m}^{-3}$.

As suggested by Levchenko et al. [345], one of the main advantages of ionized gas-based processes is the increased energy of the incident particles (e.g., ions), compared with the energies of atoms/molecules in neutral gas. An increased energy of the particle-surface interaction provides a range of beneficial effects, such as heating of the surface, increasing the species reactivity, control of the surface reactivity via dynamic termination/activation of surface dangling bonds and several others. Some of these important plasma-related controls have already been mentioned in this book. As a result, the nanowalls grown by using plasma may feature an improved structure, crystallinity, and ordering. Additionally, the ionized gas processes make it possible to generate reactive species (such as atomic hydrogen) that can cause amorphous to crystalline structural transformations. Thus, the use of plasma-based processes for the surface treatment is in most cases more advantageous than the neutral gas-based techniques.

Nevertheless, simplicity, convenience and safety remain the advantages of a gas process, which make it quite difficult to make the final choice of the optimum nanofabrication environment. However, poor "geometrical" controllability in gas-based processes, due to the diffusion-like, undirected flux of the neutral atoms to the nanostructures, may be the main decisive factor in favor of the plasma-based techniques. Moreover, in neutral gas-based processes it appears difficult to control the selective delivery of neutral particles to the nanostructures, which in turn compromises the ability to effectively control the quality of developed nanopatterns.

Here again we stress that in plasma-based processes, a better controllability can be achieved on account of the presence of ionized species in the gas phase and of patterned (focused by the nanostructures) electric fields in the immediate vicinity of the nanopattern. This enables one to selectively control the building block delivery processes. Indeed, it becomes possible to use the microscopic electric field to deliver the building material directly to the lateral surfaces of the nanowalls, which results in a significant increase of the direct (from the gas phase) particle incorporation into the nanostructures compared with the neutral gas processes [345]. Moreover, the electric field becomes an efficient control in re-distribution of the ion flux deposited to different trenches between the nanowalls.

The main distinguishing feature of the nanowalls nano-topology observed in the numerical experiments of Levchenko *et al.* [345] is the broadening of nanostructure size distributions with time (increasing mean width). This phenomenon can be understood through a careful examination of diffusion-caused re-organization of the adatom density field on the surface. It is possible that small nanostructures located near larger ones can get into the surface areas with depleted adatom densities. This depletion effectively decreases the diffusion influx to the borders of nanowalls and hence retards their growth. Thus, it turns out that the mutual interaction of growing nanostructures is essential even when the surface coverage is small and should always be taken into account.

In the neutral gas process the atom influx to the substrate surface is uncontrollable and the atoms are deposited uniformly on the entire surface. In this case the above adatom density depletion process strongly affects the growth of smaller nanostructures. Moreover, narrower nanostructures grown in the plasma create a stronger electric field, which forces the ions to deflect while traversing the sheath and eventually deposit on the substrate surface near the narrower nanowall. Thus, an effective re-distribution of the diffusion fluxes takes place, which gives an extra boost for the growth of the narrower nanostructures. As a result of the plasma process, the width distribution function equalizes, and eventually a nanopattern of very similar nanowalls is developed [345].

Using quite similar arguments, one can predict the evolution of nanosized cavities, pores and trenches in solid surfaces exposed to low-temperature plasmas. Using ion fluxes offers several additional possibilities for processing the surfaces of such voided structures. The processes involved in the profile evolution of such structures are in most cases very complex and involve deposition of "sticky" and "non-sticky" species in

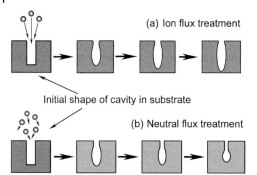

Figure 7.26 Representative screenshots of the development of cavities in (a) plasma-based and (b) neutral-based processes [173].

their neutral and ionic forms, chemical etching and physical sputtering of the pore surface material, and several others. While "sticky" species would normally incorporate into the film developing on the pore surface, "non-sticky" species may need to travel, and sometimes quite a significant distance, before they can directly participate in the growth process.

Two different possibilities in the development of cavities in a neutral and ionized gas-based processes are illustrated in Figure 7.26 [173]. From this figure one notices that the ion fluxes can penetrate deeper into the cavity and so deepen it through reactive ion etching, as commonly used in microelectronic manufacturing of tiny trenches on silicon microchips. Deposition of ionic species (and also their contribution to the etching) on sidewalls leads to the reshaping of the cavity, which is the case considered became smoother. However, if the ion species are "non-sticky", their reactivity and energy are sufficient for the etching of the substrate material, and their flux is highly-anisotropic; consequently, the cavity may deepen without any significant reshaping.

On the other hand, the flux of neutral species is isotropic and if they are sufficiently "sticky", predominant deposition near the cavity opening may lead to significant clogging of the cavity and eventually result in the formation of a closed pore in the substrate as shown in Figure 7.26(b). The interaction of ionic and neutral species with nanoporous structures and dense arrays of closely-packed nanostructures will be considered in more detail in Chapter 8.

7.7
Concluding Remarks

One of the main conclusions of this chapter is that in a low-temperature plasma, in contrast to the neutral gas-based processes, one can synthesize nanostructures of various dimensionality and shapes with a larger surface density, desired geometrical parameters, and narrower size distributions. This effect is largely attributed to strong ion focusing by irregular electric fields in the nanopatterns, which effectively redistributes the influxes of plasma-generated building units and thus provides a selective control on their delivery to the growing nanostructures [173].

In Section 7.1 we have considered some of the effects related to the presence of electric fields in the vicinity of one-dimensional nanostructures. As we have seen, plasma-related electric fields turn out to be very important for various aspects of the nanostructure growth. The most striking example of this effect is the alignment of surface-bound nanostructures along the direction of the electric field. A comparative analysis of the growth conditions has been made for nanotubes developing in three different cases; namely, surface-bound carbon nanotubes (CNTs) in the top and base growth modes.

In the above cases, we have focused primarily on the relations between the magnitudes and distributions of ion fluxes and the nanotube growth rates. It turns out that a variation of the plasma parameters affects the nanotube growth on the surface and in arc discharge plasmas in quite different ways. We have also stressed that the total fluxes to nanotubes and their distributions are different enough to cause very different growth rates and, moreover, to provoke transitions between the nanotube growth modes. As a result, the ion fluxes to a single nanotube on the surface and in the arc plasma can differ by a factor of 10^6 giving rise to the difference in the corresponding growth rates by three orders of magnitude [351].

The conditions of the metal catalyst also appear to be very different with respect to melting and poisoning. An interesting possibility in this regard is that the catalyst poisoning, which is almost inevitable for surface-bound nanotubes, can be significantly reduced or even completely avoided in arc discharge plasmas. One of the reasons is that the plasma ions effectively participate in catalyst "conditioning" and so reduce its oversaturation with carbon.

One can conclude that the quite different growth conditions for the nanotubes on the surface and in the bulk of the arc discharge plasma are intimately related to the effects exerted by the electric field and the plasma ions, and appear to be among the major factors in favor of the selection of the nanotube growth mode as well as its final structure (single-wall, double/few and multi-wall, chirality, and so on) via kinetics of carbon supply, which eventually determines the diffusion-limited and supply-limited scenarios if the nanotube growth [351].

One of the most remarkable effects caused by the presence of electric fields in the plasma sheath and in the vicinity of nanostructured surfaces is vertical alignment of one-dimensional nanostructures. This alignment markedly differs the plasma-based processes from neutral gas-based ones. Even though a large number of such nanoscale objects show this effect, the actual reasons behind their alignment along the direction of the electric field are not completely understood, and more importantly, may be different for different nanostructures.

For example, for carbon nanofibers grown in the tip growth mode, the alignment is due to the electric field-assisted, self-maintained uniform precipitation of carbon material through the catalyst particle. On the other hand, similar nanofibers but with the catalyst particle at their base do not show vertical alignment. Amazingly, single-walled carbon nanotubes, which also develop in the base growth mode, do align in the direction of the electric field.

This difference can be explained by metallic features of SWCNTs which can be either metallic or semiconducting depending on their chirality. For many vertically aligned nanostructures arranged in arrays, similar charging and direct electrostatic repulsion, complemented by a strong electric pull from the electric field in the sheath, appears to be an important factor in ensuring their ordering both in the growth direction and over the substrate. For other nanostructures with different dimensionalities, elemental composition and internal structure, the electric field-related effects may manifest quite differently.

The topic of electric field- and ion-related effects on the growth of carbon nanotubes and related structures was continued in more detail in Section 7.2 where two examples of carbon nanofibers and single-walled carbon nanotubes were considered. In particular, ion-assisted dissociation of a hydrocarbon precursor on the open surface of the catalyst particle located on top of the nanostructure appears to be one of the main processes that lead to the creation of the required building units of the CNFs. Moreover, the catalyst nanoparticle creates a unique channel to transport the as-created BUs to the graphene sheets of the CNFs, namely, over its top surface. In this way some carbon atoms may completely avoid the necessity to diffuse through the nanoparticle bulk and this may make the surface diffusion a main contributor to the plasma-assisted growth of CNFs. However, one should be very careful not to overdeposit carbon material onto the catalyst particle's open surface as this may terminate the growth process completely. Amazingly, once again the plasma can help! Indeed, atomic fluxes of hydrogen and inert gases can be used to remove unwanted deposits from the surface and thus keep the nanoparticle catalytically active.

Models based on surface diffusion turn out to be very useful to desribe the growth of single walled carbon nanotubes in plasma environments. In Section 7.2.2 we have introduced the model of Denysenko *et al.* [386], which accounts for the interactions of hydrocarbons and carbon atoms with the etching gas, dissociation of adsorbed hydrocarbon neutrals due to ion bombardment, and decomposition of hydrocarbon ions and thermal dissociation of hydrocarbons on SWCNT surfaces. This model also accounts for those effects of particular importance for the CNT synthesis such as growth and etching of the film between the SWCNTs, carbon sputtering and some others.

Using the model the conditions under which a plasma environment can be beneficial for the SWCNT growth can be predicted. The main advantages of ion and atomic hydrogen deposition and interaction on the CNT surfaces are [386]:

- Ion-induced dissociation of CH_3 radicals and decomposition of hydrocarbon ions on CNT surfaces is an essential source of carbon atoms required for their growth. At low processing temperatures these two processes, unavailable in thermal CVD, become dominant. This makes it possible to grow SWCNTs at much higher rates than that in CVD at low temperatures. This conclusion is quite similar to the case of carbon nanofiber growth discussed in Section 7.2.1.

- Reasonably strong fluxes of atomic hydrogen can completely etch away the amorphous carbon films developing between the nanotubes. This keeps the anchored catalyst particles accessible to the carbon adatoms which then eventually reach the SWCNT base and are incorporated into the growing nanostructures.

- The impinging atomic hydrogen can condition the CNT walls from unwanted adsorbates and facilitate the diffusion of carbon adatoms towards the catalyst nanoparticles.

- A reasonable increase (e.g., by varying the substrate bias) of the ion energy leads to a larger supply of atomic carbon to the CNT walls and thus faster SWCNT growth.

- In combination with the results discussed in Section 5.1, one can conclude that by varying the bias voltage within certain limits (which depend on the process and SWCNT parameters), micro-scopic ion fluxes can be concentrated in the lower CNT sections;

this leads to substantially reduced distances the carbon adatoms have to migrate from the point of deposition to the catalyst particle. This can further boost the growth process. It also explains the very high growth rates of ultra-long plasma-synthesized carbon nanotubes.

- A suitable combination of the above factors can enable rapid growth of very-high-aspect-ratio SWCNTs at surface temperatures substantially lower than that in thermal CVD.

However, as the processing temperatures decrease, one should minimize a number of adverse effects by carefully balancing the incoming fluxes of the plasma ions and the etching gas. For example, excessive production of atomic hydrogen and/or other adsorbates can significantly affect the diffusion and residence of carbon adatoms on the CNT surface. This can result in a significantly slower CNT growth since less carbon adatoms can reach the anchored catalyst particle. On the other hand, if the ion flux is too high and the etching process is not strong enough, amorphous carbon deposition can occur and eventually block the access of carbon adatoms to the catalyst particles. This can eventually terminate the growth process [386].

It is important to emphasize that the plasma-related effects are particularly important for surface temperatures $T_s \leq 1000\,\mathrm{K}$. At high substrate temperatures thermal processes dominate over the ion-induced processes, so that the benefits (except perhaps the vertical CNT alignment by the sheath electric field discussed in Section 7.1 and the radical- and excited-species production in the plasma, see Sections 4.1 and 4.2) of the latter diminish.

The results and the models described in Section 7.2 can be used for optimizing carbon nanofiber and nanotube synthesis and eventually bring it to the as yet elusive deterministic level. More importantly, the main conclusions are not restricted to CNTs or CNFs and may be relevant to the plasma-assisted catalyzed growth of a broader range of nanoassemblies, which makes these results even more valuable.

In Section 7.3, our focus was on conical nanostructures completely filled with carbon material. These objects can be crystalline, amoprphous or mixed-phase. Very different arrangements of atoms inside the nanostructures makes it possible to reshape them using plasma-based processes. And what is even more interesting is that this reshaping can be dynamic and self-organized.

For example, conical carbon nanocones with a platelet-like structure considered in Section 7.3.1, continue their growth until all available avenues are exhausted. At early stages, their growth proceeds via the addition of new carbon sheets (platelets) at the base where the catalyst is located. As the structures grow, their bases cover the catalyst and this "base" growth is discontinued. Despite this, the nanocones continue their development using the open top surface.

More importantly, the nanocone growth rates and the radius of its top surface can be precisely controlled by adjusting the ion fluxes which are progressively focused closer to the tip as the nanocone grows. This results in plasma-controlled self-sharpening of the nanocones which eventually develop into sharp, high-aspect-ratio nanoneedles.

Furthermore, multi-scale hybrid numerical simulations presented in Section 7.3.2 suggest that, in principle, nanotip-like structures can be "moulded" into any desired shape by precisely controlling the distribution of microscopic ion fluxes over their lateral surfaces. Manipulation of the main plasma parameters appears to be an efficient "turning knob" in the deterministic synthesis of the multi-purpose nanotip arrays. It is remarkable that the strong and plasma parameter-dependent DC electric field in the sheath is a crucial factor (absent in neutral gas routes) that enables a certain degree of deterministic synthesis of the desired nanopatterns. By using multistage processes and adopting specific sequences of the plasma parameters, one can create a vitually unlimited continuum of exotic shapes [100]. This approach is also generic and may be applicable to a much wider range of nanoassembly synthesis and post-processing.

In Section 7.4 we continued to explore the effect of the electric field and ion fluxes on the shapes of nanocone-like structures arranged in a microscopic pattern on a nickel-catalyzed silicon substrate. We also posed the question, why should one use ionized gas environments to synthesize arrays of carbon nanotip microemitters? The answer is that the plasma-aided process, in contrast to the neutral flux deposition, is a very efficient tool to control the nanotip aspect ratio, a critical factor in microemitter array applications. Moreover, the plasma-aided process offers a greater level of uniformity of carbon nanotip base radii within the entire nanopattern.

These two important factors can be controlled by adjusting the plasma parameters such as the degree of ionization, plasma density, electron temperature, and so on. Physically, a certain degree of determinism in the plasma-assisted synthesis of carbon nanotips can be achieved by properly manipulating the two building block delivery channels: via surface migration and directly from the ionized gas phase. It becomes evident that an increased influx and controllable deposition of ionic building units directly onto the nanotip lateral surfaces can be used to deterministically control the geometric shape of the nanotips.

In Section 7.5, we considered a more complex case of three-dimensional self-organization in arrays of carbon conical nanostructures synthesized in high-density, low-temperature plasmas. The experiments reveal a complicated scenario of the nanocone pattern development. The results of Tsakadze *et al.* [95] suggest that highly-uniform arrays of carbon nanocones can be synthesized via three-stage self-organization which involves the growth of the primary nanocone pattern on large catalyst islands followed by the growth of the secondary nanocones between the primary ones. Eventually, the nanostructures equalize in size and shape. Here we stress that this does not happen in a neutral gas-based process with very similar process parameters.

Therefore, the plasma environment plays a unique role in self-organized growth of highly-uniform carbon nanocone arrays. Moreover, plasma-based processes can be used to improve the uniformity of nanostructure growth which often suffers from non-uniform catalyst fragmentation during the preliminary growth stage. Raman spectra of as-grown nanostructures show that the developed nanocone arrays exhibit a high G:D ratio which is indicative of fairly large sizes of sp^2 crystallites. At the same time, the XRD analysis shows excellent crystallinity of the nanostructures. The semi-empirical growth scenario is justified by numerical experiments that adequately describe all the stages of the nanocone array development. This effort should be extended to incorporate more complex plasma and surface-related phenomena and involve other carbon nanostructures, such as very-high-aspect-ratio single-walled carbon nanotubes.

The final section of this chapter, Section 7.6, was devoted to a brief discussion of plasma-related effects on other nanoscale objects with the ability to focus ion fluxes. Two examples of quasi-two-dimensional nanowall-like structures and nano-sized pores were considered. In the first example, the comparison of the development of surface nanotopology in neutral and ionized gas-based processes shows that only a slightest broadening of the nanostructure width distribution takes place when a dense plasma is used. The square deviation of the nanostructures widths increases to 50% in less dense plasmas, and reaches 90% in the neutral gas process. Moreover, the plasma-aided process strongly influences the width of the nanowall-like structures due to the delivery of the building units directly to the nanoassembly sites.

The second example of ion-focusing nanostructures briefly considered at the end of Section 7.6 is related to possible implications of using ion fluxes to control the feature development in solid substrates facing the plasma. Due to faster and highly-directional penetration of ionic species into ultra-small cavities, such features may develop quite differently in ionized and neutral gas-based processes. Indeed, poorly controlled fluxes of "sticky" neutral species may cause substantial clogging of the developing nanofeatures. Moreover, under conditions of excessive attachment of neutral species in the upper areas of a cavity, a closed nanopore can be formed. The discussion of interactions of neutral and ionic species with nanopores and nanostructures arranged in dense arrays will be continued in Chapter 8.

To conclude this chapter, we stress that low-temperature plasma-based processes, in contrast to the neutral flux deposition techniques, provide a very effective control of the nanoscale surface topology, which may include various nanostructures of different dimensionality erected on the surface and impression-like features created in the substrate material. As we have seen in the cases of one-dimensional conical nanotips and two-dimensional nanowall-like structures, the use of the ion flux in addition to the neutral one ensures a much better uniformity of the nanostructure sizes. This indicates a superior potential of the plasma-based processes to achieve the as yet elusive deterministic shape control of selected nanostructures.

The results presented in this chapter allow us to confidently state that in the plasma-based process one can control the surface nanoscale morphology by changing, for example, the plasma density and degree of ionization. Physically, this control can be achieved by varying relative intensities of the direct (from the gas phase) and diffusion (over the surface) fluxes to the nanostructure surfaces, which is impossible in the neutral gas. Besides, it is worth mentioning that the structure and geometrical shape of the nanostructures also depend on the ratio of direct and surface diffusion fluxes. In most cases, an increased influx of the species from the plasma directly to the nanostructures and their subsequent redistribution about their lateral surfaces lead to notable changes in the surface nanoscale morphology.

The importance of ion-assisted processes becomes even higher in treatment of temperature-sensitive materials, which have to be processed at reasonably low surface temperatures. Under such low substrate temperature conditions, the surface diffusion may become negligible and the growth is mainly due to the direct influx of ions and other reactive species from the plasma.

These processes, largely uncommon to neutral gas-based routes, demonstrate the outstanding promise of the plasma nanofabrication processes to achieve the as yet elusive deterministic control of nanostructure growth processes and, ultimately, their physical characteristics and functionalities so much needed for the development of the next generation of nanodevices.

8
Building and Working Units at Work:
Applications in Nanoscale Materials Synthesis
and Processing

This chapter intends to show some examples of the use of specific building and working species generated in a plasma in representative nanoscale applications. The examples presented in this chapter will involve a variety of plasma-created (or more precisely, created with substantial plasma assistance) species such as ions, neutral radicals and atoms, as well as neutral and charged nanoclusters. The role of any particular species is determined by the envisaged application. In some applications a species can play the role of a building unit and that of a working unit in some others. For example, plasma-generated ions can be used as building material but can just as satisfactorily, serve a variety of other purposes ranging from surface conditioning and site preparation to reactive etching and physical sputtering.

In Sections 8.1 and 8.2 the main focus will be on showing further advantages offered by ion-assisted processes compared to neutral gas-based routes. In particular, it will be shown that ions, driven by electric fields in the vicinity of nanostructured surfaces, have a greater ability to penetrate into very dense arrays of nanostructures and nanoporous templates. This ability can be used to enable a range of post-processing options such as overcoating, functionalization, and reshaping. This wide range of sophisticated processes cannot be implemented in neutral gas-based processes due to the intrinsic randomness of motion of neutral species. Section 8.1 also shows different possibilities to control ion flux penetration into arrays of vertically aligned one-dimensional nanostructures by varying the plasma process parameters.

In Section 8.2 we will consider a very interesting example of controlled synthesis of an ordered array of gold nanodots using ion-assisted deposition through a nanoporous template. By adjusting the ion energy and fluxes, one can manage to effectively deposit nanodot structures without any significant clogging of nanopores in the template.

In the following example (Section 8.3) we will see how plasma-created working units (in this case it will be oxygen atoms) can interact with the surfaces of thin metal foils and create the most essential building units

Plasma Nanoscience: Basic Concepts and Applications of Deterministic Nanofabrication
Kostya (Ken) Ostrikov
Copyright © 2008 WILEY-VCH Verlag GmbH & Co. KGaA, Weinheim
ISBN: 978-3-527-40740-8

and also suitable surface conditions for their assembly. In this case, reactive oxygen plasmas are used to extract building units from a metal foil and cause solid-liquid-vapor growth of metal oxide nano-pyramids and nanowires.

In Section 8.4, it will be shown that plasma-generated TiO_2 nanoclusters can play a significant role in the synthesis of nanocrystalline titania films with varying presence of anatase and rutile phases. The resulting nanocrystalline structure plays a major role in enhancing biomimetic response of titania bioimplant coatings. As usual, this chapter concludes with a summary of important findings and, in particular, plasma-related effects that benefit the nanoscale assembly processes concerned.

8.1
Plasma-Based Post-Processing of Nanoarrays

A very large number of vertically aligned nanostructures have recently been studied in order to discover their highly unusual properties and convert them, via research and develoment, into commercial nanotechnology-enhanced products. Examples of relevant applications include, but are not limited to, chemical sensors and catalysts, reinforced nanocomposites, numerous optical and photonic functionalities and devices, viable alternatives to silicon-based microelectronic devices and circuitry, electron field microemitters, advanced bioscaffolds and biosensors, protein immobilization, and cell proliferation arrays [4,34,96,451–458]. These structures can be made of a variety of metallic, semiconducting, and dielectric materials including carbon, silicon as well as a broad range of pure, binary, ternary, and quarternary (ZnO, SiCN, SiCAlN, etc.) materials [459,460]. Their composition and other properties are tailored to meet the requirements of their ultimate function in nanodevices.

Moreover, there has been a rapidly increasing interest in post-processing (e.g. nanofilm coating, surface functionalization, and doping) of various nanoassemblies to enable new functionalities (e.g. biomimetic response or photoluminescence) and/or enhance their performance (e.g. field emission intensity). For example, the efficiency of electron emission from silicon nanotips can be greatly improved by coating them with carbon [459], whereas coating of Si nanotips with TiO_2 and capping with SiC leads to a substantial improvement of their wettability shown by increased water contact angles [461]. The latter property is crucial for advanced biomedical applications.

Meanwhile, doping ZnO nanorods with Ga or Al atoms not only improves electron field emission and photoluminescent properties but also significantly aids nanorod growth [462]. In another example, carbon nanotubes can be functionalized with various sugars and phosphocholine polymeric structures to become water soluble; this feature opens up a new avenue for applications in biosensing, controllable drug delivery, and gene therapy [463].

Recent research suggests that post-processing (e.g. coating, doping, or functionalization) of nanotube surfaces can significantly improve several structural, electronic, mechanical and other CNT properties and dramatically expand the field of CNT aplications. This opens broad avenues for the use of CNTs in various advanced devices that require controlled electric capacitance, thermal resistivity, hydrophobic properties, or inter-connection between different nanotubes in nanoelectronics [464–467]. Relevant examples include coating of carbon nanotube surfaces by amorphous SiO_2 for better integration in silicon-based ULSI microelectronic technology, deposition of non-wetting polymer layers for biodevice applications, or tungsten disulfide films for the development of new-generation light-emitting devices [468]. Meanwhile, functionalization of nanotube surfaces is commonly achieved, for example by using neutral fluxes of atomic hydrogen, organic molecules or fluorine-based species [172].

Despite all the research, uniformly processing large and/or dense arrays of nanostructures still proves a major challenge [34]. The main issue is to achieve a high level of control of the fluxes of reactive species and to direct them onto specific surface areas of the nanoassemblies. The latter areas can be post-processed selectively. Post-processing can include coating (e.g. monolayer coating), functionalization, doping, etching of selected areas, and some other technologically important processes. The existing techniques such as atomic layer deposition (ALD) [469] plasma enhanced CVD (PECVD) [461] and ion beam assisted deposition [470] are some of the possibilities in achieving this purpose. The ALD and its ion-assisted equivalent plasma enhanced ALD (PEALD) produce high-quality monolayer coatings.

However, a very limited choice of materials and reactive precursors make these (and many other) techniques material- and process-specific, which greatly limits their generic applicability and cost efficiency [471]. These methods also have a limited applicability for post-processing of dense nano-arrays where penetration of neutral species in narrow inter-nanostructure gaps is insufficient to coat all sections of the nanostructures [172].

In the following two subsections, we will consider two examples of uniform and selected-area post-processing of dense forests of carbon nanotubes and also nanorod arrays with different surface density. In both cases, plasma-based techniques offer significant advantages compared to thermal CVD processes.

8.1.1
Post-Processing of Nanotube Arrays

The primary aim of this subsection is to show some advantages of the plasma-based post-processing of dense arrays/forests of carbon nanotubes, as compared to neutral gas-based CVD. To do this, we will summarize the results of the original report [172] where such advantages were shown by means of multi-scale hybrid numerical simulations. More specifically, by controlling plasma-extracted ion fluxes and varying the plasma and sheath parameters, one can selectively coat, dope, or functionalize different areas on nanotube surfaces.

Interestingly, optimum conditions for the uniform deposition of ion fluxes over the entire nanotube surfaces appear to be very different for the arrays of different density. The most apparent *advantage of the plasma route is the possibility of uniform processing of lateral nanotube surfaces in very dense arrays*. This is very difficult, if possible at all, in neutral gas-based processes where radical penetration into the inter-nanotube gaps is insufficient.

Levchenko *et al.* [172] simulated a process of nanotube array coating/treatment in neutral and ionized gas-based environments and considered a uniform array of carbon nanotubes of 2 μm in height and 100 nm in diameter. The nanotubes were arranged in a hexagonal pattern where the spacing Δ between the nanostructures was varied from 500 nm in a dense nanotube forest to 4 μm in a sparse CNTs array. The spacing between the nanotubes is commonly controlled by pre-patterning of Ni/Fe/Co catalysts. In a neutral gas-based process, species of mass m_n move in all directions and their averaged thermal velocity is $V_n = (2k_B T_g/m_n)^{1/2}$, where k_B is Boltzmann's constant, and T_g is the gas temperature (in K).

In the plasma-based process, the ions are extracted from the plasma bulk and appear at the sheath border with their velocities directed normally to the substrate surface. For simplicity, only neutral Si and single-charged Si^+ silicon species with the atom mass of 28 were considered. The potential drop U_s across the near-substrate sheath was varied from the floating potential to a typical DC substrate bias of $-50\,V$, and the electron temperature T_e was varied from 2 to 5.0 eV. These are typical conditions for the synthesis of defect-free, undamaged carbon nanotubes in low-temperature plasmas [172].

The simulation was made for the case of elongated cylindrical nanotubes with hemispherical caps on top. Similar to Section 5.1, the total electric potential created by an array of N nanotubes

$$\phi(r) = \phi_w + \Sigma_1^N \phi_{S,i} + \Sigma_1^N \phi_{c,i}$$

was decomposed into the potential of the flat substrate surface ϕ_w, potentials created by cylindrical lateral surfaces of each (i-th) nanotube $\phi_{S,i}$, and potentials of the caps of each (i-th) nanotube $\phi_{c,i}$. The distributions of the atom/ion fluxes over the nanotube surfaces were computed using the Monte Carlo technique, with the total number of the species traced up to 10^5 (see Section 5.1 for more details).

The results of the atom/ion motion simulations were used to plot the atom and ion flux distributions along the nanotube length. The simulation area in the substrate surface was $25 \times 25\,\mu m$. The nanotubes were arranged into hexagonal patterns, in which the number of the CNTs varied from 200 (rarefied array) to approximately 10^4 (dense array). The study was limited to delicate post-processing such as deposition of ultra-thin (a few monolayer) coatings onto nanotube surfaces, using a dose that was varied from 1 to 5 monolayers (ML), where 1 ML is ca. $4 \times 10^{18}\,m^{-2}$ for silicon. The nett dose actually required to deposit a single monolayer over all nanotube surfaces (excluding the open surface areas) varied from 0.05 ML for a rarefied array ($\Delta = 4\,\mu m$) to 3 ML for a high-density array ($\Delta = 0.5\,\mu m$). The other parameters in the numerical experiments of Levchenko et al. [172] are: plasma density $n_p = 10^{17}$–$5 \times 10^{18}\,m^{-3}$, gas temperature $T_g = 1000\,K$, and gas pressure $p_0 = 1\,Pa$.

The most important observation is that the trajectories of neural and ionic species in dense nanotube arrays are very different. Indeed, neutral species were randomly distributed in the velocity space and mainly deposited onto the upper parts of the nanotube surface, showing a shallow penetration into the inter-nanotube gaps (Figure 8.1(a)). In contrast, the ions were effectively focused by the electric field and drawn deeper into the voids between the nanotubes. Afterwards, the ions were deflected by the electric field created by the cylindrical nanotube segments and eventually deposited on their lateral surfaces as shown in Figure 8.1(b).

Moreover, the penetration of silicon atoms into inter-nanotube gaps is more or less satisfactory only when the spacing is large enough (i.e. when the spacing is equal to the nanotube height), and significantly decreases when the inter-nanotip gaps become smaller. It is remarkable that the non-uniformity of the neutral flux deposition reaches 50% for $\Delta = 1\,\mu m$ and even 98% for $\Delta = 500\,nm$. Keeping in mind that a CNT array with the spacing ratio Δ/h (inter-nanotube spacing to nanotube height) of 0.5 and below is a typical example of a dense nanotube forest,

(a)   (b)

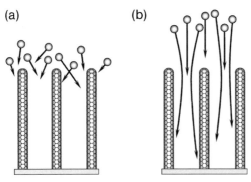

Figure 8.1 Neutral (a) and ion (b) fluxes in a dense nanotube array [172].

one can conclude that the neutral gas-based coating/functionalization in this case is very non-uniform due to unsatisfactory neutral flux penetration into the inter-nanotube space. In this case, the atoms may reach the lower parts of the nanotube only via migration over the lateral surfaces, a process with a quite limited controllability [172].

The results of computations made for the ion flux extracted from the plasma under various parameters are shown in Figures 8.2(a) and (b), for the two different inter-nanotube spacings of 500 nm and 2 μm, respectively. It is clearly seen that the ion flux distribution over the nanotube lateral surface is quite similar for small and large inter-nanotube gaps. However, the effect of the plasma-surface sheath width appears to be

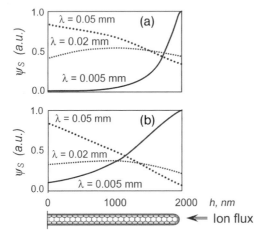

Figure 8.2 Normalized distribution of the ion flux over the nanotube lateral surface with the sheath thickness as a parameter [172]. The spacing between the nanotubes is (a) 500 nm; and (b) 2 μm.

very strong in both cases. Indeed, when the plasma sheath is thin (which is the case for the higher plasma density or electron temperature and/or lower substrate bias), the ion flux is predominantly deposited onto the upper sections of the nanotubes.

When the sheath is enlarged (e.g. by decreasing the plasma density or electron temperature and/or increasing the substrate bias), the maximum of the ion flux distribution is shifted towards the substrate surface. In this case the flux covers the entire lateral surface of the nanotubes, and becomes very uniform for the sheath of the 0.02 mm width, as can be seen in Figure 8.2. A further increase of the sheath width causes more ions to deposit in the areas closer to the substrate surface (bold dotted lines in Figure 8.2). When the sheath width increases to $\lambda = 0.05$ mm (e.g. when $T_e = 5.0$ eV, $n_p = 2 \times 10^{18}$ m^{-3}, and $U_s = -50$ V), one obtains a triangle-like ion flux distribution with the maximum near the substrate surface [172].

Thus, the use of the plasma-based process is expected to lead to a better controllability of the distribution of the working units required for post-processing of dense nanotube arrays. Indeed, in contrast to the neutral gas route, the ion current can be selectively directed to the upper or lower sections of the nanotube lateral surfaces, and, if necessary, can be uniformly distributed over the entire CNT surface with the non-uniformity over the entire nanotube length (2 µm) not exceeding 20% (short dashed lines in Figure 8.2). It is noteworthy that the uniform ion flux distribution can be obtained even for rather dense nanotube forests.

The possibility of selective coating and functionalization of different parts (e.g. of upper or lower sections of the cylindrical surfaces or caps) of CNTs arranged in dense arrays is undoubtedly very promising for complex post-processing. For example, ultra-thin films of highly-emissive materials may be required at the upper parts of the nanotubes to enhance the electron field emission, whereas surface coating by highly-conducting materials or introduction of special dopants can be useful to enhance the conductivity of the CNT-based microemitters. Such a complex treatment can be achieved by using a two-stage plasma-based process, whereby the species of a low-work-function material can be directed to the upper sections, followed by the uniform deposition of the ions of a highly-conductive material over the lateral nanotube surface.

This effect can also significantly improve the controllability of the fabrication of the post-processed arrays of carbon nanotubes. Indeed, by manipulating the plasma parameters, ionic building units can be selectively delivered to metal catalyst particles located either on top or at the base of the nanostructures depending on the prevailing CNT growth mode. This can also affect the thermokinetic growth mode selection, which in turn controls the nanotube structure (single- or multi-walled, capped or open-ended, chirality, etc.). To this end, selective and highly-controlled delivery of reactive species from the plasma to specified areas of dense-arrayed CNTs can be instrumental for the deterministic tuning of nanotube properties required in the envisaged applications [172].

8.1.2
Functional Monolayer Coating of Nanorod Arrays

As we have mentioned in the previous subsection, uniformity of post-processing of large-area, dense nanostructure arrays is currently one of the greatest challenges in nanotechnology. One of the major issues is to achieve a high level of control in specie fluxes to specific surface areas of the nanostructures. As suggested by the numerical experiments [34], this goal can be achieved by manipulating microscopic ion fluxes through varying the plasma sheath and nanorod array parameters.

Using a multi-scale hybrid numerical simulation, Tam *et al.* [34] simu-lated the dynamics of ion-assisted deposition of functional monolayer coatings onto two-dimensional carbon nanorod arrays in a hydrogen plasma. The numerical results show evidence of a strong correlation between the aspect ratios and nanopattern positioning of the nanorods, plasma sheath width, and densities and distributions of microscopic ion fluxes. Specifically, when the spacing between the nanorods and/or their aspect ratios are larger, and/or the plasma sheath is wider, the density of the microscopic ion current flowing to each of the individual nanorods increases, thus reducing the time required to apply a functional mono-layer coating down to 11 s for a 7 μm-wide sheath, and to 5 s for a 50 μm-wide sheath.

Numerical experiments from Tam *et al.* [34] aimed to quantify the de-pendence of the microscopic ion fluxes in the vicinity of ordered hexag-onal/square arrays of carbon nanorods on the nanopattern and plasma sheath parameters and elucidate the temporal dependence of the surface coverage of nanorods by an atomic hydrogen monolayer. Their numeri-cal results provided the optimized process parameters that minimize the time required to deposit a monolayer coating onto a nanorod array.

The computed monolayer development times have been found to be in remarkable agreement with previous experimental results on hydrogen plasma-based functionalization of related carbon nanostructures [406, 472]. The results considered in this subsection are generic in that they can be applied to a broader range of plasma-based processes and nanos-tructures, and contribute to the development of deterministic strategies

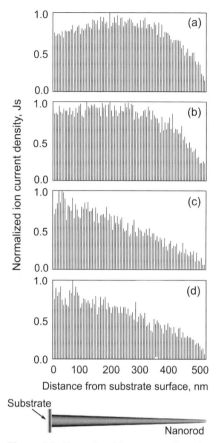

Figure 8.3 Normalized ion current densities to lateral surface of the nanorods with different parameters: (a) $R = 20$ nm, $\Delta = 100$ nm, $\lambda_s = 7$ μm; (b) $R = 40$ nm, $\Delta = 100$ nm, $\lambda_s = 7$ μm; (c) $R = 20$ nm, $\Delta = 500$ nm, $\lambda_s = 50$ μm; (d) $R = 40$ nm, $\Delta = 500$ nm, $\lambda_s = 50$ μm [34].

of post-processing and functionalization of various nano-arrays for nanoelectronic, biomedical and other emerging applications [34].

Figure 8.3 shows the normalized ion current density distribution along the lateral surfaces of the nanorods at different inter-nanorod spacing Δ, plasma sheath width λ_s and nanorod radius R. One can see that when the inter-nanorod spacing or the sheath width are small, a larger proportion of ions deposit on the upper sections of the nanorods. On the other hand, when Δ or λ_s are increased, the ions land closer to the substrate level. Moreover, the maxima of normalized ion current density distributions are shifted upwards from the substrate level (Figures 8.3(a) and (b)). These distributions make it possible to work out how a monolayer

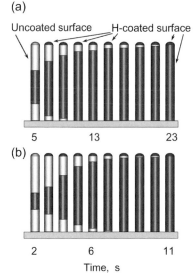

Figure 8.4 Temporal dynamics of the development of the monolayer coating to lateral surfaces of the nanorods of radius $R = 40\,\text{nm}$ (a) and $R = 20\,\text{nm}$ (b). Other parameters: $\lambda_s = 7\,\mu\text{m}$ and $\Delta = 100\,\text{nm}$ [34].

overcoat develops in time and gradually covers the entire nanorod surfaces [34].

Figures 8.4 and 8.5 show the temporal dynamics of the development of a monolayer coating over the surface of a representative nanorod. These figures show that carbon nanorods with smaller radii are coated with a hydrogen monolayer faster than their wider counterparts. Indeed, thinner nanorods are usually coated in approximately half of the time required to coat thicker nanostructures.

Another important conclusion is that when the inter-nanorod spacing is smaller, more time is required to coat the nanostructures. Moreover, in high-density arrays immersed in a plasma with a narrower sheath (Figure 8.4 with $\Delta = 100\,\text{nm}$ and $\lambda_s = 7\,\mu\text{m}$), the middle section of the nanorod is coated earlier than its other surface areas. Conversely, in low-density arrays post-processed in a plasma with a wider sheath, such as the one in Figure 8.5 with $\Delta = 500\,\text{nm}$ and $\lambda_s = 50\,\mu\text{m}$, the areas closer to nanorod bases are coated first. Therefore, *by varying the sheath width, one can deposit hydrogen monolayer coatings over selected surface areas of the nanorods.* This can be achieved, for example, by changing the plasma density and/or electron temperature.

Let us now continue discussing the issue of ion penetration into small inter-nanorod gaps. As we have seen in Section 8.1.1 and the original re-

(a)

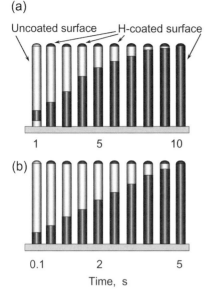

(b)

Figure 8.5 Same as in Figure 8.4 for $\lambda_s = 50\,\mu m$ and $\Delta = 500\,nm$ [34].

port [172], when an array becomes more dense, the ion penetration into it deteriorates. One can expect that in this case fewer ions reach the lower sections of the nanorods. This implies that the ion penetration into the arrays improves when the inter-nanorod spacing is increased; accordingly, this obvious conclusion leads to an increased coating thickness at the base sections of the nanorods.

From Figure 8.3, one can see that when Δ is small, there is a relative reduction in the ion current densities towards the lower sections of the nanostructures. In fact, it can be seen (with the exception of the very upper sections) that the ion current is evenly distributed over the lateral nanorod surfaces. Likewise, when the nanorod radii are larger, there is a reduction in the ion penetration into the arrays. In this case the densities of ion currents flowing onto the nanostructure areas closer to the substrate decrease. In this case, the largest fraction of the ion current is deposited onto the upper sections of the nanorods which face the plasma [34].

A similar argument helps to explain the effect of the plasma sheath thickness. It appears that increasing λ_s is an efficient means of reducing the ion deflections from straight downfall trajectories and eventually improving the ion penetration into the spaces between the nanorods. In contrast, enlarging the plasma sheath results in a major increase in the number of ions landing on the upper parts of the nanostructures.

The effects of changing the plasma process and nanopattern parameters on the dynamics of monolayer coating deposition is visualized in Figures 8.4 and 8.5. A striking observation is that the monolayer coverage of the nanorods by hydrogen atoms does not necessarily start at the substrate level as would be expected for a single nanostructure on the surface. This conclusion is consistent with the results presented in Figure 8.3. Both Figures 8.4 and 8.5 also suggest that thinner nanorods can be fully coated faster than the thicker ones. Since we already know that the time it takes to coat the nanorods is proportional to the ion flux density, one can expect that less dense arrays of higher-aspect-ratio nanorods need the shortest amount of time to be fully coated. Thus, examining Figures 8.4 and 8.5 it follows that the time needed to coat the nanorods in a denser array is longer compared to a sparse array. This can be understood by noting that in this case there are more nanorods in the array, whereas the total incoming flux remains the same [34].

It is imperative to point out that the computed time required to deposit a monolayer coating is consistent with the experimental observations, which suggest that it takes approximately 30 s to functionalize single-walled carbon nanotubes with hydrogen [472]. The small difference in time was explained by noting that the solid nanorods simulated by Tam *et al.* [34] are very different from hollow carbon nanotubes post-processed in the experiments of Khare *et al.* [472].

Here we recall that the nanotubes are hollow tubules without dangling bonds on lateral surfaces, while the nanorods usually have a crystalline structure and therefore have more readily available bonding sites on their surfaces. In the latter case the ions can bond onto the nanorods when they land on the lateral surface. On the other hand, the ions that come into contact with carbon nanotubes may not bond at the point of impact; moreover, they need to migrate over lateral nanotube surfaces to be able to insert into the nanostructure through a metal catalyst particle on top or at the base of the nanotube. Hence, one can expect that nanorod-like structures can be functionalized by atomic hydrogen faster than nanotubes. This conclusion is also applicable to other material systems [34].

However, even though less dense arrays can be coated faster when the plasma sheath thickness is larger (Figure 8.5), there is a danger that a strong non-uniformity in ion flux distributions over the lateral nanorod surfaces (Figures 8.4(c) and (d)) can result in non-even coatings, with thicker layers forming closer to the nanostructure bases. This effect is consistent with the results in Figure 8.3, which clearly show that very few ions land at the upper areas of the nanorods when Δ is large. This effect can be even stronger for heavier ions, which gain a larger momentum (compared to H^+ ions) while traversing the plasma sheath.

From an applications point of view, uniform monolayer coatings should be applied at the highest possible deposition rate. However, minimizing the time to coat the nanorods (e.g. by enlarging the plasma sheath) can somewhat compromize the coating uniformity. Moreover, non-even ion current distribution along the lateral surfaces of the nanorods can result in a buildup of undesired amorphous layers.

Nonetheless, by appropriately balancing the effects of variation of the plasma and array parameters λ_s, Δ, and R, one can work out the optimized process parameters to apply uniform monolayer coatings over the entire nanorod surfaces with reasonably high deposition rates. To this end, the numerical experiments of Tam *et al.* [34] are particuarly important since they can be used to predict the distributions of ion fluxes over the surfaces of vertically aligned nanostructures under different process conditions with a sub-nanometer precision.

In the following section, we will continue to show examples how manipulation of plasma-generated ion fluxes can be used for deposition of metal nanodots using a nanoporous template. In this process, the issue of how the ions pass through the holes and where exactly they land is particularly important to minimize the undesirable template clogging and maximize the nanodot quality and growth rates.

8.2
i-PVD of Metal Nanodot Arrays Using Nanoporous Templates

In this section we will follow the original report [473] and present the results of numerical simulation of plasma-based, porous template-assisted nanofabrication of Au nanodot arrays on highly-doped silicon substrates. In particular, we will consider the three-dimensional microscopic topography of ion flux distribution over the outer and inner surfaces of the nanoporous template in the close proximity of, and inside the nanopores. In this case, by manipulating the electron temperature, the cross-sheath potential drop, and additionally altering the structure of the nanoporous template one can control the ion fluxes within the nanopores and eventually maximize the ion deposition onto the top surface of the developing crystalline Au nanodots and minimize amorphous deposits on the sidewalls that clutter and may eventually close the nanopores thus disrupting the nanodot growth process.

Arrays of metallic (e.g. Au and Ag) nanodots, the main focus of this section, have recently shown an exceptional promise for nanofabrication of nanoplasmonic arrays and photonic crystals [474,475]. The nanodots represent individual nanostructures arranged in nanopatterns; therefore, the main problem in their fabrication is in maintaining a reasonable degree of control over the formation of highly ordered patterns of the nanodots of the required size, shape, composition, ordering and other parameters [473].

At present, three main techniques of nanodot assembly are commonly used: strain-driven Stranski–Krastanov (SK) segregation [477], catalytic growth [4], and porous template (PT) techniques [478]. Each method has specific advantages and disadvantages [479]. As we have already discussed in Chapter 6, the Stranski–Krastanov mode is essentially limited to moderately lattice-mismatched systems and is usually not applicable to material systems with a lattice mismatch less than 2–3% or exceeding 7–8%. The catalyzed growth technique is more flexible to the material properties since it involves an intermediate catalyst layer. Several examples of catalytic growth have already been considered in Chapter 7.

However, both the SK and the catalytic growth techniques exhibit a relatively poor spatial ordering and size distribution of the developed nanodot patterns. Therefore, these approaches may prove to be impractical for large-area arrays. Moreover, they may even lead to a quite poor uniformity of the nanodot shapes and sizes [477].

On the other hand, nanofabrication approaches based on nanoporous templates provide a much better degree of nanodot ordering [476], compared with the above two methods. In this case the accuracy in the nanodot positional ordering is only limited by template imperfections, which usually do not exceed a few percent. This degree of accuracy significantly exceeds the typical disorder levels which account for 50 to 100% in the SK and catalyzed growth methods. Besides, the PT technique is a cost-efficient method appropriate for the synthesis of large-area nanoarrays. A variety of materials, including metals Au, Ag, Ni, Fe, and Co, can be deposited through nanopores in the template.

Numerous deposition techniques such as the molecular beam epitaxy (MBE), chemical vapor deposition (CVD), plasma-enhanced chemical vapor deposition (PECVD), ionized physical vapor deposition (i-PVD), thermal evaporation (TE), plasma-asssited sputtering, and pulsed laser deposition (PLD) can be used to create and deliver the atomic/radical building units to the nanodot growth sites within the nanopores. However, this process is extremely delicate and case-specific because of the ultra-small size of the nanopores.

In Section 8.1, we have stressed that neutral species have a significant difficulty with penetrating deep inside features of even micrometer sizes; this is a common yet not completely resolved problem in copper metallization of micron-sized trenches in silicon wafers in ULSI microelec-tronic technology. Ionizing metal vapor and using electric field-directed ion fluxes (basic principle of the i-PVD approach) has been an effective solution of the above problem in the micrometer size range.

However, application of i-PVD for templated nanodot synthesis still remains in its infacy despite numerous reports on successful uses of this approach for the synthesis of self-assembled nanoparticles on solid sur-faces. Therefore, there is a vital demand to develop suitable i-PVD-based nanodot synthesis processes and explore viable parameter ranges. This is particularly important in view of the excellent prospects of improved nanodot crystallinity due to higher energies of ionic building units com-pared to their neutral counterparts [473].

Yuan *et al.* [473] reported on the optimized i-PVD-based process con-ditions for the synthesis of ordered arrays of gold nanodots on a highly-doped silicon substrate covered by a nanoporous conducting template. In particular, it was shown that manipulating the process parameters and adjusting the aspect ratio of the nanopores and the spacing between them makes it possible to develop a highly-ordered array of cylindrical gold nanodots. Moreover, using energetic ion fluxes appears to be favor-able for achieving a very high degree of nanodot crystallization and their structural/compositional uniformity.

Here we consider a conducting system that consists of a nanoporous template on top of the silicon substrate. This system is exposed to a low-temperature plasma with typical electron density in the range of 10^{17}–10^{18} m^{-3} and electron temperature of 2.0 to 5.0 eV, separated from the solid surface by the plasma sheath. This approach is applicable to a broad range of i-PVD systems, whereby positively charged metal ions (Au$^+$ in this case) can be generated. For simplicity, only single-charged ions are considered. It is assumed that collisions of Au$^+$ ions with gas molecules can be neglected.

A highly-doped (either n- or p-type) silicon substrate is covered with a porous mask of 300 to 500 nm height that has holes (nanopores) ar-ranged in a hexagonal pattern with fixed spacing between the nanopores $a = 100$ nm and the nanopore diameter varied from 40 to 80 nm (Fig-ure 8.6). The template material can even be dielectric as is the case for porous anodized alumina templates commonly used for templated syn-thesis of ordered arrays of carbon nanotubes and related nanostructures. For consistency, it was assumed that if the template itself is dielectric or semiconducting, its surface is pre-coated by an ultra-thin metal layer (as is commonly done for scanning electron microscopy of insulating mate-rials).

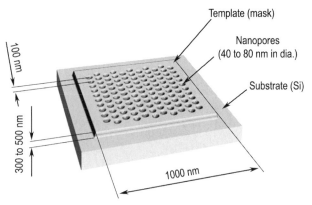

Figure 8.6 Schematics of the mask/substrate system (MSS) [473].

Alternatively, this ultra-thin conducting layer can be formed during a preliminary process that can also be conducted in a high-vacuum i-PVD environment. Therefore, the whole surface system that consists of the nanoporous mask and the silicon substrate is conducting and is hereinafter referred to as the mask/substrate system (MSS) [473]. In this setup, an ion flux is extracted from the plasma bulk and accelerated in a thin sheath between the MSS and the plasma, and is then deposited on the mask and substrate surfaces, including nanopores.

The ion motion was traced using the same Monte Carlo approach as in Sections 5.1 and 8.1. During the simulation, the coordinates of the ion impact on the MSS surface (including the interior of the nanopores) were recorded, and thus the microscopic topography of the ion current over the MSS surface was obtained. The parameters used for the simulation are listed in Table 8.1. For more details of the modeling of the electric field produced by the template and the ion motion in it please refer to the original publication [473].

The dependencies of the relative deposition rate of Au^+ ions as a function of the electron energy and MSS potential are shown in Figure 8.7. Here, the relative deposition rate $\mu = j_S/j_I$ is defined as the ratio of the density of ion current to the initially open areas of Si substrate inside the

Table 8.1 Parameters used to simulate the i-PVD of Au nanodots [473].

Group	h, nm	d, nm	U_s, V
A	500	40	25
B	500	40	75
C	500	80	75
D	300	40	25

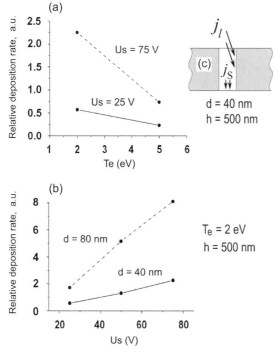

Figure 8.7 Dependence of the relative deposition rate on the (a) electron temperature; and (b) substrate potential for plasma density $n_p = 1.0 \times 10^{18}\,\mathrm{m}^{-3}$ [473].

nanopores j_S (this surface becomes the surface of predominant growth of the nanodots) to the density of the ion current on the lateral surface of the nanopore in porous template j_l, see Figure 8.7(b).

It is apparent that the higher the relative deposition rate, the more ions can be deposited onto the bottom surface of nanopore. In Figure 8.7 one can also see that the relative deposition rate decreases with the electron temperature T_e; this effect becomes stronger at higher bias potentials U_s. On the other hand, the deposition rate increases with U_s and is stronger for larger nanopore diameters, that is, for the mask holes with a smaller aspect ratio $\kappa = d/h$ [473].

Let us now consider the computed profiles of the ion deposition over nanopore surfaces. Here we make an assumption which is supported by the experimental works on quantum dot deposition through the mask [480]. It was assumed that only the ion flux deposited on the substrate surface in the nanopores, that is on the bottom surface inside the nanopore, can form a nanodot [473]. In this case all the material deposited on the mask surfaces (on the upper and lateral mask surfaces, including nanopore sidewalls) is not involved in the nanodot formation and is completely removed together with the mask during later process stages.

There is a number of reasons for this assumption. Firstly, the material deposited on the upper surface of the mask (top deposit) is apparently not bound to the nanostructures formed in the nanopores, and hence can be removed together with the mask. Secondly, the ions that deposit on the substrate surface, that is, on the bottom surface inside of the nanopore (on the surface of a developing nanodot) impact the growth surface at nearly the right angle, thus providing a strong and spatially localized energy exchange with the nanodot atoms.

This in turn leads to their fast incorporation into the crystalline structure of the growing nanodots. In contrast, ions that impact (mostly amorphous) deposits on the nanopore lateral surfaces (side deposit) contact the lateral growth surface at a very shallow angle. As a result, these ions are set to diffuse about the surface of the side deposit and eventually form an amorphous film on the nanopore side walls. This film may only be weakly bound to the nanodot crystalline structure and can thus be easily removed from the substrate during the mask removal process.

Using this assumption, Yuan *et al.* [473] simulated the nanodot growth profiles by integrating the ion current density over the entire MSS surface for various plasma and template parameters (Figure 8.8), for the electron temperature $T_e = 2.0\,\text{eV}$ and plasma density $n_p = 1.0 \times 10^{18}\,\text{m}^{-3}$ (which are typical for low-temperature plasma-based i-PVD). The results of the calculation (profiles of developing nanostructures, as well as side and top deposits) are shown for the four groups of plasma parameters which are listed in Table 8.1.

Figure 8.8 (group A) shows that it is not possible to form nanodots inside nanopores under process conditions of group A (mask height 500 nm, nanopore diameter 40 nm, bias 25 V). In this case the rate of deposition to the nanodot growth surface (nanopore bottom) is much lower than the growth rates of side deposits; hence, the nanopore "closes" well before a nanostructure is formed inside it. The deposition under conditions of group B (bias voltage increased to 75 V) exhibits a faster nanodot growth. However, the nanostructures still develop too slowly to compete with the side deposit growth. Thus, in this case nanodot formation is still substantially hampered due to the rapid increase of the side deposit thickness [473].

The best result was obtained for group C (nanopore diameter increased to 80 nm), which enables nanodot formation with growth rates comparable with those of the side deposits. This can clearly be seen from the bottom left panel in Figure 8.8. The deposition under conditions of group

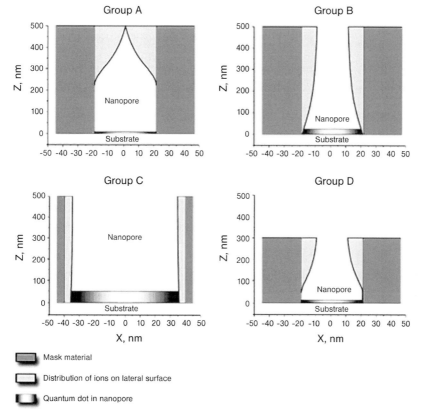

Mask material

Distribution of ions on lateral surface

Quantum dot in nanopore

Figure 8.8 Ion flux distributions inside the nanopore for the electron temperature $T_e = 2.0\,\text{eV}$ and plasma density $n_p = 1.0 \times 10^{18}\,\text{m}^{-3}$, calculated for the different groups of initial parameters [473].

D (decreased mask height at a low bias voltage of 25 V) also appears to be impractical, since the growth rate of the side deposit again exceeds the nanodot growth rate. Visualization of the growth process suggests that nanodots of group C appear to be tall and well-shaped, whereas the nanodots of group B are shorter and have the shape of truncated cones, with flat tops and radii that are larger at the base than near the top.

Therefore, the results presented in this section suggest that variation of the parameters of the plasma and nanoporous template can indeed be used to deterministically control the shape of metallic nanodots that develop inside the pores. Moreover, one could expect a higher relative ion deposition rate and a more uniform lateral distribution of the ion flux density at a lower electron temperature T_e and higher cross-sheath potential drop U_s [473]. Furthermore, the geometrical characteristics of the template also influence the ion current deposition profiles. For example, increasing the nanopore diameter d leads to better uniformity and higher deposition rates, whereas a decrease in the mask thickness h results only in higher relative deposition rates.

One can thus conclude that using a combination of lower electron temperatures, higher cross-sheath potential drops, and larger nanopore diameters may be the optimum way to achieve the ideal relative ion deposition rates and lateral ion flux distributions. Future work in this direction should be focused on the refinemnet of the nanodot growth model and its extension to more complex cases that involve different plasmas, a variety of plasma species (e.g. neutral and cationic radicals and negative ions), dielectric templates, temporal dynamics of the growth process, ion collisions with neutrals, as well as several other gas-phase and surface processes.

In the following section, we will consider a very different approach for the assembly of nanostructures on solid surfaces. The uniqueness of this approach is in that it does not involve creation of some building units in the gas phase and their subsequent transport to the growth surface. In fact, as we will see from Section 8.3, one sort of BUs is effectively created by another one right on the growth surface! This highly-unusual approach makes it possible to synthesize a variety of metal oxide nanostructures.

8.3
Metal Oxide Nanostructures: Plasma-Generated BUs Create Other BUs on the Surface

This section is based on the original report [481] on highly-unusual plasma-assisted synthesis of metal oxide nanostructures on metal foils. More specifically, we will consider the synthesis of nanostructured cadmium oxide films on Cd substrates in a reactive oxygen plasma environment. In this case, oxygen atomic building units are created in the ionized gas phase whereas cadmium species are extracted from a plasma-exposed metal foil as a result of interaction of oxygen atoms with the surface. Thus, plasma-generated building units create metal atoms (other BUs!) and combine with them to form metal oxide continuous films or well-resolved nanostructures on the surface.

However, the nanostructures, such as crystalline CdO nano-/micropyramids emerge only under very specific conditions determined by the plasma-surface interactions. These nanostructures grow via direct oxidation of Cd foil exposed to inductively coupled plasmas with the electron energy 3.0–7.0 eV, ion density 1.6×10^{16} m^{-3}, neutral oxygen atom density 1.4×10^{21} m^{-3}, and molecular gas temperature of 370 K [481].

The growth of the CdO pyramidal nanostructures takes place in the solid-liquid-solid phase, with the rates determined by the interaction of neutral oxygen atoms with cadmium BUs on the surface. Moreover, the size of the pyramidal structures can be effectively controlled by the dose of neutral oxygen atoms impinging on the metal surface. The experiments of Cvelbar *et al.* [481] suggest that the reactive plasma environment plays a crucial role in the controlled fabrication of CdO nano-pyramidal structures.

In recent years, cadmium oxide (CdO) has found applications in various fields, and is particularly attractive for optoelectronic devices, gas sensors, photodiodes, phototransistors and transparent coatings and functional films, such as transparent conducting oxide layers in photovoltaic solar cells [482–484]. Under non-stoichiometric conditions (e.g. in the presence of interstitial cadmium atoms or oxygen vacancies), cadmium oxide shows pronounced n-type semiconducting properties.

Meanwhile, crystalline CdO is a direct bandgap semiconductor, with the bandgap energy E_g varying in the 2.3–2.5 eV range depending on the grain size. The high electrical conductivity and excellent optical transmittance properties of CdO predominantly in the visible region of solar spectrum along with a moderate refractive index, makes cadmium oxide-based materials most suitable for advanced technologies, including low-emissivity windows, wear resistant coatings, flat panel displays, and thin film resistors for humidity sensing and other applications. However, the electrical and optical properties of CdO are very sensitive to deviations of its elemental composition from the stochiometric value, film thickness, microstructure, surface morphology and some other factors [481].

Despite a variety of techniques to synthesize CdO with different microstructure and surface morphology, such as wet chemical processes, vapor-phase deposition (CVD, PVD and their modifications), chemical synthesis in reactive oxygen environment, sol-gel methods, spray pyrolysis, electron beam evaporation, as well as several others [485–497], dry processes based on reactive plasmas have not received the attention they merit. Below, we will consider an advanced oxygen plasma-assisted nanofabrication method of nanostructured CdO from a bulk cadmium foil [96,481].

This method is based on the "self-catalytic" nanostructure growth, which does not require any complex metal-organic gas feedstock or additional surface catalysts. The latter was the case for carbon nanotubes and related nanostructures considered in Chapter 7. This growth is highly unusual, and at present there is no adequate explanation of the relation between the reactive plasma parameters and the surface nanostructures grown during the plasma process.

Cvelbar *et al.* [481] have shown that under certain conditions of the reactive plasma-based process, cadmium surfaces may feature intricate nano-scaled structures such as crystalline nano-pyramid arrays. The clue to explaining these unique and highly unusual surface structures is in the understanding of the role of neutral reactive species and their interactions with the solid surfaces.

The cadmium oxide nanostructures of interest to us here were fabricated via direct exposure of a high-purity cadmium foil to oxygen plasmas. Commercially available high-purity (99.9%) cadmium foil of thickness 0.3 mm was cut into rectangular pieces of $\sim 5 \times 5 \, mm^2$, and exposed to RF oxygen plasmas created in a low-pressure glow discharge [498]. The experiments [481] were performed in a vacuum system evacuated using a two-stage oil rotary pump with an ultimate pressure of approximately 0.1 Pa. After evacuating the glass reactor chamber to the base pressure, oxygen feedstock gas was continuously released into the reactor chamber.

The plasma was created by a 27.12 MHz RF generator with a variable power up to 2 kW. The temperature of the sample surface was measured through an infrared (IR) transparent window with Raytec IR camera Raynger MX. The plasma parameters were monitored using Langmuir and fiber optics catalytic probes [499, 500]. The plasma parameters and the gas temperature were controlled by varying the discharge power and oxygen gas pressure inside the reactor chamber. The plasma parameters are as follows [481]: the electron temperature $T_e \sim 3.0$–7.0 eV, ion density n_i ca. $10^{16} \, m^{-3}$, and neutral oxygen density $n_O \sim 10^{20}$–$10^{22} \, m^{-3}$. The molecular gas temperature T_g was in the range 300 to 450 K. The samples were treated in the reactive plasma environment for different durations, depending on the neutral atom dose needed to create the nanostructures. After the plasma treatment, the samples were analyzed by scanning electron microscopy Microscopy

Figure 8.9 shows the evolution of the Cd foil temperature during the first 20 s of treatment in oxygen plasmas. Depending on the equilibrium surface temperature of the foil, the surface morphology of the plasma-exposed layer turns out to be quite different. Three distinctive surface morphologies of the nanostructured CdO are shown in Figure 8.10. When the surface temperature is low, a flat-grain surface morphology is formed (Figure 8.10(a)).

When the temperature is increased slightly above the melting point of bulk cadmium (594 K), pyramid-like nano- and microstructures arise

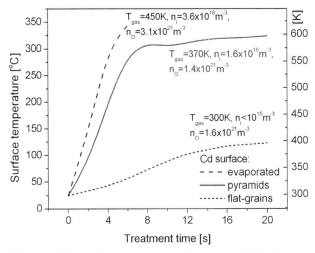

Figure 8.9 The surface temperature of cadmium foil during exposure to oxygen plasmas with different parameters [481].

as suggested by the electron micrograph in Figure 8.10(b–c). However, at very high surface temperatures cadmium material evaporates and no nanostructures are formed. The temporal dynamics of the surface temperature shown in Figure 8.9 corresponds to the above three (flat-grains, pyramids, and evaporation) typical cases.

The main tendencies observed by Cvelbar *et al.* [481] are:

- when the equilibrium temperature is too high (long dashed curve in Figure 8.9), the foil material is evaporated;

- when the surface temperature reaches the melting point of cadmium (594 K), pyramid-like nanostructures emerge as can be seen in Figure 8.10(b–c); in this temperature range a thin surface layer of the cadmium foil experiences a phase transition from the solid to the liquid state;

- a flat-grained structure with small oxide grains and no clear nanostructures (e.g. nanobelts, nanowires or nanopyramids) is formed at lower equilibrium temperatures (300–400 K). For simplicity, we will further refer to the flat-grained structure as the flat oxide.

When exposed to reactive plasmas, metal materials are normally heated up via physical (energy transfer) and chemical (chemical reactions) interactions of ions, atoms and gas molecules and radicals with the surface. Therefore, surface heating via low-energy (a few eV) ion recombination and/or energy exchange with the feedstock gas appears

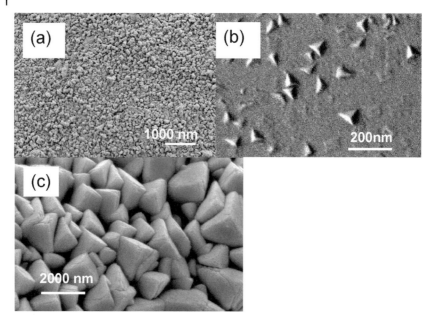

Figure 8.10 SEM images of cadmium oxide formed on the surface after oxygen plasma treatment with different plasma parameters. (a) The standard CdO in the form of grains is fabricated when the surface temperature does not reach the cadmium melting point. (b) The nano- pyramids are created after the surface receives the dose $1.5 \times 10^{24}\,\mathrm{m}^{-2}$ of oxygen atoms. (c) When the treatment is continued and the dose is increased to $7.3 \times 10^{24}\,\mathrm{m}^{-2}$, larger micro-pyramids are created over the whole surface sample area [481].

to be crucial for the growth of the nanostructures. These conditions were met when the molecular gas temperature was ca. 370 K, and the ion and oxygen atom densities were 1.6×10^{16} and $1.4 \times 10^{21}\,\mathrm{m}^{-3}$, respectively.

In this case, the solid-liquid (S-L) phase was created approximately 6 s into the discharge run; this eventually led to the growth of small CdO nanopyramids as shown in Figure 8.10(b). It would appear that, after transition of the cadmium material to the liquid state the interaction of the plasma with the surface changes. Moreover, this liquid phase can serve as an effective nanostructure growth catalyst, similar to the case of plasma-assisted growth of vertically aligned nanostructures of Chapter 6.

The growth of pyramidal nanostructures is an essentially three-dimensional process, in which nanosized nuclei evolve into micron-sized objects. Moreover, CdO crystals ($S2$ phase in the $S1 - L - S2$ system, where $S1$ corresponds to the cadmium foil) grow from the $S - L$ phase due to interactions of neutral oxygen BUs with Cd atoms. First, oxygen atoms

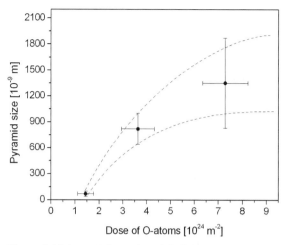

Figure 8.11 Lateral dimension of CdO pyramids versus the dose of neutral oxygen atoms deposited onto the sample surface under the solid-liquid phase conditions. The growth rates in each case can be deduced by dividing the nanocrystal sizes shown in this figure by the process time, which was approximately 20 s [481].

are dissolved inside the cadmium melt. Thereafter, the interfacial surface tension pushes cadmium and oxygen atoms (as well as any CdO molecules formed in the liquid phase) to the outer surface, on which tiny nuclei of pyramidal nanocrystals are eventually formed.

Once the nuclei are formed, further growth of the pyramidal nanostructures occurs through basal attachment of the solute material to the nucleus surface. This is quite similar to what happens during the growth of platelet-structured carbon nanocone-like structures of Chapter 7. The as-nucleated crystals can only grow vertically outwards from the solvent if the lateral growth is restricted by strong interfacial tension between the solute and the solvent.

The effect of the dose of oxygen atoms (solutes) supplied to the surface is quantified in Figure 8.11. One can see that as the dose of neutral atoms increases (this dose is controlled by the prevailing rates of dissociation of oxygen molecules in the discharge), the nanostructure sizes increase very rapidly until the whole surface is covered with CdO crystals. Then the growth process slows down, as the surface is covered and there is no more space to support three-dimensional spread. When this occurs, the lateral size of the pyramids still increases through basal attachment of solutes, albeit rather slowly. The two dashed curves in Figure 8.11 show the lower and upper limits of the CdO pyramid lateral size at different doses of oxygen atoms.

An interesting observation is that the pyramid size deviation increases with the dose of oxygen atoms impinging onto the surface. Probable reasons for the observed increase in size non-uniformity may be related to increasingly irregular (both in time and space) nanocrystal nuclei formation as well as higher rates of pyramid growth in the lateral direction giving rise to a much earlier overlap of individual nanostructures [481].

Thus, the above plasma-aided fabrication process is a very effective tool for controlling the surface growth of cadmium oxide nano- and microstructures by varying the dose of neutral oxygen atoms impinging onto the surface. However, some other conditions need to be fulfilled to make the observed cadmium oxide nano- or micro-pyramid surface structures a reality [481]. The initial heating of the metal foil to reach the solid-liquid phase needs to be quick yet moderate to avoid unwanted evaporation of the foil material. This can be controlled by adjusting the plasma parameters such as the plasma density, electron and ion temperature or their energy, neutral atom density, as well as the gas feedstock temperature. Under certain oxygen plasma conditions crystal pyramids can grow from the cadmium solid-liquid phase [481].

To conclude this section, we stress that the formation of a liquid phase during the plasma treatment is not always required for the synthesis of metal oxide nanostructures via direct oxidative treatment of metal foils. In fact, niobium oxide, vanadium oxide and iron oxide nanowires can grow from the supersaturated solid-vapor phase. For more details, please refer to the orioginal reports [96,501].

In the following section, we will consider an example of where different plasma-generated building units are used to synthesise nanostructured films with unique properties. It will be shown how different ultra-small nanoclusters created in a plasma-based magnetron sputtering process lead to very different nanocrystalline structures, and eventually, a biomimetic response of nanostructured titanium dioxide films.

8.4
Biocompatible TiO$_2$ Films: How Building Units Work

In this section, following the original work [502], we show how nanocrystallinity, phase composition, and biomimetic response of TiO$_2$ can be controlled in plasma-assisted nanofabrication. In other words, a genuine manipulation of the plasma-generated building units to fabricate the desired nanostructured coatings appears to be a smart way to enhance biomimetic response.

Briefly, the original work [502] sheds light on the role of crystal size and phase composition in inducing biomimetic apatite growth on the surface of nanostructured titania films synthesized by reactive magnetron sputtering of titanium targets in $Ar + O_2$ plasmas. Unlike most existing techniques, this method enables one to deposit highly-crystalline titania films with a wide range of phase composition and nanocrystal size without any substrate heating or post-annealing. By using this dry plasma-based method one can avoid surface hydroxylation at the deposition stage, almost inevitable in wet chemical processes. Moreover, high phase purity and optimum crystal size appear to be the essential requirements for efficient apatite formation on magnetron plasma-fabricated bioactive TiO_2 coatings.

The formation of biologically active layers on bioimplant surfaces is crucial for their integration with bone. To this end, materials with the capability of inducing calcium phosphate layer formation on their surfaces are often referred to as bioactive materials. Every specific bioactive material has different factors that induce and control their bioactive response. Generally, such factors include reactivity, morphology, roughness of the surface, crystal size, phase composition, stoichiometry and some others.

Zhou *et al.* [502] answered the question which has puzzled the minds of researchers in the last few decades: What is the role of crystal size and phase composition in sustaining the outstanding biomimetic response of titania coatings? It is generally believed that the remarkable biomimetic response of TiO_2 is attributed to the existence of surface hydroxyl (Ti–OH) groups and the induced negative charges on the TiO_2 surface, which, in turn, draw calcium- and phosphorus-based cations from the simulated body fluid (SBF) to the implant surface [503–505]. However, this growth model does not explain frequent observations of different biomimetic responses of amorphous, rutile- and anatase-rich titania films derived from conventional wet methods, such as sol-gel and other chemical methods [505–517].

These techniques almost inevitably lead to pronounced surface hydroxylation during the film synthesis stage. In these cases the TiO_2 surface is grafted with hydroxyl groups even before immersion in the SBF, which make it extremely difficult to elucidate the effects of crystal size and phase composition on the TiO_2 biomimetic response in the SBF. Moreover, post-annealing of hydroxyl-terminated titania films often results in a different abstraction of hydroxyl groups from the anatase and rutile phases, leaving some of the sites on the TiO_2 surface activated and some passivated. Indeed, higher post-annealing temperatures are needed to convert amorphous TiO_2 films (synthesized by using wet chemical methods) into the rutile phase rather than into the anatase phase.

Thus, at lower temperatures, when the anatase phase is enriched, more hydroxyl groups remain on the surface than at higher temperatures when the rutile phase is formed. The excessive Ti–OH groups might be the main reason for the frequently reported stronger biomimetic responses of the anatase-rich titania coatings in the SBF *in vitro* tests [507, 509, 510]. Therefore, the question of whether the anatase is more bioactive than the rutile phase in the SBF environment still remains essentially open. It becomes evident that reliable and unambiguous information about the effects of crystal size and phase composition on the biomimetic responses can be obtained if the as-grown films with different crystal size or phase composition are free of surface hydroxyl radicals.

In this section, following the original work [502], we will discuss the application of a dry magnetron plasma-based method to fabricate nanocrystalline and hydroxyl-free titania films with the controlled anatase and rutile phase composition and nanocrystal size. We also comment on the role of manipulation of plasma-grown building units in the fabrication of nanostructured titania films with specific required attributes. The technique in question has been discussed in Chapters 5 and 6 of the related monograph [1] in relation to the plasma-assisted fabrication of biocompatible hydroxyapatite and calcium phosphate coatings as well as semiconducting nanostructured films and low-dimensional quantum confinement structures (see also [4] and references therein).

We emphasize that *in vitro* bio-activity of TiO_2 films deposited by reactive magnetron sputtering is a new and challenging aspect of modern biomaterials research. Zhou *et al.* reported on the reactive DC magnetron sputtering deposition of titania films in $Ar + O_2$ plasmas on *unheated* substrates. Moreover, the degree of crystal size and phase composition can be controlled by the process parameters and are unambiguously related to the biomimetic response in the SBF test environment [502].

8.4.1
TiO_2 Film Deposition and Characterization

A reactive DC magnetron sputtering technique without any external substrate heating was used to grow TiO_2 films. The distance between the high-purity Ti sputtering target and deposition substrates was 90 mm. High-purity argon (99.999%) and oxygen (99.5%) were used as sputtering and reactive gases, respectively. A summary of the process conditions, crystal sizes of anatase $a[An(101)]$ and rutile $a[An(101)]$ phases, and film thickness d are given in Table 8.2. Here, p_0, P_D, and V_s are the

Table 8.2 Process conditions, crystal sizes, and film thickness [502].

Samples	p_0 (Pa)	P_D (W)	V_s (V)	a[An(101)]	a[Ru(110)]	d (nm)
TOV	1.2	40	0	6.8	-	65
T80W	1.2	80	-100	18.6	16.3	79
T36P	3.6	40	-100	19.1	-	75
T12P	1.2	40	-100	-	-	72
T03P	0.3	40	-100	-	3.2	63

working gas pressure, power supplied to the DC magnetron, and substrate bias voltage, respectively.

The "deterministic" choice of deposition conditions was motivated by the aim to synthesize nanocrystalline titania films with the required crystal size and rutile/anatase phase composition, and ultimately, to relate the varied film attributes to their biomimetic response. In a sense, a number of possible "bioactivity turning knobs" of the plasma-assisted process have been explored. Likewise, the appropriate choice of sputtering and reactive gases effectively excludes any hydrogen source and thus prevents the formation of hydroxyl groups on specimens surfaces and therefore is favorable for a hydroxyl-free deposition environment. In this set of experiments, the O_2 and Ar flow rates were maintained at 5 and 30 sccm, respectively. Furthermore, the titania films were synthesized under five different sets of process conditions. Other details of the film deposition, biocompatibility analysis, and characterization can be found elsewhere [502].

Five TiO_2 samples deposited under different process conditions summarized in Table 8.2, have been analyzed. Samples T03P, T12P, and T36P are synthesized under the same deposition (sputtering) power $P_D = 40$ W and negative DC bias on the substrate $V_s = -100$ V and different working pressures ($p_0 = 0.3$, 1.2, and 3.6 Pa, respectively). Sample TOV was fabricated under the same pressure and sputtering power as T12P but without any external biasing of the substrate. Specimen T80W was deposited under the same working pressure and substrate bias as T12P and a higher sputtering power of 80 W.

This choice of process parameters made it possible to synthesize titania films with very different crystal size and phase composition [502]. Table 8.2 also summarizes the average film thickness and sizes of An(101) and Ru(110) crystals in each specimen. X-ray diffraction (XRD) patterns show that film T03P (deposited at $p_0 = 0.3$ Pa, $P_D = 40$ W, and $V_s = -100$ V) features a wide diffraction peak of Ru(110) originating from very small rutile nanocrystals with an estimated grain size of ca. 3.2 nm. Since XRD analysis of sample T03P does not reveal any other diffraction peaks, it is reasonable to assume that this film is composed of either a pure rutile phase or a mixture of crystalline rutile and amorphous titanium oxide.

When the total pressure increases from 0.3 Pa to 1.2 Pa, no diffraction peaks are detected in film T12P (deposited at $p_0 = 1.2$ Pa, $P_D = 40$ W, and $V_S = -100$ V). Thus, the film T12P is either purely amorphous or contains ultra-small nanocrystallites undetectable by the XRD. However, diffraction patterns of the film T36P deposited at a higher working pressure of 3.6 Pa and the same V_s and P_D show pronounced peaks corresponding to An(101), An(103), An(004), An(200), An(211), and An(204) crystalline planes (details can be seen in Figure 1 of Reference [502]).

In this case TiO_2 has an anatase structure with an estimated nanocrystal size of 19.1 nm (Table 8.2). Furthermore, the intensity of the An(101) peak is the highest, which is indicative of preferential crystal growth along the (101) crystallographic direction. An increase of magnetron sputtering power results in dramatic changes in film composition and structure. Indeed, sample T80W deposited at a higher ($P_D = 80$ W) magnetron sputtering power and with all other parameters identical as T12P, shows a mixed-phase crystalline structure with estimated anatase and rutile nanocrystal sizes of 18.6 and 16.3 nm, respectively. Amongst the variety of diffraction peaks detected, the An(101) and Ru(110) peaks appear to be the strongest.

Thus, the most efficient growth of anatase crystals proceeds along the (101) direction, similar to sample T36P. In contrast, the preferential growth direction of rutile crystals in specimen T80W is quite different from their growth under T03P conditions. The sole An(101) peak also persists in the XRD spectrum from film T0V synthesized without any external DC biasing of the substrate. However, this peak is much weaker and wider and suggests the presence of anatase nanocrystals with a size of ca. 6.8 nm.

FTIR transmission spectra of the TiO_2 films deposited on unheated silicon substrates under the deposition conditions of Table 8.2 suggest that the strong absorption peak at ca. 610 cm^{-1} and two weak absorption peaks around 740 cm^{-1} and 815 cm^{-1} are present in all samples, including the pure Si(111) test sample, and are thus attributed to infrared absorption by the substrate. On the other hand, strong absorption of Ti–O vibrations in the 400–600 cm^{-1} spectral range suggests the presence of Ti–O bonds in the deposited material.

Interestingly, specimen T36P and T80W with larger crystal sizes exhibit a much stronger infrared absorption in this spectral region compared with films T0V and T03P featuring smaller crystal sizes. Moreover, the amorphous titania film of sample T12P shows almost no infrared absorption in the 400–600 cm^{-1} range, very similar to the silicon test sam-

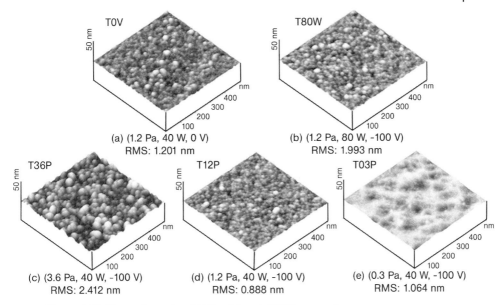

Figure 8.12 Three-dimensional (0.5×0.5 μm) AFM images of surface morphology of TiO₂ films deposited on unheated substrates under different deposition conditions [502].

ple. One can thus conclude that TiO₂ films with larger nanocrystal sizes feature stronger bonding between titanium and oxygen atoms, which is reflected by intense infrared absorption peaks in the relevant infrared spectral range.

More importantly, the FTIR spectra do not show any infrared absorption lines characteristic of hydroxyl group vibrations. Indeed, typical FTIR spectra of titania films synthesized by wet chemical methods feature absorption bands around $3400 \, \text{cm}^{-1}$ and $3700\text{–}3800 \, \text{cm}^{-1}$, which are completely absent in the experiments discussed here. Therefore, the TiO₂ films synthesized by plasma-assisted reactive magnetron sputtering without any external substrate heating and post-annealing *are free of surface hydroxyl (Ti–OH) groups* [502].

Figure 8.12 shows three-dimensional surface morphology of the TiO₂ films imaged by the AFM in a tapping mode over the surface area 0.5×0.5 μm. The process parameters are shown separately for each of the samples. The surface roughness of samples T36P and T80W is the largest, followed by T0V, T03P, and amorphous T12P.

Moreover, there is a remarkable correlation between the surface roughness and the estimated size of the nanocrystals [502]. Specifically, films T36P and T80W with the largest nanocrystal size are the roughest, whereas the films T0V and T03P with smaller crystal sizes are smoother.

Meanwhile, the amorphous T12P film appears to be the smoothest. Generally speaking, a larger crystalline size corresponds to a rougher surface morphology. However, for all samples the surface roughness remains smaller than the estimated nanocrystal size, indicating the presence of amorphous overcoats covering the nanocrystalline matter.

The surface roughness is intimately related to the wettability of the coating, which is characterized by the water contact angle. Interestingly, the measured values of the contact angle for all of the specimens fall within the range of 87–95°. Reasonably high values of the contact angle suggest low surface energy of the films synthesized by the plasma-assisted magnetron sputtering technique.

8.4.2
In Vitro Apatite Formation

We now discuss the bioactivity properties of TiO_2 films evidenced by the apatite crystal ingrowth during specimen immersion in the simulated body fluid. The XRD patterns of the samples after soaking in the SBF for 7 days feature the most prominent (211) diffraction peak at $2\Theta = 32°$, attributed to the apatite crystalline structure.

This peak has been observed in the XRD spectra of the anatase TiO_2 films of samples T36P and T0V. Moreover, the intensity of the apatite diffraction peak is higher for the T36P film with the largest anatase nanocrystal sizes. No other calcium phosphate phases have been observed in the XRD spectra [502]. FTIR transmission spectra of the TiO_2 films after soaking in the SBF for 7 days show the presence of new materials that contain the PO_4 groups and grow on the surfaces of the TiO_2 films in the SBF environment.

Furthermore, apart from the anatase TiO_2 films of samples T36P and T0V, the rutile TiO_2 film of sample T03P and the anatase/rutile mixed TiO_2 film of sample T80W also induce nucleation (in the SBF) of materials containing PO_4 groups on the surface. More importantly, no changes have been found in the FTIR spectra of the amorphous TiO_2 film of sample T12P immersed in SBF for 7 days.

Figure 8.13 shows SEM micrographs of the surfaces of TiO_2 films deposited at different deposition conditions and soaked in SBF for 7 days. A notable apatite precipitation has been observed on the surfaces of the three samples; two of them have an anatase structure (T0V, Figure 8.13(a) and T36P, Figure 8.13(c)) and the third one has a rutile structure (T03P, Figure 8.13(e)). From Figure 8.13(b), related to sample T80W with the presence of mixed anatase/rutile phases, one can notice a low degree of

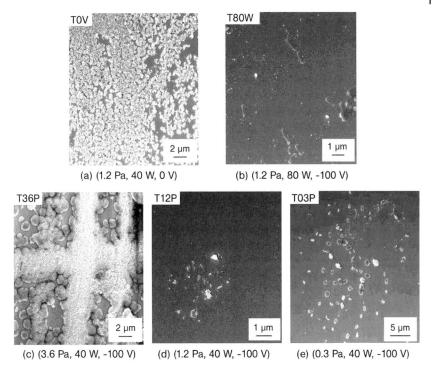

Figure 8.13 SEM micrographs of the surfaces of TiO$_2$ films after immersion in SBF for 7 days [502].

surface coverage by irregularly shaped particles. Presumably, such particles contain the PO$_4$ groups indicated by the FTIR analysis.

The most striking observation is the complete absence of any precipitation from the SBF onto the surface of the amorphous film T12P (Figure 8.13(d)). In this case, only loosely attached contaminant particles of irregular shapes have been observed [502]. On the rutile surface of sample T03P with a small crystal size of 3.2 nm, apatite particles with an average size of approximately ca. 1 μm partially cover the surface (with the average surface coverage not exceeding 20 %) as shown in Figure 8.13(e). As be seen in Figure 8.13(c), the anatase film of sample T36P with a large crystal size of 19.1 nm induces the strongest apatite formation.

Higher-resolution SEM images in Figure 8.14(b) suggest that the thickness of the apatite layer formed on the surface of the anatase film T36P exceeds 1 μm. Moreover, the porous microstructure of the apatite layer in Figure 8.14(b) is very similar to that reported elsewhere [359,505,507, 511,516,517] (see also Chapter 6 of the related monograph [1]).

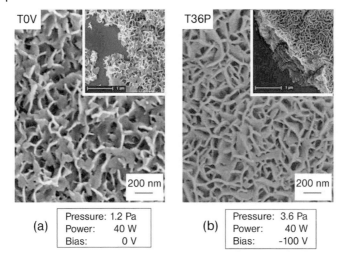

Figure 8.14 High-resolution SEM images of the surface morphology of the apatite layers deposited on anatase TiO$_2$ films of sample T0V and sample T36P after immersion in SBF for 7 days [502].

A continuous layer of porous apatite also covers the surface of the anatase film of sample T0V with a smaller crystal size of 6.8 nm. However, this layer is thinner and less dense as can be seen in Figure 8.14(a). It is also worth noting that scanning electron microscopy (Figures 8.13 and 8.14) reveals similar trends suggested by the XRD and FTIR results (not shown here) [502].

8.4.3
Growth Kinetics: Building Units at Work

Let us now discuss the growth kinetics of titania films synthesized by the plasma-assisted reactive magnetron sputtering process. To elucidate the elementary processes that occur in the ionized gas phase and on the deposition surfaces, we adopt the generic "plasma-building unit" approach that is applicable to a wide variety of plasma-assisted materials synthesis and processing applications [4] (see also Chapters 1–3).

The overwhelming complexity of possible sequences of multiple elementary reaction steps makes the analysis particularly difficult. However, by organising the possible elementary building blocks and their dynamics in the plasma sheath and on the solid surfaces one can develop reasonable scenarios for virtually any process of synthesis of various thin films and nanoassemblies.

Following the original work [502] and using the available literature on plasma-assisted synthesis of TiO$_2$ clusters, films, and nanoparticles, we consider the following range of building and other functional units (species) that originate in the ionized gas phase: Ti atoms; positive ions Ti$^+$, Ti^{2+}, and Ti^{3+}; oxygen atoms, molecules, and ions (including anions); non-reactive argon neutrals and energetic argon ions; amorphous titania clusters with the sizes less than 2 nm and crystalline clusters larger than 3 nm.

In the above, we have noted that the amorphous-to-crystalline transition in TiO$_2$ nanoclusters takes place when the cluster size exceeds ca. 2 nm [518]. Each of the species listed above serves a specific purpose. Specifically, Ti, Ti$^+$, Ti^{2+}, and Ti^{3+} are the titanium source species; oxygen species serve as oxidizing reagents needed for the synthesis of titania; TiO$_2$ nanoclusters are important building units that directly incorporate into growing films; energetic argon atoms activate and heat the surface. Likewise, low-energy (room temperature) neutral argon atoms are non-reactive and while they do not participate directly in the film growth process, they strongly affect (via elementary ion-neutral and neutral-neutral collision processes) the transport of titanium species from the sputtering target to the deposition surface.

Following the charged cluster theory (CCT) [65, 267, 268], one can assume that the nanocluster charge is positive in the specified size range. The presence of charged clusters in the few nanometers size range has been surmised by comparing the deposition conditions with the available results reporting the size dependence of the plasma-grown TiO$_2$ nanoclusters on the process parameters in a fairly similar plasma-assisted reactive magnetron sputtering process [267, 268]. The assumption of positive charge is also consistent with other reports on plasma-generated positively charged clusters and nanocrystallites of other materials in this size range [92]. Another important point frequently sidestepped in plasma-assisted materials synthesis and processing works is the fact that the plasma bulk is separated from the solid surface by the non-neutral plasma sheath area, which is collisionless under the low-pressure conditions of the experiments of Zhou *et al.* [502].

The nanofilm growth kinetics are sketched in Figure 8.15, which illustrates the effect of changing process conditions on the structure and phase composition of the films of our interest here. We recall that the T03P sample has been synthesized under a low pressure (0.3 Pa), "normal" (40 W) deposition power and "normal" (−100 V) bias.

In this case, sketched in the top left drawing in Figure 8.15, energetic argon ions heat and activate the surface. The background pressure is rather low for the formation of titania molecules and clustering in the

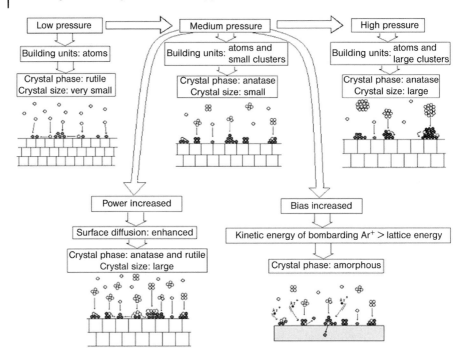

Figure 8.15 Growth kinetics of TiO$_2$ films in the plasma-assisted reactive magnetron sputtering deposition [502].

gas phase. Since the pressure of bombarding argon ions is low, the influx of titanium atoms and ions onto the deposition surface is also low; upon adsorption they become adatoms and diffuse, together with oxygen atoms, about the silicon surface, to nucleate and form nano-sized islands on the surface. In this case the growth islands develop into small rutile nanocrystallites.

Under the low-pressure conditions of T03P, formation of rutile appears to be energetically favorable. It is instructive to mention that rutile is the most stable form of TiO$_2$ and its formation is preferred under equilibrium conditions. Moreover, the critical nucleus size needed for crystallization is smaller for the rutile phase than for the anatase phase. In this case the supply of titanium-bearing BUs is too low to create the critical nucleates suitable for crystallization of the anatase phase. This conclusion is consistent with other reports on preferential synthesis of rutile under very similar pressures, for example 0.27 Pa [519].

Upon a four-fold increase of the working pressure, the density of argon and oxygen neutral species increases accordingly in the depositing sample T12P. Since the heat generated in the plasma process could be efficiently transferred out of the deposition chamber by the water-cooling system, the increase of the process pressure does not lead to any significant changes in the temperature of the gas (which thus remains at room temperature) in thermodynamic equilibrium. According to Pashen's curves in this pressure range [520] the density of the energetic argon ions that are mainly responsible for the sputtering of the titanium target also increases.

Thus, sputtering of the titanium target becomes more efficient giving rise to a larger number of titanium-containing building units supplied to the deposition site. This effect is somewhat hindered by the collisions of Ti atoms and ions with the argon gas maintained at a higher (than in the case T03P) pressure. However, since the argon gas temperature does not significantly change at higher pressures, the titanium species arrive at the silicon surface with the same kinetic energy as in case T03P owing to the rapid thermalization in the ambient room-temperature gas.

Therefore, the supply of titanium-bearing building units to the growth surface in the case T12P is likely to be excessive, which results in purely amorphous films. In this case the rates of surface diffusion are insufficient to redistribute the deposited adatoms over the surface; accordingly, excessive piling up of the building material results in a significant disorder in the films.

It is quite possible that the pressure (1.2 Pa) is too low to induce any noticeable growth of larger crystalline clusters (> 3 nm) in the gas phase and there are mainly small (< 2 nm) positively charged amorphous TiO₂ clusters in the gas phase. Such clusters contribute to the higher content of the amorphous phase. Moreover, all crystalline clusters with cohesive energies less than 100.0 eV most likely break into smaller fragments upon deposition on the biased Si(111) surface. This disintegration of crystalline nanoclusters further contributes to the buildup of the amorphous phase.

When the pressure is increased to 3.6 Pa (sample T36P, top right sketch in Figure 8.15), the supply of titanium-bearing building material further increases and becomes sufficient to generate larger (supercritical) nucleates both in the gas phase and on the surface. Crystalline clusters with cohesive energies exceeding 100.0 eV are likely to form in the gas phase; they do not break upon landing on biased silicon surfaces.

In this case the film can contain a large amount of anatase nanoclusters embedded in the amorphous matrix. It is remarkable that other authors also reported purely anatase films in the pressure range exceeding 2.7 Pa [519]. Specimen T0V is synthesized under the same magnetron sputtering power and gas pressure as the purely amorphous sample T12P, but without any external biasing of the substrate (sketch in the middle of the top row in Figure 8.15). In this case, the argon ion bombardment is significantly reduced and the surface temperature becomes lower than in cases T03P and T12P.

This condition becomes even less favorable for the rutile phase formation (compared to conditions T12P and T03P) as higher surface temperatures are usually required for the synthesis of rutile crystalline films. Since the substrate is unbiased, the gas-phase grown charged clusters do not disintegrate upon landing and the film contains a notable amount of small nanocrystallites, which is quite similar to T36P. Under such conditions, the surface temperature becomes lower than is necessary to form rutile crystals via adatom migration about the surface (as was the case for T03P).

This lower surface temperature reduces the critical nucleate size for the anatase phase crystallization, and it becomes possible to synthesize smaller anatase crystals (as can be seen in case T0V). However, it is already not possible to grow the rutile crystals, as the temperature is lower and the amount of supplied Ti atoms and ions is approximately the same as in case T12P.

In the T80W case, the gas pressure and substrate bias are the same as in T12P, but the magnetron sputtering power is doubled, from 40 to 80 W. In this case, depicted in the bottom left drawing in Figure 8.15, the plasma density substantially increases, as do the electron-impact ionization rates of argon, oxygen, and titanium species. Stronger fluxes of Ar^+ ions result in the highest (among the 5 specimens of Table 8.2) surface temperatures.

This makes it possible to significantly enhance the rates of surface diffusion and thus avoid piling-up of deposited building material, as was the case under process conditions T12P. This enhances the nucleation of islands on the surface (Volmer–Weber growth mode in lattice mismatched systems [179]). If the island growth is unobstructed by the piling up of undesired amorphous deposits, the resulting island size distribution is usually Gaussian-like [179].

Therefore, in this case there should be a sufficient number of nucleates of a suitably small size to act as growth seeds of rutile crystallites. Thus, rutile nanocrystals are also present in the film, similar to case T03P. On the other hand, small charged nanoclusters grown in the gas phase also break upon deposition, as happens under the T12P process conditions. However, the surface temperature is already high enough and there is a sufficient amount of supercritical nucleates for the efficient adatom diffusion-driven crystallization on the surface.

It is important to note that an excessive increase of the bias should increase the likelihood of energetic-ion induced disorder in the TiO_2 films. Indeed, if the energy of the impinging argon ions exceeds the lattice cohesive energy, the crystalline structure can be damaged and the films can become amorphous [518].

This situation is depicted in the bottom right drawing in Figure 8.15. However, if the rates of crystallization are larger than those of the crystalline structure damage, crystalline structures are still possible. Therefore, there is a tradeoff between higher surface temperatures and crystalline structure destruction caused by energetic argon ions.

Furthermore, heating of the surface to 915 °C (by either applying a larger bias or an external substrate heating) results in the phase transformation of any existing anatase crystals into the more stable rutile form. Finally, we note that these observations of structural and phase transformations in titania films are consistent with the report of Zeman *et al.* [521].

8.4.4
Building Units *In Vitro*: Inducing Biomimetic Response

We now comment on the roles of crystal size, and phase purity/composition in inducing bioactivity of the titania films synthesized by the plasma-assisted reactive magnetron sputtering deposition [502]. It is commonly believed that the existence of surface hydroxyl (Ti–OH) groups is crucial for apatite precipitation on the surfaces of titanium oxide films in the simulated body fluid [504,505].

After immersion in the SBF with neutral pH, acidic reaction instead of basic reaction dominates on the surface of TiO$_2$ film since the acidic hydroxides are deprotonated via the following reaction [511,522]

$$\text{Ti} - \text{OH} + \text{H}_2\text{O} \leftrightarrow [\text{Ti} - \text{O}]^- + \text{H}_3\text{O}^+$$

which results in an induced negative charge on the TiO$_2$ surface. The negatively charged Ti–OH$^-$ groups serve as precipitation sites for positive calcium and calcium oxide cations to form calcium titanate or calcium titanate oxide. As a result of positive charge buildup, the surface gradually becomes positively charged. In turn, a positively charged surface attracts negatively charged phosphate ions that combine with calcium-bearing species to form amorphous calcium phosphate with a low [Ca]/[P] ratio.

It is also commonly presumed that the amorphous calcium phosphate formed as a result of surface hydroxyl-induced precipitation, spontaneously transforms into bone-like apatite (which is essentially crystalline) by continuously consuming calcium and phosphate ions in the SBF [503]. Therefore, the existence of Ti–OH groups on the surface is widely considered to be the most essential prerequisite for the successful formation of all kinds of bioactive TiO$_2$ films [503].

For this reason, substantial international research efforts have been focused on intentional grafting of Ti–OH groups onto titania surfaces, for example, by sol-gel chemical methods, heat treatment in water vapor, or hydrogen plasma treatment, and so on. [505, 512, 514]. However, the existing nucleation models merely based on grafted Ti–OH groups do not include any charge neutral pathways and also cannot explain why amorphous TiO_2 films produced by existing wet methods do not show any significant biomimetic response (there is generally a large amount of Ti–OH groups on the surface of TiO_2 films produced by wet methods).

Moreover, none of these models shed any light on how specifically and how fast the disordered amorphous precipitates transform into crystalline apatite to stand a chance of mimicking natural bone apatites. From a clinical applications point of view, crystallization should be induced instantaneously, without any intermediate resorption frequently observed in *in vitro* SBF tests involving hydroxyapatite bioceramics [359] (see also Chapter 6 of the related monograph [1]).

To single out the effect of crystal size and phase composition on the bioactivity of TiO_2 films, during the synthesis stage the formation of Ti-OH surface groups, almost unavoidable in wet chemical methods, was *intentionally* avoided. Nevertheless, several samples exhibited a notable bioactivity in the SBF, as is clearly seen in Figures 8.13(a,c,e). We emphasize that without the interference from Ti–OH surface groups grafted during the film synthesis stage one can unambiguously investigate the effect of crystal size and phase composition on the bioactivity of the deposited TiO_2 films.

Among 5 remarkably different samples, two are purely anatase and contain nanocrystals of different sizes (T36P and T0V) [502]. Sample T03P is a phase-pure rutile, whereas sample T80W is a mixture of rutile and anatase phases. Finally, sample T12P is amorphous. We recall that the best biomimetic response is generated by the purely anatase samples T36P and T0V (Figures 8.13(a) and (c)). Interestingly, the TiO_2 surface coverage by the apatite material is higher in the T0V case (Figure 8.13(a)), when the estimated anatase nanocrystal size is smaller (ca. 6.8 nm). On the other hand, the film thickness and surface morphology feature sizes are larger under process conditions T36P.

This observation can be explained in two respects [502]. Firstly, with the decrease of crystal size, there are more crystal boundaries, defects and monoatomic terraces that can serve as sites for the apatite nucleation and subsequent crystallization in SBF. Lattices of smaller crystals are strained and are generally more energetically favorable for new building units to insert and recrystallize in the existing crystal structure. However, it is more difficult to insert into larger crystals with established and strain-relaxed lattices.

This could be the reason why (nucleation-site-free) single-crystalline anatase exhibits no bioactivity [504]. Secondly, the lattice match (both lattice parameters and orientation) is an important factor that controls epitaxial growth and recrystallization of apatite crystals from initial nucleation sites on TiO$_2$ surfaces [507,509,510,515]. When the crystallinity is low and the nanocrystal size is small, the mutual disorientation of the apatite and titania lattices increases, mostly because of the higher structural disorder on the surface. This disorder is highest in amorphous titania featuring chaotically oriented lattice nanofragments.

Moreover, the surface strain becomes so irregular that the Volmer–Weber mechanism applicable to the crystal growth in lattice-mismatched crystalline systems can no longer sustain the apatite crystalization. This explains, in part, why amorphous TiO$_2$ layers always fail to induce any apatite formation during immersion in the SBF, which can be seen in Figure 8.13(d) and was reported elsewhere [507–510]. Therefore, there is a tradeoff between crystal size, surface defects and structural disorder of anatase TiO$_2$ films to induce the best in vitro biomimetic response [502].

The phase purity of the titania films is an important element in sustaining the required apatite precipitation. Indeed, the phase-mixed TiO$_2$ film of specimen T80W did not induce any significant apatite ingrowth. This observation can be interpreted, in part, in terms of an increased structural disorder on the surfaces of phase-mixed systems. However, it still remains unclear as to why the apatite does not follow the poly-nanocrystalline growth scenario peculiar to the mixed-phase anatase/rutile film of sample T80W.

Although only a little apatite formation is observed after the 7-day soaking of sample T03P in the SBF, this observation does not necessarily indicate the conclusion that the biomimetic response of rutile phase is worse than that of the anatase phase, as several existing reports suggest. Perhaps it is the very small nanocrystal size (ca. 3.8 nm) of film T03P that led to such weak biomimetic response. One can thus speculate that only when the crystal size of the TiO$_2$ films is similar and the interference of initial Ti–OH groups on their surface is excluded, it becomes possible to compare biomimetic responses between anatase phase and rutile phase not only *in vitro* but also *in vivo*.

To conclude this section, we stress that the plasma-based magnetron sputtering process turned out to be extremely useful in synthesizing biocompatible titanium dioxide films with the tailored nanocrystalline structure and phase composition. More importantly, using the arguments that involve the plasma-generated nanocluster building units and the generic nanofabrication approach of Chapter 2, one can explain the relation between the plasma-based process parameters and the film microstructure. This ability is one of the most important milestones on the way to deterministic nanofabrication.

8.5
Concluding Remarks

The main focus of this chapter has been on continuing to demonstrate the advantages and benefits of using low-temperature plasmas for nanofabrication. While the pervious sections were mostly related to the growth processes of various nanostructures, this chapter dealt with other aspects of nanoscale processing. In terms of the "cause and effect" approach introduced in Chapter 2, the species involved have also been from the working unit rather than merely from the building unit category. The examples considered in this chapter ranged from post-processing of surfaces of one-dimensional vertically aligned nanostructures to using plasma-generated nanocluster BUs to synthesize biocompatible titania films.

In Section 8.1, we considered two examples of plasma-assisted post-processing of vertically aligned nanostructures arranged in arrays of different densities. These processes are usually subdivided into thin layer (e.g. monolayer) coating, functionalization, and doping. In many applications, it is a requirement that only selected areas are processed. Moreover, different areas of nanostructures may need to be processed differently. For instance, to enhance field emission properties of carbon nanotips, one might think about depositing a thin layer of emitting material with a very low work function onto the top, implant atoms of rare earth metals over the entire lateral surface, and deposit a "collar" made of a sticky material around the base. The first two actions would enhance the yield of field emission whereas the last one is intended to improve the nanostructure stability on the substrate.

In Section 8.1.1, the advantages of low-temperature plasmas for post-processing of dense arrays of vertically aligned carbon nanotubes have been summarized. Specifically, by using plasma-extracted ion fluxes, carbon nanotubes can be uniformly coated and treated along their entire length. Moreover, the uniformity of the ion flux deposition is optimal when the thickness of the plasma sheath is approximately one order of magnitude larger than the nanotube length. More importantly, by manipulating the plasma parameters, one can direct the ion flux to pre-selected areas on the nanotube surfaces. Interestingly, this effect can also be used for deterministic low-temperature plasma-assisted synthesis of dense arrays of carbon nanotubes [172].

This topic was continued in Section 8.1.2 for carbon nanorods arranged in microscopic arrays of different density. Similar to the conclusions of Section 8.1.1, the ion trajectories and deposition can be effectively controlled by the electrostatic potential drop across the plasma sheath. Likewise, by changing the spacings between the nanorods, the nanoarray type, or aspect ratio of the nanostructures, it is possible to steer, with sub-nanometer precision, the ion flux over the surfaces of the nanorods. In particular, this makes it possible to optimize the process of carbon nanorod coating with a hydrogen monolayer [34].

More specifically, the main findings presented in Section 8.1.2 can be summarized as [34]:

- by altering the spacing between the nanorods one can control the ion penetration into the arrays, the relative ion deflections and the magnitude of the microscopic ion current. An increase of the inter-nanorod spacing leads to better ion penetrations, larger ion currents to the nanostructures, increases the coating thickness at the base sections of the nanorods, as well as minimizes the time (being approximately 11 and 5 s for a plasma sheath with a thickness of 7 and 50 μm, respectively) required to coat the structures with a functional monolayer;

- adjusting the plasma sheath width provides a high degree of control of the ion deflections from straight downfall paths and eventually their penetration into the array. By enlarging the sheath, one can reduce the ion deflections and eventually achieve a better and deeper ion penetration into the arrays. However, this also results in an increase in the time required to fully coat the nanorods due to a decrease in the ion current to the nanostructures;

- the aspect ratio of the nanorods strongly affects the ion current to the nanostructures. Indeed, by reducing the radii of the nanorods, the ion flux to their surfaces can be greatly increased. The ion penetration is also affected by the radii of the nanorods; however, this effect is particularly important in sufficiently dense nanoarrays;

- there is a tradeoff between the time needed to deposit a monolayer coating over the entire surface of the nanorods and the uniformity of such a coating; for every specific nanorod array, the plasma process parameters can be adjusted to produce uniform ion flux distributions over the nanorod lateral surfaces, yet maintaining reasonably high deposition rates;

- finally, the time required to deposit a hydrogen monolayer over the entire surface of carbon nanorods is in encouragingly reasonable agreement with the experimental results on plasma-assisted functionalization of related carbon nanostructures in hydrogen-based plasmas.

The original model of Tam *et al.* [34] should be refined by inclusion of more details of the plasma-, nanostructure-, and surface-related phenomena, for example surface charging by microscopic electron and ion fluxes. This improvement will eventually make it possible to develop higher-fidelity, microscopic models of the interaction of the plasma-generated species with the nanostructure surfaces, and so more realistically quantify the growth kinetics of the nanostructures and their functional overcoats.

Nevertheless, the results presented in Section 8.1 give a clear indication of how and where exactly the plasma-generated ionic building units are deposited. The results of such numerical experiments can serve as input conditions for microscopic models of surface/interface phenomena on the surfaces of a large number of arrayed nanorod-like structures. Most importantly, this approach is generic, can be applied to a broader range of nanostructures and materials, and is directly relevant to the development of deterministic strategies towards precise and cost-efficient plasma-aided nanofabrication.

In Section 8.2, we presented the results of Monte Carlo numerical simulations of transport and deposition of Au^+ ions onto the three-dimensional nanostructured mask/substrate system. The investigation into the three-dimensional microscopic topography of the ion flux distributions over the surfaces of the nanoporous template (including the inner surfaces of the pores) helps one to optimize the plasma process conditions that make it possible to grow cylindrical metallic nanodots within the nanopores. Moreover, these numerical experiments can also be used to predict the nanodot shapes. The relative ion deposition rates increase and the distributions of ion fluxes over the lateral nanopore surfaces become more uniform when the electron temperature or the template height decrease.

Additionally, increasing the cross-sheath potential drop or the nanopore diameter leads to the same result. The nano-scaled microscopic topography of the electric field is an important factor which causes these changes. Therefore, by controlling and steering the ion fluxes about the template, nanopore, and nanodot growth surfaces, one can achieve the as yet elusive deterministic level in plasma-based, porous template-assisted nanofabrication of ordered arrays of metallic nanodots.

Section 8.3 primarily aimed at showing advantages offered by plasma-aided direct nanofabrication of metal oxides in controlling the growth and size of nano- and micro-structures on their surfaces. The salient features of this method are its simplicity and very high growth rates compared to wet chemical and other neutral gas-based techniques. In fact, pyramidal nanostructures emerge within ca. 10 s from the discharge ignition.

Other benefits of this plasma-aided fabrication include a complete absence of additional surface catalyst, extra flexibility in manipulation of precursor species in the near-surface area, and several others. The method is based on the self-organization of atoms on the material surface, which is controlled by the process parameters, most effectively by the density of neutral oxygen atoms and the surface temperature. More specifically, atomic oxygen building units are created in the plasma and then interact with the surface (together with oxygen ions) to create suitable conditions for the growth of nanostructures. When the surface of the metal foil melts locally, oxygen BUs penetrate into the liquid phase and combine with metal atoms giving rise to basal growth of platelet structured nano-pyramids made of metal oxide. These structures grow in a rather similar way to single-crystalline platelet-structured carbon nanotips of Chapter 7. The elegant experiment of Cvelbar *et al.* [481] has demonstrated the usefulness of the reactive plasma-based approach in nanofabrication of arrays of cadmium oxide nano- and micron-sized pyramidal structures, which are very promising for advanced optoelectronic and other applications.

In Section 8.4 we considered an advanced plasma-assisted DC magnetron sputtering deposition of nanostructured titanium dioxide films on unheated substrates. These films feature different nanocrystal size, anatase/rutile phase composition, and are free of *surface hydroxyl groups* [502]. By using the "plasma-building unit" approach of Chapter 2, one can explain the growth kinetics of the films with the required crystal size and phase purity and select appropriate sets of process parameters. For example, by increasing the working gas pressure from 0.3 to 3.6 Pa, it is possible to switch the process output from the pure-phase ultrananocrystalline rutile to pure-phase nanocrystalline anatase.

By analyzing the biomimetic responses of all of the specimens in the simulated body fluid (SBF) environment, one can confidently surmise that phase-pure anatase films feature the best biomimetic response, which is proven by the pronounced apatite precipitation, nucleation, and crystallization in the SBF and also depends on the TiO_2 nanocrystal size. Bioactive responses of mixed-phase anatase/rutile and ultrananocrystalline rutile appear to be weaker.

Furthermore, since neither amorphous nor single-crystalline anatase can induce the apatite formation in the SBF, one can conclude that high phase purity and structural order, as well as optimum nanocrystal size and surface density of nucleation sites appear to be the essential requirements for efficient apatite crystallization on magnetron plasma-fabricated bioactive titania coatings.

All the TiO_2 films of Section 8.4 have almost no initial hydroxyl groups on their surfaces, but several of them still exhibit intense biomimetic response in SBF. Therefore, one can speculate that the existence of initial hydroxyl groups on the surface *may not be the main decisive factor* in inducing the positive bioactive response of biocompatible titania films, as is commonly believed.

Let us now summarize the main advantages of the dry plasma-based method of fabrication of nanostructured biocompatible TiO_2 films compared with most of the conventional wet chemical methods [502]:

- without interference from initial surface hydroxyl groups that are almost unavoidable in many wet chemical processes, it is possible to study more easily the effect of crystal structure (such as crystal size and phase composition) of the TiO_2 films on their bioresponse not only *in vitro* but also *in vivo*;

- in the plasma-based process, one can deposit TiO_2 films with a wide range of crystal size and phase composition by varying the main process parameters such as the deposition power, total pressure, oxygen partial pressure, substrate bias voltage, distance between the substrate and titanium target, and so on;

- since the deposition can be operated at room temperatures without any additional substrate heating or post annealing through the use of plasma-based methods, it becomes possible to deposit bioactive titania films on many kinds of temperature-sensitive materials.

These results have an enormous potential to boost the interest of the research and development community and biomedical industry towards the plasma-assisted fabrication of biomaterials, in particular, temperature-sensitive and cost-efficient materials such as polymers and plastics.

Finally, the examples presented in this chapter present interesting possibilities arising from the unique and highly-unusual properties of low-temperature plasma environments, where a range of suitable building and working units can be generated and gainfully used.

9
Conclusions and Outlook

This monograph aimed at introducing plasma nanoscience as a distinctive research area, introduce its main aims and approaches and show how plasma-based environments work in assembling a variety of nanoscale objects in very different settings ranging from stellar outflows in astrophysics to laboratory experiments and applications in nanotechnology. The main focus of this work has been on presenting specific case studies that examplify salient benefits and advantages of plasma-based approaches in various forms of nanoscale assembly and processing.

Moreover, we attempted to highlight the most important aspects and physical phenomena that make plasma-aided nanoassembly so markedly different from other, in particular, neutral gas-based processes. We did not try to provide detailed practical recipes for the synthesis of specific nanoassemblies or give specific suggestions to researchers with either experimental or theoretical/computational profiles. In part, this is because of large number of such recipes and suggestions given in our earlier related monograph [1].

Neither did we attempt to provide an exhaustive coverage of all (or even most) uses of low-temperature plasmas for nanoscale processing and synthesis. Indeed, due to the very large and continuously increasing number of publications, it would be a completely futile attempt to even briefly mention them all. As we have seen from the previous chapters, the range of related processes begins from traditional processes in present-day microelectronics such as material ashing and stripping, surface conditioning, passivation/activation, deposition of conformal coatings, reactive chemical etching, physical sputtering on one end, extends to more common nanostructure synthesis and post-processing (e.g., doping, functionalization, monolayer coating), and evolves into the as yet elusive domain of plasma-controlled self-assembly.

The range of nanoscale objects and nanostructured materials with highly unusual properties that can be created using plasma-based approaches is also very wide. The examples considered include common metals, oxides, nitrides, carbon, silicon, various semiconducting and insulating materials and several others. Interestingly, the properties and attributes the materials gain through plasma synthesis or treatment appear to be remarkably different compared with the same materials obtained through other, e.g., neutral gas-based, routes.

We hope that the reader has already received some idea on what plasma nanoscience is about. It is a common knowledge nowadays that nanoscale synthesis is object- and process-specific and, taken the emergenece of a huge number of nanoscale objects and features, the number of specific techniques to fabricate them is equally large. Plasma nanoscience research aims at finding, via better understanding of the elementary processes involved, common features in this overwhelming variety of objects and approaches and elaborate effective "turning knobs" to control the processes involved, and eventually create physical foundations for the as yet elusive deterministic nanoassembly. This particular point has been stressed in Chapter 2.

Every particular chapter aimed at stressing the importance of specific plasma-related phenomena and at identifying how exatly nanoscale objects can benefit from the ionized gas environment. For instance, nanoparticles in stellar outflows nucleate (and eventually develop) faster in the presence of ions created via photoionization of neutral gas; nanoscale features on biopolymer surfaces can be created at room temperatures which is not possible in thermal CVD; plasma-synthesized vertically aligned nanostructures not only grow faster but also have better attributes, such as purity, stability, aspect ratio, alignment, and so on – these examples can go on until all cases of successful uses of low-temperature plasmas for nanoscale assembly and processing have been covered. As already stressed above, the number of such cases is extensive, which is reflected by the exponential growth in the number of publications in the area.

Despite the very large number of such examples and specific manifestations of gainful uses of ionized gas environments, in the end, any remarkable differences from neutral gas-based approaches are in some way related to those plasma features that differ them from neutral gases. The first thing that comes to our mind is that the plasma is an *ionized* gas, and the main difference from the neutral gases is therefore in the ionization, the presence of electric charges, electric fields, polarization, as well as other unusual attributes such as those related to the ability of plasmas to dissociate neutral gases into reactive radicals, create a variety of species, sustain clustering and polymerization, and so on.

Thus, the plasma-related features are what makes all these seemingly completely different phenomena quite similar. Plasma nanoscience tries to elucidate how these nanoscale (and even atomic) features are related to the available plasma-related controls such as electric charges, electric fields, ionized species, plentiful radicals, clusters and other particles and some others. Moreover, as we have seen from a variety of examples presented above, each of the above "turning knobs" needs to be used with due care. For instance, ion bombardment can substantially improve crystalline properties and growth rates of nanostructured films. On the other hand, energetic ions can easily destroy some delicate nanostructures and in some cases even damage the substrate. Thus, reliable knowledge is required to fully enjoy the benefits of using plasmas for nanoscale synthesis and processing.

Related examples are numerous and even a simple search for "plasma and nano" through any reliable database immediately returns several thousand entries. And one could assume that many other reports could still be found upon a more careful search. Therefore, it would be a futile attempt to try to cover all the examples where low-temperature plasmas have been used for nanoscale materials synthesis and processing. For this reason we decided not to attempt to provide a full coverage of all relevant processes, features, and phenomena and concentrated instead on those examples that best illustrate the amazing physics behind plasma-based nanoassembly and show how the building unit-based "cause and effect" generic approach works in most common cases.

Nevertheless, we do provide a brief overview of a large number (yet still very far from being exhaustive) of relevant examples in Appendix B. Due to significant space constraints, we have adopted a very concise "telegraphic" style when mentioning any particular work. Thus, if the reader has reached this far, now would be the right time to check the overview in Appendix B. We believe that a relatively large number of works mentioned there, in addition to those already discussed throughout the monograph, will boost the reader's confidence that plasma nanoscience is not only an interesting research topic but also a potential area for substantial research, development and capital investment in the coming years.

In the rest of this concluding chapter, we will briefly stress the importance of the fundamental issues of determinism and complexity, summarize some of the most important features and competitive advantages of plasma-based processes for nanoscale synthesis and also attempt to highlight some of the most interesting topics for the future research. In Section 9.1, the fundamental issues of determinism and complexity will be revisited with the aim of relating the processes of plasma-assisted nanoscale synthesis in the universe to that in a terrestrial lab and to bring them even closer than has already been done in the introductory chapter. Section 9.2 is a brief "dot-point" summary of some of the most important plasma-related features of nanoscale materials synthesis and processing. The final section of this monograph, Section 9.3, gives an optimistic outlook for the future development of this undoubtedly interesting multidisciplinary research area.

9.1
Determinism and Higher Complexity

Let us now consider an interesting example illustrating the usefulness of using more complex ionized gas environments to assemble highly-unusual, exotic nanoassemblies in natural and laboratory (and potentially, industrial) environments. To do so, one might pose an obvious question about the relationship between the vertically aligned carbon nanostructures and nature's mastery in the synthesis of cosmic dust discussed in Section 1.3? To clarify this issue, let us examine the energetic properties of size-reduced atomistic models of carbon nanotips with different aspect ratios, one of them grown by the plasma technique (bottom left panel in Figure 9.1) and the other one by the neutral gas process (top left panel in Figure 9.1).

We stress that one of requirements of efficient electron field emission is that the nanotips should ideally be highly-conductive and also have as large an aspect ratio as possible (see Chapter 7 for more details). Interestingly, the atomistic nanotip models with smaller apex angles (and hence, larger aspect ratios) feature a negligibly small energy bandgap, which make them perfect conductors.

One such model composed of 18 carbon atoms terminated at the periphery by 18 hydrogen atoms ($C_{18}H_{18}$) is shown in the bottom right panel in Figure 9.1. The cohesive energy E_c of this atomistic structure (the amount of energy that holds the atoms together) is ca. 363.200×10^{-19} J (227 eV) and the energy bandgap $\mathcal{E}_g$ (the difference between the highest occupied molecular orbital (HOMO) and the lowest unocuppied molecular orbital (LUMO)) is 0.016×10^{-19} J (0.01 eV). The apex angle of this atomistic model is ca. $43°$.

In comparision, shorter and wider atomistic nanotip models exhibit semiconducting properties and larger bandgaps. However, such nanostructures have a somewhat better cohesion and structural stability. Indeed, the structure $C_{51}H_{27}$ in the top right panel in Figure 9.1 features E_c ca. 924.160×10^{-19} J (577.6 eV) and $\mathcal{E}_g$ ca. 0.210×10^{-19} J (0.131 eV). The apex angle of this atomistic model is significantly larger at approximately $87°$ as can be seen in Figure 9.1.

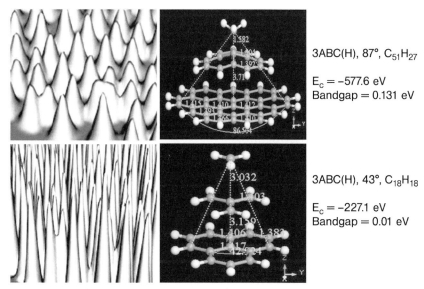

3ABC(H), 87°, $C_{51}H_{27}$

$E_c = -577.6$ eV
Bandgap $= 0.131$ eV

3ABC(H), 43°, $C_{18}H_{18}$

$E_c = -227.1$ eV
Bandgap $= 0.01$ eV

Figure 9.1 Carbon nanotips synthesized through the neutral (top row) and ionized (bottom row) gas routes and their downscaled atomistic models [1,23].

Extrapolating the above results into downscaled atomistic models (with a few tens to a few hundred atoms) to real nanostructures (which in the case considered may contain up to 10^5–10^6 atoms and even more), one can conclude that taller and narrower plasma-synthesized carbon nanocones are less stable than shorter and wider nanostructures grown by neutral-gas CVD. However, the high-aspect-ratio geometry of the plasma-grown nanocones strongly affects their electronic spectra, which improves their conductivity and eventually makes them more suitable as electron microemitters compared to the nanostructures synthesized by thermal CVD.

Therefore, without pursuing the specific purpose of enhancing the field emission, it is worthwhile to synthesize simpler but more stable (better satisfying the energy minimum principle) structures, with larger apex angles. More importantly, for this purpose one should use simpler, neutral gas-based fabrication routes, which are more appropriate for the synthesis of the wide-base nanotips shown in Figure 9.1. In fact, this is the "minimum-energy-minimum-effort" scenario advocated by nature's mastery (Chapter 1) [23].

However, should one want to synthesize sharper and conducting carbon nanotip microemitters, one should use more complex, plasma-based fabrication routes, which lead to low-aspect-ratio structures shown in the bottom left panel of Figure 9.1. The nanocone structures synthesized via a plasma route have higher energy minima (less structurally stable) and also require a higher-level-complexity fabrication environment, which contains an ionized gas component (which, in turn, also needs extra energy for ionization). Therefore, special-purpose nanostructures of a higher complexity, for example microemitters, should be fabricated using higher-complexity nanofabrication environments! Generalizing this conclusion, one can state that *nanofabrication environments of a higher complexity should be used to synthesize more unusual, exotic nanoscale objects.*

A similar situation takes place in nature's nanofab (see Section 1.3 of this monograph). The degree of ionization ($\equiv$ complexity) of the gas environment in the red giant star exteriors is what controls the rates of dust nucleation. For example, if the demand, for example to maintain the chemical balance in the universe, for the dust is low, it is worthwhile to create it in a cold and rarefied neutral gas, for example area III in Figure 1.7. On the other hand, if the demand for the dust is higher, then higher (ion-induced) nucleation rates, and hence, higher ionization degrees (area II in Figure 1.7) are required.

In a sense, nature's nanofab acts as a complex system, in which parts of different complexity come into play depending on the balance between the dust demand and supply. We emphasize that in this complex system both the neutral- and ionized gas-based nanoassembly routes have their place and a synergy would cater to all necessary requirements. A similar synergetic approach should also be used in terrestrial laboratory experiments [23].

Speaking of higher-complexity environments, we should recall our discussion about self-organized quantum dot arrays in Chapter 6, where we have stressed that such "more complex" process environments are prone to numerous additional controls that are not available in similar environments of a lower complexity. For example, a comparison of a plasma with a neutral gas leads us to the appreciation that electric fields, charges, and ionized ion and electron components are what actually make the plasma environment more complex. Some may even stress that since most of the low-temperature plasmas currently used for nanoscale assembly are weakly ionized, plasma environment is just a little bit more complex than the equivalent neutral gas. And despite this minor "complexity upgrade", plasma offers so many new and exciting possibilities to control self-organized processes at nanometer and sub-nanometer levels. Some examples have been considered in Chapter 6 (see also examples of plasma-based nanoassembly in Chapters 7 and 8).

9.2
Plasma-Related Features and Areas of Competitive Advantage

The qualitative arguments of the previous section make us believe that in lab-based nanotechnology it should be possible to use low-temperature plasma environments to synthesize higher-complexity nanoassemblies and to do so more efficiently than is presently possible using other techniques. The only thing that remains to decipher is how exactly to do that, and this is one of the main roles of plasma nanoscience [23].

Having said that, we should now mention some of the most important advantages of plasma-aided nanofabrication over non-plasma-based approaches and techniques, and wherever applicable relate these to the most important features of low-temperature plasmas [5].

A particular advantage of thermally non-equilibrium low-temperature plasmas is the possibility of directed motion of charged species under the influence of the electric field. In particular, this allows for highly-anisotropic etching of substrates as well as controlled, with subnanometer precision, deposition of nanoassembly building units, including in very narrow gaps in dense arrays of vertically aligned nanostructures and tiny pores in the substrate. This function is particularly strong at low gas pressures when the mean free path of the BUs in collisions with other particles is larger than the distance they have to cross before insertion into nanoscale objects. In high-density (e.g., atmospheric pressure thermal) plasmas, collision rates may be much higher, thus the directed motion of BUs is affected to a much larger extent.

In another aspect, low-temperature (in particular, thermally non-equilibrium plasmas where electrons are hot enough to sustain the ionization at the required level whereas the ions and neutrals can be maintained at room temperatures [1]) can be advantageously used for a range of processes that involve temperature-sensitive materials such as polymers, plastics, and ultra-thin nanolayers.

If necessary, one can also access very high temperatures using low-temperature thermal plasmas of transferred arcs and inductively-coupled RF plasma torches, where the temperatures of all three classes of discharge species (electrons, neutrals, and ions) can reach ca. 20 000 and 10 000 K, respectively. These high temperatures cannot be achieved in conventional, combustion-based systems and allow the use of solid, liquid, vapor, gas, suspension and solution precursors, with the possibility of complete dissociation of chemical reactants into atoms [5].

Further, high operation pressures (e.g., comparable with the atmospheric pressures) usually mean very high densities of plentiful discharge species. Moreover, the plasma temperature is largely independent of the chemical reactions, and there is a very wide choice of both reactive and inert atmospheres. In contrast, combustion-based technologies rely on chemical reactions to heat the gas, and so the atmosphere necessarily contains combustible gases and the products of combustion. As such, it is easier to control the process temperature in a plasma environment, with the added advantage that the plasma environment is much cleaner.

The high temperature of thermal plasmas further means that very steep temperature gradients can be obtained. Under such conditions, gas-phase nanoparticle nucleation becomes pronounced, leading to high degree of supersaturation and therefore *high nucleation rates*, which allow for high-yield synthesis of nanoparticles and nanostructures, something not achievable by other methods including thermally non-equilibrium plasma-based ones [5].

Specific examples of situations in which low-temperature plasmas can be advantageously and purposely used for nanoscale materials synthesis and processing include those in which [5]:

- specific (e.g., vertical) alignment or electric field-based control of preferential growth directions of nanostructures is required. This applies not only to common carbon nanotubes but also to a range of materials and high-aspect-ratio nanostructures;

- directionality of fluxes and penetration of species is an issue. In these cases electric field-driven ion fluxes can be a significant help; for example, in the cases of nanofeature metallization, filling nanopores, and post-processing of dense nanotube arrays. Low-pressure operation conditions are more appropriate in this case;

- processes are sensitive to high temperatures of the fabrication environments and substrates can easily melt or be otherwise damaged. This is the case with polymer processing and deposition of ultra-thin (as thin as a few atomic layers) conformal layers; therefore maintaining process temperatures below the melting points of metallic interconnects is one of the pressing issues in micro/nanoelectronics. Thermally non-equilibrium plasmas (with hot electrons and cold ions and radicals) are best suited for this purpose and have an indisputable advantage over thermal plasmas and many other methods and techniques;

- enhanced crystallization is needed without increasing temperature. In such cases plasma-generated crystallization agents (e.g., reactive radicals or subplanted ions) can be particularly useful;

- higher dissociation rates of precursor gases and large, yet controlled, amounts of specific radicals are required;

- pre-formed building units such as nanoclusters or nanocrystallites are needed; polymorphous silicon films is one such example;

- otherwise neutral species (e.g., radicals) can be ionized for better control of their motion and interaction with nanostructured surfaces. This is the case in various modifications of ionized physical vapor deposition (i-PVD);

- specific (passivation versus activation) surface preparation is needed. Low-temperature plasmas can offer a great deal of surface preparation processes assisted by a large variety of working units (WUs) including atoms, ions and radicals;

- it is necessary to control/enhance catalyst activity but use of other methods is not straightforward. For example, satisfactory fragmentation of catalyst interlayers can be achieved via reactive chemical etching or ion bombardment, a process which is otherwise unavailable or inefficient;

- both high gas temperatures and high densities of ionized species are required. This is a definite advantage of thermal plasmas in, for example synthesis of single-walled carbon nanotubes in the gas phase;

- condensed precursors are required. High-density (e.g., thermal) plasmas can provide the energy flux necessary for complete vaporization of liquids and solids; morever, intense ion-assisted processes also make non-equilibrium plasmas an efficient source of ionized physical vapor of liquid and solid precursors;

- high-rate production or deposition of nanostructures is required. High-density and higher-temperature (e.g., thermal) plasmas have a clear advantage in the synthesis of large amounts of nanoparticles in the gas phase, whereas thermally non-equilibrium plasmas are better suited for high-rate fabrication of delicate nanoassemblies on solid substrates.

This list represents only a few examples of many different possibilities offered by low-temperature plasmas and is by no means exhaustive. These theoretical capacities have been proven by a number of successful experiments in the last decade. Some of the advantages related to effective control of self-organizing processes involved in the synthesis of delicate nanoassemblies on plasma-exposed surfaces predicted via theoretical/numerical analysis (see, e.g., Chapter 6), are still awaiting their experimental realization. These and some other exciting prospects for plasma nanoscience research in the coming years are discussed in the following section.

9.3
Outlook for the Future

In this section we will highlight some of the pressing challenges and unresolved issues, and map potentially attractive directions for future endeavours in this exciting and commercially attractive research field [5]. Due to the overwhelming variety of possible nanoassemblies, processes, tools and techniques, it would be futile to try to prescribe what should be done in the field in the near future. It is nevertheless possible to highlight some of the most important challenges, unresolved issues, and opportunities in plasma-aided nanofabrication that arise from the topics discussed in this monograph.

The ultimate crux of research in plasma nanoscience might be summarized as to *fully understand the most effective controls of plasma-based nanoscale processes with the long-term aim of achieving a deterministic level in nanofabrication*. Rephrasing the famous Richard Feynman statement [2],

> *one should learn how to create and control the nanoworld, in a plasma environment, plasma species by plasma species, the way we want them.*

In fact, this is what nature's nanofab does, as we have seen from Chapter 1 (see also Reference [23])!

More specifically, one needs to address the yet unresolved problem of self-assembly/self-organization of nanoscale objects on plasma-exposed surfaces (as is the case of surface-bound nanoarrays) or in the ionized gas phase (e.g., growth of single-walled carbon nanotubes between the anode and cathode in arc discharges) and elucidate the roles of charges on plasma-exposed surfaces and plasma-nucleated nanoparticles, electric potentials (or currents) in the vicinity and across the nanoassemblies, ion fluxes and ion-flux-related temperature effects. This will not be possible without precise balancing of delivery and consumption of plasma-generated building units.

To this end, one needs to properly identify exactly which species are building units and which ones are working units and thus serve for other relevant purposes, such as surface preparation [5]. The latter point is probably one of the most difficult milestones on the way to deterministic nanofabrication. Indeed, relying on and trying to control wrongly chosen plasma-generated building units will lead us nowhere. However, even an incorrect choice is not automatically a dead-end, and lessons learned through such trial and error may eventually lead to the optimum process.

From the point of view of nanotool/process design, those plasma tools and processes that can provide a measured and controlled delivery of building units (and any required working units) have the potential to become major competitors to extremely expensive nanofabrication tools such as metal-organic vapor phase epitaxy (MOVPE) and molecular beam epitaxy (MBE); the latter presently remains among the most effective tools for the assembly of arrays of epitaxial semiconducting quantum dots. Moreover, there is a great potential for plasma nanotools used in combination with other nanoassembly tools such as laser-assisted methods, chemical vapor deposition, atomic layer deposition (ALD), and even MBE! It is likely that nanofabrication techniques with the prefix "plasma-assisted/enhanced" will become even more widespread than they are now; the existing examples include PE-CVD, PE-ALD, PE-PLD, PE-MBE, and i-PVD. Further exploration of the opportunities for hybrid plasma–laser systems to selectively control processes in the ionized gas phase and on solid surfaces should also be undertaken [5].

Interestingly, a metal surface exposed to a gas discharge is an example of a hybrid system where a gaseous plasma meets a plasma of free electrons. The plasma frequency associated with electron oscillations is determined by the electron density in metals and the effective electron mass (which is different from the fundamental constant m_e) and can be in the terahertz, infrared, and optical ranges. Collective responses of free electrons to electromagnetic fields (e.g., laser excitation) can lead to the excitation of surface plasmons, in other words, electromagnetic surface waves. These waves can alter the energetic states of the surface on which adatom self-assembly takes place and thus significantly affect the nanoscale assembly. Moreover, surface plasmons can be used for *in situ* real-time monitoring of the nanoarray development process. These and numerous other exciting possibilities are offered by the new and rapidly expanding research field of plasmonics [523].

The field of plasma-based synthesis of biomimetic materials and devices and the processing of temperature-sensitive polymer bio-interfaces is one of the main areas where plasma nanotools display a competitive advantage. It is imperative, however, to be able to identify, generate and control suitable functional plasma species; that is, some species for nanostructure formation and some for surface tethering to attract biomolecules and ultimately improve the bioresponse [5].

In the area of plasma etching of nanoassemblies and nanofeatures, an important aim is to continue to increase resolution and selectivity and so ultimately be able to etch ultra-deep trenches at and beyond the limits of present-day lithographic tools. It would be wise here to pose a deceptively simple question:

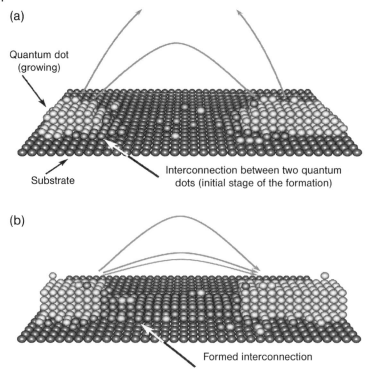

(a)

Quantum dot
(growing)

Substrate

Interconnection between two quantum
dots (initial stage of the formation)

(b)

Formed interconnection

Figure 9.2 Electric field-guided self-assembly of interconnected quantum dots on a plasma-exposed surface.
Courtesy of I. Levchenko and K. Ostrikov (unpublished).

When the capacity of the highest-resolution lithography is reached
(problems are already experienced in the sub-100 nm domain), how can
ultra-nanofeatures be etched on essentially self-organized patterns? Are
self-organized ultra-nano-trenches/pores filled with metal (a nanoscale
analogue of copper metallization of silicon trenches) a myth or reality [5]?

One can thus expect a much stronger push towards fabricating functional elements of nanodevices relying merely on plasma-controlled self-
assembly of individual nanostructures and their self-organization into
ordered and interconnected networks. Advanced numerical simulations
suggest that such may become a reality in the near future. Figure 9.2
shows an example of how a nano-scale electric field in the vicinity of a
plasma-exposed Si surface with two growing quantum dots can help to
overcome uncontrollable stress-driven assembly and lead to the creation
of a perfectly aligned interconnection between the QD's. The experimental confirmation of this theory is still awaiting its realization. Neverteless,
pilot experiments by Mariotti *et al.* [524] on the plasma-based synthesis of

rather disordered networks of gold nanoparticles interconnected by carbon leads, suggest that with some more effort this predicted effect may
become a reality within the next few years.

One can thus anticipate that the nanoscale self-organization of nanoassemblies on plasma-exposed surfaces and in the ionized gas phase will
be a fertile topic for major advances in the coming years. In this regard, it
is imperative that we first understand and can then create the capabilities
to [5]:

- properly identify, generate and then deliver suitable plasma-generated building units and properly position them, with atomic precision, for further insertion/self-assembly into nanostructures and
  nanoscale assemblies;

- precisely control surface conditions over ultra-small areas, for example, by localized energy transfer to specific areas;

- control crystallographic growth directions (e.g., pull nanowires or
  deterministically shape nanocrystals) by varying the plasma process parameters;

- predict and control catalyst nucleation, self-organization, and saturation with building materials (e.g., nickel catalyst with carbon
  atoms) which is essential for the growth of ultra-dense arrays of
  nanotube and other nanostructures;

- be able to predict, with subnanometer precision, where a nanoassembly will nucleate on a range of nanostructured, rough, and
  atomically smooth surfaces, and ultimately, learn how to control
  nucleation sites;

- remove unwanted amorphous deposits, for example, via plasma
  etching, and improve nanostructure crystallization, for example,
  via atomic/radical crystallization agents or energetic ion bombardment and subplantation;

- properly manipulate energetic ion beams to improve the quality
  of nanostructures without causing any substantial structural damage. As stated by Kato et al. [370], to improve the quality of single-
  walled carbon nanotubes for future nanoelectronic applications,
  precise control of ion energy will be a crucial issue in any future
  plasma nanotechnology;

- precisely manipulate various nanoparticles in the gas phase, in the
  vicinity of and upon deposition onto solid surfaces. This is required

in a broad range of problems spanning from carbon nanotube collection to management of particulate contamination in microelectronics;

- remove unwanted contaminants and enable delivery of, for example crystalline nanoparticles to specified nanodevice locations by using different plasma forces;

- deterministically control self-organization of developing nanostructures into ordered nanopatterns with pre-determined properties;

- same as above for the plasma-guided, self-organized assembly of interconnects between the individual nanostructures arranged into ordered arrays, as well as leads and other connections between the nanoscale objects, their functionalities and circuit elements;

- balance different scenarios of nanoparticle nucleation in the plasma, such as homogeneous nucleation and ion-induced nucleation, the latter being unique to ionized gas environments; and

- implement high-precision, selective-area post-processing, for example, functionalization, of a range of nanoscale assemblies to enable new and unique functions tailored for specific applications, e.g., in *in vivo* biosensors.

The above list is far from being exhaustive and mentions just a few of the most obvious conclusions that directly arise from the examples given elsewhere in this monograph.

In the area of fundamental plasma nanoscience research, one should continue to generate a new understanding which will allow us to identify common effects exerted by a selected plasma-related feature (e.g., the sheath electric field or ion flux density) on a range of different nanoscale objects and both mark and explain any possible similarities or differences. This will eventually lead to the development of deterministic strategies and approaches for nanoscale materials synthesis and processing. From the practical, and ultimately commercial perspectives, one should achieve highly-controlled, predictable, reproducible, and cost-effective plasma-assisted nanofabrication processes and design the most appropriate and effective plasma nanotools.

Ostrikov and Murphy [5] stressed that in trying to achieve the above specific nanoassembly-related aims, low-temperature plasmas should be made stable, reproducible, and feature an excellent uniformity of the densities, fluxes, and temperatures of the required species. To this end, the controlled feed of gaseous, liquid, and solid precursors is essential.

Plasmas offer a range of means (from "simple" thermal vaporization to "complex" polymerization) by which to generate the entire spectrum of building units discussed in detail in Chapters 2–4.

Moreover, the residence time of building units, catalysts and nanoparticles in the plasma reactor should be adjusted to optimize the processes for which they are intended. For example, to deterministically control gas-phase synthesis of single-walled carbon nanotubes, metal catalyst nanoparticles should be confined in the plasma long enough to sustain the nanotube growth up to the desired length. In this case, the (usually negative) electric charge the particles acquire in the plasma makes it possible to levitate them in the vicinity of negatively charged surfaces thus increasing the nanotube growth time [5].

In most cases, researchers need to make the right choice of the suitable plasma (and an efficient plasma reactor to generate and use such plasmas) in view of their future prospects in the nanoscience research field. The ideal combination is to have a range of versatile plasma tools in the lab; if this is not possible, the best advice would be to try to find the plasma that best suits the needs of the specific nanoassembly problems that are aimed for. Whatever the final choice, the list of intriguing problems related to various aspects of plasma-aided nanofabrication is virtually unlimited (as is the number of different possibilities), just to mention two extreme problems in carbon nanotube synthesis that need two different plasmas to resolve. Indeed, the problem of finding the ultimate physical limits of nanoparticle production rates in the gas phase will almost certainly require thermal plasmas. Furthermore, in trying to find the absolute minimum surface temperatures that can still sustain carbon nanotube growth, one needs to deal with thermally non-equilibrium plasmas. Thus, both plasmas have tantalizing future prospects for specific and, more importantly, relevant purposes [5]!

One of the major issues that still needs a clear, quantitative validation is the effect of ions, as well as electromagnetic fields, on the formation of nanoassemblies. Recent results on plasma-generated electrtic field-assisted growth of zinc oxide nanorods and numerical simulation of plasma-assisted growth of ion-focusing nanostructures of different dimensionality (including nanoporous structures), ion-assisted post-processing of dense carbon nanotube arrays (impossible or inefficient via neutral gas-based routes), specific roles of plasma-generated species in controlling surface conditions for precise nanoscale processing in microelectronics shed more light on this problem.

It should however be stressed that existing knowledge on the enhancement of nanoparticle and nanostructure nucleation rates due to ionization, charge separation, and electric fields needs to be complemented by studies of the effect of plasma turbulence, instabilities, and temperature gradients, which have been known to increase the nucleation and agglomeration rates of dust particles in astrophysical, space and laboratory plasmas. These important issues are still awaiting a conclusive answer.

Even though the ongoing research efforts are still a long way from fully bridging the 9-order-of-magnitude gap mentioned in Chapter 1 (see also Figure 1.11), there are several encouraging computational and experimental results relevant to each specific spatial sub-scale. The relevance of such advances varies from one sort of nanostructures to another. At present, it is possible to establish a range of critical dependencies of the nanostructure sizes, shapes, compositions and other characteristics on the plasma process parameters (e.g., plasma density, substrate bias, electron temperature, surface temperature) and to explain such dependencies by following the "cause and effect" approach based on building unit generation in the plasma, their delivery to nanoassembly sites, and incorporation into developing nanostructures (please refer to Chapter 2 for more details).

Due to the limited space of this monograph (and also because of the limited time available to prepare it), we did not touch on an important class of problems related to the possible role of the plasma environment in the creation of the building blocks of life in the primordial earth. In the last 50 years, there have been extensive debates on whether or not life could have originated by chemical processes involving nonbiological components.

There is a possibility that simple building blocks can arrange into more complex biochemical assemblies, which in turn lead to living cell creation. It is accepted that there is a range of common organic molecules that define life, for example amino acids, nucleotides and some others; furthermore, under certain environment-related conditions, these organic molecules can assemble into macromolecules, for example proteins and nucleic acids. Organic chemistry suggests that effective macromolecule formation in most cases requires some sort of catalyst or process accelerators. As we have seen from previous chapters of this monograph, some of the features of the environment (such as the electric charges or fields) may be able to function as such. In this regard, one of the greatest puzzles of modern science has been centered around the possible synthesis of basic organic molecules and also suitable macromolecules in the atmosphere of primordial earth, when the first primitive forms of life were created.

It is natural to pose a simple question: What does this have to do with plasma nanoscience, the main focus of this monograph? Similar to how we have proceeded in the previous chapters, let us examine the environment where the first building blocks of life could have been created. At that time the earth's atmosphere was oxygen-deficient and contained water, methane (CH_4), ammonia (NH_3) and hydrogen. Importantly, the frequency of violent electric discharges (lightnings) in the primordial earth atmosphere was very high. Thus, low-temperature plasmas played a very significant role in chemical processes in the atmosphere. Experiments conducted under conditions intended to resemble those present on primordial earth have resulted in the production of some of the chemical components of proteins, DNA, and RNA [525].

Scientists have concluded that these building blocks could have been available early in the earth's history [525]. It has also been suggested that the synthesis of the "building blocks of life" was most effective on the surfaces of rocks, sand grains and so on, whose most essential component is silicon. Does that not resemble some of the problems we have considered in this monograph? For example, self-assembly of nanostructures on silicon surfaces exposed to low-temperature plasmas. For a recap and in-depth analysis of the relevant efforts and possible pathways of transformations of organic macromolecules (which have been successfully synthesisized using low-temperature plasmas) please refer to the available literature (see, e.g., Reference [526] and references therein).

There are many great challenges and unresolved issues in this area, such as which pathway (among a large number of possible options) would nature choose to synthesize the self-reproducible building blocks of life? For a plasma nanoscience researcher, however, the most important issue is to improve our existing knowledge on a possible role of the plasma environment in prebiotic synthesis in nanoassembly of nanometer-sized macromolecules under primordial earth conditions. This is another area where our knowledge of plasma-controlled molecular assembly and polymerization could prove indispensable.

There are many other unresolved and unclear issues in plasma nanoscience area, with only a few representative examples given below [5]:

- Can plasma-aided nanofabrication be sufficiently developed to enable *atomic-scale* processing of surface features with at least one dimension in the atomic size range such as zero-dimensional clusters of very small number of atoms, one-dimensional atomic chains, and two-dimensional monoatomic terraces?

- Can plasma-based devices be made fully compatible with the high-precision ultra-high-vacuum tools of surface science?

- Is it possible to use low-temperature plasmas to fabricate structures containing just a few atoms (at any specified position on or inside the surface) for quantum computing applications via controlled manipulation of small (compared to typical surface densities) numbers of atoms/ions?

- Is it possible to use plasma-related "turning knobs" to control the chirality of single-walled nanotubes, which is currently believed to be merely determined by the catalyst properties?

- How can one control the nanostructure nucleation sites without pre-patterning? How can one use plasma-related effects to control the distribution of surface stresses and temperatures, as well as any other parameters which in turn determine where exactly the nanostructures will nucleate?

- Can one use plasma-related effects to control atomic spin directions and flip rates for quantum computing applications?

- What is the influence of nanoparticle charging and charged species in plasmas and on solid surfaces and how can they be utilized for controlled production of a wider range of nanoassemblies?

- How can one control charging states of plasma-generated building units on the surface? For instance, how can one preserve or dissipate the charge of ions or clusters when they land on a solid surface?

- In which aspects of nanofabrication are particular plasmas better than any others? For example, thermal plasmas or thermally non-equilibrium low-temperature plasmas?

- Is there "universal" plasma that could be used for each of the most important nanoscale processes and as such adopted for the next-generation plasma-based nanotechnology?

- Which plasma-based facilities and associated nanotools should one adopt to enable next-generation nanoscale and atomic-level processing?

The number of these and similar questions is much larger than the number of the available answers and solutions. Future work in the next few years should be focused on numerical and experimental identification of the most efficient "turning knobs" of plasma-based nanoscale processes and nanotools and on unlocking their enormous potential for large-area deterministic nanoassembly [5].

The outlook for research in this important research field is very positive; moreover, this topic is full of exciting opportunities for scientific research and techology-oriented research and development in the near future. As suggested by Ostrikov and Murphy, the scale of this research ranges from undergraduate projects to major national and international research programs [5]. More importantly, this field offers an exciting synergy of experimental, theoretical, and computational research in a number of diverse disciplines ranging from astrophysics to structural chemistry and from surface science to atomic physics.

9.4
Final Remarks

In this monograph, we have explored various possibilities for achieving a reasonable level of deterministic control of the quality of various nanoscale assemblies through specific manipulation of the plasma-related process parameters. For example, the assembly and ordering of semiconductor nanodots can be effectively controlled by adjusting the surface temperatures and incoming fluxes of building units from the plasma. This can be achieved, for example, by varying the degree of ionization in the plasma discharge, external bias applied to the surface, partial pressures of working gases, using surface temperature controllers, and some other means.

Most of the nanoassemblies considered in this monograph were chosen to represent the most common classes of nanoscale objects. For example, quantum dots are typical examples of low-dimensional nanostructures with outstanding prospects for optoelectronic, biosensing and other applications. Beyond these classes of nanoassemblies, the scope of deterministic plasma-aided nanofabrication is set to launch into more complicated nanostructures, their ordered arrays and interconnected networks and eventually fully operational nanodevices. Here we should stress that the level of determinism that can be achieved is limited by the possible combinations and sequences of the process and environmental adjustments. Finding suitable combinations of the process parameters to create the desired nanoassemblies in a plasma environment is one of the main objectives of plasma nanoscience.

The examples presented in this monograph suggest that the plasma-based route is a promising deterministic fabrication approach which shows good potential in terms of commercial feasibility. The need for these new techniques and nanodevices is critical as the current cost of lithographic manufacture increases, and the operational limits of traditional semiconducting devices are approached. Here we stress that plasma nanoscience contributes to the improvement of present-day capabilities of plasma nanotools, with the ultimate goal of achieving fully deterministic, yet cost-efficient and microelectronic industry-compatible, plasma-aided nanofabrication and the development of plasma-based nanotechnologies.

In many of the examples we have shown some important advantages offered by the plasma-aided nanoassembly in controlling the growth and various parameters of typical nanostructures and their arrays. One of the challenges for the future endeavors is to explore the detailed parameter ranges where such plasma-based control is most effective and efficient. We have also highlighted some other benefits of plasma-aided nanofabrication compared to neutral gas-based techniques, such as generation of a variety of building units in the required energetic and chemical states, nanomanipulation of building units by using various forces in the plasma sheath (such as the electric field in the plasma sheath, ion drag force, etc.), electric field-controlled alignment of nanoassemblies, surface-charge controlled self-organization on plasma-exposed surfaces, and several others. From the fundamental point of view, plasma-aided nanoassembly simultaneously involves two main approaches of modern nanoscience, namely, nanomanipulation (e.g., ions by the electric field in the plasma sheath) and self-organization (including migration on plasma exposed surfaces and self-assembly) of building units.

Owing to the exciting research opportunities in the area, a broad international research community should concentrate efforts on *deterministic* plasma-assisted synthesis and post-processing of nanoscale assemblies (from individual nanoclusters to nanostructured biomaterials and photonic nanocavities) and on better understanding the role and purpose of the plasma environments and optimizing relevant plasma processes and nanotools. Future work in the area should be focused on numerical and experimental identification of the most efficient "turning knobs" of plasma-based nanotools and of unravelling their enormous potential for large-area deterministic nanoassembly.

The next important step will be to translate the knowledge obtained via plasma nanoscience research into industrially-viable process control strategies for nanomanufacturing. This is important for the delivery of sustained economic benefits and societal impacts as well as substantial wealth creation. Fundamental approaches of plasma nanoscience can be very useful in this regard.

It is now the best time to pose one final "fundamental" question: What is the plasma nanoscience anyway [23]? Is this surface science of plasma-exposed surfaces, physics/chemistry of building blocks in the plasma, physics/chemistry of plasma-surface interactions, physics/chemistry of nucleation of cosmic dust, plasma engineering of nanostructures and nanoassemblies, or plasma-based nanoscience? The answer would be not only "all of the above", but also any relevant sub-area/sub-field that comes along as the role of the plasma in nanoscience and nanotechnology becomes clearer and the plasma-based nanoscale processes gradually reach the as yet elusive deterministic level. Rapid advances in the area make us very optimistic in this regard.

10
Appendix A. Reactions and Rate Coefficients in Low-Temperature PECVD of Carbon Nanostructures

10.1
Plasmas of Ar + H$_2$+ CH$_4$ Gas Mixtures (Section 4.1)

Rate coefficients for electron-neutral collisions in Equations (4.2)–(4.6) of Section 4.1 have been calculated by integrating the relevant collisional cross-sections [53]. Table 10.1 summarizes the main electron-impact reactions involved in the creation of a variety of species in low-pressure inductively coupled plasmas [527] of Ar $+$ H$_2$ $+$ CH$_4$ gas mixtures [53]. Two vibrational excitation reactions with the thresholds of 0.259×10^{-19} J (0.162 eV) and 0.578×10^{-19} J (0.361 eV) [528, 529] for the electron collisions with CH$_4$ molecules needed to be considered in Section 4.1. For the inelastic electron – C$_2$H$_2$ collisions 3 vibrational excitations with the threshold energies ca. 0.144×10^{-19} J (0.09 eV), 0.408×10^{-19} J (0.255 eV) and 0.651×10^{-19} J (0.407 eV) [528,530] need to be accounted for. Two vibrational excitations of C$_2$H$_4$ with the threshold energies of 0.160×10^{-19} J (0.1 eV) and 0.576×10^{-19} J (0.36 eV) [528,530] are also included in the model of Section 4.1.

Likewise, to describe electron – H$_2$ collisions, two rotational excitations with the threshold energies of 0.070×10^{-19} J (0.044 eV) and 0.117×10^{-19} J (0.073 eV) and three vibrational excitation with the thresholds of 0.826×10^{-19} J (0.516 eV), 1.600×10^{-19} J (1.0 eV), and 2.400×10^{-19} J (1.5 eV) need to be taken into consideration [53,528]. Electronic excitations of Ar, CH$_4$, C$_2$H$_2$, C$_2$H$_4$, H$_2$, and H are accounted in the power balance equation (4.6) of Section 4.1. The relevant reaction cross-sections can be found elsewhere (see, e.g., References [53,528,531] and the references therein). The number of different vibrational and rotational excitation reactions (noted "vib" and "rot" in Table 10.1) is also shown in Table 10.1.

A list of the neutral-neutral reactions and reaction rate coefficients compiled using the available data [191, 192, 536, 537] are given in Table 10.2 (T_g in Table 10.2 is the gas temperature in K). Full data on the

Plasma Nanoscience: Basic Concepts and Applications of Deterministic Nanofabrication
Kostya (Ken) Ostrikov
Copyright © 2008 WILEY-VCH Verlag GmbH & Co. KGaA, Weinheim
ISBN: 978-3-527-40740-8

Table 10.1 Electron reactions with atoms and molecules [53].

Reaction	Chemical reaction	Reference
Ar, excitation	$e^- + Ar \rightarrow Ar^* + e^-$	[528]
Ar, ionization	$e^- + Ar \rightarrow Ar^+ + e^-$	[528]
CH_4, excitation	$e^- + CH_4 \rightarrow CH_4^* + e^-$ (2 vib)	[529]
CH_4, ionization	$e^- + CH_4 \rightarrow CH_4^+ + 2e^-$	[528,529]
CH_4, ionization	$e^- + CH_4 \rightarrow CH_3^+ + H + 2e^-$	[529]
CH_4, dissociation	$e^- + CH_4 \rightarrow CH_3 + H + e^-$	[532]
CH_4, dissociation	$e^- + CH_4 \rightarrow CH_2 + 2H + e^-$	[532]
H, ionization	$e^- + H \rightarrow H^+ + e^-$	[533]
H, excitation	$e^- + H \rightarrow H^* + e^-$	[531]
H_2, excitation	$e^- + H_2 \rightarrow H_2^* + e^-$ (2 rot, 3 vib)	[528]
H_2, ionization	$e^- + H_2 \rightarrow H_2^+ + 2e^-$	[533]
H_2, dissociation	$e^- + H_2 \rightarrow 2H + e^-$	[534]
C_2H_4, excitation	$e^- + C_2H_4 \rightarrow C_2H_4^* + e^-$ (2 vib)	[528,530]
C_2H_4, ionization	$e^- + C_2H_4 \rightarrow C_2H_4^+ + 2e^-$	[528,530]
C_2H_4, dissociation	$e^- + C_2H_4 \rightarrow C_2H_2 + 2H + e^-$	[535]
C_2H_2, excitation	$e^- + C_2H_2 \rightarrow C_2H_2^* + e^-$ (3 vib)	[528,530]
C_2H_2, ionization	$e^- + C_2H_2 \rightarrow C_2H_2^+ + 2e^-$	[535]
C_2H_6, ionization	$e^- + C_2H_6 \rightarrow C_2H_6^+ + 2e^-$	[535]
C_2H_6, ionization	$e^- + C_2H_6 \rightarrow C_2H_5^+ + H + 2e^-$	[535]
C_2H_6, dissociation	$e^- + C_2H_6 \rightarrow C_2H_5 + H + e^-$	[535]
C_2H_6, dissociation	$e^- + C_2H_6 \rightarrow C_2H_4 + 2H + e^-$	[535]
C_2H_5, ionization	$e^- + C_2H_5 \rightarrow C_2H_5^+ + 2e^-$	[535]
C_2H_5, ionization	$e^- + C_2H_5 \rightarrow C_2H_4^+ + H + 2e^-$	[535]
C_2H_5, dissociation	$e^- + C_2H_5 \rightarrow C_2H_4 + H + e^-$	[535]
CH_2, ionization	$e^- + CH_2 \rightarrow CH_2^+ + 2e^-$	[535]
CH_2, dissociation	$e^- + CH_2 \rightarrow CH + H + e^-$	[535]
CH_3, ionization	$e^- + CH_3 \rightarrow CH_3^+ + 2e^-$	[535]
CH_3, dissociation	$e^- + CH_3 \rightarrow CH_2 + H + e^-$	[535]
CH_3, dissociation	$e^- + CH_3 \rightarrow CH + 2H + e^-$	[535]

rate coefficients for the ion-neutral reactions can be found in other relevant papers [190, 528, 537–539]. The corresponding rate constants are shown in Table 10.3.

The set of equations (4.2)–(4.6) of Section 4.1 has been solved by the time evolution method [53]. To start a numerical cycle an initial estimate of the effective electron temperature is needed. This allows one to obtain the average electron energy $\langle \varepsilon \rangle = (3/2) T_{\text{eff}}$ and the rate coefficients entering (4.2)–(4.6). We note that Equations (4.2)–(4.6) of Section 4.1 are nonlinear and time-dependent equations with respect to the time-varying number densities of different species $n_i(t)$ and $n_j(t)$. As a

Table 10.2 Neutral-neutral reactions [53].

Reaction	Rate constant k (cm^{3}s^{-1})	Reference
$H + CH_4 \rightarrow CH_3 + H_2$	$2.2 \times 10^{-20} T_g^3 \exp(-4045/T_g)$	[536]
$H + CH_3 \rightarrow CH_2 + H_2$	$1 \times 10^{-10} \exp(-7600/T_g)$	[536]
$H + CH_2 \rightarrow CH + H_2$	$1 \times 10^{-11} \exp(900/T_g)$	[536]
$H + C_2H_6 \rightarrow C_2H_5 + H_2$	$2.4 \times 10^{-15} T_g^{1.5} \exp(-3730/T_g)$	[536]
$H + C_2H_5 \rightarrow 2CH_3$	6×10^{-11}	[537]
$H + C_2H_5 \rightarrow C_2H_4 + H_2$	5×10^{-11}	[536]
$H + C_2H_4 \rightarrow C_2H_3 + H_2$	$9 \times 10^{-10} \exp(-7500/T_g)$	[536]
$H + C_2H_2 \rightarrow C_2H + H_2$	$1 \times 10^{-10} \exp(-14000/T_g)$	[536]
$CH_3 + CH_3 \rightarrow C_2H_6$	6×10^{-11}	[536]
$CH_3 + CH_3 \rightarrow C_2H_5 + H$	$5 \times 10^{-11} \exp(-6800/T_g)$	[536]
$CH_3 + CH_3 \rightarrow C_2H_4 + H_2$	$1.7 \times 10^{-8} \exp(-16000/T_g)$	[536]
$CH_3 + CH_2 \rightarrow C_2H_4 + H$	7×10^{-11}	[536]
$CH_3 + CH \rightarrow C_2H_3 + H$	5×10^{-11}	[536]
$CH_2 + CH_2 \rightarrow C_2H_4$	1.7×10^{-12}	[536]
$CH_2 + CH_2 \rightarrow C_2H_2 + H_2$	$2 \times 10^{-10} \exp(-400/T_g)$	[536]
$CH + CH_4 \rightarrow C_2H_4 + H$	1×10^{-10}	[536]
$CH + CH_2 \rightarrow C_2H_2 + H$	6.6×10^{-11}	[536]
$CH + CH \rightarrow C_2H_2$	2×10^{-10}	[536]
$C_2H_5 + CH_3 \rightarrow C_3H_8$	4.2×10^{-12}	[192]
$CH + C_2H_6 \rightarrow C_3H_8$	4×10^{-10}	[191]

result of the linearization of (4.2)–(4.6), the products $n_i(t)n_j(t)$ can be re-placed by $n_i(t)n_j(t - \Delta)$, where $n_j(t - \Delta)$ is the neutral/charged particle density at the previous moment of time $(t - \Delta)$.

Here, the time step Δ can be chosen empirically to enable the best con-vergence of the numerical routine [53]. Therefore, the l.h.s. of (4.2)–(4.4) can be presented in the algebraic form $[n_\alpha(t) - n_\alpha(t - \Delta)]/\Delta$. At the ini-tial stage one can use an estimated T_{eff} and "plasma-off" densities of Ar, CH$_4$, and H$_2$. The densities of other non-radical and radical neutrals have been assumed to be zero at this stage.

For the assumed T_{eff}, the number densities of the species have been computed from the linearized equations (4.2)–(4.6), the latter set being solved by the conventional Gauss method as detailed in the original work [53]. At the next temporal step it is assumed that $n_\alpha(t - \Delta)$ is equal to the as-calculated value from the previous step, and repeat the cycle at fixed T_{eff} until the absolute value of $[n_\alpha(t) - n_\alpha(t - \Delta)]/n_\alpha(t)$ becomes of the order of 10^{-3}.

This routine usually requires approximately 250 computation cycles. Thereafter, solving (4.2)–(4.6), the corrected densities of H$_2$ and CH$_4$, as well as the densities of other species have been calculated [53]. This rou-tine is followed by probing into the plasma quasineutrality condition of

Table 10.3 Ion-neutral reactions [53].

Reaction	k (cm^{3}s^{-1})	Reference
$CH_4^+ + CH_4 \rightarrow CH_5^+ + CH_3$	1.5×10^{-9}	[537]
$CH_4^+ + H_2 \rightarrow CH_5^+ + H$	3.3×10^{-11}	[537]
$CH_3^+ + CH_4 \rightarrow CH_4^+ + CH_3$	1.36×10^{-10}	[537]
$Ar^+ + H_2 \rightarrow ArH^+ + H$	1.6×10^{-9}	[538]
$Ar^+ + H_2 \rightarrow Ar^+ H_2^+$	2.7×10^{-10}	[538]
$Ar^+ + CH_4 \rightarrow CH_3^+ + H + Ar$	1.05×10^{-9}	[539]
$H_2^+ + H_2 \rightarrow H + H_3^+$	2.5×10^{-9}	[537]
$CH_3^+ + CH_4 \rightarrow C_2H_5^+ + H_2$	1.2×10^{-9}	[537]
$H_3^+ + C_2H_6 \rightarrow C_2H_5^+ + 2H_2$	2.0×10^{-9}	[190]
$H_3^+ + CH_4 \rightarrow CH_5^+ + H_2$	1.6×10^{-9}	[537]
$H_3^+ + C_2H_4 \rightarrow C_2H_5^+ + H_2$	1.9×10^{-9}	[537]
$H_3^+ + C_2H_2 \rightarrow C_2H_3^+ + H_2$	1.94×10^{-9}	[537]
$C_2H_2^+ + CH_4 \rightarrow C_2H_3^+ + CH_3$	4.1×10^{-9}	[537]
$C_2H_2^+ + CH_4 \rightarrow C_3H_4^+ + H_2$	6.25×10^{-10}	[537]
$C_2H_2^+ + CH_4 \rightarrow C_3H_5^+ + H$	1.44×10^{-9}	[537]
$C_2H_4^+ + C_2H_4 \rightarrow C_3H_5^+ + CH_3$	3.9×10^{-10}	[537]
$C_2H_4^+ + C_2H_4 \rightarrow C_4H_8^+$	4.3×10^{-10}	[537]
$CH_5^+ + C_2H_6 \rightarrow C_2H_5^+ + H_2 + CH_4$	5.0×10^{-10}	[190]
$C_2H_4^+ + C_2H_6 \rightarrow C_3H_6^+ + CH_4$	2.03×10^{-13}	[537]
$C_2H_4^+ + C_2H_6 \rightarrow C_3H_7^+ + CH_3$	1.32×10^{-11}	[537]
$C_2H_5^+ + C_2H_2 \rightarrow C_4H_7^+$	6.7×10^{-10}	[537]
$C_2H_5^+ + C_2H_4 \rightarrow C_3H_5^+ + CH_4$	3.1×10^{-10}	[537]
$C_2H_5^+ + C_2H_4 \rightarrow C_4H_9^+$	3.0×10^{-10}	[537]

Section 4.1, proper adjustment of the effective electron temperature and subsequent repetition of any number of cycles until the above condition has been satisfied [53].

10.2
Plasmas of Ar + H$_2$ + C$_2$H$_2$ Gas Mixtures (Section 4.2)

The two-dimensional fluid code used for the numerical simulation of number densities and surface fluxes of the main building and associated working (surface preparation) units in an inductively coupled plasma-based nanofabrication facility is based on a fluid plasma model and consists of a set of fluid equations for electrons, ions, and neutrals, which are solved self-consistently with Poisson's equation in a specified spatial grid in r-z cross-section of a cylindrical chamber described in Section 4.2. The Ar, H$_2$, and C$_2$H$_2$ gas feedstock is assumed uniformly distributed in the plasma reactor. Their densities are calculated from the input gas pressure and temperature by using the ideal gas law [54].

Table 10.4 Reactions involving hydrogen and argon species [54].

Reaction	Class of reactions	References
$H + e \rightarrow H^+ + 2e$	Ionization	[540]
$H_2 + e \rightarrow H_2^+ + 2e$	Ionization	[540]
$H_2 + e \rightarrow H^+ + H + 2e$	Ionization	[540]
$H_2 + H_2 \rightarrow H_2^+ + H_2 + e$	Dissoc. ionization	[237,541]
$H_2 + H_2 \rightarrow H^+ + H + H_2 + e$	Dissoc. ionization	[237,541]
$H + H_2 \rightarrow H^+ + H_2 + e$	Dissoc. ionization	[237,541]
$H_2^+ + H_2 \rightarrow H^+ + H + H_2^+ + e$	Dissoc. ionization	[237,541]
$H^+ + e \rightarrow H$	Recombination	[540]
$H^+ + 2e \rightarrow H + e$	Recombination	[540]
$H_3^+ + e \rightarrow H_2 + H$	Recombination	[540]
$H_3^+ + e \rightarrow H + H + H$	Recombination	[237,541]
$H^+ + H_2 + H_2 \rightarrow H_3^+ + H_2^+$	Recombination	[541]
$H_2 + e \rightarrow 2H + e$	Dissociation	[540]
$H_2 + H_2 \rightarrow H_2 + 2H$	Dissociation	[540]
$H_2 + H \rightarrow 3H$	Dissociation	[540]
$H_2^+ + H_2 \rightarrow H^+ + H + H_2$	Collision-induced dissoc.	[540]
$H_3^+ + H_2 \rightarrow H^+ + H_2 + H_2$	Collision-induced dissoc.	[540]
$H_3^+ + H_2 \rightarrow H_2^+ + H + H_2$	Collision-induced dissoc.	[540]
$H_2^+ + e \rightarrow H^+ + H + e$	Collision-induced dissoc.	[540]
$H_2 + 2H \rightarrow 2H_2$	Association	[540]
$2H + H \rightarrow H_2 + H$	Association	[540]
$H_2^+ + H \rightarrow H^+ + H_2$	Charge exchange	[540]
$H_2^+ + H_2 \rightarrow H_3^+ + H$	Charge exchange	[540]
$H^+ + H_2 \rightarrow H + H_2^+$	Charge exchange	[237,541]
$Ar + e \rightarrow Ar^+ + 2e$	Ionization	[542–544]
$Ar + e \rightarrow Ar^* + e$	Excitation	[542–544]
$Ar + e \rightarrow Ar + e$	Elastic scattering	[542–544]

The electron and ion densities are computed from the continuity equation

$$\frac{\partial n_j}{\partial t} + \nabla \cdot \vec{\Gamma}_j = R_{crea} - R_{dest}, \tag{10.1}$$

where n_j and $\vec{\Gamma}_j$ are the density and flux of the electrons/ionic species j, respectively; R_{crea} and R_{dest} are the total creation and destruction rates for the electron and ion species.

The creation and destruction processes of the species involved for the various collisions are summarized in Tables 10.4–10.8. [237, 540–545]. The continuity equation is solved by the alternative direction implicit (ADI) method [54]. The drift-diffusion momentum transfer equation for electron and ion species is

$$\vec{\Gamma}_j = sgn(q_j)\mu_j n_j \vec{E} - D_j \nabla n, \tag{10.2}$$

where μ_j and D_j are the mobility and diffusion coefficients for the charged species (which can be obtained by using Einstein's relation) and

Table 10.5 Electron impact reactions [54].

Reactions	Class of reaction	References
$e + CH_3 \rightarrow CH_3^+ + 2e$	Ionization	[545]
$\rightarrow CH_2^+ + H + 2e$	Ionization	[545]
$e + CH_2 \rightarrow CH_2^+ + 2e$	Ionization	[545]
$\rightarrow CH^+ + H + 2e$	Ionization	[545]
$e + CH \rightarrow CH^+ + 2e$	Ionization	[545]
$\rightarrow C^+ + H + 2e$	Ionization	[545]
$e + C_2H_2 \rightarrow C_2H_2^+ + 2e$	Ionization	[545]
$\rightarrow C_2H^+ + H + 2e$	Ionization	[545]
$e + CH_2 \rightarrow C_2H^+ + 2e$	Ionization	[545]
$\rightarrow C^+ + C + H + 2e$	Ionization	[545]
$e + CH_3 \rightarrow CH_2 + H + e$	Dissociation	[545]
$\rightarrow CH + 2H + e$	Dissociation	[545]
$e + CH_2 \rightarrow CH + H + e$	Dissociation	[545]
$\rightarrow C + 2H + e$	Dissociation	[545]
$e + CH \rightarrow C + H + e$	Dissociation	[545]
$e + CH_3^+ \rightarrow CH_2 + H$	Dissoc. recombination	[545]
$e + CH_2^+ \rightarrow CH + H$	Dissoc. recombination	[545]
$e + CH^+ \rightarrow C + H$	Dissoc. recombination	[545]
$e + C_2H_2^+ \rightarrow 0 \cdot 33 C_2H + H$		
$\rightarrow 0 \cdot 33 CH + CH$	Dissoc. recombination	[545]
$\rightarrow 0 \cdot 33 C + C + H + H$		
$e + C_2H^+ \rightarrow 0 \cdot 5 C + C + H$	Dissoc. recombination	[545]
$\rightarrow 0 \cdot 5 CH + C$	Dissoc. recombination	[545]

Table 10.6 Proton impact ionization reactions [54].

Reaction	Class of reaction	References
$H^+ + CH_3 \rightarrow 0 \cdot 5 CH_3^+ + H$	Proton impact ionization	[545]
$\rightarrow 0 \cdot 5 CH_2^+ + H_2$	Proton impact ionization	[545]
$H^+ + CH_2 \rightarrow 0 \cdot 5 CH_2^+ + H$	Proton impact ionization	[545]
$\rightarrow 0 \cdot 5 CH^+ + H_2$	Proton impact ionization	[545]
$H^+ + CH \rightarrow CH^+ + H$	Proton impact ionization	[545]
$H^+ + C_2H_2 \rightarrow 0 \cdot 5 C_2H_2^+ + H$	Proton impact ionization	[545]
$\rightarrow 0 \cdot 5 C_2H^+ + H$	Proton impact ionization	[545]
$H^+ + C_2H \rightarrow C_2H^+ + H$	Proton impact ionization	[545]

$\vec{E}$ is the electric field. Equation (10.2) is solved by using the Scharfetter–
Gummel scheme [546], which can handle large density gradients. It is
assumed that the ion-neutral collision rates and ion/neutral temperatures remain constant within the reactor chamber. The drift-diffusion
momentum transfer equation for the neutrals does not include the first
term that contains the electric field $\vec{E}$.

Table 10.7 Reactions involving argon and hydrogen species [54].

Reaction	Class of reaction	References
$H + Ar \rightarrow H^+ + Ar + e$	Ionization	[237,541]
$H^+ + Ar \rightarrow H + Ar^+$	Charge exchange	[237,541]
$H_2^+ + Ar \rightarrow H_2 + Ar^+$	Charge exchange	[237,541]
$Ar^+ + H_2 \rightarrow Ar + H_2^+$	Charge exchange	[53,237,541,545]
$H_2^+ + Ar \rightarrow ArH^+ + H$	Proton transfer	[53,237,541,545]
$H_3^+ + Ar \rightarrow ArH^+ + H$	Proton transfer	[237,541]
$Ar^+ + H_2 \rightarrow ArH^+ + H$	H-atom transfer	[237,541]
$ArH^+ + H_2 \rightarrow H_3^+ + Ar$	Proton transfer	[237,541]
$ArH^+ + e \rightarrow Ar + H$	Recombination	[237,541]
$H_2 + Ar \rightarrow H + H + Ar$	Dissociation	[237,541]
$H + H + Ar \rightarrow H_2 + Ar$	Association	[237,541]

Table 10.8 Miscellaneous electron impact, neutral-neutral, and neutral-ion reactions [54].

Reaction	Class of reaction	References
$Ar^+ + C_2H_2 \rightarrow C_2H_2^+ + Ar$	Charge exchange	[137]
$CH + CH \rightarrow C_2H + H$		[137]
$Ar + C_2H_2 + H \rightarrow C_2H + H_2 + Ar$	Abstraction	[137]
$C_2H + H \rightarrow C_2H_2$		[137]
$Ar + C_2H + H \rightarrow C_2H_2 + Ar$		[137]
$C_2H_2 + CH_2 \rightarrow CH_3 + CH$		[137]
$C_2H_2 + H_2 \rightarrow CH_3 + H$		[137]
$C_2H + H_2 \rightarrow CH_3$		[137]
$CH_2 + H \rightarrow CH + H_2$	Abstraction	[137]
$CH + H_2 \rightarrow CH_2 + H$		[137]
$CH + H \rightarrow C + H_2$	Abstraction	[137]
$H + CH_3 \rightarrow CH_2 + H_2$	Abstraction	[53]
$H + C_2H_2 \rightarrow C_2H + H_2$	Abstraction	[53]
$CH_2 + CH_2 \rightarrow C_2H_2 + H_2$		[53]
$CH + CH \rightarrow C_2H_2$		[53]
$C_2H + C_2H \rightarrow C_2 + C_2H_2$		[53]
$C_2H + H \rightarrow C_2 + H_2$	Abstraction	[53]
$C_2H_2 + Ar \rightarrow C_2 + H_2 + Ar$	Abstraction	[53]
$e + C_2 \rightarrow e + C_2(a^3\Pi_u)$	Excitation	[137]
$C_2 + H_2 \rightarrow C_2H + H$		[137]
$C_2(a^3\Pi_u) + H_2 \rightarrow C_2H + H$		[137]

The electron temperature is determined by solving the energy balance equation

$$\frac{\partial}{\partial t}\left(\frac{3}{2}n_e k T_e\right) + \nabla \cdot \vec{q}_e + e\vec{\Gamma}_e \cdot \vec{E} + P_{coll} = P_{abs},\qquad(10.3)$$

where D_e, T_e, and $\vec{\Gamma}_e$ are the diffusion coefficient, temperature and flux of electrons, respectively. Here, P_{coll} is the energy loss rate per unit volume due to the electron-neutral collisions [543]. Furthermore,

$$\vec{q_e} = -\frac{5}{2} n_e D_e \nabla (kT_e) + \frac{5}{2} kT_e \vec{\Gamma}_e \tag{10.4}$$

is the energy flux [54] written in a form similar to that of the fluxes of charged particles by using the exponential scheme proposed by Scharf-fetter and Gummel [546].

In Equation (10.3), the absorbed energy is [54]

$$P_{abs} = Re(\sigma_p |E_\theta(r, z)|^2), \tag{10.5}$$

where $\sigma_p = \epsilon_0 w_{pe}^2 / (v_{en} - jw)$ is the conductivity of the cold plasma, w_{pe} is the plasma frequency, and v_{en} and w are the frequencies of electron-neutral collisions and RF generator, respectively. Here, $E_\theta(r, z)$ is the azimuthal electric field induced in the plasma by an external flat spiral antenna and can be calculated from

$$\nabla^2 E_\theta(r, z) + \frac{w^2}{c^2} K E_\theta(r, x) = -i w \mu_0 J_{ext,\theta} \tag{10.6}$$

which relates the induced electric field to the RF current through the inductive coil $J_{ext,\theta}$, where $K = 1 + j\sigma_p / w\epsilon_0$ is the plasma tensor and ϵ_0 is the dielectric constant [54].

The electrostatic field in the plasma reactor is obtained from Poisson's equation

$$\nabla^2 \phi = \frac{e}{\epsilon_0} (n_e - n_+), \tag{10.7}$$

where n_e is the electron density and n_+ is the combined number density of all positive ions included in the simulation (see Section 4.2). Equation (10.7) is solved by using the successive-over-relaxation (SOR) algorithm with Chebyshev acceleration.

The time-averaged inductive power deposited to the plasma is given by Equation (8) of Reference [547]. In this simulation, standard electrodynamic boundary conditions on the dielectric and metal surfaces of the plasma reactor have been used. Further details of the basic version of the numerical code can be found elsewhere [548].

Finally, Tables 10.4–10.8 provide the details of chemical reactions in the ionized gas phase included in our discharge model, with the references to original sources of reaction rate coefficients [54].

11
Appendix B. Why Plasma-based Nanoassembly: Further Reasons

The range of nanofabrication processes that can benefit from plasma-based processes is in fact much broader than we have considered so far in this monograph. We have only addressed some of the typical advantages of using plasmas in nanotechnology and related them to specific properties, such as the presence of electric fields, ionized component, polarization, and some others.

However, due to significant space (and also time) constraints, it was not practical to provide exhaustive coverage of all the main plasma-assisted nanoscale processes and related issues. The material in this appendix is intended to partially mediate this problem by giving the reader a succinct description of some of the main problems and achievements to be found in a number of published reports. Some of the works have already been cited in the main part of this monograph.

We emphasize that the statements that accompany each of the reports do not aim at describing everything or even the most important results and conclusions of the relevant work. They rather serve as prompts on what sort of research has been conducted and to guide the interested reader to relevant publications. As we have already stressed above, the succinct, prompt-like descriptions and linked references serve to further show the widespread use of plasma nanotools and nanoscale processes.

For the convenience of the reader, the references will be split into several main categories for easier reference. These categories have been selected to make them consistent with the main part of this monograph.

11.1
Carbon Nanotubes and Related Structures

Bell *et al.* [549] present a number of plasma-related factors important to the growth of carbon nanofibers and multiwalled carbon nanotubes by PECVD. Some of the issues considered are the effect of the sheath electric field on nanostructure vertical alignment, low-temperature deposition in plasma environments, and nanostructure reshaping by varying the ratio

Plasma Nanoscience: Basic Concepts and Applications of Deterministic Nanofabrication
Kostya (Ken) Ostrikov
Copyright © 2008 WILEY-VCH Verlag GmbH & Co. KGaA, Weinheim
ISBN: 978-3-527-40740-8

of carbon feedstock gas to the etchant gas. Related articles: [21,231,550–555].

A large number of carbon nanostructures can be synthesized using gas-phase processes in atmospheric-pressure *thermal plasmas*. Gonzalez-Aguilar *et al.* [556] reviewed specific roles of thermal plasmas in the gas-phase synthesis of carbon black and carbon-based nanostructures (fullerenes and nanotubes) and highlights the advantages and disadvantages of plasma-based techniques compared to other existing technologies. Related articles: [557–569].

As we have stressed in Chapter 7, thermal plasmas of arc discharges are particularly suitable for high-rate gas-phase synthesis of high-quality carbon nanotubes. References [272, 570–581] provide further evidence to this effect. Many of the above works pay special attention to single-walled carbon nanotubes.

In addition to the above well-established methods of nanocarbon synthesis, in recent years there has been a strong trend towards carbon nanostructure and nanoparticle production using low-temperature plasmas of submerged arc discharges in liquids such as water, liquid hydrogen, molten chloride, toluene, benzene, and so on. This nanoscale synthesis environment is becoming increasingly attractive owing to its apparent simplicity, low-cost, high yield, and relatively simple system design. This environment also provides an effective and fast cooling and condensation of reactants, which may lead to novel nanocarbon materials with highly-unusual properties [582]. Related articles: [583–589].

In Section 7.3, we considered some examples of ion-assisted reshaping of carbon nanostructures. Wang *et al.* [590, 591] experimentally demonstrated that the shape of conical carbon nanotips sensitively depends on ion bombardment and proposed a model based on the competition of processes of ion deposition and ion-assisted etching and sputtering of the surface. Related articles: [592–595].

Microwave plasma-based CVD is one of the most effective approaches for the synthesis of various forms of nanocarbons. For instance, carbon nanoparticles, nanodiamond, carbon nanotubes and DLC films were grown using a microwave, most commonly used 2.45 GHz, plasma reactor [393, 596]. Interestingly, substrate biasing using a 13.56 MHz RF source often appears useful to control plasma-generated building units.

Further examples of various aspects (e.g., effect of variation of different process parameters) of plasma-assisted growth of single- and multiwalled carbon nanotubes are given in References [597–604]. Related articles: [605–607].

Reports [608,610–612] highlight the importance of an appropriate balance of carbon-bearing building units. In practice, single-walled nanotubes may be grown under low building material supply conditions (supply-limited growth); whereas MWCNTs appear under much higher supply conditions (diffusion-limited growth, oversupply of BUs). Several papers [609–611,613–620] discuss the role of metal catalyst nanoparticles in the controlled carbon nanotube synthesis. The importance of these processes has been highlighted in Chapters 2, 3, and 7.

A very large number of articles report on various aspects (effect of process parameters, feedstock gases, catalyst, substrate, etc.) of the plasma-assisted growth (e.g., different modifications of PECVD techniques) of carbon nanofibers (CNFs, considered in Section 7.2) and related nanostructures [621–645]. In addition, articles [646, 647] report on effective ways of post-processing (e.g., post-growth catalyst removal and functionalization) of carbon nanostructures. For a more detailed discussion of the synthesis and properties of carbon nanofibers and related nanostructures please refer to the comprehensive review [22].

11.2
Semiconductor Nanostructures and Nanomaterials

Shieh *et al.* [342] summarized recent studies on the quantum dot, nanograss, porous film, and nanocones that were synthesized using plasma methods. It contains discussions of the issues of generation and energy of building units and of the significant reduction of incubation time of nanograss. Most of the processes appear to be compatible with silicon-based nanotechnology. Related articles: [313,648–651].

Bapat *et al.* [85] described a new plasma process for the synthesis of cubic-shaped silicon nanocrystals for nanoelectronic device applications. The advantages of using plasma processes for building nanoelectronic devices are discussed. Related articles: [47,48,84,126,652].

Roca i Cabarrocas *et al.* [348] studied the synthesis of silicon nanocrystals in RF silane-based plasmas with the aims of: (i) production of devices based on quantum size effects associated with small dimensions of silicon nanocrystals and (ii) synthesis of polymorphous and polycrystalline silicon using silicon nanocrystals as building units. Related articles: [80,92,653–655].

Koga *et al.* [656] studied different regimes of silicon nanoparticle synthesis in capacitively coupled RF plasmas. These nanoparticles are very promising for fabricating nanoporous materials for future low-k inter-level dielectrics (ILDs) with the dielectric constant $\varepsilon_{IDL} < 2.0$, expected by the International Technology Roadmap for Semiconductors by the year 2012. Related articles: [77,123,657,658].

Cheng *et al.* [659] performed detailed investigations on the growth dynamics and characterization of SiC quantum dots synthesized by low-frequency inductively coupled plasma assisted RF magnetron sputtering. This technique leads to the formation of nanoislanded nanocrystalline SiC films discussed in Chapter 3 of this monograph. Related articles: [129,660,661].

An intersting case of plasma-based synthesis of core-shell Si/Ge nanoparticles was reported [662]. These particles ranged in size from 5 to 15 nm, contained a crystalline core and a very thin (typically 1–2 nm) outer shell. For more discussion on plasma-assisted synthesis of core-shell nanoparticles, please refer to Section 6.3 of this monograph. A quite similar plasma-based technique was used to synthesize SiC/SiN nanopowders and make unique nanocomposites with oxides, or TiN/AlN [663]. Furthermore, by intermixing SiC and SiN nanolayers, it is possible to fabricate homogeneous Si_3N_4/SiC nanocomposites [664].

SiCN nanocrystals and nanotubes with interesting properties can be grown in a VLS (Vapor-Liquid-Solid) mode from supersaturated Fe catalyst using microwave plasma-enhanced chemical vapor deposition [665].

Hori *et al.* [666] reported on the possibility of growing GaN quantum dots on $Al_xGa_{1-x}N$ layers by plasma-assisted molecular beam epitaxy.

Shiratani *et al.* [667] recently introduced a new single-step method to deposit Si quantum dot films using plasmas of very high frequency discharges in $H_2 + SiH_4$ gas mixtures. This technique may significantly improve the electron mobility in Si quantum dot-enhanced solar cells of the third generation. Related articles: [668,669].

Articles [670–685] show further examples of various aspects of plasma-assisted synthesis of low-dimensional semiconducting nanostructures for a broad range of applications. These structures are made of a range of materials and their synthesis involves various plasma-based approaches.

11.3
Other Nanostructures and Nanoscale Objects

Komatsu [686] showed an interesting example of a successful synergy of plasma- and laser-assisted processes to synthesize various nanostructured forms of boron nitride (BN) such as microcones for electron field emission applications. It is stressed that thermodynamically non-equilibrium forms of BN can be grown under far-from-equilibrium plasma conditions. Examples of plasma-based processes include plasma-assisted laser chemical vapor deposition (PALCVD) and plasma-assisted pulsed laser deposition (PAPLD). Related articles: [687–695].

Valsesia *et al.* [696] presented a low-cost and effective technique to fabricate intricate nanopatterns on polymer surfaces for advanced applications in protein chips, cell immobilization arrays, and eventually integrated bio-nanodevices. This technique (commonly termed nanosphere lithography) relies on a top-down nanofabrication approach and uses oxygen plasma etching through a monolayer of polystyrene nanospheres as an etching mask. Related articles: [697–706].

Polymeric nanospheres can be synthesized using a rapid, repetitive-burst, continuous wave plasma polymerization process described in details elsewhere [707]. If properly functionalized, these nanospheres can be used to fabricate superhydrophobic surfaces for a variety of biomedical applications.

Thermal plasmas have been successfully used for high-rate synthesis of TiO_2 [708] and SiC [709] nanoparticles. The experimental findings are consistent with the results of numerical modeling. In some cases plasma-grown nanoparticles can be used as effective building units, quite similar to what was discussed in Chapters 2 and 3. Related articles: [710–716].

Atmospheric pressure inductively coupled plasmas have also shown outstanding performance in highly-efficient synthesis of functional silicon-based nanoparticles such as dicilicides of molybdenum $MoSi_2$ and titanium TiS_2 [717]. Inductively coupled thermal plasma established itself as a powerful tool for fabrication of functional nanoparticles with precise control of particle size distribution and stoichiometric composition. Predictive numerical modeling of thermal plasma characteristics and nanoparticle nucleation and growth are becoming increasingly important. Related articles: [718–725].

Article [286], already mentioned in this monograph in respect to the most important practical requirements for the synthesis of arrays of quantum confinement structures, also gives several interesting examples of the plasma-assisted synthesis of low-dimensional nanomaterials. These examples include self-assembled size-uniform ZnO nanoparticles, ultra-high-aspect-ratio Si nanowires, vertically aligned cadmium sulfide (CdS) nanostructures, and quarternary semiconducting SiCAlN nanomaterial, which were synthesized using inductively coupled plasma-assisted RF magnetron sputtering deposition. The observed increase in crystallinity and growth rates of the nanostructures are explained by using a model of plasma-enhanced adatom surface diffusion under conditions of local energy exchange between the ion flux and the growth surface. Some of these effects have been discussed in Chapters 6 and 7.

Low-temperature plasmas have been successfully used for nanoassembly of a range of one-dimensional nanostructures, such as ZnO nanorods [726], silica fibers with embedded gold nanoparticles [727], and Ag/TiO$_2$ core-shell nanofibers [728,729].

RF and microwave plasmas have been applied for the synthesis of a range of nanoparticles and nanoparticle-made nanomaterials such as TiO$_2$ [730], iron oxide nanoparticles coated with various types of polymer films including polystyrene, polymethyl methacrylate, polycarbonate and several others [731]. Iron nanopowders can be encapsulated *in situ* into a macromolecular thin film structure generated by plasma polymerization [732]. For more details related to the plasma source and plasma parameters please refer to Reference [733].

11.4
Materials with Nanoscale Features

Li *et al.* [734] reported on the plasma-assisted cathodic vacuum arc deposition of nanostructured zirconia (ZrO$_2$) films. The structure and other properties of these films have been related to the ability to induce the formation of apatite (See Section 8.4 of this monograph for apatite formation on surfaces of bioactive materials) in a simulated body fluid. This ability is crucial for clinical applications such as ball heads in total hip replacements. Related articles: [735–737].

Cheng *et al.* [661] has shown the possibility of highly-controlled synthesis of nanocrystalline SiC films using inductively coupled plasma-assisted RF magnetron sputtering deposition. The mechanism of the formation of nanocrystallites at low process temperatures (ca. 400 °C) is discussed. This work is part of a focused effort to fabricate nanostructured silicon carbide films with different structure and morphology and using different plasma-based approaches. For an example of a nanocrystalline structure of SiC films please refer to Figure 3.4 in Chapter 3.

Using low-temperature reactive plasmas of silane and methane gases diluted in argon and/or hydrogen, one can synthesize, at low process temperatures, nanocrystalline SiC (*nc*-SiC) with a range of excellent parameters (e.g., nanocrystalline fraction, stoichiometric elemental composition) suitable for the future nanodevice applications [739]. By appropriately controlling the process parameters, one can obtain very homogeneous nanocrystalline cubic silicon carbide films [740, 741]. The reactive plasma-based approach makes it possible to synthesize amorphous, polymorphous, and polycrystalline SiC with different attributes and properties [740]. In particular, the film properties can be controlled by varying the degree of hydrogen dilution [742]. Related articles: [129, 743–755].

An RF plasma CVD reactor has been successfully used to fabricate Si-C-N-based nanoparticles [756]. This process enables the precise control of nanoparticle size distribution, elemental composition, and surface texture; consequently, this may result in improved properties relative to the SiC and SiN nanomaterials. Furthermore, SiN_x films have been synthesized on SiO_2 buffer layers by plasma-assisted radio frequency magnetron sputtering [757, 758]. These films show a strong photoluminescence yield which further improves under conditions of sufficient oxidation and moderate nitridation of SiN_x/SiO_2 films during the plasma-based growth process.

Nanoscale plasma processing becomes increasingly important in many different areas such as nanomaterials for nuclear fusion devices [759–762], proton exchange membrane (PEM) fuel cells [763–769], and a range of advanced biomedical applications (surface treatment, bio-active materials, sterilization, biosensors, etc.) [770–777]. For more details on these and other relevant applications of low-pressure PECVD, please refer to the review article [778]. Recent and expected roles of reactive plasmas and plasma-polymerized films in a broad range of biomedical applications have also been reviewed [779].

11.5
Plasma-Related Issues and Fabrication Techniques

Anders [347] reviewed the issues of metal plasma production in ionized physical vapor deposition (i-PVD) systems, the energetic condensation of metal plasmas, metallization of ULSI features in microelectronics, and the formation of various nanostructures using such plasmas. Selected examples include top-down nanofabrication via a combination of lithographic pattern delineation and conformal coating as well as self-organized nanocomposites and nanoporous materials. Related articles: [780–782].

Cvelbar and Mozetic [501] investigated the behavior of oxygen atoms in the vicinity of nanostructured niobium pentoxide (Nb_2O_5) surfaces. Recombination of oxygen atoms plays a crucial role in the development of Nb_2O_5 nanowires. This mechanism is similar to the one discussed in Section 8.3 of this monograph. Related articles: [783–785].

As we have mentioned several times in this monograph, it is not possible to achieve highly-controlled nanoscale processing without reliable knowledge of the interactions of building and working units with solid surffaces. This is a vast area of research represented by a substantial number of publications that clarify the mechanisms of interactions of specific species (e.g., radicals) with the surfaces being processed. These studies make it possible to identify the major precursors of nanostructures being grown or otherwise predict the outcomes of interactions of WUs with the surface. Review articles and theoretical and experimental reports [786,787,789–802] clarify the main issues involved. Recently, Hori and Goto [803] reviewed the status of ongoing investigations into sticking of radicals on solid surfaces for smart plasma-based nanoscale processing.

Finally, we stress that the number of relevant publications is large and our attempt to give some additional (to the main chapters of this monograph) coverage in this appendix is just the tip of the iceberg ... However, we sincerely appreciate and are interested in the efforts of everyone who is involved in any aspects of research related to this monograph. To appreciate these efforts, we have lumped all unnamed relevant references (together with our sincere gratitide) in one concluding reference [804].

References

1 K. Ostrikov and S. Xu, *Plasma-Aided Nanofabrication: From Plasma Sources to Nanoassembly* (Wiley-VCH Verlag GmbH, Weinheim Germany, 2007).

2 R. P. Feynman, *There is Plenty of Room at the Bottom*, Paper presented at the American Physical Society Annual Meeting, 29 December 1959. Reprinted in: *Miniaturization*, edited by H. D. Gilbert (Reinhold, New York, 1961). See also http://www.zyvex.com/nanotech/feynman.html

3 S. Iijima, Nature (London) **354**, 56 (1991).

4 K. Ostrikov, Rev. Mod. Phys. **77**, 489 (2005).

5 K. Ostrikov and A. B. Murphy, J. Phys. D: Appl. Phys. **40**, 2223 (2007).

6 C. P. Poole and F. J. Owens, *Introduction to Nanotechnology* (John Wiley and Sons Ltd., New York, 2003), and the references therein.

7 J. H. Fendler, *Nanoparticles and Nanostructured Films: Preparation, Characterization and Applications* (Wiley-VCH Verlag GmbH, Weinheim, Germany, 1998).

8 G. S. Oehrlein, *Plasma Processing of Electronic Materials* (Springer, Berlin, 2003)

9 *Molecular Building Blocks for Nanotechnology: From Diamondoids to Nanoscale Materials and Applications.*, edited by G. A. Mansoori, Th. F. George, L. Assoufid, and G. Zhang. Series: Topics in Applied Physics, vol. 109 (Springer, Berlin, 2007).

10 *Advances in Low Temperature RF Plasmas: Basis for Process Design*, edited by T. Makabe (Elsevier, Amsterdam, 2002).

11 *Ionized Physical Vapor Deposition*, edited by J. A. Hopwood (Academic Press, San Diego, 2001).

12 *Handbook of Plasma Immersion Ion Implantation and Deposition*, edited by A. Anders (John Wiley and Sons, Canada, Montreal, 2000).

13 *Dusty Plasmas: Physics, Chemistry, and Technological Impacts in Plasma Processing*, edited by A. Bouchoule (J. Wiley and Sons, Chicester, UK, 1999).

14 S. V. Vladimirov, K. Ostrikov, and A. Samarian, *Physics and Applications of Complex Plasmas* (Imperial College Press, Singapore, London, 2005).

15 F. F. Chen and J. P. Chang, *Principles of Plasma Processing: A Lecture Course* (Kluwer Academic Publishers, Amsterdam, 2002).

16 *Nanoparticles and Nanostructured Films: Preparation, Characterization and Applications*, edited by J. H. Fendler (Wiley-VCH, Weinheim, 1998).

17 *Nano-Architectured and Nanostructured Materials: Fabrication, Control and Properties*, edited by Y. Champion and H.-J. Fecht (Wiley-VCH, Weinheim, 2004).

18 G. Schmid, *Nanoparticles: From Theory to Application* (John Wiley and Sons, New York, 2004).

19 S. Reich, *Carbon Nanotubes: Basic Concepts and Physical Properties* (John Wiley and Sons, New York, 2004).

20 *Encyclopedia of Nanoscience and Nanotechnology*^{TM}, edited by H. S. Nalva (American Scientific Publishers, New York, 2004).

21 M. Meyyappan, L. Delzeit, A. Cassell, and D. Hash, Plasma Sources Sci. Technol. **12**, 205 (2003), and the references therein.

22 A. V. Melechko, V. I. Merkulov, T. E. McKnight, M. A. Guillorn, K. L. Klein, D. H. Lowndes, and M. L. Simpson, J. Appl. Phys. **97**, 041301 (2005).

23 K. Ostrikov, IEEE Trans. Plasma Sci. **35**, 127 (2007).

24 C. Helling, R. Klein, P. Woitke, U. Nowak, and E. Seldmayr, Astron. Astrophys. **423**, 657 (2004).

25 D. A. Williams and E. Herbst, Surf. Sci. **500**, 823 (2002).

26 Y. J. Pendleton and L. J. Allamandolla, Astrophys. J. Suppl. Ser. **138**, 75 (2002).

27 S. I. Popel and A. A. Gisko, Nonlin. Proc. Geophys. **13**, 223 (2006).

28 S. V. Vladimirov and K. Ostrikov, Phys. Repts. **393**, 175 (2004).

29 M. Keidar, Y. Raitses, A. Knapp, and A. M. Waas, Carbon **44**, 1013 (2006).

30 A. Chhowalla and G. A. J. Amaratunga, Nature (London) **407**, 164 (2000).

31 S. Sriraman, S. Agrawal, E. S. Aydil, and D. Maroudas, Nature (London) **418**, 62 (2002).

32 E. Ott, C. Grebogi, and J. A. Yorke, Phys. Rev. Lett. **64**, 1196 (1990).

33 J. Li, C. Papadopoulos, J. M. Xu, and M. Moskovits, Appl. Phys. Lett. **75**, 367 (1999).

34 E. Tam, I. Levchenko, K. Ostrikov, M. Keidar, and S. Xu, Phys. Plasmas **14**, 033503 (2007).

35 M. P. Stoykovich and P. F. Nealey, Mater. Today **9**, 20 (2006).

36 L. Xu, S. C. Vemula, M. Jain, S. K. Nam, V. M. Donnelly, D. J. Economou, and P. Ruchhoeft, Nano Lett. **5**, 2563 (2005).

37 D. M. Gruen, MRS Bulletin **6**, 771 (2001).

38 D. M. Gruen, P. C. Redfern, D. A. Horner, P. Zapol, and L. A. Curtis, J. Phys. Chem B **103**, 5459 (1999).

39 R. Doering, *Societal Implications of Scaling to Nanoelectronics*, in: *Societal Implications of Nanoscience and Nanotechnology*, Nanoscale Science, Engineering, and Technology (NSET) Subcommittee Workshop Report, Edited by M. C. Roco and W. S. Bainbridge (National Science Foundation, Arlington, Virginia, 2001).

40 Y. J. T. Lii, *Etching*, in *ULSI Technology*, edited by C. Y. Chang and S. M. Sze (McGraw Hill, N.Y., 1996), p. 329–370.

41 H. O. U. Fynbo, *et al.*, Nature (London) **433**, 136 (2005).

42 M. S. Povich, J. C. Raymond, G. H. Jones, M. Uzzo, Y. K. Ko, P. D. Feldman, P. L. Smith, B. G. Marsden, and T. N. Woods, Science **302**, 1949 (2003).

43 S. Kempf, R. Srama, M. Horanyi, M. Burton, S. Helfert, G. Moragos-Klostermeyer, M. Roy, and E. Grun, Nature (London) **433**, 289 (2005).

44 M. S. Tillack, D. W. Blair, and S. S. Harilal, Nanotechnology **15**, 390 (2004).

45 M. C. Roco, S. Williams, and A. P. Alivisatos, *Nanotechnology Research Directions: Vision for Nanotechnology Research and Development in the next Decade* (Kluwer Academic, Amsterdam, 1999). See also: US National Nanotechnology Initiative, http://www.nano.gov

46 A. McWilliams, *GB-290 Nanotechnology: A Realistic Market Evaluation* (Business Comm. Co., 2004), http://www.bccresearch.com/

47 A. Bapat, C. Anderson, C. R. Perrey, C. B. Carter, S. A. Campbell, and U. Kortshagen, Plasma Phys. Control. Fusion **46**, B97 (2004).

48 L. Mangolini, E. Thimsen, and U. Kortshagen, Nano Lett. **5**, 655 (2005).

49 S. Xu, K. Ostrikov, J. D. Long, and S. Y. Huang, Vacuum **80**, 621 (2006).

50 K. Ostrikov, J. D. Long, P. P. Rutkevych, and S. Xu, Vacuum **80**, 1126 (2006).

51 I. Levchenko, K. Ostrikov, M. Keidar, and S. Xu, J. Appl. Phys. **98**, 064304 (2005).

52 I. Levchenko, K. Ostrikov, M. Keidar, and S. Xu, Appl. Phys. Lett. **89**, 033109 (2006).

53 I. B. Denysenko, S. Xu, P. P. Rutkevych, J. D. Long, N. A. Azarenkov, and K. Ostrikov, J. Appl. Phys. **95**, 2713 (2004).

54 K. Ostrikov, H. J. Yoon, A. E. Rider, and S. V. Vladimirov, Plasma Proc. Polym. **4**, 27 (2007).

55 K. Ostrikov, Z. Tsakadze, I. Denysenko, P. P. Rutkevych, J. D. Long, and S. Xu, Contr. Plasma Phys. **45**, 514 (2005).

56 K. Ostrikov and S. Xu, *Nanofabrication of Single-Crystalline Flat-Panel Display Microemitters: a Plasma-Building Unit Approach*, SPIE Proceedings "Microelectronics, MEMS and Nanotechnology" (SPIE, Bellingham WA, USA, 2005), vol. 6037, 6037-32 (2005).

57 A. Fridman and L. A. Kennedy, *Plasma Physics and Engineering* (Taylor & Francis, New York, 2004).

58 J. Perrin and Ch. Hollenstein, in *Dusty Plasmas: Physics, Chemistry and Technological Impacts in Plasma Processing*, edited by A. Bouchoule (John Wiley and Sons Inc., New York, 1999), p.77–180.

59 C. Hollenstein, Plasma Phys. Control. Fusion **42**, R93 (2000).

60 D. S. Bethune, C. H. Kiang, M. S. DeVries, G. Gorman, R. Savoy, and R. Beyers, Nature (London) **363**, 605 (1993).

61 J. Stangl, V. Holy, and G. Bauer, Rev. Mod. Phys. **76**, 725 (2004).

62 V. Shchukin, N. N. Ledentsov, and D. Bimberg, *Epitaxy of Nanostructures* (Springer, Berlin/Heidelberg, 2003).

63 M. S. Dresselhaus, G. Dresselhaus, and P. C. Eklund, *Science of Fullerenes and Carbon Nanotubes* (Academic, San Diego, CA, 1996).

64 N. R. Franklin and H. Dai, Adv. Mater. **12**, 890 (2002).

65 N. M. Hwang and D. Y. Kim, Int. Mater. Rev. **49**, 171 (2004).

66 N. M. Hwang and D. Y. Yoon, J. Cryst. Growth **143**, 103 (1994).

67 I. D. Jeon, *et al.*, J. Cryst. Growth **223**, 6 (2001).

68 M. A. Lieberman and A. J. Lichtenberg, *Principles of Plasma Discharges and Materials Processing* (John Wiley and Sons Inc., New York, 1994).

69 H. Yasuda, *Plasma Polymerization* (Academic, New York, 1985).

70 N. M. Hwang, J. H. Hahn, and D. Y. Yoon, J. Cryst. Growth **162**, 55 (1996).

71 N. M. Hwang, J. Cryst. Growth **204**, 85 (1999).

72 I. D. Jeon, C. J. Park, D. Y. Kim, and N. M. Hwang, J. Cryst. Growth **213**, 79 (2000).

73 P. Gerhardt and K. H. Homann, Combust. Flame **81**, 289 (1990).

74 N. M. Hwang, W. S. Cheong, D. Y. Yoon, and D. Y. Kim, J. Cryst. Growth **218**, 33 (2000).

75 A. V. Rode, *et al.*, Phys. Rev. B **70**, 054407 (2004).

76 M. Tanda, M. Kondo, and A. Matsuda, Thin Sol. Films **427**, 33 (2003).

77 M. Shiratani, K. Koga, and Y. Watanabe, Thin Sol. Films **427**, 1 (2003).

78 K. Wegner, P. Piseri, H. V. Tafreshi, and P. Milani, J. Phys. D: Appl. Phys. **39**, R439 (2006).

79 E. Magnano, *et al.*, Phys. Rev. B **67**, 125414 (2003).

80 G. Viera, M. Mikikian, E. Bertran, P. Roca i Cabarrocas, and L. Boufendi, J. Appl. Phys. **92**, 4684 (2002).

81 Y. Poissant, P. Chatterjee, and P. Roca i Cabarrocas, J. Appl. Phys. **94**, 7305 (2003).

82 A. Fontcuberta i Morral, P. Roca i Cabarrocas, and C. Clerc, Phys. Rev. B **69**, 125307 (2004).

83 V. Suendo, A. Kharchenko, and P. Roca i Cabarrocas, Thin Sol. Films **451–452**, 259 (2004).

84 A. Bapat, C. R. Perrey, S. A. Campbell, C. B. Carter, and U. Kortshagen, J. Appl. Phys. **94**, 1969 (2003).

85 A. Bapat, M. Gatti, Y.-P. Ding, S. A. Campbell, and U. Kortshagen, J. Phys. D.: Appl. Phys. **40**, 2247 (2007)

86 P. Mulvaney, MRS Bulletin **6**, 1009 (2001).

87 B. Gilbert, F. Huang, H. Zhang, G. Waychunas, and J. F. Banfield, Science **305**, 651 (2004).

88 H. Kersten, *et al.*, New J. Phys. **5**, 93.1 (2003).

89 M. Chhowalla, K. B. K. Teo, C. Dukati, N. L. Rupersinghe, G. A. J. Amaratunga, A. C. Ferrari, D. Roy, J. Robertson, and W. I. Milne, J. Appl. Phys. **90**, 5308 (2001).

90 C. Bower, W. Zhu, S. Jin, and O. Zhou, Appl. Phys. Lett. **77**, 830 (2000).

91 J. Frantz and K. Nordlund, Phys. Rev. B **67**, 075415 (2003).

92 P. Roca i Cabarrocas, N. Chaabane, A. V. Kharchenko, and S. Tchakarov, Plasma Phys. Control. Fusion **46**, B235 (2004).

93 Z. L. Tsakadze, K. Ostrikov, J. D. Long, and S. Xu, Diam. Relat. Mater. **13**, 1923 (2004).

94 Z. L. Tsakadze, K. Ostrikov, and S. Xu, Surf. Coat. Technol. **191/1**, 49 (2005).

95 Z. L. Tsakadze, I. Levchenko, K. Ostrikov, S. Xu, Carbon **45**, 2022 (2007).

96 M. Mozetic, U. Cvelbar, M. K. Sunkara, and S. Vaddiraju, Adv. Mater. **17**, 2138 (2005).

97 W. W. Stoffels, E. Stoffels, G. H. P. M. Swinkels, H. Videlot, M. Boufnichkel, and G. M. W. Kroesen, Phys. Rev. E **59**, 2302 (1999).

98 J. Perrin, M. Shiratani, P. Kae-Nune, H. Videlot, J. Jolly, and J. Guillion, J. Vac. Sci. Technol. **16**, 278 (1998).

99 M. Frenklash and H. Wang, Phys. Rev. B **43**, 1520 (1991).

100 E. Tam, I. Levchenko, and K. Ostrikov, J. Appl. Phys. **100**, 036104 (2006).

101 T. Baron, P. Gentile, N. Magnea, and P. Mur, Appl. Phys. Lett. **79**, 1175 (2001).

102 K. Tsubouchi and K. Masu, J. Vac. Sci. Technol. A **10**, 856 (1992).

103 H. M. Thomas and G. E. Morphill, Nature (London) **379**, 806 (1996).

104 R. Merlino and J. Goree, Phys. Today, July 2004, 32 (2004).

105 P. Rutkevych, K. Ostrikov, S. Xu, and S. V. Vladimirov, J. Appl. Phys. **96**, 4421 (2004).

106 J. Moran, K. Ostrikov, B. W. James, and A. Samarian, *Investigation Into the Deposition of Ordered Structures in an Ionized Granular Gas*, Interim Report on research project "Complex Ionized Gas Systems: Pattern Transfer From Gas to Surface", School of Physics, The University of Sydney, Australia (2006).

107 P. Hohenberg and W. Kohn. Phys. Rev. **136**, B864 (1964).

108 W. Kohn and L. J. Sham, Phys. Rev. **140**, A1133 (1965).

109 S. Lunqvist and N. H. March, *The Theory of the Homogenous Electron Gas* (Plenum Press, New York, 1983).

110 http://www.accelrys.com/

111 B. Delley, J. Chem. Phys. **113**, 7756 (2000).

112 *Handbook of Nanophase and Nanostructured Materials. Volume II: Characterization*, edited by Z. L. Wang, Y. Liu, and Z. Zhang, (Springer, Berlin, 2002).

113 U. V. Bhandarkar, M. T. Swihart, S. L. Girshik, and U. Kortshagen, J. Phys. D: Appl. Phys. **33**, 2731 (2000).

114 S.-M. Suh, S. L. Girshick, U. R. Kortshagen, and M. R. Zachariah, J. Vac. Sci. Technol. A **21**, 251 (2003).

115 K. De Bleecker, A. Bogaerts, R. Gijbels, and W. Goedheer, Phys. Rev. E **69**, 056409 (2004).

116 K. Ostrikov, I. B. Denysenko, S. V. Vladimirov, S. Xu, H. Sugai, and M. Y. Yu, Phys. Rev. E **67**, 056408 (2004).

117 I. B. Denysenko, K. Ostrikov, S. Xu, M. Y. Yu, and C. H. Diong, J. Appl. Phys. **94**, 6097 (2003).

118 H. Sugai, I. Ghanashev, M. Hosokawa, K. Mizuno, K. Nakamura, H. Toyoda, and K. Yamauchi, Plasma Sources Sci. Technol. **10**, 378 (2001).

119 A. A. Fridman, L. Boufendi, T. Hbid, B. N. Potapkin, and A. Bouchoule, J. Appl. Phys. **79**, 1303 (1996).

120 P. Roca i Cabarrocas, S. Hamma, S. N. Sharma, G. Viera, E. Bertran, and J. Costa, J. Non-Cryst. Sol. **227–230**, 871 (1998).

121 N. Chaabane, A. V. Kharchenko, H. Vach, and P. Roca i Cabarrocas, New J. Phys. **3**, 37.1 (2003).

122 N. Chaabane, P. Roca i Cabarrocas, and H. Vach, J. Non-Cryst. Solids **338–340**, 51 (2004).

123 M. Shiratani, S. Maeda, K. Koga, and Y. Watanabe, Jpn. J. Appl. Phys., Part 1 **39**, 287 (2000).

124 A. Fontcuberta i Morral and P. Roca i Cabarrocas, Thin Sol. Films **383**, 161 (2001).

125 J. Costa, in *Handbook of Nanostructured Materials*, edited by H. S. Nalva (Academic, New York, 2000), vol. 1, p. 57.

126 P. Cernetti, R. Gresback, S. A. Campbell, and U. Kortshagen, Chem. Vap. Deposition **13**, 345 (2007).

127 M. L. Ostraat, *et al.*, Appl. Phys. Lett. **79**, 433 (2001).

128 G. Conibeer, M. Green, R. Corkish, Y. Cho, E. C. Cho, C. W. Jiang, T. Fangsuwannarak, E. Pink, Y. Huang, T. Puzzer, T. Trupke, B. Richards, A. Shalav, and L. Lin, Thin Sol. Films **511–512**, 645 (2006).

129 Q. J. Cheng, S. Xu, J. D. Long, and K. Ostrikov, Appl. Phys. Lett. **90**, 173112 (2007).

130 K. Ostrikov, I. Denysenko, M. Y. Yu, and S. Xu, Phys. Scripta **72**, 277 (2005).

131 T. Kuykendall, P. J. Pauzauskie, Y. Zhang, J. Goldberger, D. Sirbuly, J. Denlinger, and P. Yang, Nature Mater. **3**, 524 (2004).

132 O. A. Louchev and J. R. Hester, J. Appl. Phys. **94**, 2002 (2003).

133 S. Hong, I. Stefanovic, J. Berndt, and J. Winter, Plasma Sources Sci. Technol. **12**, 46 (2003).

134 E. Kovacevic, I. Stefanovic, J. Berndt, and J. Winter, J. Appl. Phys. **93**, 2924 (2003).

135 G. Gebauer and J. Winter, New J. Phys. **5**, 38.1 (2003).

136 S. Stoykov, C. Eggs, and U. Kortshagen, J. Phys. D: Appl. Phys. **34**, 2160 (2001).

137 F. J. Gordillo-Vazques and J. M. Albella, Plasma Sources Sci. Technol. **13**, 50 (2004).

138 K. Ostrikov, H.-J. Yoon, A. E. Rider, and V. Ligatchev, Phys. Scripta **76**, 187 (2007).

139 B. J. Hrivnak and S. Kwok, Astroph. J. **513**, 869 (1999).

140 J. D. Long, S. Xu, S. Y. Huang, P. P. Rutkevych, M. Xu, and C. H. Diong, IEEE Trans. Plasma Sci. **33**, 240 (2005).

141 V. I. Merkulov, D. H. Lowndes, Y. Y. Wei, G. Eres, and E. Voelkl, Appl. Phys. Lett. **76**, 3555 (2000).

142 D. B. Hash and M. Meyyappan, J. Appl. Phys. **93**, 750 (2003).

143 S. Reich, L. Li, and J. Robertson, Chem. Phys. Lett. **421**, 469 (2006).

144 S. Hofmann, R. Sharma, C. Ducati, G. Du, C. Mattevi, C. Cepek, M. Cantoro, S. Pisana, A. Parvez, F. Cervantes-Sodi, A. C. Ferrari, R. Dunin-Borkowski, S. Lizzit, L. Petaccia, A. Goldoni, and J. Robertson, Nano Lett. **7**, 602 (2007).

145 T. Nozaki, K. Ohnishi, K. Okazaki, and U. Kortshagen, Carbon **45**, 364 (2007).

146 T. Saito, S. Ohshima, W. C. Xu, H. Ago, M. Yumura, and S. Iijima, J. Phys. Chem. B **109**, 10647 (2005).

147 M. Paillet, J. C. Meyer, T. Michel, V. Jourdain, P. Poncharal, J.-L. Sauvajol, N. Cordente, C. Amiens, B. Chaudret, S. Roth, and A. Zahab, Diam. Relat. Mater. **15**, 1019 (2006).

148 J. Y. Raty, F. Gygi, and G. Galli, Phys. Rev. Lett. **95**, 096103 (2005).

149 B. Q. Wei, J. D'Arcy-Gall, P. M. Ajayan, and G. Ramanath, Appl. Phys. Lett. **83**, 3581 (2003).

150 T. Hirata, N. Satake, G. H. Jeong, T. Kato, R. Hatakeyama, K. Motomiya, and K. Tohji, Appl. Phys. Lett. **83**, 1119 (2003).

151 D. Cai, J. M. Mataraza, Z. H. Qin, Z. Huang, J. Huang, T. C. Chiles, D. Carnahan, K. Kempa, Z. Ren, Nature Methods **2**, 449 (2005).

152 P. P. Rutkevych, K. Ostrikov, and S. Xu, Phys. Plasmas **12**, 103507 (2005).

153 P. P. Rutkevych, K. Ostrikov, and S. Xu, Phys. Plasmas **14**, 043502 (2007).

154 S. Walch and R. Merkle, Nanotechnology **9**, 1998 (1998).

155 A. T. H. Chuang, J. Robertson, B. O. Boskovic, K. K. K. Koziol, Appl. Phys. Lett. **90**, 123107 (2007).

156 M. Hiramatsu, K. Shiji, H. Amano, and M. Hori, Appl. Phys. Lett. **84**, 4708 (2004).

157 N. Marks, N. C. Cooper, D. R. McKenzie, D. G. McCulloch, P. Bath, and S. P. Russo, Phys. Rev. B **65**, 075411 (2002).

158 N. Marks, J. M. Bell, G. K. Pearce, D. R. McKenzie, and M. M. M. Bilek, Diam. Relat. Mater. **12**, 2003 (2003).

159 I. Levchenko, M. Korobov, M. Romanov, and M. Keidar, J. Phys. D: Appl. Phys. **37**, 1690 (2004).

160 V. I. Merkulov, A. V. Melechko, M. A. Guillorn, D. H. Lowndes, and M. L. Simpson, Appl. Phys. Lett. **79**, 2970 (2001).

161 J. F. AuBuchon, L-H.Chen, and S. Jin, J. Phys. Chem. B **109**, 6044 (2005).

162 J. Robertson, Mater. Today **10**, 36 (2007).

163 E. Tam, K. Ostrikov, I. Levchenko, M. Keidar, and S. Xu, *Multi-Scale Hybrid Numerical Simulation of the Growth of High Aspect Ratio Nanostructures*, Comput. Mater. Sci. in press, DOI:10.1016/j.commatsci.2008.01.048 (2008).

164 H. Kanzow and A. Ding, Phys. Rev. B **60**, 11180 (1999).

165 S. Helveg, C. Lopez-Cartez, J. Sehested, P. L. Hansen, B. S. Clausen, J. R. Rostrup-Nielsen, F. Abild-Pedersen, and J. Norskov, Nature (London) **427**, 426 (2004).

166 J. Goree and T. E. Sheridan, J. Vac. Sci. Technol. A **10**, 3540 (1992).

167 I. Denysenko and K. Ostrikov, Appl. Phys. Lett. **90**, 251501 (2007).

168 A. N. Obraztsov, I. Pavlovsky, A. P. Volkov, E. D. Obraztsova, A. L. Chuvilin, and V. L. Kuznetsov, J. Vac. Sci. Technol. B **18**, 1059 (2000).

169 S. H. Tsai, F. K. Chiang, T. G. Tsai, F. S. Shieu, and H. C. Shih, Thin Sol. Films **366**, 11 (2000).

170 S. Hofmann, C. Dukati, J. Robertson, and B. Kleinsorge, Appl. Phys. Lett. **83**, 135 (2003).

171 K. Ostrikov, I. Levchenko, and S. Xu, Comput. Phys. Commun. **177**, 110 (2007).

172 I. Levchenko, K. Ostrikov, and E. Tam, Appl. Phys. Lett. **89**, 223108 (2006).

173 I. Levchenko and K. Ostrikov, J. Phys. D **40**, 2308 (2007).

174 A. Rider, I. Levchenko, and K. Ostrikov, J. Appl. Phys. **101**, 044306 (2007).

175 I. Levchenko, A. Rider, and K. Ostrikov, Appl. Phys. Lett. **90**, 193110 (2007).

176 A. Rider, I. Levchenko, K. Ostrikov, and M. Keidar, Plasma Proc. Polym. **4**, 638 (2007).

177 J. C. Ho, I. Levchenko, and K. Ostrikov, J. Appl. Phys. **101** 094309 (2007).

178 I. Levchenko and K. Ostrikov, *Numerical Simulation of Self-Organized Nanoislands in Plasma-Based Assembly of Quantum Dot Arrays*, SPIE Proceedings "Microelectronics, MEMS and Nanotechnology" (SPIE, Bellingham WA, USA, 2005), vol. 6039, 6039-24 (2005).

179 F. Rosei, J. Phys. Condens. Matter **16**, S1373 (2004).

180 I. H. Hutchinson, *Principles of Plasma Diagnostics*, Second Edition (Cambridge University Press, Cambridge, UK, 2002).

181 H. R. Griem, *Principles of Plasma Spectroscopy* (Cambridge University Press, Cambridge, UK, 2005).

182 M. Su, B. Zheng, and J. Liu, Chem. Phys. Lett. **322**, 321 (2000).

183 L. Delzeit, B. Chen, A. Cassell, R. Stevens, C. Nguyen, and M. Meyyappan, Chem. Phys. Lett. **348**, 368 (2001).

184 L. Delzeit, I. McAninch, B. A. Cruden, D. Hash, B. Chen, J. Han, and M. Meyyappan, J. Appl. Phys. **91**, 6027 (2002).

185 Y. M. Shyu and F. C. N. Hong, Diam. Relat. Mater. **10**, 1241 (2001).

186 M. Chen, C. M. Chen, S. C. Shi, and C. F. Chen, Jpn. J. Appl. Phys **42**, 614 (2003).

187 Y. S. Woo, I. T. Han, Y. J. Park, H. J. Kim, J. E. Jung, N. S. Lee, D. Y. Jeon, and J. M. Kim, Jpn. J. Appl. Phys Part 1 **42**, 1410 (2003).

188 T. Hirao, K. Ito, H. Furuta, Y. K. Yap, T. Ikuno, S. Honda, Y. Mori, T. Sasaki, and K. Oura, Jpn. J. Appl. Phys. Part 1 **40**, L631 (2001).

189 S. Xu, K. N. Ostrikov, Y. Li, E. L. Tsakadze, and I. R. Jones, Phys. Plasmas **8**, 2549 (2001).

190 D. Herrebout, A. Bogaerts, M. Yan, R. Gijbels, W. Goedheer, and E. Dekempeneer, J. Appl. Phys. **90**, 570 (2001).

191 S. F. Yoon, K. H. Tan, Rusli, and J. Ahn, J. Appl. Phys. **91**, 40 (2002).

192 V. Ivanov, O. Proshina, T. Rakhimova, A. Rakhimov, D. Herrebout, and A. Bogaerts, J. Appl. Phys. **91**, 6296 (2002).

193 K. Bera, B. Farouk, and Y. H. Lee, Plasma Sources Sci. Technol. **10**, 211 (2001).

194 K. Bera, B. Farouk, and P. Vitello, J. Phys. D: Appl. Phys. **34**, 1479 (2001).

195 D. Hash, D. Bose, T. R. Govindan, and M. Meyyappan, J. Appl. Phys. **93**, 6284 (2003).

196 M. Camero, F. J. Gordillo-Vazques and C. Gomez-Aleixandre, Chem. Vap. Deposition **13**, 326 (2007).

197 S. Xu, K. N. Ostrikov, W. Luo, and S. Lee, J. Vac. Sci. Technol. A, **18**, 2185 (2000).

198 K. N. Ostrikov, S. Xu, and A. B. M. Shafiul Azam, J. Vac. Sci. Technol. A **20**, 251 (2002).

199 K. N. Ostrikov, S. Xu, and M. Y. Yu, J. Appl. Phys. **88**, 2268 (2000).

200 K. Ostrikov, E. Tsakadze, N. Jiang, Z. Tsakadze, J. Long, R. Storer, and S. Xu, IEEE Trans. Plasma Sci. **30**, 128 (2002).

201 K. Ostrikov, E. Tsakadze, S. Xu, S. V. Vladimirov, and R. Storer, Phys. Plasmas **10**, 1146 (2003).

202 K. N. Ostrikov, I. B. Denysenko, E. L. Tsakadze, S. Xu, and R. G. Storer, J. Appl. Phys. **92**, 4935 (2002).

203 V. Schulz-von der Gathen, J. Roepcke, T. Gans, M. Kaning, C. Lukas, and H. F. Doebele, Plasma Sources Sci. Technol. **10**, 530 (2001).

204 C. Riccardi, R. Barni, M. Fontanesi, P. Tosi, Chem. Phys. Lett. **329**, 66 (2000).

205 T. Nozaki, K. Okazaki, U. Kortshagen, and J. Heberlein, Bull. Amer. Phys. Soc. **48**, No. 6, 13 (2003).

206 E. Gogolides, D. Mary, A. Rhallabi, and G. Turban, Jpn. J. Appl. Phys., Part 1 **34**, 261 (1995).

207 J. Geddes, R. W. McCullough, A. Donnelly, and H. B. Gilbody, Plasma Sources Sci. Technol. **2**, 93 (1993).

208 H. Amemiya, J. Phys. Soc. Japan **66**, 1335 (1997).

209 J. T. Gudmundsson, Plasma Sources Sci. Technol. **10**, 76 (2001).

210 T. Chevolleau and W. Fukarek, Plasma Sources Sci. Technol. **9**, 568 (2000).

211 N. Sadeghi, M. van de Grift, D. Vender, G. M. W. Kroesen, and F. J. de Hoog, Appl. Phys. Lett. **70**, 835 (1997).

212 H. Kojima, H. Toyoda, and H. Sugai, Appl. Phys. Lett. **55**, 1292 (1989).

213 C. Hopf, K. Letourneur, W. Jacob, T. Schwarz-Selinger, and A. Von Keudell, Appl. Phys. Lett. **74**, 3800 (1999).

214 A. von Keudell, T. Schwarz-Selinger, M. Meier, and W. Jacob, Appl. Phys. Lett. **76**, 676 (2000).

215 N. Mutsukura, S. Inoue, and Y. Machi, J. Appl. Phys. **72**, 43 (1992).

216 C. Hopf, T. Schwarz-Selinger, W. Jacob, and A. Von Keudell, J. Appl. Phys. **87**, 2719 (2000).

217 C. Lee and M. A. Lieberman, J.Vac. Sci. Technol. A **13**, 368 (1995).

218 V. A. Godyak, *Soviet Radio Frequency Discharge Research* (Delphic, Falls Church, VA, 1986).

219 M. Kawase, T. Nakai, A. Yamaguchi, T. Hakozaki, and K. Nashimoto, Jpn. J. Appl. Phys, **36**, Part 1, 3396 (1997).

220 I. Peres, M. Fortin, and J. Margot, Phys. Plasmas, **3**, 1754 (1996).

221 M. Keidar and A. M. Waas, Nanotechnology **15**, 1571 (2004).

222 D. B. Hash, M. S. Bell, K. B. K. Teo, B. A. Cruden, W. I. Milne, and M. Meyyappan, Nanotechnology **16**, 925 2005

223 M. J. Kushner, J. Appl. Phys. **95**, 846 (2004).

224 K. de Bleecker, A. Bogaerts, and W. Goedheer, Appl. Phys. Lett. **88**, 151501 (2006).

225 W. B. Choi, D. S. Chung, J. H. Kang, H. Y. Kim, Y. W. Jin, I. T. Han, Y. H. Lee, J. E. Jung, N. S. Lee, G. S. Park, and J. M. Kim, Appl. Phys. Lett. **75**, 3129 (1999).

226 C. L. Tsai, C. W. Chao, C. L. Lee, H. C. Shih, Appl. Phys. Lett. **74**, 3462 (1999).

227 C. L. Tsai, C. F. Chen, and L. K. Wu, Appl. Phys. Lett. **81**, 721 (2002).

228 S. B. Lee, A. S. Teh, K. B. K. Teo, M. Chhowalla, D. G. Hasko, W. I. Milne, G. A. J. Amaratunga, and H. Ahmed, Nanotechnology **14**, 192 (2003).

229 S. Bhattacharyya, A. Granier, and G. Turban, J. Appl. Phys. **86**, 4668 (1999).

230 A. von Keudell and W. Jacob, Progr. Surf. Sci. **76**, 21 (2004).

231 K. B. K. Teo, D. B. Hash, R. G. Lacerda, N. L. Rupesinghe, M. S. Bell, S. H. Dalal, D. Bose, T. R. Govindan, B. A. Cruden, M. Chhowalla, G. A. J. Amaratunga, M. Meyyappan, and W. I. Milne, Nano Lett. **4**, 921 (2004).

232 E. L. Tsakadze, K. Ostrikov, Z. L. Tsakadze, and S. Xu, J. Appl. Phys. **97**, 013301 (2005).

233 J. Perrin, C. Bohm, R. Etemadi, and A. Lloret, Plasma Sources Sci. Technol. **3**, 252 (1994).

234 A. Gallagher, Phys. Rev. E **62**, 2690 (2000).

235 A. A. Howling, L. Sansonnens, J.-L. Dorrier, and Ch. Hollenstein, J. Phys. D: Appl. Phys. **26**, 1003 (1993).

236 Ch. Hollenstein, J.-L. Dorier, J. Dutta, L. Sansonnens, and A. A. Howling, Plasma Sources Sci. Technol. **3**, 278 (1994).

237 A. Bogaerts, K. de Bleecker, V. Georgieva, I. Kolev, M. Madani, E. Neyts, Plasma Proc. Polym. **3**, 110 (2006).

238 L. Boufendi, J. Herman, A. Bouchoule, B. Dubreuil, E. Stoffels, W. W. Stoffels, and M. L. de Giorgi, J. Appl. Phys. **76**, 148 (1994).

239 P. Cernetti, R. Gresback, S. A. Campbell, and U. Kortshagen, Chem. Vap. Depos. **13**, 345 (2007).

240 U. Kogelschats, Plasma Phys. Contr. Fusion **46**, B63 (2004).

241 S. P. Fisenko, D. B. Kane, and M. S. El-Shall, J. Chem. Phys. **123**, 104704 (2005).

242 M. N. Mautner, *et al.*, Faraday Discuss. **133**, 103 (2006).

243 S. P. Fisenko, private communication (2007).

244 S. P. Fisenko, Appl. Surf. Sci. **106**, 94 (1996).

245 T. Seto, Y. Kawakami, N. Suzuki, M. Hirasawa, S. Kano, N. Aya, S. Sasaki, and H. Shimura, J. Nanopart. Res. **3**, 185 (2001).

246 K. De Bleecker, A. Bogaertz, and W. Goedheer, Phys. Rev. E **73**, 026405 (2006).

247 Ch. Deschenaux, *Etude de l'Origine et de la Croissance de Particules Sub-micrometriques dans des Plasmas Radiofrequence Reactifs*, Ph. D. Thesis, Centre de Recherches en Physique des Plasmas, Association Euratom, Switzerland (2002).

248 K. De Bleecker, *Modeling of the Formation and Behavior of Nanoparticles in Dusty Plasmas*, Ph. D. Thesis, University of Antwerpen, Beligium (2006).

249 Market research by iSuppli/Stanford Resources, El Segundo, CA; http://www.isuppli.com

250 W. P. Kang, T. Fisher, and J. L. Davidson, New Diam. and Frontier Carbon Technol. **11**, 129 (2001).

251 E. T. Thostenson, Z. Ren, and T. W. Chou, Compos. Sci. Technol. **61**, 1899 (2001).

252 K. B. K. Teo, R. G. Lacerda, M. H. Yang, A. S. Teh, L. A. W. Robinson, S. H. Dalal, N. L. Rupesinghe, M. Chhowalla, S. B. Lee, D. A. Jefferson, D. G. Hasko, G. A. J. Amaratunga, W. L. Milne, P. Legagneux, L. Gangloff, E. Minoux, J. P. Schnell, and D. Pribat, IEE Proc. Circuits Devices and Systems **151**, 443 (2004).

253 L. Nilsson, O. Groening, O. Kuettel, P. Groening, and L. Schlapbach, J. Vac. Sci. Technol. **20**, 326 (2002).

254 M. Mauger, V. T. Bihn, A. Levesque, and D. Guillot, Appl. Phys. Lett. **85**, 305 (2004).

255 Y. Shiratori, H. Hiraoka, Y. Takeuchi, S. Itoh, and M. Yamamoto, Appl. Phys. Lett. **82**, 2485 (2003).

256 C. Ducati, I. Alexandrou, M. Chhowalla, J. Robertson, G. A. J. Amaratunga, J. Appl. Phys. **95**, 6387 (2004).

257 M. S. Bell, R. G. Lacerda, K. B. K. Teo, N. L. Rupesinghe, G. A. J. Amaratunga, W. I. Milne, and M. Chhowalla, Appl. Phys. Lett. **85**, 1137 (2004).

258 I. Levchenko, K. Ostrikov, J. D. Long, and S. Xu, Appl. Phys. Lett. **91**, 113115 (2007).

259 H. L. Chua and S. Xu, *Ab initio Density Functional Theory Simulations of Single-Crystalline Carbon Nanotip Structures*, Internal Report 2678/2005, National Institute of Education, Singapore.

260 I. Levchenko and K. Ostrikov, Int. J. Nanosci. **5**, 621 (2006).

261 Q. Y. Zhang and P. Chu, Surf. Coat. Technol. **158**, 247 (2002).

262 Y. Watanabe, M. Shiratani, and K. Koga, Plasma Sources Sci. Technol. **11**, A229 (2002).

263 E. I. Waldorff, A. M. Waas, P. P. Friedmann, and M. Keidar, J. Appl. Phys. **95**, 2749 (2004).

264 E. Barborini, I. N. Kholmanov, P. Piseri, C. Ducati, C. E. Bottani, and P. Milani, Appl. Phys. Lett. **81**, 3052 (2002).

265 F. Di Fonzo, A. Gidwani, M. H. Fan, D. Neumann, D. I. Iordanoglou, J. V. R. Heberlein, P. H. McMurry, S. L. Girshick, N. Tymiak, W. W. Gerberich, and N. P. Rao, Appl. Phys. Lett. **77**, 910 (2000).

266 M. C. Barnes, A. R. Gerson, S. Kumar, L. Green, and N. M. Hwang, Thin Sol. Films **436**, 181 (2003).

267 M. C. Barnes, A. R. Gerson, S. Kumar, and N. M. Hwang, Thin Sol. Films **446**, 29 (2004).

268 M. C. Barnes, S. Kumar, L. Green, N. M. Hwang, and A. R. Gerson, Surf. Coat. Technol. **190**, 321 (2005).

269 S. V. Vladimirov and N. F. Cramer, Phys. Rev. E **62**, 2754 (2000).

270 K. N. Ostrikov, S. Kumar, and H. Sugai, Phys. Plasmas **7**, 3490 (2001).

271 O. Havnes, T. Nitter, V. Tsytovich, G. E. Morfill, and T. Hartquist, Plasma Sources Sci. Technol. **3**, 448 (1994).

272 M. Keidar, J. Phys. D **40**, 2388 (2007).

273 I. Levchenko and O. Baranov, Vacuum **72**, 205 (2004).

274 S. Nunomura, M. Kita, K. Koga, M. Shiratani, and Y. Watanabe, Jpn. J. Appl. Phys. **40**, L1509 (2005).

275 E. V. Shevchenko, D. V. Talapin, N. A. Kotov, S. O'Brien, and C. B. Murray, Nature **439**, 55 (2006).

276 P. P. Rutkevych, K. Ostrikov, and S. Xu, Int. J. Nanosci. **5**, 465 (2006).

277 J. A. Venables, *Introduction to Surface and Thin Film Processes* (Cambridge Univ. Press, Cambridge, UK, 2000).

278 F. Rosei and R. Rosei, Surf. Sci. **500**, 395 (2002).

279 K. Ostrikov, *Surface Science of Plasma Exposed Surfaces: a Challenge for Applied Plasma Science*, Vacuum, in press, DOI: 10.1016/ j.vacuum.2008.03.051 (2008).

280 R. Rohrer, Surf. Sci. **299–300**, 956 (1994).

281 C. F. Quate, Surf. Sci. **299–300**, 980 (1994).

282 Y. Arakawa and H. Sakai, Appl. Phys. Lett. **40**, 939 (1982).

283 G. M. Whitesides, J. P. Mathias, and C. T. Seto, Science **254**, 1312 (1991).

284 A. V. Krasheninnikov and F. Banhart, Nature Mater. **6**, 723 (2007).

285 X. Michalet, *et al.*, Science **307**, 538 (2005).

286 K. Ostrikov, I. Levchenko, S. Xu, S. Y. Huang, Q. J. Cheng, J. D. Long, and M. Xu, Thin Sol. Films, in press, DOI:10.1016/ j.tsf.2007.11.045 (2008).

287 J. Osaka, M. S. Kumar, H. Toyoda, T. Ishijima, H. Sugai, and T. Mizutani, Appl. Phys. Lett. **90** 172114 (2007).

288 E. A. Fitzgerald Jr. and D. G. Ast, US Patent 5156995 (1992).

289 S. Kicin, A. Pioda, T. Ihn, M. Sigrist, A. Fuhrer, K. Ensslin, M. Reinwald and W. Wegscheider, New. J. Phys. **7**, 185 (2005).

290 S. J. Choi and M. J. Kushner, J. Appl. Phys. **74**, 853 (1993).

291 F. M. Ross, J. Tersoff, and R. M. Tromp, Phys. Rev. Lett. **80**, 984 (1998).

292 G. Fuchs, P. Melinon, F. Santos Aires, M. Treilleux, B. Cabaud and A. Hoareau, Phys. Rev. B **44**, 3926 (1991).

293 G. Fuchs, M. Treilleux, F. Santos Aires, B. Cabaud, P. Melinon and A. Hoareau, Phys. Rev. A **40**, 6128 (1989).

294 P. Jensen, Rev. Mod. Phys. **71**, 1695 (1999).

295 T. Ishikawa, S. Kajii, K. Matsunaga, T. Hogami, Y. Kohtoku, and T. Naga-sawa, Science **282**, 1295 (1998).

296 D. Nakamura, I. Gunjishima, S. Yam-aguchi, T. Ito, A. Okamoto, H. Kondo, S. Onda, and K. Takatori, Nature (London) **430**, 1009 (2004).

297 F. Liao, S. L. Girshick, W. M. Mook, W. W. Gerberich, and M. R. Zachariah, Appl. Phys. Lett. **86**, 171913 (2005).

298 S. Kerdiles, A. Berthelot, F. Gourbil-leau, and R. Rizk, Appl. Phys. Lett. **76**, 2373 (2000).

299 S. Y. Huang, S. Xu, J. D. Long, Z. Sun, and T. Chen, Phys. Plasmas **13**, 023506 (2006).

300 M. Xu, V. M. Ng, S. Y. Huang, J. D. Long, and S. Xu, IEEE Trans. Plasma Sci. **33**, 242 (2005).

301 J. Mi, R. Johnson, and W. J. Lackey, J. Am. Ceram. Soc. **89**, 519 (2006).

302 J. Y. Fan, X. L. Wu, H. X. Li, H. W. Liu, G. S. Huang, G. G. Siu, and P. K. Chu, Appl. Phys. A **82**, 485 (2006).

303 H. Colder, R. Rizk, M. Morales, P. Marie, J. Vicens, and I. Vickridge, J. Appl. Phys. **98**, 024313 (2005).

304 L. J. Wang and F. C. N. Hong, Micropor. Mesopor. Mater. **77**, 167 (2005).

305 X. L. Wu, Y. Gu, S. J. Xiong, J. M. Zhu, G. S. Huang, X. M. Bao, and G. G. Siu, J. Appl. Phys. **94**, 5247 (2003).

306 F. A. Reboredo, L. Pizzagalli, and G. Galli, Nano Lett. **4**, 801 (2004).

307 J. Y. Fan, X. L. Wu, H. X. Li, H. W. Liu, G. G. Siu, and P. K. Chu, Appl. Phys. Lett. **88**, 41909 (2006).

308 H. C. Lo, D. Das, J. S. Hwang, K. H. Chen, C. H. Hsu, C. F. Chen, and L. C. Chen, Appl. Phys. Lett. **83**, 1420 (2003).

309 K. Chew, Rusli, S. F. Yoon, J. Ahn, V. Ligatchev, E. J. Teo, T. Osipowicz, and F. Watt, J. Appl. Phys. **92**, 2937 (2002).

310 A. R. Oliveira and M. N. P. Carreno, Mat. Sci. Eng. B **128**, 44 (2006).

311 T. Miyasato, Y. Sun, J. K. Wigmore, J. Appl. Phys. **85**, 3565 (1999).

312 H. J. Kim, Z. M. Zhao, J. Liu, V. Ozolins, J. Y. Chang, and Y. H. Xie, J. Appl. Phys. **95**, 6065 (2004).

313 J. Shieh, T. S. Ko, H. L. Chen, B. T. Dai and T. C. Chu, Chem. Vap. Dep. **10**, 265 (2004).

314 T. P. Munt, D. E. Jesson, V. A. Shchukin, and D. Bimberg, Appl. Phys. Lett. **85**, 1784 (2004).

315 F. Gibou, C. Ratsch, and R. Caflisch, Phys. Rev. B. **67**, 155403 (2003).

316 T. Doi, M. Ichikawa, S. Hosoki, and K. Ninomiya, Surf. Sci. **343**, 24 (1995).

317 J. D. Gale and J. M. Seddon, *Thermodynamics and Statistical Mechanics* (Wiley-Interscience, New York, 2002).

318 "Properties of Solids", in *CRC Handbook of Chemistry and Physics, Internet Version*, 87th Edition, edited by D. R. Lide, (Taylor and Francis, Boca Raton, FL, 2007).

319 M. S. Silverberg, *Chemistry: The Molecular Nature of Matter and Change*, 3rd Ed.(McGraw-Hill, New York, 2003), p. 339.

320 T. L. Cottrell, *The Strength of Chemical Bonds*, 2nd Ed., (Butterworths Publications Ltd., London, 1958), p. 246.

321 X. Y. Tian, C. J. Ruan, P. Cui, W. T. Liu, J. Zheng, X. Zhang, X. Y. Yao, K. Zheng, and Y. Li, J. Macromol. Sci. Phys. **45**, 835 (2006).

322 H.-Y. Fan, K. Yang, D. M. Boye, T. Sigmon, K. J. Malloy, H. Xu, G. P. López, and C. J. Brinker, Science **304**, 567 (2004).

323 F. X. Redl, K.-S. Cho, C. B. Murray and S. O'Brien, Nature (London) **423**, 968 (2003).

324 R. E. Bailey and S. Nie, J. Am. Chem. Soc. **125**, 7100 (2003).

325 N. Onojima, J. Suda, T. Kimoto, H. Matsunami, Appl. Phys. Lett. **83**, 5208 (2003).

326 T. M. Jovin, Nature Biotech. **21**, 32 (2003).

327 C. Lang, D. Nguyen-Manh, and D. J. H. Cockayne, J. Appl. Phys. **94**, 7067 (2003).

328 A. I. Yakimov, A. V. Dvurechenskii, A. I. Nikiforov, and Y. Y. Proskuryakov, J. Appl. Phys. **89**, 5676 (2001).

329 J. L. Gray, R. Hull, and J. A. Floro, J. Appl. Phys. **100**, 084312 (2006).

330 P. A. Cain, H. Ahmed, D. A. Williams, J. Appl. Phys. **92**, 346 (2002).

331 M. Brust, Nature Mater. **4**, 364 (2005).

332 A. Karmous, A. Cuenat, A. Ronda, and I. Berbeziera, Appl. Phys. Lett. **83**, 6401 (2004).

333 R. Noetzel, Z. Niu, M. Ramsteiner, H. P. Schoenherr, A. Tranpert, L. Daeweritz, and K. H. Ploog, Nature **392**, 56 (1998).

334 V. Ligatchev, Rusli, and Z. Pan, Appl. Phys. Lett. **87**, 242903 (2005).

335 V. Ligatchev, T. K. S. Wong, and S. F. Yoon, J. Appl. Phys. **95**, 7681 (2004).

336 E. Kasper and S. Heim, Appl. Surf. Sci. **224**, 3 (2004).

337 A. Karmous, A. Cuenat, A. Ronda, I. Berbeziera, S. Atha, and R. Hull, Appl. Phys. Lett. **85**, 6401 (2004).

338 A. Pascale, P. Gentile, J. Eymery, J. Meziere, A. Bavard, T. U. Schulli, and F. Fournel, Surf. Sci. **600**, 3187 (2006).

339 M. De Seta, G. Capellini, and F. Evangelisti, Cryst. Res. Technol. **40**, 942 (2005).

340 C. Zhao, Y. H. Chen, C. X. Cui, B. Xu, L. K. Yu, W. Lei, J. Sun, and Z. G. Wang, Solid State Commun. **137**, 630 (2006).

341 T. Takaki, T. Hasebe, and Y. Tomita, J. Cryst. Growth **287**, 495 (2006).

342 J. Shieh, C. H. Lin, and M. C. Yang, J. Phys. D: Appl. Phys. **40**, 2242 (2007).

343 K. Ostrikov, I. Levchenko, and S. Xu, *Self-Organized Nanoarrays: Plasma-Related Controls,* Pure Appl. Chem., in press (2008).

344 N. Motta, A. Sgarlata, F. Rosei, P. D. Szkutnik, S. Nufris, M. Scarselli, and A. Balzarotti, Mater. Sci. Eng. B **101**, 77 (2003).

345 I. Levchenko, K. Ostrikov, A. E. Rider, E. Tam, S. V. Vladimirov, and S. Xu, Phys. Plasmas **14**, 063502 (2007).

346 Q. J. Cheng, S. Xu, and J. D. Long, J. Appl. Phys. **101**, 094304 (2007).

347 A. Anders, J. Phys. D: Appl. Phys **40**, 2272 (2007).

348 P. Roca i Cabarrocas, Th. Nguyen-Tran, Y. Djeridane, A. Abramov, E. Johnson and G. Patriarche, J. Phys. D: Appl. Phys. **40**, 2258 (2007).

349 A. V. Dvurechenski, J. V. Smagina, R. Groetzschel, V. A. Zinovyev, V. A. Armbrister, P. L. Novikov, S. A. Teys and A. K. Gutakovskii, Surf. Coat. Technol. **196**, 25 (2005).

350 T. J. Krinke, K. Deppert, M. H. Magnusson, F. Schmidt and H. Fissan, J. Aerosol Sci. **33**, 1341 (2002).

351 I. Levchenko, K. Ostrikov, and M. Keidar, *Plasma-Assembled Carbon Nanotubes: Electric Field - Related Effects*, J. Nanosci. Nanotechnol., in press, DOI: 10.1166/jnn.2008.010 (2008).

352 H. Huang, O. K. Tan, Y. C. Lee, M. S. Tse, J. Guo, and T. White, Nanotechnology **17**, 3668 (2006).

353 X. S. Peng, L. D. Zhang, G. W. Meng, X. Y. Yuan, Y. Lin, and Y. T. Tian, J. Phys. D: Appl. Phys. **36**, L35 (2003).

354 S. Valizadeh, M. Abid, F. Hernandez-Ramirez, A. R. Rodriguez, K. Hjort, and J. A. Schweitz, Nanotechnology **17**, 1134 (2006).

355 J. C. Johnson, H. Q. Yan, R. D. Schaller, L. H. Haber, R. J. Saykally, and P. D. Yang, J. Phys. Chem. B **105**, 11387 (2001).

356 Y. Wu, P. Qiao, T. Chong, and Z. Shen, Adv. Mater. **14**, 64 (2002).

357 F. Liu, P. J. Cao, H. R. Zhang, J. Q. Li, and H. J. Gao, Nanotechnology **15**, 949 (2004).

358 Y. Zhang, X. Song, J. Zheng, H. Liu, X. Li, and L. You, Nanotechnology **17**, 1916 (2006).

359 S. Xu, J. D. Long, L. Sim, C. H. Diong, and K. Ostrikov, Plasma Proc. Polym. **2**, 373 (2005).

360 K. Jiang, C. Feng, K. Liu, and S. Fan, J. Nanosci. Nanotechnol. **7**, 1494 (2007).

361 V. L. Kuznetsov, A. N. Usoltseva, A. L. Chuvilin, E. D. Obraztsova, and J.-M. Bonard, Phys. Rev. B **64**, 235401 (2001).

362 R. G. Lacerda, A. S. Teh, M. H. Yang, K. B. K. Teo, N. L. Rupesinghe, S. H. Dalal, K. K. K. Koziol, D. Roy, G. A. J. Amaratunga, W. I. Milne, M. Chhowalla, F. Wyczisk and P. Legagneux, Appl. Phys. Lett. **84**, 269 (2004).

363 T. Kato, G.-H. Jeong, T. Hirata, R. Hatakeyama, K. Tohji, K. Motomiya, Chem. Phys. Lett. **381**, 422 (2003).

364 D. B. Geohegan, A. A. Puretzky, I. N. Ivanov, S. Jesse, G. Eres, and J. Y. Howe, Appl. Phys. Lett. **83**, 1851 (2003).

365 X. Zhang, A. Cao, B. Wei, Y. Li, J. Wei, C. Xu, and D. Wu, Chem. Phys. Lett. **362**, 285 (2002).

366 A. Moisala, A. G. Nasibulin, and E. I. Kauppinen, J. Phys.: Condens. Matter. **15**, S3011 (2003).

367 O. Jasek, M. Elias, L. Zajickova, Z. Kucerova, J. Matejkova, A. Rek, and J. Bursik, J. Phys. Chem. Sol. **68**, 738 (2007).

368 M. Cantoro, S. Hofmann, S. Pisana, C. Ducati, A. Parvez, A. C. Ferrari, and J. Robertson. Diam. Relat. Mat. **15**, 1029 (2006).

369 Q. Bao, H. Zhang, and C. Pan, Comput. Mater. Sci. **39**, 616 (2007).

370 T. Kato, R. Hatakeyama, and K. Tohji, Nanotechnology **17**, 2223 (2006).

371 Y. Zhang, A. Chang, J. Cao, Q. Wang, W. Kim, Y. Li, N. Morris, E. Yenilmez, J. Kong, and H. Dai, Appl. Phys. Lett. **79**, 3155 (2001).

372 L. D. Landau, E. M. Lifshits, and L. P. Pitaevskii, *Electrodynamics of Continuous Media* (Pergamon, Oxford, UK).

373 S. Hofmann, G. Csanyi, A. C. Ferrari, M. C. Payne, and J. Robertson, Phys. Rev. Lett. **95**, 036101 (2005).

374 N. Grobert, Mater. Today **10**, 28 (2007).

375 A. Okita, Y. Suda, A. Ozeki, H. Sugawara, Y. Sakai, A. Oda, and J. Nakamura, J. Appl. Phys. **99**, 014302 (2006).

376 S. H. Lim, H. S. Yoon, J. H. Moon, K. C. Park, and J. Jang, Appl. Phys. Lett. **88**, 033114 (2006).

377 A. Anders, Appl. Phys. Lett. **80**, 1100 (2002).

378 O. A. Louchev and Y. Sato, Appl. Phys. Lett. **74**, 194 (1999).

379 O. A. Louchev, Y. Sato, and H. Kanda, Appl. Phys. Lett. **80**, 2752 (2002).

380 J. C. Charlier, A. de Vita, X. Blase, and R. Car, Science **275**, 647 (1997).

381 C. Ducati, I. Alexandrou, M. Chhowalla, G. A. J. Amaratunga, and J. Robertson, J. Appl. Phys. **92**, 3299 (2002).

382 M. Tanemura, K. Iwata, K. Takahashi, Y. Fujimoto, F. Okuyama, H. Sugie, and V. Filip, J. Appl. Phys. **90**, 1529 (2001).

383 O. A. Louchev, T. Laude, Y. Sato, and H. Kanda, J. Chem. Phys. **118**, 7622 (2003).

384 S. Hong, *et al.*, Jpn. J. Appl. Phys. **41**, 6142 (2002).

385 N. V. Mantzaris, E. Gogolides, A. G. Boudouvis, A. Rhallabi, and G. Turban, J. Appl. Phys. **79**, 3718 (1996).

386 I. Denysenko, K. Ostrikov, M. Y. Yu, and N. A. Azarenkov, J. Appl. Phys. **102**, 074308 (2007).

387 Y. Li *et al.*, Nano Lett. **4**, 317 (2004).

388 M. R. Maschmann, P. B. Amama, A. Goyal, Z. Iqbal, R. Gat, and T. S. Fisher, Carbon **44**, 10 (2006).

389 W. L. Wang, X. D. Bai, Z. Xu, S. Liu, and E. G. Wang, Chem. Phys. Lett. **419**, 81 (2005).

390 G. Zhong, T. Iwasaki, K. Honda, Y. Furukawa, I. Ohdomari, and H. Kawarada, Jpn. J. Appl. Phys., Part 1 **44**, 1558 (2005).

391 A. Gohier, T. M. Minea, M. A. Djouadi, and A. Granier, J. Appl. Phys. **101**, 054317 (2007).

392 S. Reich, L. Li, and J. Robertson, Chem. Phys. Lett. **421**, 469 (2006).

393 Y. Y. Wang, S. Gupta, and R. J. Nemanich, Appl. Phys. Lett. **85**, 2601 (2004).

394 G. Zhang, D. Mann, L. Zhang, A. Javey, Y. Li, E. Yenilmez, Q. Wang, J. P. McVittie, Y. Nishi, J. Gibbons, and H. Dai, Proc. Natl. Acad. Sci. **102**, 16141 (2005).

395 O. A. Louchev, Y. Sato, and H. Kanda, J. Appl. Phys. **89**, 3438 (2001).

396 L. T. Chadderton and Y. Chen, Phys. Lett. A **263**, 401 (1999).

397 Y. H. Lee, S. G. Kim, and D. Tomanek, Phys. Rev. Lett. **78**, 2393 (1997).

398 O. A. Louchev, C. Dussarrat, and Y. Sato, J. Appl. Phys. **86**, 1736 (1999).

399 O. A. Louchev, Y. Sato, and H. Kanda, Phys. Rev. E **66**, 011601 (2002).

400 A. V. Krasheninnikov, K. Nordlund, P. O. Lehtinen, A. S. Foster, A. Ayuela, and R. M. Nieminen, Phys. Rev. B **69**, 073402 (2004).

401 A. V. Krasheninnikov, K. Nordlund, P. O. Lehtinen, A. S. Foster, A. Ayuela, and R. M. Nieminen, Carbon **42**, 1021 (2004).

402 Y.-K. Kwon, Y. H. Lee, S. G. Kim, P. Jund, D. Tomanek, and R. E. Smalley, Phys. Rev. Lett. **79**, 2065 (1997).

403 N. Kitamura and A. Oshiyama, J. Phys. Soc. Jpn. **70**, 1995 (2001).

404 O. A. Louchev, H. Kanda, A. Rosen, and K. Bolton, J. Chem. Phys. **121**, 446 (2004).

405 P. Reinke, W. Jakob, and W. Moller, J. Appl. Phys. **74**, 1354 (1993).

406 B. N. Khare, M. Meyyappan, A. M. Casell, C. V. Nguyen, and J. Han, Nano Lett. **2**, 73 (2002).

407 K. S. Kim, D. J. Bae, J. R. Kim, K. A. Park, S. C. Lim, J. J. Kim, W. B. Choi, C. Y. Park, and Y. H. A. Lee, Adv. Mater. **14**, 1818 (2002).

408 P. Ruffieux, O. Groning, M. Bielmann, P. Mauron, L. Schlapbach, P. Groning, Phys. Rev. B **66**, 245416 (2002).

409 A. Nikitin, H. Ogasawara, D. Mann, R. Denecke, Z. Zhang, H. Dai, K. Cho, A. Nilsson, Phys. Rev. Lett. **95**, 225507 (2005).

410 Q. Wan, E. N. Dattoli, W. Y. Fung, W. Guo, Y. B. Chen, X. Q. Pan, and W. Lu, Nano Lett. **6**, 2909 (2006).

411 F. J. Gordillo-Vazquez, A. Perea, A. P. McKiernan, and C. N. Afonso, Appl. Phys. Lett. **86**, 181501 (2005).

412 D. Riabinina, M. Chaker, and F. Rosei, Appl. Phys. Lett. **89**, 131501 (2006).

413 B. S. Kang, S. J. Pearton, and F. Ren, Appl. Phys. Lett. **90**, 083104 (2007).

414 C.-H. Shen, H.-Y. Chen, H.-W. Lin, S. Gwo, A. A. Klochikhin, and V. Yu. Davydov, Appl. Phys. Lett. **88**, 253104 (2006).

415 T. Yamashita, S. Hasegawa, N. Nishida, M. Ishimaru, Y. Hirotsu, and H. Asahi, Appl. Phys. Lett. **86**, 082109 (2005).

416 S. McCaldin, M. Bououdina, D. M. Grant, and G. S. Walker, Carbon **44**, 2273 (2006).

417 P. E. Nolan, M. J. Schabel, D. C. Lynch, and A. H. Cutler, Carbon **33**, 79 (1995).

418 Z. L. Wang, X. Y. Kong, and J. M. Zuo, Phys. Rev. Lett. **91**, 185502 (2003).

419 M. Tanemura, T. Okita, J. Tanaka, M. Kitazawa, K. Itoh, L. Miao, S. Tanemura, S. P. Lau, H. Y. Yang, L. Huang, IEEE Trans. Nanotech. **5**, 587 (2006).

420 A. Hellemans, Science **273**, 1173 (1996).

421 B. T. Liu, *et al.*, Appl. Phys. Lett. **80**, 4801 (2002).

422 Q. Wang, J. J. Li, Y. J. Ma, X. D. Bai, Z. L. Wang, P. Xu, C. Y. Shi, B. G. Quan, S. L. Yue, and C. Z. Gu, Nanotechnology **16**, 2919 (2005).

423 C. H. Oon, J. T. L. Thong, Y. Lei, and W. K. Chim, Appl. Phys. Lett. **81**, 3037 (2002).

424 H. C. Lo, D. Das, J. S. Hwang, K. H. Chen, C. H. Hsu, C. F. Chen, and L. C. Chen, Appl. Phys. Lett. **83**, 1420 (2003).

425 J. Zhou, L. Gong, S. Z. Deng, J. Chen, J. C. She, N. S. Xu, R. Yang, and Z. L. Wang, Appl. Phys. Lett. **87**, 223108 (2005).

426 K. Okano, S. Koizumi, S. R. P. Silva, and G. A. J. Amaratunga, Nature **381**, 140 (1996).

427 K. Kang, L. H. Lewis, and A. R. Moodenbaugh, Appl. Phys. Lett. **87**, 062505 (2005)

428 H. C. Lo, D. Das, J. S. Hwang, K. H. Chen, C. H. Hsu, C. F. Chen, and L. C. Chen, Appl. Phys. Lett. **83**, 1420 (2003)

429 Y. H. Wang, M. J. Kim, H. W. Shan, C. Kittrell, H. Fan, L. M. Ericson, W. F. Hwang, S. Arepalli, R. H. Hauge, R. E. Smalley, Nano Lett. **5**, 997 (2005)

430 J. V. Barth, Surf. Sci. Repts. **40**, 75 (2000).

431 T. Kyotani, Bull. Chem. Soc. Jap. **79**, 1322 (2006).

432 P. X. Hou, T. Yamazaki, H. Orikasa, and T. Kyotani, Carbon **43**, 2624 (2005).

433 V. Ligatchev and B. Gan, Diam. Relat. Mater. **15**, 410 (2006).

434 U. Cvelbar, B. Markoli, I. Poberaj, A. Zalar, L. Kosec, and S. Spaic, Appl. Surf. Sci. **253**, 1861 (2006).

435 G. Y. Zhang, X. Jiang, and E. G. Wang, Science **300**, 472 (2003).

436 M. Mannsberger, A. Kukovecz, V. Georgakilas, J. Rechthaler, F. Hasi, G. Allmaier, M. Prato, and B. Kuzmany, Carbon **42**, 953 (2004).

437 X. Sun, R. Li, B. Stansfield, J. P. Dodelet, G. Menard, and S. Desilets, Carbon **45**, 7323 (2007).

438 A. Okita, Y. Suda, A. Oda, J. Nakamura, A. Ozeki, K. Bhattacharyya, H. Sugawara, and Y. Sakaia, Carbon **45**, 1518 (2007).

439 A. Burian, J. C. Dore, T. Kyotani, and V. Honkimaki, Carbon **43**, 2723 (2005).

440 B. B. Wang and B. Zhang, Carbon **44**, 1949 (2006).

441 Y. C. Sui, B. Z. Cui, R. Guardian, D. R. Acosta, L. Martinez, and R. Perez, Carbon **40**, 1011 (2002).

442 S. H. Jeong, H. Y. Hwang, and K. H. Lee, Appl. Phys. Lett. **78**, 2052 (2001).

443 Y. J. Kim, T. S. Shin, H. D. Choi, J. H. Kwon, Y. C. Chung, and H. G. Yoon, Carbon **43**, 23 (2005).

444 K. Bradley, J. Gabriel, A. Star, and G. Gruener, Appl. Phys. Lett. **83**, 3821 (2003).

445 C. Merino, P. Soto, E. Vilaplana-Ortego, J. M. Gomez de Salazar, F. Pico, and J. M. Rojo, Carbon **43**, 551 (2005).

446 C. K. Tan, K. P. Loh, J. T. L. Thong, C. H. Sow, and H. Zhang, Diam. Relat. Mater. **14**, 902 (2005).

447 P. K. Chuang, I. J. Teng, W. H. Wang, and C. T. Kuo, Diam. Relat. Mater. **14**, 1911 (2005).

448 J. Yu, E. G. Wang, and X. D. Bai, Appl. Phys. Lett. **78**, 2226 (2001).

449 P. C. Eklund, J. M. Holden, and R. A. Jishi, Carbon **33**, 959 (1995).

450 A. Anders, Appl. Phys. Lett. **85**, 6137 (2004).

451 K. Bradley, J. C. P. Gabriel, A. Star, and G. Grüner, Appl. Phys. Lett. **83**, 3821 (2003).

452 G. Du, F. Xu, Z. Yuan, and G. V. Tendeloo, Appl. Phys. Lett. **24**, 243101 (2006).

453 J. W. Park, J. K. Kim, and K. Y. Suh, Nanotechnology **17**, 2631 (2006).

454 M. Bockrath, D. H. Cobden, P. L. McEuen, N. G. Chopra, A. Zettl, A. Thess, and R. E. Smalley, Science **275**, 1922 (1997).

455 Y. C. Hong and H. S. Uhm, Phys. Plasmas **12**, 053504 (2005).

456 W. Zhou, X. X. Zhong, X. C. Wu, L. Q. Yuan, Z. C. Zhao, H. Wang, Y. X. Xia, Y. Y. Feng, J. He, and W. T. Chen, Surf. Coat. Technol. **200**, 6155 (2006).

457 Y. C. Hong, J. H. Kim, S. C. Cho, and H. S. Uhm, Phys. Plasmas **13**, 063506 (2006).

458 Y. C. Hong, J. H. Kim, C. U. Bang, and H. S. Uhm, Phys. Plasmas **12**, 114501 (2005).

459 Q. Wang, J. Li, Y. Ma, X. Bai, Z. Wang, P. Xu, C. Shi, B. Quan, S. Yue, and C. Gu, Nanotechnology **16**, 2919 (2005).

460 F. Xu, K. Yu, G. Li, Q. Li, and Z. Zhu, Nanotechnology **17**, 2855 (2006).

461 I. Levchenko, K. Ostrikov, and A. B. Murphy, J. Phys. D: Appl. Phys. **41**, 092001 (2008).

462 M. Yan, H. T. Zhang, E. J. Widjaja, and R. P. H. Chang, J. Appl. Phys. **94**, 5240 (2003).

463 R. Narain, A. Housni, and L. Lane, J. Poly. Sci. A - Poly. Chem. **44**, 6558 (2006).

464 M. Terrones, N. Grobert, J. Olivares, J. P. Zhang, H. Terrones, K. Kordatos, W. K. Hsu, J. P. Hare, P. D. Townsend, K. Prassides, A. K. Cheetham, H. W. Kroto, D. R. M. Walton, Nature (London) **388**, 52 (1997).

465 Y.-T. Kim, Y. Ito, K. Tadai, T. Mitani, U.-S. Kim, H.-S. Kim, B.-W. Cho, Appl. Phys. Lett. **87**, 234106 (2005).

466 S. Shenogin, A. Bodapati, L. Xue, R. Ozisik, and P. Keblinski, Appl. Phys. Lett. **85**, 2229 (2004).

467 P. W. Chiu, G. S. Duesberg, U. Dettlaff-Weglikowska, and S. Roth, Appl. Phys. Lett. **80**, 3811 (2002).

468 V. Stolojan, S. R. P. Silva, M. J. Goringe, R. L. D. Whitby, W. K. Hsu, D. R. M. Walton, H. W. Kroto, Appl. Phys. Lett. **86**, 063112 (2005).

469 D. Wang and H. Dai, Appl. Phys. A **85**, 217 (2006).

470 P. Paredez, M. Marchi, M. M. da Costa, C. Figuero, M. Kleinke, C. Ribeiro, J. Sanchez-Lopez, T. Rojas, and F. Alvarez, J. Non-Cryst. Sol. **352**, 1303 (2006).

471 K. E. Elers, T. Blomberg, M. Peussa, B. Aitchison, S. Haukka, and S. Marcusi, Chem. Vap. Deposition **12**, 13 (2006).

472 B. N. Khare, M. Meyyappan, J. Kralj, P. Wilhite, M. Sisay, H. Imanaka, J. Koehne, C. W. Baushchlicher, Jr. Appl. Phys. Lett. **81**, 5237 (2002).

473 L. Yuan, X. Zhong, I. Levchenko, Y. Xia, and K. Ostrikov, Plasma Proc. Polym. **4**, 612 (2007).

474 J. Stodolka, D. Nau, M. Frommberger, C. Zanke, H. Giessen, and E. Quandt, Microelectron. Eng. **78–79**, 442 (2005).

475 W. Wang and S. A. Asher, J. Am. Chem. Soc. **123**, 12528 (2001).

476 H. Masuda, K. Yasui, and K. Nishio, Adv. Mater. **12**, 1031 (2000).

477 R. Wetzler, R. Kunert, A. Wacker, and E. Schoell, New. J. Phys. **6**, 81 (2004).

478 T. Gao, J. C. Fan, G. W. Meng, Z. Q. Chu, and L. D. Zhang, Thin Sol. Films **401**, 102 (2001).

479 D. Crouse, Yu-Hwa Lo, A. E. Miller, and M. Crouse, Appl. Phys. Lett. **76**, 49 (2000).

480 M. Kokonou, A. G. Nassiopoulou and K. P. Giannakopoulos, Nanotechnology **16**, 103 (2005).

481 U. Cvelbar and K. Ostrikov, *Plasma-Assisted Nanofabrication of CdO and the Role of Reactive Plasma Environment* (unpublished).

482 C. Champeness and C. Chan, Sol. Energy Mater. Sol. Cells **37**, 75 (1995).

483 X. Liu, C. Li, S. Han, J. Han, and C. Zhou, Appl. Phys. Lett. **82**, 1 (2003).

484 E. Martin, M. Yan, M. Lane, J. Ireland, C. Kannewurf, and R. H. Chang, Thin Sol. Films **371**, 105 (2001).

485 H. M. Ali, H. A. Mohamed, M. M. Wakkad, and M. F. Hasaneen, Thin Sol. Films **515**, 3024 (2007).

486 K. Gurumurgan, D. Mangalaray, and Sa. K. Narayandass, J. Cryst. Growth **147**, 355 (1995).

487 K. T. R. Reddy, C. Sravani, and R. W. Miles, J. Cryst. Growth **184**, 1031 (1998).

488 T. K. Subramanyam, G. M. Rao, and S. Uthanna, Mater. Chem. Phys. **69**, 133 (2001).

489 M. Ghosh and C. N. R. Rao, Chem. Phys. Lett. **393**, 493 (2004).

490 J. A. Gerbec, D. Magana, A. Washington, and G. F. Strouse, J. Amer. Chem. Soc. **127**, 15791 (2005).

491 N. Singh, S. Charan, K. R. Patil, A. K. Viswanath, and P. K. Khanna, Mater. Lett. **60**, 3492 (2006).

492 X. S. Peng, X. F. Wang, Y. W. Wang, C. Z. Wang, G. W. Meng, and L. D. Zhang, J. Phys. D: Appl. Phys. **35**, L101 (2002).

493 T. Ghoshal, S. Kar, and S. Chaudhuri, Appl. Surf. Sci. **253**, 7578 (2007).

494 C. Yan and D. Xue, J. Phys. Chem. B **110**, 1582 (2006).

495 S. D. Bunge, *et al.*, J. Mater. Chem. **13**, 1705 (2003).

496 H. F. Yang, *et al.*, Adv. Func. Mater. **15**, 1377 (2005).

497 W. D. Shi, C. Wang, H. S. Wang, and H. J. Zhang, Cryst. Growth and Design **6**, 915 (2006).

498 U. Cvelbar, S. Pejovnik, M. Mozetic, and A. Zalar, Appl. Surf. Sci. **210**, 255 (2003).

499 M. Mozetic, U. Cvelbar, A. Vesel, A. Ricard, D. Babic, and I. Poberaj, J. Appl. Phys. **97**, 103308 (2005).

500 M. Mozetic, A. Vesel, U. Cvelbar, and A. Ricard, Plasma Chem. Plasma Process. **26**, 103 (2006).

501 U. Cvelbar and M. Mozetic, J. Phys. D **40**, 2300 (2007).

502 W. Zhou, X. Zhong, X. Wu, L. Yuan, Q. Shu, Y. Xia, and K. Ostrikov, J. Biomed. Mater. Res. **81A**, 453 (2007).

503 T. Kokubo, H. M. Kim, and M. Kawashita, Biomaterials **24**, 2161 (2003).

504 P. J. Li, C. Ohtsuki, T. Kokubo, K. Nakanishi, N. Soga, and K. de Groot, J. Biomed. Mater. Res. **28**, 7 (1994).

505 P. Li, I. Kangasniemi, K. de Groot, and T. Kokubo, J. Amer. Ceram. Soc. **77**, 1307 (1994).

506 T. Peltola, M. Patsi, H. Rahiala, I. Kangasniemi, and A. Yli-Urpo, J. Biomed. Mater. Res. **41**, 504 (1998).

507 M. Uchida, H. M. Kim, T. Kokubo, S. Fujibayashi, and T. Nakamura, J. Biomed. Mater. Res. Part A **64A**, 164 (2003).

508 N. Moritz, S. Areva, J. Wolke, and T. Peltola, Biomaterials **26**, 4460 (2005).

509 X. X. Wang, S. Hayakawa, K. Tsuru, and A. Osaka, J. Biomed. Mater. Res. **52**, 171 (2000).

510 M. Wei, M. Uchida, H. M. Kim, T. Kokubo, and T. Nakamura, Biomaterials **22**, 167 (2002).

511 R. Rohanizadeh, M. Al-Sadeq, and R. Z. LeGeros, J. Biomed. Mater. Res. Part A **71A**, 343 (2004).

512 X. Y. Liu, X. B. Zhao, R. K. Y. Fu, J. P. Y. Ho, C. X. Ding, and P. K. Chu, Biomaterials **26**, 6143 (2005).

513 W. H. Song, Y. K. Jun, Y. Han, and S. H. Hong, Biomaterials **25**, 3341 (2004).

514 B. Feng, J. Y. Chen, S. K. Qi, L. He, J. Z. Zhao, and X. D. Zhang, J. Mater. Sci.: Mater. Med. **13**, 457 (2002).

515 J. M. Wu, S. Hayakawa, K. Tsuru, and A. Osaka, J. Amer. Ceram. Soc. **87**, 1635 (2004).

516 W. Q. Yan, T. Nakamura, M. Kobayashi, H. M. Kim, F. Miyaji, and T. Kokubo, J. Biomed. Mater. Res. **37**, 267 (1997).

517 X. X. Wang, W. Yan, S. Hayakawa, K. Tsuru, and A. Osaka, Biomaterials **24**, 4631 (2003).

518 S. Monticone, R. Tufeu, A. V. Kanaev, E. Scolan, and C. Sanchez, Appl. Surf. Sci. **162**, 565 (2000).

519 K. Okimura, N. Maeda, and A. Shibata, Thin Sol. Films **281-282**, 427 (1996).

520 R. Bartnikas and E. J. McMahon, *Engineering Dielectrics: Corona Measurement and Interpretation.* (ASTM Intnl., New York, 1979), p. 25.

521 R. Zeman and S. Takabayashi, Surf. Coat. Technol. **153**, 93 (2002).

522 D. M. Brunette, P. Tengvall, M. Textor, and P. Thomsen, *Titanium in Medicine* (Springer, New York, 2001), p. 204.

523 M. L. Brongersma, R. Zia, and J. A. Schuller, Appl. Phys. A **89**, 221 (2007).

524 D. Mariotti, V. Svrcek, and D. G. Kim, Appl. Phys. Lett. **91**, 183111 (2007).

525 *Science and Creationism: A View from the National Academy of Sciences,* Second Edition, http:// books.nap.edu/ html/ creationism/ index.html, 28 July 1999.

526 S. L. Miller and L. E. Orgel, *The Origins of Life on the Earth* (Prentice-Hall, Englewood Cliffs, NJ, 1974).

527 S. Ashida, C. Lee, and M. A. Lieberman, J. Vac. Sci. Technol. A **13**, 2498 (1995).

528 http://www.kinema.com/ download.htm

529 D. K. Davies, L. E. Kline, and W. E. Bies, J. Appl. Phys. **65**, 3311 (1989).

530 M. Hayashi, *Electron Collision Cross Sections Determined From Beam and Swarm Data by Boltzmann Analysis,* in *Nonequilibrium Processes in Partially Ionized Gases*, edited by M. Capitelli and J. N. Bardsley (Plenum Press, New York, 1990).

531 W. L. Fite and R. T. Brackmann, Phys. Rev. **112**, 1151 (1958).

532 T. Nakano, H. Toyoda, and H. Sugai, Jpn. J. Appl. Phys. **30**, 2912 (1991).

533 Y.-K. Kim and M. E. Rudd, Phys. Rev. A **50**, 3954 (1994).

534 A. G. Engelhardt and A. V. Phelps, Phys. Rev. **131**, 2115 (1963).

535 D. A. Alman, D. N. Ruzic, and J. N. Brooks, Phys. Plasmas, **7**, 1421 (2000).

536 M. Heintze, M. Magureanu, and M. Kettlitz, J. Appl. Phys. **92**, 7022 (2002).

537 K. Tachibana, M. Nishida, H. Harima, and Y. Urano, J. Phys. D: Appl. Phys. **17**, 1727 (1984).

538 R. L. Mills, P. C. Ray, B. Dhandapani, R. M. Mayo, and J. He, J. Appl. Phys. **92**, 7008 (2002).

539 H. Chatham, D. Hils, R. Robertson, and A. C. Gallagher, J. Chem. Phys. **79**, 1301 (1983).

540 J. L. Giuliani, V. A. Shamamian, R. E. Thomas, J. P. Apruzese, M. Mulbrandon, R. A. Rudder, R. C. Hendry, and A. E. Robson, IEEE Trans. Plasma Sci. **27**, 1317 (1999).

541 T. G. Beuthe and J. S. Chang, Jpn. J. Appl. Phys. Part 1 **38**, 4576 (1999).

542 R. A. Stewart, P. Vitello, and D. B. Graves, J. Vac. Sci. Technol. B **12**, 478 (1994).

543 R. A. Stewart, P. Vitello, and D. B. Graves, Plasma Sources Sci. Technol. **4**, 36 (1995).

544 D. P. Lymberopoulos and D. J. Economou, J. Res. Natl. Stand. Technol. **100**, 473 (1995).

545 D. A. Alman, D. N. Ruzic, and J. N. Brooks, Phys. Plasmas **7**, 1421 (2000).

546 D. L. Scharfetter and H. K. Gummel, IEEE Trans. Electron. Devices **16**, ED-64 (1967).

547 J. D. Bukowski and D. B. Graves, J. Appl. Phys. **80**, 2614 (1996).

548 H. J. Yoon and T. J. Chung, J. Kor. Phys. Soc. **34**, 29 (1999).

549 M. S. Bell, K. B. K. Teo, and W. I. Milne, J. Phys. D: Appl. Phys. **40**, 2285 (2007).

550 Z. F. Ren, Z. P. Huang, J. W. Xu, J. H. Wang, P. Bush, M. P. Siegal, and J. C. Provencio, Science **282**, 1105 (1998).

551 L. Delzeit, C. V. Nguyen, R. M. Stevens, J. Han, and M. Meyyappan, Nanotechnology **13**, 280 (2002).

552 A. M. Rao, D. Jacques, R. C. Haddon, W. Zhu, C. Bower, and S. Jin, Appl. Phys. Lett. **76**, 3813 (2002).

553 B. O. Boskovic, V. Stolojan, R. U. A. Khan, S. haq, and S. R. P. Silva, Nature Mater. **1**, 165 (2002).

554 C. Bower, O. Zhou, W. Zhu, D. J. Werder, and S. Jin, Appl. Phys. Lett. **77**, 2767 (2000).

555 M. S. Bell, K. B. K. Teo, R. G. Lacerda, W. I. Milne, D. B. Hash, and M. Meyyappan, Pure Appl. Chem. **78**, 1117 (2006).

556 J. Gonzalez-Aguilar, M. Moreno, and L. Fulcheri, J. Phys. D: Appl. Phys. **40**, 2361 (2007).

557 G. N. Churilov, Instr. Exper. Techniques **43**, 1 (2000).

558 G. A. Dyuzhev, Plasma Devices and Operations **10** 63 (2002).

559 R. E. Smalley, Acc. Chem. Res. **25**, 98 (1992).

560 K. Yoshie, S. Kasuya, K. Eguchi, and T. Yoshida, Appl. Phys. Lett. **61**, 2782 (1992).

561 S. Xie, R. Huang, L. Yu, J. Ding, and L. Zheng, Appl. Phys. Lett. **75**, 2764 (1999).

562 T. Alexakis, P. G. Tsantrizos, Y. S. Tsantrizos, and J. L. Meunier, Appl. Phys. Lett. **70**, 2102 (1997).

563 J. Hahn, J. H. Han, J. E. Yoo, H. Y. Jung, and J. S. Suh, Carbon **42**, 877 (2004).

564 S. I. Choi, J. S. Nam, J. I. Kim, T. H. Hwang, J. H. Seo, and S. H. Hong, Thin Sol. Films **506–507**, 244 (2006).

565 O. Smiljanic, B. L. Stansfield, J. P. Dodolet, A. Serventi, and S. Desilets, Chem. Phys. Lett. **356**, 189 (2002).

566 H. Okuno, E. Grivei, F. Fabry, T. M. Gruenberger, J. Gonzalez-Aguilar, A. Palnichenko, L. Fulcheri, N. Probst, and J. C. Charlier, Carbon **42**, 2543 (2004).

567 J. Gavillet, A. Loiseau, F. Ducastelle, S. Thair, P. Bernier, O. Stephan, J. Thibault, and J.-C. Charlier, Carbon **40**, 1649 (2002).

568 T. M. Gruenberger, J. Gonzalez-Aguilar, L. Fulcheri, H. Okuno, J. C. Charlier, F. Fabry, E. Grivei, and N. Probst, Nanotubes and Carbon Nanostr. **13**, 67 (2005).

569 L. Fulcheri, T. M. Gruenberger, J. Gonzalez-Aguilar, F. Fabry, E. Grivei, N. Probst, G. Flamant, H. Okuno, and J. C. Charlier, High Tech. Plasma Proc. **8**, 119 (2004).

570 K. S. Kim, G. Cota-Sanchez, C. T. Kingston, M. Imris, B. Simard, and G. Soucy, J. Phys. D: Appl. Phys. **40**, 2375 (2007).

571 Y. Ando, X. Zhao, K. Hirahara, K. Suenaga, S. Bandow, and S. Iijima, Chem. Phys. Lett. **323**, 580 (2000).

572 D. Harbec, J. L. Meunier, L. Guo, R. Gauvin, and N. E. Mallah, J. Phys. D: Appl. Phys. **37**, 2121 (2004).

573 S. I. Choi, J. S. Nam, C. M. Lee, S. S. Choi, J. I. Kim, J. M. Park, and S. H. Hong, Curr. Appl. Phys. **6**, 224 (2006).

574 G. Cota-Sanchez, G. Soucy, A. Huczko, and H. Lange, Carbon **43**, 3153 (2005).

575 K. S. Kim, J. H. Seo, J. S. Nam, W. T. Ju, and S. H. Hong, IEEE Trans. Plasma Sci. **33**, 813 (2005).

576 M. Meyyappan, *Carbon Nanotubes: Science and Applications* (CRC Press, Boca Raton, 2005).

577 C. Journet, W. K. Maser, P. Bernier, A. Loiseau, M. L. de la Chapelle, S. Lefrant, P. Deniard, R. Lee, J. E. Fischer, Nature **388**, 756 (1997).

578 T. Zhao and Y. Liu, Carbon **42**, 2735 (2004).

579 X. Lv, F. Du, Y. Ma, Q. Wu, and Y. Chen, Carbon **43**, 2020 (2005).

580 D. Tang, L. Sun, J. Zhou, W. Zhou, and S. Xie, Carbon **43**, 2812 (2005).

581 M. Yao, B. Liu, Y. Zou, L. Wang, D. Li, T. Cui, G. Zou, B. Sundqvist, Carbon **43**, 2894 (2005).

582 H. Lange, M. Sioda, A. Huczko, Y. Q. Zhu, H. W. Kroto, and D. R. M. Walton, Carbon **41**, 1617 (2003).

583 H. Kawamura, K. Moritani, and Y. Ito, Plasmas Ions **1**, 29 (1998).

584 M. T. Beck, Z. Dinya, S. Keki, and L. Papp, Tetrahedron **49**, 285 (1993).

585 M. Ishigami, J. Cumings, A. Zettl, and S. Chen, Chem. Phys. Lett. **319**, 457 (2000).

586 M. V. Antisari, R. Marazzi, and R. Krsmanovic, Carbon **41**, 2393 (2003).

587 Y. L. Hsin, K. C. Hwang, F. R. Chen, and J. J. Kai, Adv. Mater. **13**, 830 (2001).

588 N. Sano, H. Wang, M. Chhowalla, M. I. Alexandrou, and G. A. J. Amaratunga, Nature **414**, 527 (2001).

589 W. H. Zhu, X. S. Li, B. Jiang, C. L. Xu, Y. F. Zhu, D. H. Wu, and X. H. Chen, Chem. Phys. Lett. **366**, 664 (2002).

590 B. Wang and B. Zhang, Diam. Relat. Mater. **16**, 1982 (2007).

591 B. Wang and B. Zhang, Appl. Surf. Sci. **253**, 6951 (2007).

592 J. Jang, S. J. Jung, H. S. Kim, S. H. Lim, and C. H. Lee, Appl. Phys. Lett. **79**, 1682 (2001).

593 C. J. Huang, Y. K. Chih, J. Hwang, A. P. Lee, and C. S. Kou, J. Appl. Phys. **94**, 6796 (2003).

594 B. B. Wang, W. L. Wang, K. J. Liao, and J. L. Xiao, Phys. Rev. B **63**, 85412 (2001).

595 Z. L. Tsakadze, K. Ostrikov, R. Storer, and S. Xu, J. Metastable and Nanocryst. Mater. **23–25**, 297 (2005).

596 S. Kumar, C. M. S. Rauthan, K. M. K. Srivasta, P. N. Dixit, and R. Bhattacharyya, Appl. Surf. Sci. **182**, 326 (2001).

597 L. Zhu, J. Xu, Y. Xiu, Y. Sun, D. W. Hess, and C.P. Wong, Carbon **44**, 253 (2006).

598 C. Daraio, V. Nesterenko, and S. Jin, Appl. Phys. Lett. **85**, 5724 (2004).

599 D.-H. Kim, H.-S. Jang, C.-D. Kim, D.-S. Cho, H.-S. Yang, H.-D. Kang, B.-K. Min, and H.-R. Lee, Nano Lett. **3**, 863 (2003).

600 H. Sato, T. Sakai, M. Matsubayashi, K. Hata, H. Miyake, K. Hiramatsu, A. Oshita, and Y. Saito, Vacuum **80**, 798 (2006).

601 Y. T. Lee, J. Park, Y. S. Choi, H. Ryu, and H. J. Lee, J. Phys. Chem. B **106**, 7614 (2002).

602 Le Thien-Nga, J.-M. Bonard, R. Gaal, L. Forro, and K. Hernadi, Appl. Phys. Lett. **80**, 850 (2002).

603 R. Hatakeyama, G.-H. Jeong, T. Kato, and T. Hirata, J. Appl. Phys. **96**, 6053 (2004).

604 T. Kato, G.-H. Jeong, T. Hirata, R. Hatakeyama, and K. Tohji, Japan. J. Appl. Phys. **43**, L1278 (2004).

605 G.-H. Jeong, R. Hatakeyama, T. Hirata, K. Tohji, K. Motomiya, and N. Sato, Appl. Phys. Lett. **79**, 4213 (2001).

606 G.-H. Jeong, R. Hatakeyama, T. Hirata, K. Tohji, K. Motomiya, T. Yaguchi, and Y. Kawazoe, Chem. Commun., Issue 1, 152 (2003).

607 T. Okada, T. Kaneko, R. Hatakeyama, and K. Tohji, Chem. Phys. Lett. **417**, 289 (2005).

608 M. Paillet, V. Jourdain, P. Poncharal, J.-L. Sauvajol, A. Zahab, J. C. Meyer, S. Roth, N. Cordente, C. Amiens, and B. Chaudret, J. Phys. Chem. B **108**, 17112 (2004).

609 Y. F. Guan, A. V. Melechko, A. J. Pedraza1, M. L. Simpson, and P. D. Rack, Nanotechnology **18**, 335306 (2007).

610 V. L. Kuznetsov, A. N. Usoltseva, A. L. Chuvilin, E. D. Obraztsova, and J. M. Bonard, Phys. Rev. B **64**, 235401 (2001).

611 C. L. Cheung, A. Kurtz, H. Park, and C. M. Lieber, J. Phys. Chem. B **106**, 2429 (2002).

612 J. H. Hafner, M. J. Bronikowski, B. R. Azamian, P. Nikolaev, A. G. Rinzler, D. T. Colbert, K. A. Smith, and R. E. Smalley, Chem. Phys. Lett. **296**, 195 (1998).

613 Y. Li, W. Kim, Y. Zhang, M. Rolandi, D. Wang, and H. Dai, J. Phys. Chem. B **105**, 11424 (2001).

614 M. P. Siegal, D. L. Overmyer, and P. P. Provencio, Appl. Phys. Lett. **80**, 2171 (2002).

615 J.-M. Bonard, P. Chauvin, and C. Klinke, Nano Lett. **2**, 665 (2002).

616 L. An, J. M. Owens, L. E. McNeil, and J. J. Liu, J. Am. Chem. Soc. **124**, 13688 (2002).

617 H. C. Choi, W. Kim, D. Wang, and H. J. Dai, J. Phys. Chem. B **106**, 12361 (2002).

618 S. Sato, A. Kawabata, M. Nihei, and Y. Awano, Chem. Phys. Lett. **382**, 361 (2003).

619 Q. Fu, S. Huang, and J. J. Liu, J. Phys. Chem. B **108**, 6124 (2004).

620 S. Han, T. Yu, J. Park, B. Koo, J. Joo, T. Hyeon, S. Hong, and J. J. Im, J. Phys. Chem. B **108**, 8091 (2004).

621 W. Hofmeister, W. P. Kang, Y. M. Wong, and J. L. Davidson, J. Vac. Sci. Technol. B **22**, 1286 (2004).

622 A. M. Cassell, Q. Ye, B. A. Cruden, J. Li, P. C. Sarrazin, H. T. Ng, J. Han, and M. Meyyappan, Nanotechnology **15**, 9 (2004).

623 H. Sato, H. Tagegawa, and Y. Saito, J. Vac. Sci. Technol. B **21**, 2564 (2003).

624 C. H. Lin, H. L. Chang, C. M. Hsu, A. Y. Lo, and C. T. Kuo, Diamond Relat. Mater. **12**, 1851 (2003).

625 C. Qin, D. Zhou, A. R. Krauss, and D. M. Gruen, Appl. Phys. Lett. **72**, 3437 (1998).

626 Q. Zhang, S. F. Yoon, J. Ahn, B. Gan, Rusli, and M. B. Yu, J. Phys. Chem. Solids **61**, 1179 (2000).

627 H. Cui, O. Zhou, and B. R. Stoner, J. Appl. Phys. **88**, 6072 (2000).

628 M. Kuttel, O. Groening, C. Emmenegger, and L. Schlapbach, Appl. Phys. Lett. **73**, 2113 (1998).

629 T. Ikuno, *et al.*, Surf. Interface Anal. **35**, 15 (2003).

630 M. Okai, T. Muneyoshi, T. Yaguchi, and S. Sasaki, Appl. Phys. Lett. **77**, 3468 (2000).

631 K. Y. Lee, *et al.*, Jpn. J. Appl. Phys., Part 2 **42**, L804 (2003).

632 N. A. Kiselev, *et al.*, Carbon **42**, 149 (2004).

633 Y. S. Woo, D. Y. Jeon, I. T. Han, N. S. Lee, J. E. Jung, and J. M. Kim, Diamond Relat. Mater. **11**, 59 (2002).

634 Y. Tu, Y. H. Lin, and Z. F. Ren, Nano Lett. **3**, 107 (2003).

635 J. Koehne, J. Li, A. M. Cassell, H. Chen, Q. Ye, H. T. Ng, J. Han, and M. Meyyappan, J. Mater. Chem. **14**, 676 (2004).

636 J. Han, W. S. Yang, J. B. Yoo, and C. Y. Park, J. Appl. Phys. **88**, 7363 (2000).

637 Y. Chen, L. P. Guo, D. J. Johnson, and R. H. Prince, J. Cryst. Growth **193**, 342 (1998).

638 K. M. Ryu, M. Y. Kang, Y. D. Kim, and H. T. Jeon, Jpn. J. Appl. Phys., Part 1 **42**, 3578 (2003).

639 Q. Yang, C. Xiao, W. Chen, A. K. Singh, T. Asai, and A. Hirose, Diamond Relat. Mater. **12**, 1482 (2003).

640 S. Honda, M. Katayama, K. Y. Lee, T. Ikuno, S. Ohkura, K. Oura, H. Furuta, and T. Hirao, Jpn. J. Appl. Phys., Part 2 **42**, L441 (2003).

641 Y. C. Choi, Y. M. Shin, S. C. Lim, D. J. Bae, Y. H. Lee, B. S. Lee, and D. C. Chung, J. Appl. Phys. **88**, 4898 (2000).

642 Y. C. Choi, D. J. Bae, Y. H. Lee, B. S. Lee, I. T. Han, W. B. Choi, N. S. Lee, and J. M. Kim, Synth. Met. **108**, 159 (2000).

643 V. I. Merkulov, D. K. Hensley, A. V. Melechko, M. A. Guillorn, D. H. Lowndes, and M. L. Simpson, J. Phys. Chem. B **106**, 570 (2002).

644 J. B. O. Caughman, L. R. Baylor, M. A. Guillorn, V. I. Merkulov, D. H. Lowndes, and L. F. Allard, Appl. Phys. Lett. **83**, 1207 (2003).

645 M. Nihei, A. Kawabata, and Y. Awano, Jpn. J. Appl. Phys., Part 2 **42**, L721 (2003).

646 V. Nguyen, L. Delziet, K. Matthews, B. Chen, and M. Meyyappan, J. Nanosci. Nanotechnol. **3**, 121 (2003).

647 M. A. Guillorn, A. V. Melechko, V. I. Merkulov, E. D. Ellis, M. L. Simpson, L. R. Baylor, and G. J. Bordonaro, J. Vac. Sci. Technol. B **19**, 2598 (2001).

648 J. Shieh, H. L. Chen, T. S. Ko, H. C. Cheng, and T. C. Chu, Adv. Mater. **16**, 1121 (2004).

649 M. J. Yang, J. Shieh, S. L. Hsu, I. Y. Huang, C. C. Leu, S. W. Shen, T. Y. Huang, P. Lehnen, and C. H. Chien, Electrochem. Solid-State Lett. **8** C74 (2005).

650 M. C. Yang, J. Shieh, C. C. Hsu, and T. C. Cheng, Electrochem. Solid-State Lett. **8** C131 (2005).

651 T. C. Cheng, J. Shieh, W. J. Huang, M. C. Yang, M. H. Cheng, H. M. Lin, and M. N. Chang, Appl. Phys. Lett. **88**, 263118 (2006).

652 S. Thompson, C. R. Perrey, C. B. Carter, T. J. Belich, J. Kakalios, and U. Kortshagen, J. Appl. Phys. **97**, 034310 (2005).

653 P. Roca i Cabarrocas, J. Non-Cryst. Solids **164-166**, 37 (1993).

654 N. Chaabane, V. Suendo, H. Vach, and P. Roca i Cabarrocas, Appl. Phys. Lett. **88**, 203111 (2006).

655 Y. Watanabe, J. Phys. D: Appl. Phys. **39**, R329 (2006).

656 K. Koga, S. Iwashita, and M. Shiratani, J. Phys. D: Appl. Phys. **40**, 2267 (2007).

657 International Technology Roadmap for Semiconductors, http://www.itrs.net

658 S. Nunomura, M. Kita, K. Koga, M. Shiratani, and Y. Watanabe, Jpn. J. Appl. Phys. **40**, L1509 (2005).

659 Q. J. Cheng, J. D. Long, and S. Xu, J. Appl. Phys. **101**, 094304 (2007).

660 Y. Sun, T. Miyasato, and J. K. Wigmore, J. Appl. Phys. **86**, 3076 (1999).

661 Q. J. Cheng, J. D. Long, Z. Chen, and S. Xu, J. Phys. D: Appl. Phys. **40**, 2304 (2007).

662 C. R. Gorla, S. Liang, G. S. Tompa, W. E. Mayo, and Y. Lu, J. Vac. Sci. Technol. **15**, 860 (1997).

663 J. Grabis, D. Jankovica, M. Berzins, and L. Chera, Sol. State Phenom. **94**, 151 (2003).

664 J. Grabis, D. Jankovica, M. Berzins, L. Chera, and I. Zalite, J. Eur. Ceram. Soc. **24**, 179 (2004).

665 H. L. Chang, C. M. Hsu, C. T. Kuo, Appl. Phys. Lett. **80**, 4638 (2002).

666 Y. Horia, O. Oda, E. Bellet-Amalric, and B. Daudin, J. Appl. Phys. **102**, 024311 (2007).

667 M. Shiratani, K. Koga, S. Ando, T. Inoue, Y. Watanabe, S. Nunomura, and M. Kondo, Surf. Coat. Technol. **201**, 5468 (2007).

668 M. Shiratani, T. Kakeya, K. Koga, Y. Watanabe, and M. Kondo, Trans. Mater. Res. Soc. Jpn. **30**, 307 (2005).

669 T. Kakeya, K. Koga, M. Shiratani, Y. Watanabe, and M. Kondo, Thin Sol. Films **506–507**, 288 (2006).

670 T. Xu, A. Yu. Nikiforov, R. France, C. Thomidis, A. Williams, and T. D. Moustakas, Phys. Stat. Sol (a) **204**, 2098 (2007).

671 S. Founta, C. Bougerol, H. Mariette, B. Daudin, and P. Vennegues, J. Appl. Phys. **102**, 074304 (2007).

672 M. Sutiknoa, U. Hashim, and Z. A. Z. Jamal, Microelectron. Journ. **38**, 823 (2007).

673 K. Makiharaa, H. Dekib, H. Murakamia, S. Higashia, and S. Miyazaki, Appl. Surf. Sci. **244**, 75 (2005).

674 Y. E. Romanyuk, R.-G. Dengel, L. V. Stebounova, and S. R. Leone, J. Cryst. Growth **304**, 346 (2007).

675 S. Y. Huang *et al.*, Physica E **31**, 200 (2006).

676 L. W. Yu, K. J. Chen, J. Song, J. M. Wang, J. Xu, W. Li, and X. F. Huang, Thin Sol. Films **515**, 5466 (2007).

677 D. Nie, T. Mei, H. S. Djie, B. S. Ooi, and X. H. Zhang, Appl. Phys. Lett. **88**, 251102 (2006).

678 A. Tanaka, G. Yamahata, Y. Tsuchiya, K. Usami, H. Mizuta, and S. Oda, Curr. Appl. Phys. **6**, 344 (2006).

679 D. Song, E. C. Cho, G. Conibeer, Y. Huang, and M. A. Green, Appl. Phys. Lett. **91**, 123510 (2007).

680 S. Y. Huang, S. Xu, J. D. Long, Q. J. Cheng, X. B. Xu, and J. B. Chu, Surf. Rev. Lett. **14**, 225 (2007).

681 F. Zhu, Z. X. Yang, W. M. Zhou, and Y. F. Zhang, Sol. State Commun. **137**, 177 (2006).

682 F. Zhu, Z. X. Yang, W. M. Zhou, and Y. F. Zhang, Appl. Surf. Sci. **252**, 7930 (2006).

683 F. Zhu, Z. X. Yang, W. M. Zhou, and Y. F. Zhang, Physica E **30**, 155 (2005).

684 Q. J. Feng, D. Z. Shen, J. Y. Zhang, H. W. Liang, D. X. Zhao, Y. M. Lua, and X. W. Fan, J. Cryst. Growth **285**, 561 (2005).

685 L. C. Chen, *et al.*, J. Phys. Chem. Sol. **62**, 1567 (2001).

686 S. Komatsu, J. Phys. D: Appl. Phys. **40**, 2320 (2007).

687 S. Komatsu, K. Okada, Y. Shimizu, and Y. Moriyoshi, J. Phys. Chem. B **103**, 3289 (1999).

688 S. Komatsu, A. Okudo, D. Kazami, D. Golberg, Y. Li, Y. Moriyoshi, M. Shiratani, and K. Okada, J. Phys. Chem. B **108**, 5182 (2004).

689 S. Komatsu, D. Kazami, H. Tanaka, Y. Shimizu, Y. Moriyoshi, M. Shiratani, and K. Okada, Appl. Phys. Lett. **88**, 151914 (2006).

690 S. Komatsu, D. Kazami, H. Tanaka, Y. Moriyoshi, M. Shiratani, and K. Okada, J. Appl. Phys. **99**, 123512 (2006).

691 S. Komatsu, Y. Shimizu, Y. Moriyoshi, K. Okada, and M. Mitomo, Appl. Phys. Lett. **79**, 188 (2001).

692 S. Komatsu, K. Kurashima, H. Kanda, M. Mitomo, Y. Moriyoshi, Y. Shimizu, K. Okada, M. Shiratani, T. Nakano, and S.Samukawa, Appl. Phys. Lett. **81**, 4547 (2002).

693 S. Komatsu, Y. Shimizu, Y. Moriyoshi, K. Okada, and M. Mitomo, J. Appl. Phys. **91**, 6181 (2002).

694 S. Komatsu, K. Kurashima, Y. Shimizu, Y. Moriyoshi, M. Shiratani, and K. Okada, J. Phys. Chem. B **108**, 205 (2004).

695 S. Komatsu, D. Kazami, H. Tanaka, Y. Moriyoshi, M. Shiratani, and K.Okada, J. Chem. Phys. **125**, 084701 (2006).

696 A. Valsesia, T. Meziani, F. Bretagnol, P. Colpo, and F. Rossi, J. Phys. D: Appl. Phys. **40**, 2341 (2007).

697 A. Valsesia, P. Colpo, P. Lisboa, M. Lejeune, T. Meziani, and F. Rossi, Langmuir **22**, 1763 (2006).

698 K. B. Lee, E. Y. Kim, C. A. Mirkin, and S. M. Wolinsky, Nano Lett. **4**, 1869 (2004).

699 H. Agheli, J. Malmstroem, E. M. Larsson, M. Textor, and D. S. Sutherland, Nano Lett. **6**, 1165 (2006).

700 M. Arnold, E. A. Cavalcanti-Adam, R. Glass, J. Blummel, W. Eck, M. Kantlehner, H. Kessler, and J. P. Spatz, Chem. Phys. Chem. **5**, 383 (2004).

701 B. D. Gates, Q. Xu, J. C. Love, D. B. Wolfe, and G. M. Whitesides, Ann. Rev. Mater. Res. **34**, 339 (2004).

702 F. Bretagnol, A. Valsesia, G. Ceccone, P. Colpo, D. Gilliland, L. Ceriotti, M. Hasiwa, and F. Rossi, Plasma Proc. Polym. **3**, 443 (2006).

703 M. Lejeune, A. Valsesia, P. Colpo, M. Kormunda, and F. Rossi, Surf. Sci. **583**, L142 (2005).

704 A. Valsesia, P. Colpo, T. Meziani, F. Bretagnol, M. Lejeune, T. Meziani, F. Rossi, A. Bouma, and M. Garcia-Parajo, Adv. Funct. Mater. **16**, 1242 (2006).

705 F. Bretagnol, L. Ceriotti, M. Lejeune, A. Papadopoulou-Bouraoui, M. Hasiwa, D. Gilliland, G. Ceccone, P. Colpo, and F. Rossi, Plasma Proc. Polym. **3**, 30 (2006).

706 A. Valsesia, M. M. Silvan, G. Ceccone, D. Gilliland, P. Colpo, and F. Rossi, Plasma Proc. Polym. **2**, 334 (2005).

707 D. Teare, C. G. Spanos, P. Ridley, E. J. Kinmond, V. Roucoules, J. Badyal, S. A. Brewer, S. Coulson, and C. Willis, Chem. Mater. **14**, 4566 (2002).

708 J.-G. Li, M. Ikeda, R. Ye, Y. Moriyoshi, and T. Ishigaki, J. Phys. D: Appl. Phys. **40**, 2348 (2007).

709 S. L. Girshick and J. Hafiz, J. Phys. D: Appl. Phys. **40**, 2354 (2007).

710 Y. L. Li and T. Ishigaki, J. Phys. Chem. B **108**, 15536 (2004).

711 S. M. Oh, J. G. Li, and T. Ishigaki, J. Mater. Res. **20**, 529 (2005).

712 X. H. Wang, J. G. Li, H. Kamiyama, M. Katada, N. Ohashi, Y. Moriyoshi, and T. Ishigaki, J. Am. Chem. Soc. **127**, 10982 (2005).

713 J. G. Li, X. H. Wang, K. Watanabe, and T. Ishigaki, J. Phys. Chem. B **110**, 1121 (2006).

714 F. Liao, S. Park, J. M. Larson, M. R. Zachariah, and S. L. Girshick, Mater. Lett. **57**, 1982 (2003).

715 F. Liao, S. L. Girshick, W. M. Mook, W. W. Gerberich, and M. R. Zachariah, Appl. Phys. Lett. **86**, 171913 (2005).

716 N. P. Rao, H. J. Lee, M. Kelkar, D. J. Hansen, J. V. R. Heberlein, P. H. McMurry, and S. L. Girshick, Nanostruct. Mater. **9**, 129 (1997).

717 M. Shigeta and T. Watanabe, J. Phys. D: Appl. Phys. **40**, 2407 (2007).

718 T. Watanabe and K. Fujiwara, Chem. Eng. Commun. **191**, 1343 (2004).

719 M. Shigeta, T. Watanabe, and H. Nishiyama, Thin Sol. Films **457**, 192 (2004).

720 T. Watanabe, A. Nezu, Y. Abe, Y. Ishii, and K. Adachi, Thin Sol. Films **435**, 27 (2003).

721 M. Shigeta and T. Watanabe, J. Mater. Res. **20**, 2801 (2005).

722 T. Watanabe and H. Okumiya, Sci. Tech. Adv. Mater. **5**, 639 (2004).

723 A. B. Murphy, J. Phys. D: Appl. Phys. **29**, 1922 (1996).

724 A. B. Murphy, J. Phys. D: Appl. Phys. **34**, R151 (2001).

725 A. B. Murphy, J. Phys. D: Appl. Phys. **37**, 2841 (2004).

726 X. Liu, X. Wu, H. Cao, and R. P. H. Chang, J. Appl. Phys. **95**, 3141 (2004).

727 M. S. Hu, H. L. Chen, C. S. Shen, L. S. Hong, B. R. Huang, K. H. Chen, and L. C. Chen, Nature Mater. **5**, 102 (2006).

728 A. Borras, A. Barranco, F. Yuberto, and A. R. Gonzalez-Elipe, Nanotechnology **17**, 3518 (2006).

729 A. Borras, A. Barranco, J. P. Espinos, J. Cotrino, J. P. Holgado, and A. R. Gonzalez-Elipe, Plasma Proc. Polym. **4**, 515 (2007).

730 M. I. Boulos, J. W. Jurewicz, C. A. Nessim, C. A. A. Messih, *Plasma Synthesis of Metal Oxide Nanopowder and Apparatus Therefore*, US Patent Application No. 20030143153 (2003).

731 R. Tannenbaum, et al., *New Technique Controls Nanoparticle Size Research News and Publication Office*, Georgia Institute of Technology (2005).

732 H. Srikanth, R. Hajndi, C. Chirios, and J. Sanders, Appl. Phys. Lett. **79**, 3503 (2001).

733 R. Kalyanaraman, S. Yoo, M. S. Krupashankara, T. S. Sudarshan, and R. J. Dowding, Nanostruct. Mater. **10**, 1379 (1999).

734 W. Li, X. Liu, A. Huang, and P. K. Chu, J. Phys. D: Appl. Phys. **40**, 2293 (2007).

735 J. Chevalier, Biomaterials **27**, 535 (2006).

736 Y. W. Gua, A. U. J. Yap, P. Cheang, and K. A. Khor, Biomaterials **26**, 713 (2005).

737 Y. T. Xie, X. Y. Liu, A. P. Huang, C. X. Ding, and P. K. Chu, Biomaterials **27**, 3904 (2006).

738 A. M. Morales and C. M. Lieber, Science **279**, 208 (1998).

739 Q. J. Cheng, S. Xu, J. D. Long, and K. Ostrikov, Chem. Vap. Deposition **13**, 561 (2007).

740 Q. J. Cheng, S. Xu, J. D. Long, S. Huang, and J. Guo, Nanotechnology **18**, 465601 (2007).

741 Q. J. Cheng and S. Xu, J. Appl. Phys. **102**, 056101 (2007).

742 Q. J. Cheng, S. Xu, J. W. Chai, S. Y. Huang, Y. P. Ren, J. D. Long, P. P. Rutkevych, and K. Ostrikov, Thin Solid Films, in press, DOI: http://dx.doi.org/ 10.1016/ j.tsf.2007.10.091

743 *Silicon Carbide and Related Materials*, Materials Science Forum, Vols. 433–436, edited by P. Bergman and E. Jantzen (Trans Tech Publications, Zurich, Switzerland, 2003).

744 H. Colder, R. Rizk, M. Morales, P. Marie, J. Vicens, and I. Vickridge, J. Appl. Phys. **98**, 024313 (2005).

745 Y. Sun, T. Miyasato, and J. K. Wigmore, J. Appl. Phys. **86**, 3076 (1999).

746 V. Cimalla, A. A. Schmidt, T. Stauden, K. Zekentes, O. Ambacher, and J. Pezoldt, J. Vac. Sci. Technol. B **22**, L20 (2004).

747 J. Y. Fan, X. L. Wu, H. X. Li, H. W. Liu, G. G. Siu, and P. K. Chu, Appl. Phys. Lett. **88**, 041909 (2006).

748 A. Fissel, K. Pfennighaus, and W. Richter, Appl. Phys. Lett. **71**, 2981 (1997).

749 A. Fissel, K. Pfennighaus, and W. Richter, Thin Sol. Films **88**, 318 (1998).

750 W. K. Choi, T. Y. Ong, L. S. Tan, F. C. Loh, and K. L. Tan, J. Appl. Phys. **83**, 4968 (1998).

751 T. Berlind, N. Hellgren, M. P. Johansson, and L. Hultman, Surf. Coat. Technol. **141**, 145 (2001).

752 K. Yamamoto, Y. Koga, and S. Fujiwara, Diam. Relat. Mater. **10**, 1921 (2001).

753 F. Liao, S. L. Girshick, W. M. Mook, W. W. Gerberich, and M. R. Zachariah, Appl. Phys. Lett. **86**, 171913 (2005).

754 Z. Hu, X. Liao, H. Diao, G. Kong, X. Zeng, Y. Xu, J. Cryst. Growth **7**, 264 (2004).

755 W. Yu, W. Lu, L. Han, and G. Fu, J. Phys. D: Appl. Phys. **37**, 3304 (2004).

756 G. Viera, E. Garcia-Caurel, J. Costa, J. L. Andujar, and E. Betran, Applied Surface Science **144**, 702 (1999).

757 M. Xu, S. Xu, Y. C. Ee, C. Yong, J. W. Chai, S. Y. Huang, J. D. Long, Mater. Sci. Eng. B **128**, 89 (2006).

758 M. Xu, S. Xu, J. W. Chai , J. D. Long, and Y. C. Ee, Appl. Phys. Lett. **89**, 251904 (2006).

759 O. I. Buzhinskij, V. G. Otroshchenko, D. G. Whyte, M. Baldwin, R. W. Conn, R. P. Doerner, R. Seraydarian, S. Luckhardt, H. Kugel, and W. P. West, J. Nucl. Mater. **313–316**, 214 (2003).

760 Y. Oya, H. Kodama, M. Oyaidzu, Y. Morimoto, M. Matsuyama, A. Sagara, N. Noda, and K. Okuno, J. Nucl. Mater. **329–333**, 870 (2004).

761 H. Kodama, M. Oyaidzu, A. Yoshikawa, H. Kimura, Y. Oya, M. Matsuyama, A. Sagara, N. Noda, and K. Okuno, J. Nucl. Mater. **337–339**, 649 (2005).

762 H. Sugai, H. Toyoda, K. Nakamuta, K. Furuta, M. Ohori, K. Toi, S. Hirokura, and K. Sato, J. Nucl. Mater. **220–222**, 254 (1995).

763 P. Brault, S. Roualdes, A. Caillard, A. L. Thomann, J. Mathias, J. Durand, C. Coutanceau, J. M. Leger, C. Charles, and R. Boswell, Eur. Phys. J. Appl. Phys. **34**, 151 (2006).

764 N. Inagaki, S. Tasaka, and T. Kurita, Polym. Bull. **22**, 15 (1989).

765 Z. Ogumi, Y. Uchimoto, K. Yasuda, and Z. I. Takehara, Chem. Lett. **19**, 953 (1990).

766 Y. Uchimoto, E. Endo, K. Yasuda, and Y. Yamasaki, J. Electrochem. Soc. **147**, 111 (2000).

767 K. Yasuda, Y. Uchimoto, Z. Ogumi, and Z. I. Takehara, J. Electrochem. Soc. **141**, 2350 (1994).

768 C. J. Brumlik, A. Parthasarathy, W. J. Chen, and C. R. Martin, J. Electrochem. Soc. **141**, 2273 (1994).

769 D. Kim, M. A. Scibioh, S. Kwak, I.-H. Oh, H. Y. Ha, Electrochem. Commun. **6**, 1069 (2004).

770 H. Liang, B. Shi, A. Fairchild, and T. Cale, Vacuum **73**, 317 (2004).

771 P. K. Chu, J. Y. Chen, L. P. Wang, N. Huang, Mater. Sci. Eng. **R36**, 143 (2002).

772 M. Laroussi, Plasma Process. Polym. **2**, 391 (2005).

773 C. P. Klages, Mat.-Wiss. u. Werkstofftech. **30**, 767 (1999).

774 R. Foerch, Z. Zhang, and W. Knoll, Plasma Process. Polym. **2**, 351 (2005).

775 A. Ohl, K. Schroeder, Surf. Coat. Technol. **116**, 820 (1999).

776 F. Z. Cui and D. J. Li, Surf. Coat. Technol. **131**, 481 (2000).

777 P. Favia and R. D'Agostino, Surf. Coat. Technol. **98**, 1102 (1998).

778 F. J. Gordillo-Vazquez, V. Herrero, and I. Tanarro, Chem. Vapor Deposition **13**, 267 (2007).

779 R. Foersh, A. N. Chifen, A. Bousquet, H. L. Khor, M. Jungblut, L. Q. Chu, Z. Zhang, I. Osey-Mensah, E.-K. Sinner, and W. Knoll, Chem. Vapor Deposition **13**, 280 (2007).

780 A. Anders, Thin Sol. Films **502**, 22 (2006).

781 P. Siemroth, C. Wentzel, W. Kliomes, B. Schultrich, and T. Schuelke, Thin Sol. Films **308**, 455 (1997).

782 D. M. Mattox, *Handbook of Physical Vapor Deposition (PVD) Processing* (Noyes Publications, Park Ridge, NJ, 1998).

783 M. K. Sunkara, S. Sharma, R. Miranda, G. Lian, and E. C. Dickey, Appl. Phys. Lett. **79**, 1546 (2001).

784 M. Mozetic *et al.*, J. Vac. Sci. Technol. A **21**, 369 (2003).

785 U. Cvelbar, M. Mozetic, and A. Ricard, IEEE Trans. Plasma Sci. **33**, 834 (2005).

786 A. von Keudell, Plasma Sources Sci. Technol. **9**, 455 (2000).

787 C. Cavallotti, M. Di Stanislao, and S. Carrà, Prog. Cryst. Growth Charact. Mater. **48**, 123 (2004).

788 A. von Keudell, T. Schwarz-Selinger, and W. Jacob, J. Appl. Phys. **89**, 2979 (2001).

789 M. Meier and A. von Keudell, J. Appl. Phys. **90**, 3585 (2001).

790 M. Meier and A. von Keudell, J. Chem. Phys. **116**, 5125 (2002).

791 C. Hopf, A. von Keudell, and W. Jacob, J. Appl. Phys. **93**, 3352 (2003).

792 C. Hopf, A. von Keudell, and W. Jacob, J. Appl. Phys. **94**, 2373 (2003).

793 M. M. Millard and E. Kay, J. Electrochem. Soc. **129**, 160 (1982).

794 R. d'Agostino, F. Cramarossa, V. Colaprico, R. d'Ettole, J. Appl. Phys. **54**, 1284 (1983).

795 J. P. Booth, G. Cunge, P. Chabert, and N. Sadeghi, J. Appl. Phys. **85**, 3097 (1999).

796 G. Cunge and J. P. Booth, J. Appl. Phys. **85**, 3952 (1999).

797 E. R. Fisher, Plasma Process. Polym. **1**, 13 (2004).

798 P. R. McCurdy, K. H. A. Bogart, N. F. Dalleska, and R. E. Fisher, Rev. Sci. Instrum. **68**, 1684 (1997).

799 I. T. Martín, J. Zhou, and E. R. Fisher, J. Appl. Phys. **100**, 013301 (2006).

800 J. P. Booth, H. Abada, P. Chabert, and D. B. Graves, Plasma Sources Sci. Tech. **14**, 273 (2005).

801 F. Gaboriau, G. Carty, M. C. Peignon, and C. Cardinaud, J. Phys. D: Appl. Phys. **39**, 1830 (2006).

802 D. Liu, I. T. Martin, J. Zhou, and E. R. Fisher, Pure Appl. Chem. **78**, 1187 (2006).

803 M. Hori and T. Goto, Appl. Surf. Sci. **253**, 6657 (2007).

804 This reference is created intentionally to show the author's sincere gratitude to anyone who has ever published or will ever publish any report on any aspect related to this monograph. The number of these reports is very large (and will undoubtedly increase in the future) and is far above what is possible to cover in a single monograph.

Index

Plasma Nanoscience: Basic Concepts and Applications of Deterministic Nanofabrication
Kostya (Ken) Ostrikov
Copyright © 2008 WILEY-VCH Verlag GmbH & Co. KGaA, Weinheim
ISBN: 978-3-527-40740-8